The physical basis of organic chemistry

The physical basis of organic chemistry

HOWARD MASKILL

Department of Chemistry,
University of Stirling

Oxford · New York
OXFORD UNIVERSITY PRESS

Oxford University Press, Walton Street, Oxford OX2 6DP

Oxford New York
Athens Auckland Bangkok Bombay
Calcutta Cape Town Dar es Salaam Delhi
Florence Hong Kong Istanbul Karachi
Kuala Lumpur Madras Madrid Melbourne
Mexico City Nairobi Paris Singapore
Taipei Tokyo Toronto

Oxford is a trade mark of Oxford University Press

Published in the United States by
Oxford University Press Inc., New York

First published 1985
Reprinted 1989 (with corrections), 1990, 1993, 1995

British Library Cataloguing in Publication Data
Maskill, Howard
The physical basis of organic chemistry.
1. Chemistry, Physical organic
I. Title
547.1'3 QD476
ISBN 0 19 855199 1 (Pbk)

Printed in Great Britain by
The Ipswich Book Co.

Preface

Traditionally, chemistry has been taught as a clearly defined subject with major subdivisions – physical, organic, and inorganic – and practitioners of the separate branches are usually seen by students not only as covering different areas of the subject, but also as having their own distinctive approaches and methods. This image of separateness is reinforced by university chemists identifying themselves as one type or another, organizing themselves into separate administrative and teaching units, and sometimes housing themselves in different buildings. Whilst we maintain that such diversity enriches the discipline as a whole, we should also acknowledge that it frequently causes difficulties for students who fail to recognize the relationships between different areas, for example, thermodynamics and synthesis. There is a responsibility, therefore, upon those of us who teach to relate our own corner of chemistry to the wider subject, and to ensure that there is overlap at each boundary rather than a gap or, even worse, a barrier. Moreover, chemistry, like all modern sciences, constitutes a growing body of knowledge, and some of the most exciting and important current advances are at the boundaries between the subdivisions, and where chemistry as a whole impinges upon other sciences traditionally regarded as different – principally biology and physics.

Organic and biological chemistry investigated by the methods and techniques of physics and physical chemistry have been particularly fruitful areas of study. And although physical chemists have always investigated the properties and reactions of organic compounds, the discipline that today we recognize as physical organic chemistry really began when organic chemists themselves became seriously interested in the relationship between molecular structure and chemical properties, and in how, at the molecular level, the reactions of organic compounds take place. This leads me to offer to the student at an early stage the important distinction between experimental facts and reaction mechanisms.

Structures of reactants and products are usually the principal features of a chemical reaction to be established. Equilibrium constants for reversible processes are also normally determinable, and a third experimental aspect of a reaction which may be quantified is its rate (or velocity). Molecular structures and reaction parameters can be expressed numerically with associated degrees of precision which are usually taken as measures of accuracy. In contrast, a mechanism – a view of how reactants become products – constitutes a set of deductions derived by theory and intuition from the experimental results. A mechanism is not, therefore, a primary experimental feature in the way that a molecular structure, an equilibrium constant, or a rate constant is, and we can never express it in fine detail and with complete confidence.

Today, a general organic chemistry textbook which made no attempt to describe how reactions take place, or to explain why one compound is more reactive than another under the same experimental conditions, would be unacceptable. But even the best available texts of general organic chemistry seldom find room to describe how the mechanism of a given reaction has been investigated, or what the experimental basis is for alleging that one compound is, for example, a more potent electrophile or nucleophile than another. The aim of this book, however, is not to demonstrate how, in detail, experimental results lead towards a mechanistic understanding of a particular reaction; that, over a wide subject area, is impracticable. It is, in the first place, to define a coherent vocabulary and grammar of what is currently recognized as physical organic chemistry and, secondly, to illustrate in a single progressive account the application of principles by which this knowledge is used to interpret experimental results within the general context of organic chemistry. (According to one's point of view, this attempted development of the interface between organic and physical chemistry may be seen either as putting organic flesh upon the dry bones of physical chemistry, or as introducing physical rigour into the flab of organic chemistry). My task has been made easier by working in the small, undivided chemistry department of a new university and, like my colleagues, in teaching a wider range of chemistry than would be regarded as normal in larger, older, and more conventional institutions.

The level and extent of coverage are dual problems for the author of any textbook. Broadly, the first five chapters deal with molecular structure and equilibrium properties of chemical compounds, and the remaining five deal with various aspects of reaction rates and mechanisms, but I have included cross-references since thermodynamic and kinetic phenomena are occasionally amenable to a common molecular analysis and interpretation. Later chapters assume an understanding of the contents of earlier ones, and throughout I have presumed that the reader has some knowledge of basic organic and physical chemistry. The subject of each chapter is developed from a level well within the grasp of an undergraduate student, but some topics are taken beyond normal chemistry degree requirements, and, in some places, I have indicated areas of current development. Consequently, the book could also serve as an introductory text for graduate studies. Either way, I would be grateful to hear from users of errors of fact or interpretation which have eluded our best detective efforts.

A wide-ranging list of topics could be included under the general title of physical organic chemistry, and such lists compiled by different chemists would vary. I am aware that some topics, principally molecular orbital theory and derived subjects such as aromaticity and symmetry-control of chemical reactions, would be on almost every list and yet are not covered in this book. I anticipate criticism on this account and offer the following explanation. I see physical organic chemistry as a range of approaches and methods concerned primarily with the quantification of experimental results and their interpretation. It includes theory or, rather, theories which are diverse and concerned with deriving results from measurements then interpreting them in the context of a wider body of knowledge; theories in this sense permeate the subject matter of this book without being filtered out and treated separately.

In contrast, that substantial and self-contained subject area based upon molecular orbital theory, whose antecedents are essentially mathematical, neither blends in readily with the range of other subjects which I wanted to cover, nor lends itself to the sort of treatment I wished to adopt for these other subjects. Nevertheless, some aspects of molecular orbital theory inevitably impinge upon some topics discussed in this book. Where this occurs, I rely upon good and well-known texts on the subject presently available at various levels, the existence of which constitutes another reason for my not including the topic for its own sake in this book.

Another criticism which I anticipate relates to the inclusion of topics such as molecular mechanics or the extended discussion of entropy and chemical equilibrium. Some may regard these as either inappropriate in a textbook at this level, or better left in mainstream physical chemistry texts. This criticism springs from different views of where the lines should be drawn which divide and bound chemistry as a whole at university degree level. Some topics are treated as essential core material in one institution but as optional extras in another; several topics will be taught by 'organic' chemists in some places and by 'physical' chemists elsewhere. By extending into certain areas conventionally regarded as physical chemistry, I hope to establish that they are neither especially difficult at this level nor irrelevant to a proper study of the chemistry of organic compounds.

Finally, I wish to thank colleagues and students at Stirling who read chapters of the manuscript. Because chemistry is an expanding subject, no one remains fully informed on more than a part and I would not wish to deny my reliance upon others. I am especially grateful to Dr. Brian G. Cox who, after reading a large proportion of the manuscript, prevented me from inflicting some of my chemical misconceptions upon a wider readership. I also thank Drs. P. Murray-Rust, F. G. Riddell, and W. V. Steele for expert guidance and patience in reading other parts. Drs. Alan A. Wilson and Stuart Corr shared the reading of the whole manuscript from the user's point of view during their time as graduate students at Stirling, and identified sections which needed clarification; to them both, I am particularly grateful.

To conclude, I must express my thanks to Mrs. Joan Weber who uncomplainingly typed the whole manuscript, most parts through several versions, and to my wife, Jean, who read every part at each stage and rooted out many grammatical errors, clichés, obscurities, and infelicities. Even with all this help, some mistakes will inevitably remain; these are my own responsibility.

Stirling
June 1984

H.M.

Contents

Abbreviations and symbols for units and properties

Units

Å	ångström
amu	atomic mass unit
atm	standard atmosphere
cm	centimetre (10^{-2} m)
cm^3	cubic centimetre (millilitre; 10^{-6} m^3)
°C	degrees centigrade (Celsius)
dm	decimetre (10^{-1} m)
dm^3	cubic decimetre (litre; 10^{-3} m^3)
h	hour
Hz	hertz
J	joule
kJ	kilojoule (10^3 J)
K	kelvin
m	metre
min	minute
mm	millimetre (10^{-3} m)
mol	mole
MHz	megahertz (10^6 Hz)
N	newton
nm	nanometre (10^{-9} m)
Pa	pascal
s	second

Properties

A	pre-exponential factor in the Arrhenius equation
a_i	activity of i
$a_i^\ominus$ (a_i°)	standard activity of i
a_i^r	relative activity of i (activity of i relative to its value in a defined standard state)
α (β)	Brönsted coefficient
α	degree of reaction
$BH^\circ(X{-}Y)$	incremental standard molar bond enthalpy term for the bond between X and Y
c_i	activity of i using the molarity scale

c_i^r	relative activity of *i* using molarity scale with standard state activity = 1 mol dm^{-3}
C	molar heat capacity
$DH^\circ(XY)$	standard molar bond dissociation enthalpy for the bond between X and Y
D_e	dissociation energy
D_0	spectroscopic bond dissociation energy
E, ε, ϵ	energy
E_a	activation energy in the Arrhenius equation
EA	electron affinity
$\Delta G^\ominus$ (ΔG°)	standard molar free-energy change
$\Delta G^{\ominus\ddagger}$ ($\Delta G^{\circ\ddagger}$)	standard molar free energy of activation
γ_i	activity coefficient of *i*
$\Delta H^\ominus$ (ΔH°)	standard molar enthalpy change
$\Delta H^{\ominus\ddagger}$ ($\Delta H^{\circ\ddagger}$)	standard molar enthalpy of activation
IP	ionization potential
k	rate constant
K_{AP}	autoprotolysis constant
K_{AH} (K_{BH})	acidity constant
K_c	practical equilibrium constant using molarity scale for concentrations
$K_a^\ominus$	general thermodynamic equilibrium constant for which the activity scale and standard state must be defined
K_P°	thermodynamic equilibrium constant using pressure in atm as the activity scale and the pure material at 1 standard atmosphere as the standard state
κ	force constant
log	logarithm to the base 10
ln	logarithm to the base e
ξ	extent of reaction
λ	wavelength
μ	reduced mass
μ_i	chemical potential of *i*
$\mu_i^\ominus$ (μ_i°)	standard chemical potential of *i*
n_i	number of moles of *i*
$\boldsymbol{n}$	reaction molecularity or order
ν	frequency
ν_A	number of moles of A in a balanced chemical equation
$\bar{\nu}$	wave number
P	pressure
PA	proton affinity
pH	$-\log[H_3O^+]$
pK	$-\log K$
q	heat

Q	molecular partition function
r	buffer ratio
r^+ (r^-)	Yukawa–Tsuno reaction parameter
r_e	equilibrium internuclear distance
ρ	Hammett reaction parameter
ρ_I	reaction parameter for correlation analysis of aliphatic (alicyclic) systems
S°	absolute standard molar entropy (standard state = pure material at 1 standard atmosphere pressure)
$\Delta S^\ominus$ (ΔS°)	standard molar entropy change
$\Delta S^{\ominus\ddagger}$ $(\Delta S^{\circ\ddagger})$	standard molar entropy of activation
σ	Hammett substituent constant
σ^+ (σ^-)	modified Hammett substituent constant
σ_I	substituent constant for correlation analysis of aliphatic (alicyclic) systems
T	(absolute) temperature
t	time
U	internal energy
V	potential energy
V	volume
w	work
X_i	mole fraction of i
$\bar{\omega}_e$	energy spacing (in cm^{-1}) between successive vibrational levels of a harmonic diatomic molecule
[Z]	molarity of Z (concentration of Z expressed in moles per dm^3)

For my mother, Rita,
and in memory of my
father, Walter Maskill.
In gratitude and with love.

1

Molecular properties

1.1 Introduction

One of the objectives of chemistry is to be able to account for the physical and chemical behaviour of chemical compounds in terms of the structures and attributed properties of discrete molecules. Over the years, various concepts and models have been developed to assist in this exercise, and individual molecules and parts of molecules have been credited with many and diverse properties.

In this chapter we shall review aspects of molecular structure with any emphasis on those facets which, in later chapters, will help us to understand reactions and reactivities of organic compounds.

1.2 Molecular energy levels and spectroscopy

1.2.1 Electronic levels

The lowest of the most widely-spaced heavy lines in the energy level diagram in Fig. 1.1 represents the electronic ground state of a molecule. If an electron from the occupied molecular orbital of highest energy is excited into the initially unoccupied molecular orbital of lowest energy, the first electronically-excited state of the molecule is produced. This state is represented in Fig. 1.1 by the next of the widely-spaced heavy lines. The electronic excitation is induced by interaction between the

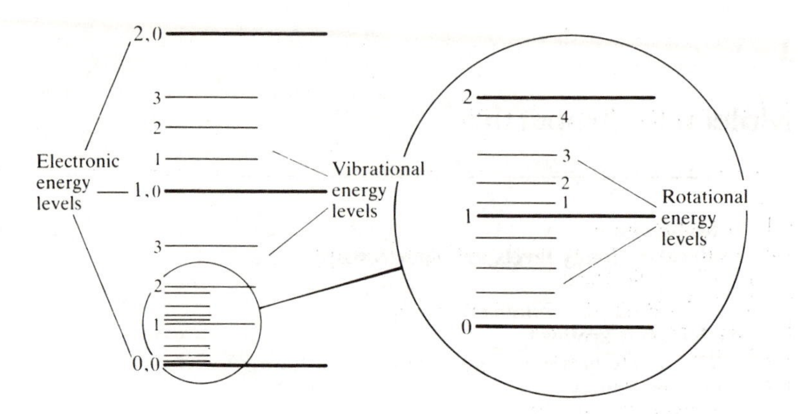

Fig. 1.1. A general molecular energy level diagram.

molecule and electromagnetic radiation in the *ultraviolet* or *visible* range, with transfer to the molecule of one photon of energy.

The energy of one photon, ϵ, is given by

$$\epsilon = h\nu$$

where h = the Planck constant (6.6262×10^{-34} J s) and
ν = the frequency of the radiation.
If the energy difference between the two molecular electronic states is $\Delta\varepsilon$

$$\Delta\varepsilon = \epsilon,$$

so

$$\Delta\varepsilon = h\nu,$$

or, on a molar basis,

$$\Delta E = N_A \,.\, \Delta\varepsilon = N_A \,.\, \epsilon = N_A \,.\, h\nu$$

where N_A = the number of molecules per mole (the Avogadro constant, 6.0220×10^{23} mol^{-1}). So, by measuring the frequency of the radiation absorbed by a compound, the energy difference between the two electronic levels involved in the excitation can be calculated.

A conventional ultraviolet and visible spectrophotometer records absorption spectra between approximately 200 and 800 nm; this corresponds to $\Delta E \sim 600$–150 kJ mol^{-1}. These are very large energy differences and virtually all stable organic molecules with complements of valence electrons exist in their electronic ground states at normal temperatures.

Electronic transitions in organic compounds which contain chromophores no more extensive than unconjugated carbon–carbon double bonds, or sigma-bonded atoms, require radiation of wavelength less than 200 nm ($\Delta E > 600$ kJ mol^{-1}). This very short wavelength region presents experimental difficulties and the recording of spectra in this range is not a routine matter.

Example. Acetaldehyde in hexane solution shows a low intensity ultraviolet absorption maximum at 293.4 nm due to excitation of a lone pair electron to an antibonding π^*-orbital (an n–π^* transition).[1] Calculate the energy per photon of radiation of this wavelength, and the energy difference per mole between the two electronic states of acetaldehyde which are involved.

$$\lambda_{\text{max}} = 293.4 \times 10^{-9}\ \text{m}$$

The relationship between frequency ν, wavelength λ, and the velocity of propagation, c, of electromagnetic radiation is $c = \nu . \lambda = 2.998 \times 10^{8}\ \text{m s}^{-1}$.

Therefore

$$\nu_{\text{max}} = \frac{c}{\lambda_{\text{max}}} = \frac{2.998 \times 10^{8}\ \text{m s}^{-1}}{293.4 \times 10^{-9}\ \text{m}}$$
$$= 1.022 \times 10^{15}\ \text{s}^{-1}.$$

The energy per photon $= h\nu$

$$= 6.626 \times 10^{-34}\ \text{J s} \times 1.022 \times 10^{15}\ \text{s}^{-1}$$
$$= \underline{6.772 \times 10^{-19}\ \text{J}}.$$

Molar energy difference between the two electronic states, ΔE, equals the energy per mole of photons:

$$\Delta E = 6.772 \times 10^{-19}\ \text{J} \times 6.022 \times 10^{23}\ \text{mol}^{-1}$$
$$= \underline{407.8\ \text{kJ mol}^{-1}}.$$

The promotion of an electron to a higher energy molecular orbital within a molecule A initially containing only fully-paired valence electrons by direct interaction with electromagnetic radiation takes place with no inversion of electronic spin and first gives an electronically-excited singlet state.[2] The lifetime of this species is exceedingly short and it rapidly undergoes one of several processes:

(a) emission of a photon to produce the electronic ground state (fluorescence),

(b) dissipation of energy non-radiatively to regenerate the electronic ground state in thermal equilibrium with its surroundings (internal conversion),

(c) spin inversion to generate the triplet electronically-excited state (intersystem crossing),

(d) transfer of singlet electronic energy to another molecule B to yield the electronic ground state of A and an electronically-excited singlet state of B, or

(e) a chemical reaction with dissipation of thermal energy.

Which of these processes take place depends upon the nature of the compound and the experimental conditions. If the electronically-excited triplet state is generated (route (c)), it too is normally short-lived (though not so short-lived as the upper singlet) and either undergoes one of the following three physical processes to regenerate the singlet ground state:

(a) emission of a photon with concurrent spin inversion (phosphorescence),

(b) non-radiative dissipation of energy with spin inversion (intersystem crossing), or

(c) transfer of electronic energy to another molecule B to give the triplet excited state of B,

or a fourth possibility,

(d) a chemical reaction with dissipation of thermal energy.

Transitions from lower to upper electronic levels with the absorption of electromagnetic radiation, and intersystem crossing from an excited singlet to a triplet state, initially produce molecules in upper vibrational and rotational levels of the photo-excited state. Much faster than such molecules can do anything else, they shed vibrational and rotational energy non-radiatively. This vibrational and rotational relaxation tends towards an equilibrium distribution of the vibrational and rotational levels within the electronically-excited state and will achieve it if the upper electronic level is long enough lived. Consequently, fluorescence and phosphorescence take place from upper electronic states with equilibrium populations of vibrational and rotational levels. These processes then give molecules in excited vibrational and rotational levels of the electronic ground state which then, very rapidly, achieve complete thermal equilibrium non-radiatively. These vibrational and rotational complications account for fluorescence and phosphorescence spectra being at longer wavelengths than, and appearing as the mirror images of, a compound's absorption spectrum.

The chemical properties of electronically-excited molecules (singlet or triplet states) are quite distinct from the same compound's ground state properties. They may undergo the same reaction, e.g. as an acid or base, but at a very different rate; or, in isomerizations or addition reactions with a common reactant, they may give quite different products. Many of these differences originate in the different symmetry properties of the highest occupied or lowest unoccupied molecular orbitals of ground and electronically-excited states.[3]

Details of how molecules interact with electromagnetic radiation and of how photo-excited molecules react chemically, especially in the biological processes of vision and photosynthesis, are still being investigated. Furthermore, some photochemical transformations are of immense value in organic synthesis and a few are used industrially.

1.2.2 Vibrational and rotational levels

A molecule is not, of course, a rigidly-bonded static assembly of atoms. It has 3 translational degrees of freedom so, if there are N atoms per molecule, it has $3N$-3 rotational and vibrational internal degrees of freedom.

A linear molecule, for example acetylene H—C≡C—H, has all its atoms on a single axis; about this axis, therefore, the molecule has no moment of inertia and, consequently, no rotational degree of freedom. Moments of inertia about the other two mutually perpendicular axes are identical, so a linear molecule has two rotational degrees of freedom which are *degenerate*; they are identical except for spatial orientation. The other $3N$-5 internal degrees of freedom of a linear molecule are vibrational.

All non-linear polyatomic molecules, regardless of symmetry, have 3 rotational and $3N$-6 vibrational degrees of freedom although, again, these may include degeneracies.

The energy levels of just one vibrational mode in both ground and first excited electronic states are included in Fig. 1.1. As indicated, smaller quanta of energy are required for vibrational compared with electronic excitations, consequently radiation of lower frequency (longer wavelength or lower wavenumber) is required. Broadly, this corresponds to the *infra-red* region of the electromagnetic spectrum. The range of a commercial infra-red spectrometer is approximately 4000 to 250 cm^{-1} which translates into energy differences of about 48 to 3 kJ mol^{-1}. We shall see later, however, that excitation of some very low energy molecular vibrations of polyatomic molecules requires radiation in the far infra-red and even microwave regions.

The wavenumber, $\bar{\nu}$, of radiation is the reciprocal of the wavelength: $\bar{\nu} = 1/\lambda$. It can be thought of as the number of wavelengths of radiation per unit length just as the frequency is the number per unit time and has the advantage over wavelength that, like frequency, it is directly proportional to the energy of the radiation since $\epsilon = h\nu = hc\bar{\nu}$. The units are usually cm^{-1} so that wavenumbers in the infra-red range are more manageable than, for example, the corresponding frequencies expressed in s^{-1}.

Example. The infra-red absorption spectrum of cyclohexanone has an intense band at 1710 cm^{-1}.[4] Calculate the energy per photon of this radiation and the energy difference per mole between the two vibrational states of cyclohexanone which are involved.

$$\bar{\nu} = 1710\ \mathrm{cm^{-1}},$$

therefore

$$\nu = c\bar{\nu} = 2.998 \times 10^{10}\ \mathrm{cm\ s^{-1}} \times 1710\ \mathrm{cm^{-1}}$$
$$= 5.127 \times 10^{13}\ \mathrm{s^{-1}}.$$

$$\text{Energy per photon} = h\nu = 6.626 \times 10^{-34}\ \mathrm{J\ s} \times 5.127 \times 10^{13}\ \mathrm{s^{-1}}$$
$$= \underline{3.397 \times 10^{-20}\ \mathrm{J}}.$$

Molar energy difference between the two states

$$= 3.397 \times 10^{-20}\ \mathrm{J} \times 6.022 \times 10^{23}\ \mathrm{mol^{-1}}$$
$$= \underline{20.5\ \mathrm{kJ\ mol^{-1}}}.$$

The very narrowly spaced lines in Fig. 1.1 which are superimposed upon two of the vibrational levels in the electronic ground state represent energy levels of one rotational degree of freedom; rotational levels within a single vibrational state are not equally spaced. Energy differences between successive rotational levels range between about 6 kJ mol^{-1} and 0.3 J mol^{-1} depending very much upon the particular compound. This wide range of energies corresponds to far infra-red and microwave radiation (approximately 500–5 and 30–0.03 cm^{-1} respectively). Pure rotational spectra allow the determination of moments of inertia which reveal symmetries, bond angles, and bond lengths frequently with great precision. However, as mentioned above, these spectral ranges also correspond to transitions between some close vibrational levels of polyatomic molecules. Consequently, the interpretation of far infra-red and microwave spectra depends to a considerable extent upon the nature of the compound under investigation and the precise experimental technique.

The chemical conversion of one organic molecule into another involves atomic reorganization with consequent re-bonding. Before we can expect to understand how such gross atomic rearrangements take place, we must have some prior comprehension of simpler atomic motions within molecules. We need to understand, therefore, how atoms and groups of atoms within a molecule can vibrate, twist, and rotate, i.e. how they move reversibly with respect to other parts of the molecule. The rest of this chapter is devoted to this subject, and we shall not restrict ourselves simply to organic molecules.

1.3 Molecular vibrations

1.3.1 The simple harmonic approximation

A diatomic molecule A–B has only one vibrational degree of freedom, the stretching mode shown in eqn (1.1):

$$\underset{r>r_e}{\text{A——B}} \rightleftharpoons \underset{r=r_e}{\text{A—B}} \rightleftharpoons \underset{r<r_e}{\text{A–B}} \qquad (1.1)$$

where r = the instantaneous internuclear distance and r_e = the equilibrium internuclear distance.

Implicit in eqn (1.1) is the notion that at the equilibrium internuclear distance, r_e, the molecule is experiencing neither compression nor stretching, and the vibration has only *kinetic energy* (not to be confused with *translational* kinetic energy). If we assume that, to a first approximation, the instantaneous force tending to restore the two atoms to their equilibrium internuclear distance is proportional to the displacement from the equilibrium internuclear distance, we have the Hooke's Law model for the vibration of a diatomic molecule. This is also called the *simple harmonic approximation.* The restoring force f at internuclear distance r is given by[5]

$$-f = \kappa x$$

and

$$\kappa x = \mathrm{d}V/\mathrm{d}x,$$

where κ = a constant of proportionality (the *force constant* of the bond),
$x = r - r_e$, and
V = the *vibrational potential energy* of the diatomic molecule,

therefore

$$V = \tfrac{1}{2}\kappa x^2. \tag{1.2}$$

The force constant measures the ability of the bond to resist a stretching or compressing force; it is also a measure of the restoring force of the bond which becomes stretched or compressed. It is independent of the masses of the atoms joined by the bond and, within the limit of the harmonic approximation, independent of the extent of the distortion from the equilibrium internuclear distance.

Equation (1.2) shows that a plot of vibrational potential energy against the internuclear distance is a parabola and this is illustrated in Fig. 1.2.

It also follows from the above equations that

$$\kappa = \mathrm{d}^2V/\mathrm{d}x^2,$$

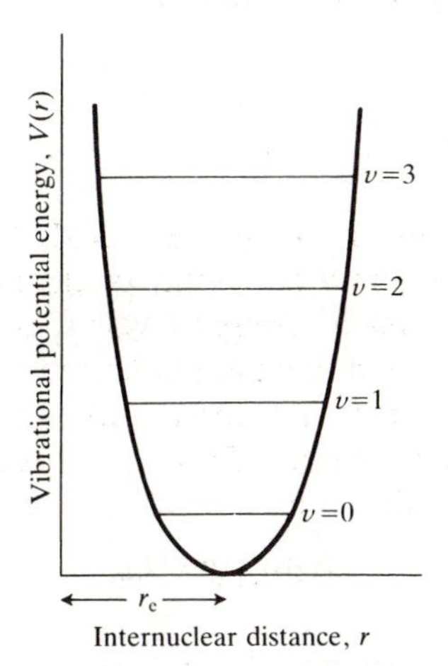

Fig. 1.2. Potential-energy diagram for a diatomic molecule according to the simple harmonic approximation.

so the force constant is the *curvature* of the parabola in Fig. 1.2. If κ is large, the parabola is narrow and a small change in the bond length requires a large force and causes a large change in the vibrational potential energy of the molecule; this is sometimes called a *hard* potential function. If κ is small, the parabola is wide, and a small force can bring about a large change in the bond length and cause only a small change in the potential energy of the system (a *soft* potential function).

The *total* vibrational energy, ε(vib) of the harmonic diatomic molecule is the sum of the vibrational potential and kinetic energies:

$$\varepsilon(\text{vib}) = \tfrac{1}{2}\kappa x^2 + \tfrac{1}{2}\mu\left(\frac{dx}{dt}\right)^2,$$

where μ is the *reduced mass* of the molecule defined by

$$\frac{1}{\mu} = \frac{1}{m_A} + \frac{1}{m_B},$$

and m_A and m_B are the masses of the two atoms. (The reduced mass is required in order that the vibrational and the translational kinetic energies may be treated as independent.)

Application of wave mechanics with quantum restrictions to the simple harmonic oscillator provides the total energy expression:[5]

$$\varepsilon(\text{vib}) = (v + \tfrac{1}{2})h\nu, \tag{1.3}$$

where h = the Planck constant,
v = a vibrational quantum number which may have values 0, 1, 2, 3, . . . , and
ν = the frequency of the vibration,

where

$$\nu = \frac{1}{2\pi}\sqrt{\frac{\kappa}{\mu}}. \tag{1.4}$$

Two important conclusions from eqns (1.3) and (1.4) have been incorporated into Fig. 1.2. First, the total energy of the system may not assume any value; only equally-spaced energy levels specified by integral quantum numbers are allowed and these are indicated by the horizontal lines.

Second, the vibration does not cease and even at absolute zero the molecule retains *zero-point energy*. This is the total vibrational energy of the ground state level with $v = 0$, and so for one molecule

$$\varepsilon(\text{vib})_{v=0} = \tfrac{1}{2}h\nu,$$

or, on a molar basis,

$$E(\text{vib})_{v=0} = \tfrac{1}{2}h\nu \,.\, N_A.$$

When a molecule becomes vibrationally excited from the level with

$v=0$ to the one with $v=1$, the amplitude of the vibration increases but the frequency remains constant.

By eqn (1.3),

$$\varepsilon(\mathrm{vib})_{v=1}=\tfrac{3}{2}h\nu,$$

therefore the molecular energy difference between the two levels is given by

$$\varepsilon(\mathrm{vib})_{v=1}-\varepsilon(\mathrm{vib})_{v=0}=\delta\varepsilon(\mathrm{vib})=h\nu. \quad (1.5)$$

Such a vibrational excitation will be induced by infra-red radiation, and one photon of energy ($\epsilon=h\nu$ where ν here is the frequency of the *radiation*) must be equal to $\delta\varepsilon(\mathrm{vib})$. Consequently, the frequency (or wavenumber) of the radiation which will induce a transition between the $v=n$ and $v=n+1$ levels is equal to the frequency (or wavenumber) of the harmonic vibration.

Example. Hydrogen iodide has an infra-red absorption band at 2230 $\mathrm{cm^{-1}}$ due to a transition between $v=0$ and $v=1$ vibrational levels.[6(a)] Calculate the energy difference per mole between the two vibrational levels, the zero-point energy, and the force constant of the bond assuming that the vibration is harmonic.

$$\bar{\nu}=2230\ \mathrm{cm^{-1}}$$
$$\nu=c\bar{\nu}$$
$$=2.998\times10^{10}\ \mathrm{cm\,s^{-1}}\times2230\ \mathrm{cm^{-1}}$$
$$=6.686\times10^{13}\ \mathrm{s^{-1}}.$$

Energy difference per mole between the two levels $=N_A\,.\,h\nu$

$$=6.022\times10^{23}\ \mathrm{mol^{-1}}\times6.626\times10^{-34}\ \mathrm{J\,s}\times6.686\times10^{13}\ \mathrm{s^{-1}}$$
$$=26.68\ \mathrm{kJ\,mol^{-1}}.$$

Molar zero-point energy $=\tfrac{1}{2}h\nu\,.\,N_A$

$$=13.34\ \mathrm{kJ\,mol^{-1}}.$$

The reduced mass of HI is given by

$$\frac{1}{\mu_{HI}}=\frac{1}{m_H}+\frac{1}{m_I}.$$

Taking $m_H=1$ and $m_I=127$ atomic mass units (amu),

$$\mu_{HI}=0.9922\ \mathrm{amu}$$

or 1.648×10^{-27} kg since 1 amu $=1.6606\times10^{-27}$ kg.
The vibrational frequency is given by

$$\nu=\frac{1}{2\pi}\sqrt{\frac{\kappa}{\mu}},$$

therefore

$$\kappa_{HI} = 4\pi^2 . \nu^2 . \mu_{HI}$$
$$= 4 \times 9.868 \times 44.70 \times 10^{26}\ s^{-2} \times 1.648 \times 10^{-27}\ kg$$
$$= 291\ kg\ s^{-2}.$$

These are not the usual units for a force constant, but since

$$1\ kg\ m\ s^{-2} = 1\ newton \quad (newton = N),$$
$$\kappa_{HI} = 291\ N\ m^{-1},$$

which is in the units of modern convention. Since 1 Nm = 1 J, the result could be expressed as $\kappa_{HI} = 291\ J\ m^{-2}$, which is more obviously consistent with the definition of κ in eqn (1.2).

Hydrogen iodide has a relatively low reduced mass and a moderately high vibrational wavenumber, therefore the vibrational energy levels are widely spaced. Consequently, the overwhelming majority of a large assembly of HI molecules will be in the ground vibrational state even at 25 °C. The equilibrium ratio of molecules in the two states corresponding to $v = 1$ and $v = 0$ can be calculated quite easily from the *Boltzmann Distribution Law.*

If n_0 and n_1 are the numbers of molecules in the ground and first excited vibrational levels,

$\delta\varepsilon$(vib) = the energy difference per molecule between the two levels, and
ΔE(vib) = the energy difference per mole between the two levels
= 26.68 kJ mol^{-1} as calculated above,

then

$$\frac{n_1}{n_0} = e^{-\delta\varepsilon(vib)/k_B T},$$

where k_B = the Boltzmann constant ($1.38066 \times 10^{-23}\ J\ K^{-1}$) and
T = the (absolute) temperature in kelvin (K).

Since $k_B . N_A = R$, the universal gas constant ($8.314\ J\ K^{-1}\ mol^{-1}$),

$$n_1/n_0 = e^{-\Delta E(vib)/RT}$$
$$= e^{-26\,680/8.314 \times 298}$$
$$= e^{-10.77},$$

therefore $n_1/n_0 = 2.1 \times 10^{-5}$ at 298 K.

A heavier diatomic molecule such as ICl ($\mu_{ICl} = 27.44$ amu using the ^{35}Cl isotope of chlorine) which has a broadly comparable force constant (236 N m^{-1} compared with 291 N m^{-1} for HI) has a much lower vibrational wavenumber as required by eqn (1.4) and found experimentally ($\bar{\nu} = 382\ cm^{-1}$).[7] The energy levels for ICl are, therefore, much closer together (ΔE(vib) = 4.57 kJ mol^{-1}) so the higher vibrational levels will be more heavily populated in a large assembly of ICI molecules than is the case for HI. Using the Boltzmann Distribution Law expression as above,

TABLE 1.1
Frequencies (ν), *molar energy differences between* $v = 0$ *and* $v = 1$ *levels* (ΔE(vib)), *and force constants* (κ) *for vibrations of diatomic molecules*

Molecule	$\bar{\nu}$/cm^{-1}	10^{-13} ν/s^{-1}	ΔE(vib)/kJ mol^{-1}	κ/Nm^{-1}
HF[a]	3962	11.878	47.40	879
HCl[b]	2886	8.652	34.52	477
HBr[b]	2559	7.672	30.61	381
HI[b]	2230	6.686	26.68	291
CO[b]	2143	6.425	25.64	1856
NO[b]	1876	5.624	22.44	1548
ICl[c]	382	1.145	4.57	236

[a] G. A. Kuipers, D. F. Smith, and A. H. Nielsen, *J. chem. Phys.* **25,** 275 (1956).
[b] Ref. 6(*a*), p. 62.
[c] Ref. 7.

we can calculate that

$$\frac{n_1}{n_0} = 0.16 \quad \text{at 298 K for ICl.}$$

Correspondingly, it follows from eqns (1.4) and (1.5) that if one of two molecules of comparable reduced mass has a high force constant (narrow potential-energy curve) and the other has a low force constant (broad curve), the former will have the higher frequency of vibration and more-widely-spaced energy levels.

Some collected results calculated from experimental wavenumbers using the harmonic approximation are shown in Table 1.1.

1.3.2 Anharmonicity

The simple harmonic description of a real vibrating diatomic molecule provides satisfactory agreement with experiment only near the minimum of the potential-energy curve where $r \sim r_e$. Spectra show that higher vibrational levels are not equally spaced as this simple model requires, but that they become increasingly close together.

If the nuclei are forced towards each other more closely, the potential energy of the system increases steeply. And, because the two nuclei cannot occupy the same space, as r approaches zero, the potential energy tends towards infinity asymptotically as shown in Fig. 1.3; when $r \ll r_e$, therefore, κ is large and increases as $r \rightarrow 0$.

Conversely, spectral results require that the gradient of the correlation between potential energy and internuclear distance becomes progressively less steep as r becomes large. In the region where $r \gg r_e$, therefore, κ is small and decreases as r increases.

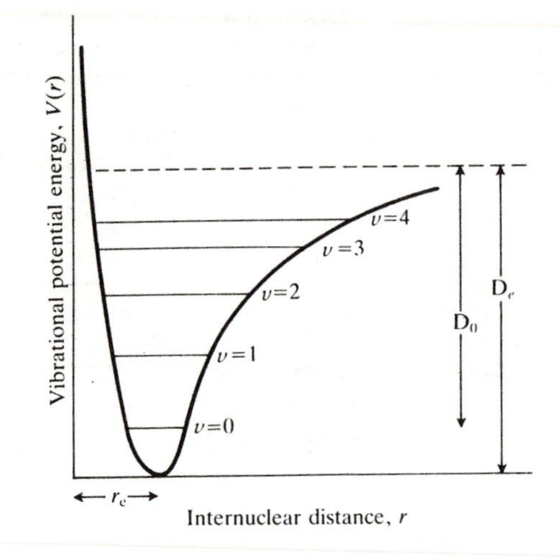

Fig. 1.3. Anharmonic potential-energy diagram for a diatomic molecule.

In the limit,

$$\frac{dV}{dr} \to 0 \quad \text{and} \quad \kappa\left(=\frac{d^2V}{dr^2}\right) \to 0 \quad \text{as } r \text{ becomes very large,}$$

so the potential-energy curve becomes horizontal. This state corresponds to dissociation of the molecule into two atoms.

A potential-energy curve which deviates from the parabolic form as shown in Fig. 1.3 is called *anharmonic* because of the inadequacy of the simple harmonic approximation except when $r \sim r_e$. And the more anharmonic the potential-energy curve of a diatomic molecule, the more the force constant depends upon the displacement from the equilibrium bond length.

It is, nevertheless, useful to assign to a *real* molecule a parameter which is the equal energy spacing that the molecule would have between successive vibrational levels if its vibration were fully harmonic with the force constant corresponding to the minimum in its actual potential-energy profile (this is close to the force constant at $r \sim r_e$ in the ground state vibration). This parameter is normally expressed as a wavenumber (units cm^{-1}) and given the symbol $\bar{\omega}_e$. Anharmonicity of the real molecule, i.e. the decreasing spacings between successive vibrational energy levels, is then expressed by an *anharmonicity constant* x_e which occurs with a squared quantum number term in a refined equation for the total vibrational energy of the real molecule (the anharmonic analogue of eqn (1.3), p. 8):

$$\varepsilon(\text{vib}) = (v+\tfrac{1}{2})h\nu_e - x_e(v+\tfrac{1}{2})^2 h\nu_e,$$

TABLE 1.2
Anharmonicity of some diatomic molecules[a]

Molecule	$\bar{\nu}$/cm^{-1} for the $v=0$ to $v=1$ transition	$\bar{\omega}_e$/cm^{-1}	$x_e\bar{\omega}_e$/cm^{-1}	x_e
HF	3961.6	4138.3	89.9	0.0217
HCl	2885.9	2989.7	52.0	0.0174
HBr	2559.3	2649.2	45.2	0.0171
HI	2230.1	2309.1	39.7	0.0172
CO	2143.2	2169.5	13.4	0.0062
NO	1875.9	1903.6	14.0	0.0073
ICl	381.5	382.2	1.45	0.0038

[a] See Table 1.1 and ref. 8.

where ν_e is the hypothetical vibrational frequency of the molecular vibration corresponding to $\bar{\omega}_e$ or, expressed as a wavenumber,

$$\bar{\varepsilon}(\text{vib}) = (v+\tfrac{1}{2}) \,.\, \bar{\omega}_e - x_e(v+\tfrac{1}{2})^2 \,.\, \bar{\omega}_e.$$

x_e is a measure of the rate at which the successive vibrational levels become closer as v increases. Some results are given in Table 1.2.

The change in potential energy from the minimum at $r=r_e$ to $r\rightarrow\infty$ in Fig. 1.3 is the *Dissociation Energy*, D_e. This parameter is really a property of the curve rather than of the molecule because the molecule never has less than zero-point energy. The *Bond Dissociation Energy* D_0 is of greater physical significance to the chemist; this is the energy required to dissociate a diatomic molecule in its ground vibrational (and rotational) level into two atoms, all species being in their electronic ground states with no translational kinetic energy. It may be expressed either in energy units per molecule (usually electron volts or, loosely, cm^{-1} by spectroscopists) or on a molar basis in kJ mol^{-1}. D_0 is a property of a molecule, related to D_e by eqn (1.6),

$$D_e = D_0 + \tfrac{1}{2}h\nu, \tag{1.6}$$

and is usually obtained from the electronic spectra of the diatomic molecule and the constituent atoms into which the molecule dissociates. (As we shall see in the next chapter, D_0 can also be measured calorimetrically). The zero-point energy is calculated from the stretching frequency which is determined by infra-red spectroscopy. Some results which were obtained spectroscopically are shown in Table 1.3.[9]

To describe the anharmonic relationship between V and r in Fig. 1.3 mathematically, a more complex expression than eqn (1.2) (p. 7) is

TABLE 1.3

Bond dissociation energies and zero-point energies of some diatomic molecules[9]

Molecule	D_0/kJ mol^{-1}	$\bar{\nu}$/cm^{-1}	Zero-point energy/ kJ mol^{-1}	D_e/kJ mol^{-1}
CO	1069.9	2143	12.8	1082.7
NO	627.1	1876	11.2	638.3
HF	563.4	3962	23.7	587.1
ICl	207.6	382	2.3	209.9

required. An empirical equation proposed by Morse is shown in eqn (1.7) and is satisfactory except at very short internuclear distances.[5,9]

$$V(r) = D_e(1 - e^{-\beta(r-r_e)})^2, \tag{1.7}$$

where $V(r)$ = the vibrational potential energy as a function of r, the internuclear distance,

D_e = the dissociation energy defined in eqn (1.6) and represented in Fig. 1.3,

r_e = the equilibrium internuclear distance, and

β = a constant for the particular molecule composed of experimentally determinable parameters defined above and fundamental constants:

$$\beta = \bar{\omega}_e \sqrt{\frac{2\pi^2 c\mu}{D_e h}}.$$

Potential-energy curves of the type shown in Fig. 1.3 which are described by eqn (1.7) are sometimes called *Morse curves.*

1.3.3 Polyatomic molecules

The simple harmonic model with an empirical correction to account for anharmonicity in the higher vibrational levels is reasonably satisfactory for diatomic molecules. But most organic molecules have more than two atoms!

A complex vibration of a molecule with N atoms can be resolved into $3N$-6 (or $3N$-5 if the molecule is linear) so-called *normal modes* of vibration. Each normal mode is described by a *normal coordinate.* A normal coordinate is not usually a simple structural parameter such as a bond length or angle, but some composite unidimensional term, q, which allows an expression in the harmonic form for the potential energy, V, of a single normal mode:

$$V_i = \tfrac{1}{2}\kappa_i q_i^2,$$

and the total vibrational potential energy, $V(\text{vib})$, is given by

$$V(\text{vib}) = \sum \tfrac{1}{2}\kappa_i q_i^2.$$

We shall encounter some vibrational modes later for which the normal coordinate q is approximately equivalent to some internal coordinate such as an intramolecular distance or angle.

The spectral absorption which corresponds to the transition from the ground state ($v = 0$) to the first excited state ($v = 1$) of a normal mode is called the *fundamental* band of that normal mode. The normal modes of hydrogen cyanide are indicated below and the wavenumbers of the fundamental bands are shown in parentheses:[6(b)]

H—C≡N ($\bar{\nu}_1 = 3312\ \text{cm}^{-1}$)   H—C≡N ($\bar{\nu}_2 = 2089\ \text{cm}^{-1}$)

H—C≡N   H—C≡N

degenerate modes, ($\bar{\nu}_3 = 712\ \text{cm}^{-1}$)

Arrows indicate motion within the plane of the paper, − and + indicate motion into and out of the plane of the paper. In each representation, the upper and lower symbols indicate the in-phase nuclear motion during the two halves of the vibrational period, respectively.

These comprise two stretching modes ($\bar{\nu}_1$ and $\bar{\nu}_2$) with force constants κ_1 and κ_2, and a doubly degenerate bending mode ($\bar{\nu}_3$) with force constant κ_3 making the required four normal modes ($3N$-5).

Fundamental frequencies of the normal modes of a relatively simple molecule (but one with more than two atoms) can usually be identified in its infra-red spectrum (or Raman spectrum for a vibration which does not involve a change in dipole moment). The force constants of the normal modes can then generally be calculated from knowledge of the fundamental frequencies, the atomic masses, and a few approximations.[6(b),10]

The force constants for the normal modes of hydrogen cyanide shown above are:[10]

$$\kappa_1 = 582\ \text{N m}^{-1}$$

$$\kappa_2 = 1788\ \text{N m}^{-1}$$

$$\kappa_3 = 20\ \text{N m}^{-1}.$$

When the relative displacements of the atoms in each of these normal modes are considered, it transpires that the first approximates the ≡C—H stretching (the nitrogen hardly moving relative to the carbon), the second the —C≡N stretching (the hydrogen hardly moving relative to the carbon), and the third doubly degenerate normal mode approximates the C—H bending out of the line of the N≡C— atoms in each of two mutually perpendicular planes.

We see from the force constants above that the —C≡N triple bond stretching is the 'stiffest' vibration, and that much less energy is required to bend the ≡C—H bond than to stretch it. These conclusions are fairly general – bending modes have smaller force constants than stretching modes, and force constants of bonds increase with the *bond order* (the number of bonding electron pairs minus the number of antibonding electron pairs).

The approximation that a particular normal mode can be associated with the stretching or bending of a particular functional group has allowed at least some bands in complicated spectra of complex organic compounds to be interpretable. Thus —C≡N, >C=O, —C—H, O—H, and other groups are regarded as having stretching and bending vibrations which give rise to absorption bands in characteristic parts of the infra-red spectrum.

A deviation from the expected wavenumber of such a group vibration is commonly interpreted in terms of special structural features of the particular compound. But because of the relative constancy of the wavenumber of many group vibrations, infra-red spectroscopy is a very useful analytical method in organic chemistry.[4]

In the case of R—H compounds, it is the very small mass of the hydrogen atom which causes some normal modes to approximate localized group vibrations. There can be no overall change in translational momentum due to a vibration, so the centre of mass of the molecule must remain stationary. (If it were otherwise, we could not treat the vibration and translation as independent.) Consequently, the amplitudes of some vibrations involve an exceedingly small movement by the group R of large mass, and a large and opposite movement by the very much lighter hydrogen atom. These vibrations, therefore, appear to be localized oscillations of only the hydrogen atom about a mean position:

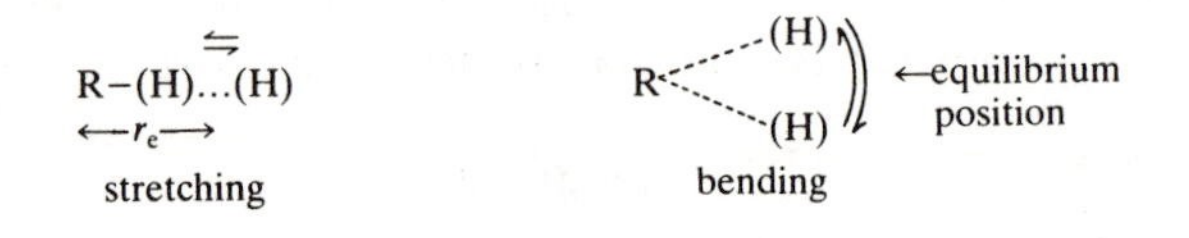

The explanation is different for compounds containing dipolar groups with multiple bonds such as >C=O and —C≡N. The force constants of the strong multiple bonds *within* these groups are very much higher than the force constants of the bonds which join these groups to the rest of the molecule – usually carbon–carbon single bonds. These weaker bonds allow only poor coupling between the vibrations of the rest of the molecule and those within the group. Consequently, localized internal vibrations of multiply-bonded groups are largely unaffected by the nature of the rest of the molecule.

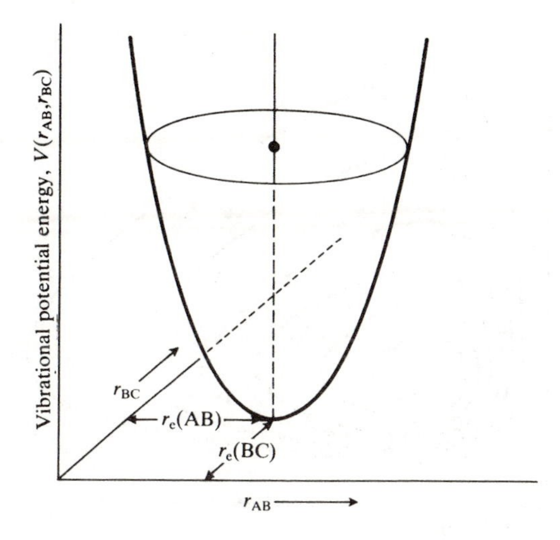

Fig. 1.4. Potential-energy surface for the two stretching modes of a linear triatomic molecule A—B═C according to the harmonic approximation.

Besides characteristic group vibrations, there are many other bands in the infra-red spectrum of a typical organic compound which are not easily interpretable. These include bands due to excitations of complex skeletal vibrations of the molecule as a whole which are usually not attributable to isolated structural features. None the less, they often prove invaluable as an empirical method of compound identification.[4]

1.3.4 Potential-energy surfaces

Vibrational modes of a molecule can be represented separately as potential-energy curves as we have already seen; alternatively, a pair can be combined to give a *potential-energy surface*. The normal modes which approximate the A—B— and —B═C stretchings in the linear molecule A—B═C are combined in Fig. 1.4. This includes the harmonic approximation and is the three-dimensional analogue of the potential-energy curve shown in Fig. 1.2. The surface is elipsoidal and vertical sections through it at r_e(AB) and r_e(BC) give the harmonic potential-energy curves for the —B═C and A—B═ bonds respectively. The former has the larger force constant as indicated by the eccentricity of the ellipsoid, and the position of the two-dimensional minimum shows that the single bond between A and B is longer than the double bond between B and C.

We can include qualitatively the effect of anharmonicity by constructing the potential-energy surface of a real molecule. This time in Fig. 1.5 we represent r(C—H) and r(C≡N) bond distances of H—C≡N along two

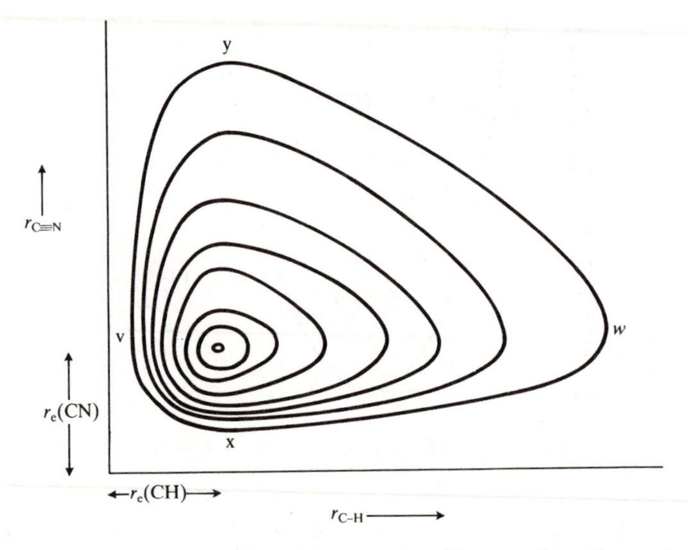

Fig. 1.5. Molecular potential-energy contour diagram for the anharmonic ≡C—H and —C≡N stretching vibrations in H—C≡N.

axes in the plane of the paper, and the molecular vibrational potential-energy axis comes vertically out of the paper. Points of equal energy are joined to give an energy contour diagram. In this way, the asymmetry of the surface due to anharmonicity is clearly shown by the different spacings between the contours for compression and stretching of the bonds. Sections through this surface perpendicular to the plane of the paper at vw and xy give the Morse molecular potential-energy curves for the single anharmonic stretching modes of the two bonds, each being similar to Fig. 1.3. Note also that as r_{CN} increases from r_e(CN) to ∞, r_e(CH) increases (from 1.066 to 1.1198 Å, 1 ångström = 10^{-10} m); correspondingly, as r_{CH} increases from r_e(CH) to ∞, r_e(CN) increases (from 1.153 to 1.175 Å).[8]

We cannot represent graphically a combination of more than two coordinates versus potential energy. Consequently, the full relationship between vibrational potential energy of a polyatomic molecule and its internal coordinates may be described only in mathematical form. Such a relationship is sometimes called a *force field.* It is a multi-dimensional equivalent of eqn (1.2) (p. 7) if it assumes harmonic approximations, or eqn (1.7) (p. 14) if anharmonicity is included. Because of the helpfulness of the graphical representations of the two- and three-dimensional potential-energy relationships (Figs 1.2–5), we can imagine that a force field expression describes a hypothetical *potential-energy hypersurface* – a multi-dimensional relationship between total potential energy and atomic configuration. These and related matters are discussed more fully in the next chapter.

1.3.5 Pyramidal inversion

A simple derivative of methane such as [1] has a unique minimum potential-energy configuration if groups X–Z have no internal structure, for example if they are single atoms. In the case of methane itself, this occurs at the tetrahedral distribution of hydrogen atoms about the central carbon, with equilibrium C–H bond lengths equal at 1.093 Å.[11]

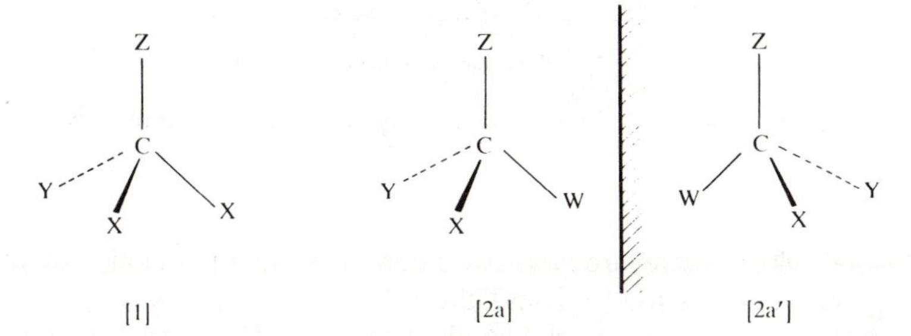

Compounds [2a] and [2a′] are structures which differ only in *chirality*. They are different configurations of five atoms with the same connectivity and therefore they correspond to different regions of a common potential-energy hypersurface. But this extended hypersurface contains an extremely high potential-energy barrier which effectively divides the whole surface into two; and one half, which corresponds to distortions of [2a], is the mirror image of the other half, which corresponds to distortions of [2a′]. This high energy barrier corresponds to all those configurations of CWXYZ which are coplanar and, therefore, achiral. The interconversion of [2a] and [2a′], regardless of mechanism, is an *inversion of configuration* and, for a saturated methane derivative, must proceed through a very high energy state.

There are chiral molecules, however, which undergo inversion of configuration without involving achiral configurations of exceedingly high potential energy. An illustration is the pyramidal molecule [3a] which may invert via the planar configuration [3b] to give [3a′], eqn (1.8), and Fig. 1.6 represents the potential-energy diagram for this process. This molecular inversion proceeds by a *pyramidal inversion* at the atom X.[12]

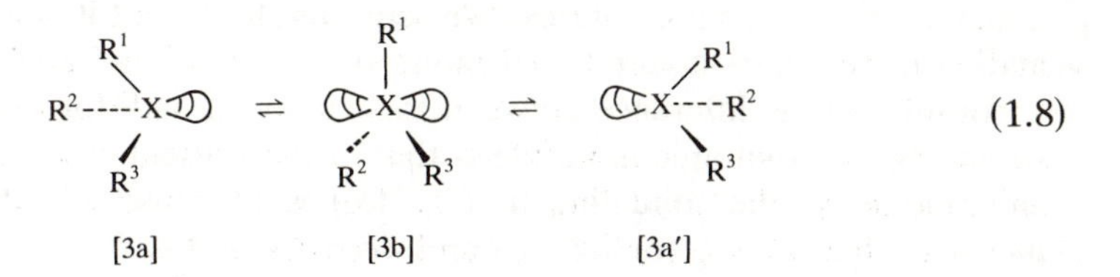

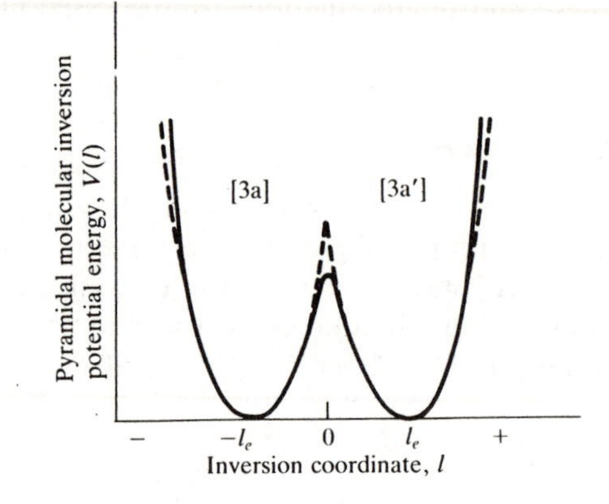

Fig. 1.6. Potential-energy curve for a pyramidal molecular inversion.

Figure 1.6 can be regarded as two interpenetrating parabolic potential-energy curves, one for [3a] and the other for [3a′], and the point of intersection corresponds to the configuration which is common to both – the planar form, [3b].

The inverson coordinate which is included in Fig. 1.6 is defined as l = the perpendicular distance from X to the plane which contains the centres of masses of the three ligands R^1, R^2, and R^3, and $l = l_e$ and $-l_e$ at the potential-energy minima of [3a′] and [3a] respectively.

If the vibrations of [3a] and [3a′] within the inversion coordinate are harmonic, the potential energy of the system is described by the broken line and, because [3a] and [3a′] are related by a mirror plane, the overall potential-energy curve for their interconversion must also contain a mirror plane.

For small deviations from either equilibrium structure, the vibration can be regarded as a bond angle bending mode with an associated force constant κ:

$$V(l) = \tfrac{1}{2}\kappa(l - l_e)^2 \quad \text{for [3a′] and}$$

$$V(l) = \tfrac{1}{2}\kappa(l + l_e)^2 \quad \text{for [3a].}$$

But at the energy maximum, the X–R bond lengths within the planar configuration may be significantly shorter than their values in the minimum energy configurations. We see, therefore, that it is an approximation to regard this vibrational mode simply as a bond angle bending.

An interesting difference between this vibration and, for example, the stretching of a diatomic molecule is that in the present case the effect of anharmonicity (the solid line in Fig. 1.6) is to cause $\kappa(= d^2V/dl^2)$ to *increase* when $l \gg +l_e$ (or $l \ll -l_e$) and *decrease* as $l \to 0$.

If the central potential-barrier in Fig. 1.6 is much higher than any appreciably occupied vibrational levels at a particular temperature, the inversion will be very slow and the molecules will simply execute a sort of 'breathing' vibration about the equilibrium pyramidal configuration. In contrast, if the potential-barrier is lower than appreciably occupied vibrational levels, the inversion will be fast and a time-averaged planar structure is observed even though the minimum energy configuration is still pyramidal.

Phosphines ([3], X = P) have high barriers to inversion and if compounds with different R^1, R^2, and R^3 groups are resolved, the enantiomers are configurationally stable at room temperature. Only by raising the temperature to 120–140 °C, and thereby increasing the population of the higher vibration levels, could compound [4] be racemized at an appreciable rate.[13] From Arrhenius plots (see p. 231) of rate constants for racemization at various temperatures, the barrier to inversion was estimated to be ~126 kJ mol^{-1}.

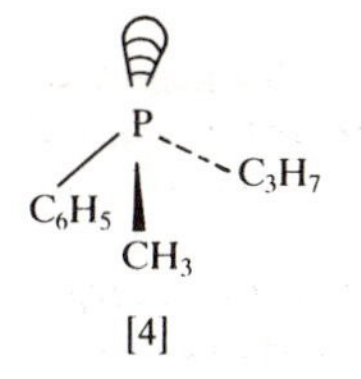

[4]

This method of estimating the barrier height involves resolving the compound (or preparing it in an enantiomerically-enriched state) and then measuring the rate of racemization (see p. 280). If the inversion barrier is too low (less than about 80 kJ mol^{-1}), the racemic modification will not be resolvable at room temperature; and if the barrier is too high (greater than about 140 kJ mol^{-1}), then the racemization will be too slow to measure accurately at reasonable temperatures. These limits for barrier heights, therefore, restrict the range of compounds amenable to investigation by this classical chemical kinetics method.

No simple acyclic amines ([3], X = N) have been resolved so their potential-energy barriers to inversion are very much lower than those of the analogous phosphines. Even below room temperature, nitrogen inversion in such compounds is usually rapid.

Molecular potential-energy barriers from less than 10 J mol^{-1} to over 70 kJ mol^{-1} have been investigated by microwave, infra-red, and Raman spectroscopy, or from the fine structure of electronic spectra. Experimentally, these techniques are quite specialized, but the usual procedure is to determine the shape of the potential-energy curve and the positions of the vibrational energy levels which best account for the observed spectra.[14] By such a method, the barrier to inversion of ammonia in the gas phase was found to be 24.16 kJ mol^{-1}.[6(b),16]

Barrier heights in an intermediate range (~30–80 kJ mol^{-1}) can be measured by dynamic nuclear magnetic resonance spectroscopy. If the rate of molecular inversion is relatively slow, the n.m.r. spectrum corresponds to the equilibrium pyramidal form. But if the inversion is very fast on the n.m.r. time-scale, the spectrum is of the time-averaged species, so the molecule appears planar. Over a particular temperature range, therefore, the spectrum is temperature-dependent. By detailed computer-assisted line-shape analysis of spectra obtained over this temperature range, a series of rate constants for the inversion process can be calculated.[17] The inversion barrier is then estimated from an Eyring plot (see p. 247) of the temperature dependence of the rate constants.

Some results for pyramidal molecular inversions by various methods are shown in Table 1.4.

The chirality of a whole molecule is inverted in the above examples by a process which is centred upon a particular atom – nitrogen or phosphorus. But if these, or any other comparable atoms, undergo their

TABLE 1.4

Potential-energy barriers for pyramidal molecular inversions

Compound	Potential-energy barrier/kJ mol^{-1}	Method
NH_3	24.16	i.r.[a]
CH_3NH_2	20.19	i.r.[b]
$C_6H_5NH_2$	5.43	resonance fluorescence spec.[c]
2-aminopyridine (ring N, NH_2)	2.99	microwave[d]
N-methyl-1,2-oxazinane (ring O–N, N–CH_3)	63.2	dynamic n.m.r.[e]
PH_3	156	calculated[f]
PF_3	189	calculated[g]
$PhP(CH_3)C_3H_7$	126	rate of racemization[h]

[a] See ref. 16.
[b] M. Tsuboi, A. Y. Hirakawa, and K. Tamigake, *J. molec. Spectrosc.* **22,** 272 (1967).
[c] M. Quack and M. Stockburger, *J. molec. Spectrosc.* **43,** 87 (1972).
[d] D. Christen, D. Norburg, D. G. Lister, and P. Palmieri, *J. chem. Soc., Faraday Trans. 2,* **71,** 438 (1975).
[e] F. G. Riddell, E. S. Turner, and A. Boyd, *Tetrahedron* **35,** 259 (1979).
[f] J. M. Lehn and B. Munsch, *Chem. Commun.* 1327 (1969).
[g] R. E. Weston, *J. Am. chem. Soc.* **76,** 2645 (1954).
[h] See ref. 13.

pyramidal inversion in a molecule which has another source of chirality, it interconverts a pair of *diastereoisomers* (stereoisomers which are not enantiomers). This is exemplified by the nitrogen inversion in the deuteriated aziridines, [5a, b], which have been investigated by dynamic

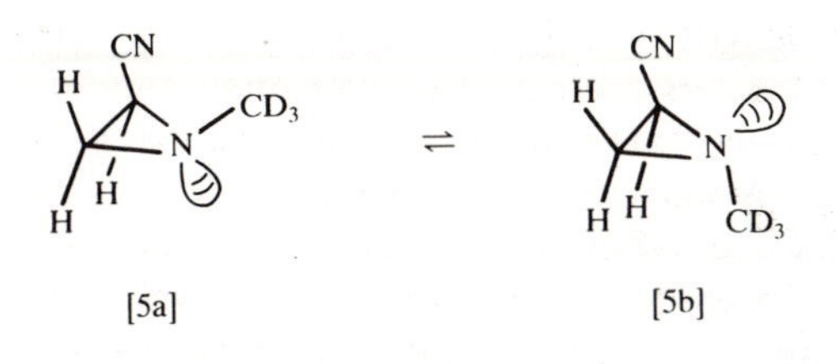

nuclear magnetic resonance spectroscopy.[18] The profile for the interconversion of [5a] and [5b] still comprises two interpenetrating vibrational potential-energy curves, one for [5a] and the other for [5b], but these are not enantiomeric. Consequently, the resultant (Fig. 1.7) lacks the symmetry of Fig. 1.6. Note that virtually the same phenomenon – pyramidal inversion – is regarded as a molecular vibration in ammonia, a mechanism for racemization of the phosphine [4], or a mechanism for equilibration of two chemically-different compounds [5a] and [5b] according to the molecular context.

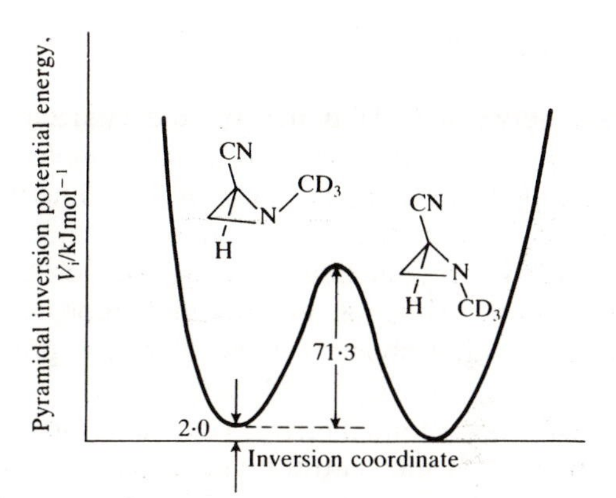

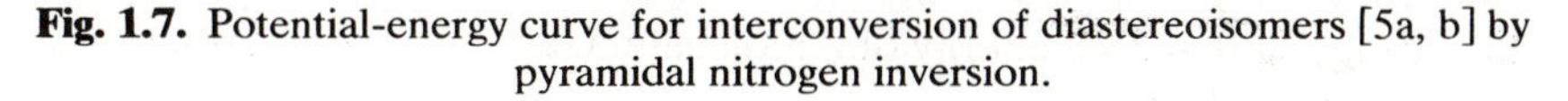
Fig. 1.7. Potential-energy curve for interconversion of diastereoisomers [5a, b] by pyramidal nitrogen inversion.

1.3.6 Internal rotation and torsional vibrations

Rotation of one part of an organic molecule with respect to another about a single bond which is not part of a cyclic structure is usually relatively free and is called an *internal rotation.*[15] If some structural feature inhibits the internal rotation without making it impossible, then it becomes a *hindered* or *restricted* rotation. These physical processes are really *torsional vibrations* (librations) with very low energy barriers, and must not be confused with the rotational modes of the molecule as a whole (see p. 4). Different molecular configurations with the same atom connectivity which are interconverted only by rotations about single bonds (with any associated bond length or angle deformations) or pyramidal inversions with low energy barriers are called *conformations*, and a molecule which has, or exists in, one conformation (a structural description) rather than another, exists as one *conformer* (a physical entity) rather than another.

Rotation of one methyl group through 120° about the central carbon–carbon bond of the conformation of ethane shown in [6a] interconverts conformations with the same minimum potential energy. These conformations are indistinguishable as we have not labelled the hydrogens, so another one, two, or more rotations of 120° will be equivalent to the first.

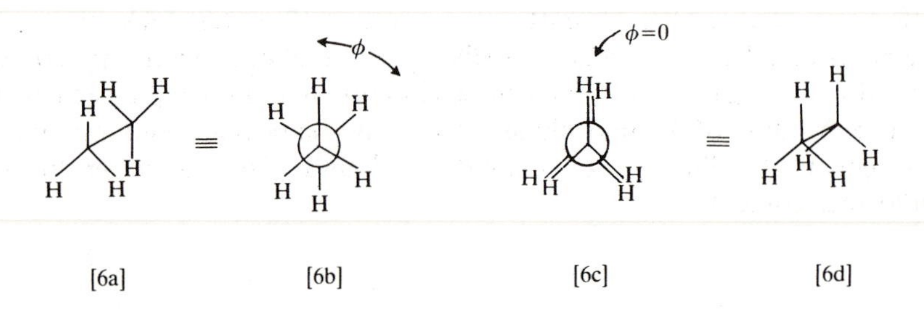

[6a] [6b] [6c] [6d]

The angle seen between C–H bonds on the adjacent carbon atoms when looking along the C–C axis, as shown for example in the *Newman projection* [6b], is called the *dihedral angle, ϕ*. The dihedral angle at the energy minimum is 60° and the C–H bonds are said to be *staggered* [6a, b]. The potential energy maximum in the rotation occurs when the dihedral angle is 0° and C–H bonds on adjacent carbons are *eclipsed.* This occurs in the conformation shown by the Newman projection [6c] of structure [6d].

A plot of the change in potential energy of a molecule of ethane against the angle of rotation about the central bond, ω, starting from an eclipsed conformation ($\omega = \phi = 0$) is shown in Fig. 1.8. The general equation for the change in potential energy of a molecule due to an internal rotation is given by eqn (1.9).

$$V(\omega) = \tfrac{1}{2}V_0(1 - \cos n\omega) \qquad (1.9)$$

where V_0 = the energy difference between minimum and maximum in the profile,
n = the number of times the profile repeats itself in 360° of bond rotation ($n = 3$ for ethane), and
ω = the angle of bond rotation.

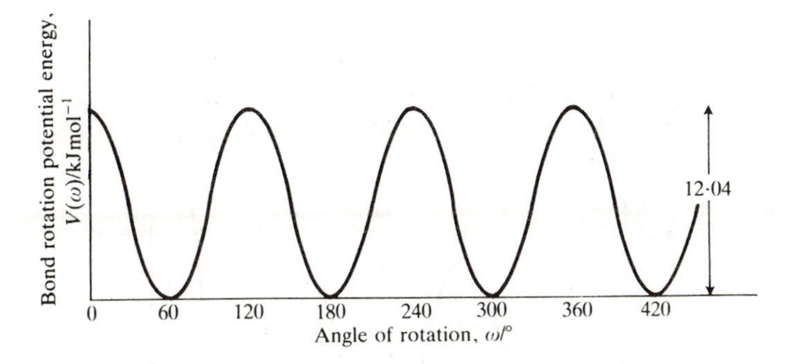

Fig. 1.8. Potential-energy profile for carbon–carbon bond rotation in ethane.

The molecular potential energy difference between eclipsed and staggered conformations is usually much smaller than that between atomic configurations which have substantially different bond angles or bond lengths. This is equivalent to stating that the force constant for an internal rotation about a single bond is usually much smaller than that for a bending or stretching vibration. The difference between maxima and minima (V_0 in eqn (1.9)) for ethane has been determined by gas-phase microwave spectroscopy and is only 12.04 kJ mol^{-1} as indicated in Fig. 1.8, so bond rotation is very rapid at room temperature.[19,15(b)] V_0 for hexachloroethane, which has been estimated from liquid-state spectroscopy, is substantially higher at 73.2 kJ mol^{-1}.[20]

The potential energy profile shown in Fig. 1.8 is *periodic* and appears very different from that, for example, in Fig. 1.3 (p. 12). Yet both represent a relationship between molecular shape and potential energy.

Figure 1.9 shows a rotation profile with potential energy plotted against ϕ, the dihedral angle between the terminal methyl groups of n-butane. This is not as simple as the one for ethane but it is still periodic and the diagram includes one complete cycle.

The two minima with methyl groups related by a dihedral angle of about 60°, corresponding to the so-called *gauche* conformations, are not as low in potential energy as the single minimum with the two methyl groups *trans* and coplanar ($\phi = 180°$). The energy difference between these minima has been measured by temperature-dependent n.m.r. spectroscopy on a deuteriated butane in solution (2.85 kJ mol^{-1})[21] and calculated from *ab initio* theory for the gas phase (4.0 kJ mol^{-1}).[22]

Correspondingly, the calculated[22] potential energy at the single maximum with the two methyl groups eclipsed ($\phi = 0°$) is higher than at the two enantiomeric maxima when both methyl groups are eclipsed with hydrogens ($\phi = 120°$). This latter barrier maximum, which separates the

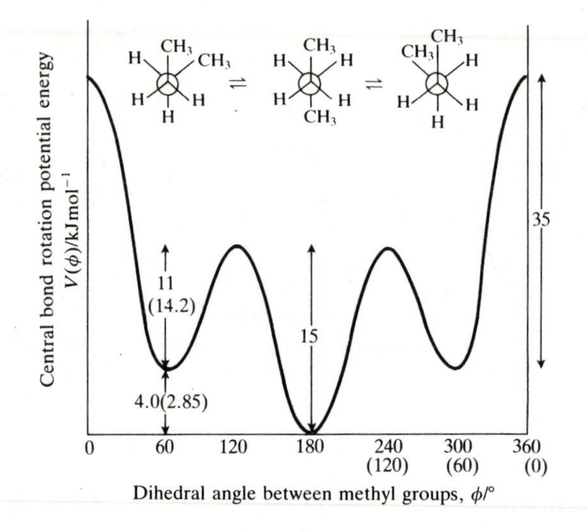

Fig. 1.9. Potential-energy profile for the central bond rotation in n-butane. (Energy differences are theoretical[22] and, in parentheses, experimental.[21,23])

gauche and *trans* conformations, has also been estimated experimentally by an ultrasonic relaxation technique (14.2 kJ mol^{-1} above the gauche form).[23]

Pairs of compounds such as *cis*- and *trans*-but-2-ene have the same atom connectivity and correspond to different regions of an extended potential-energy hypersurface. Fig. 1.10 is the potential energy curve for their *torsional vibrations* (the reversible twisting of one $=CHCH_3$ group with respect to the other i.e. partial rotation about a carbon–carbon double bond), and the internal coordinate is the dihedral angle between the methyl groups.

Again the profile is periodic, and in order to identify *cis* and *trans* stereoisomers easily, more than just one cycle has been included in Fig. 1.10. But in this example, the potential-energy barrier separating *cis* and *trans* structures is very high[24] and thermal interconversion is not possible at normal temperatures. Consequently, the two diastereoisomers are distinct isolable compounds and both undergo torsional vibrations described by the common internal coordinate. Quantized energy levels are indicated in Fig. 1.10.

The difference between Figs. 1.8 and 1.9 on the one hand, and Fig. 1.10 on the other is one of degree rather than fundamental type. Consequently, the sharp distinction which is commonly held to exist between rapidly interconverting conformers and separable stereoisomers

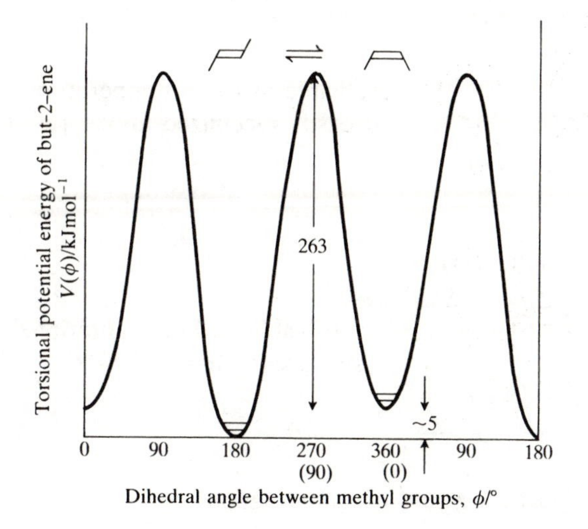

Fig. 1.10. Potential-energy profile for the torsional vibrations of *cis*- and *trans*-but-2-enes.

is not always justifiable. For example, the thermal interconversion of the enantiomeric biphenyls [7a] and [7b] is easy or difficult according to the size of the substituents adjacent to the central carbon–carbon bond.[25]

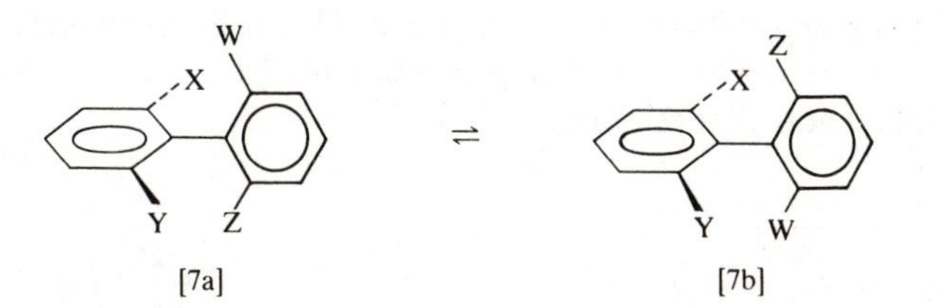

Alternatively, we can say that a given enantiomeric pair [7a, b] will be rapidly interconverting conformers of a single achiral compound, or a potentially-resolvable racemic modification according to the temperature. Examples are shown below:

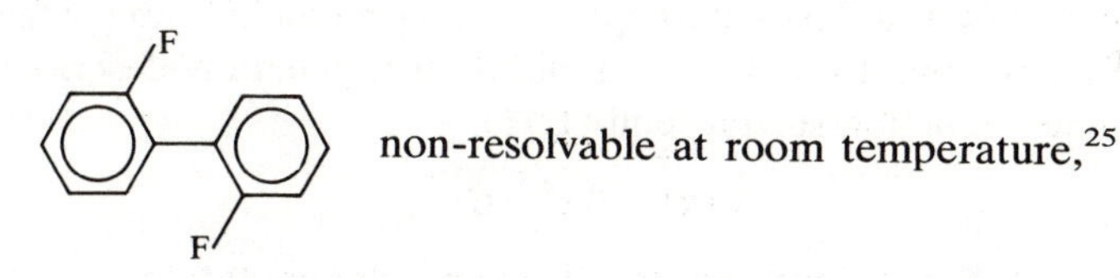

non-resolvable at room temperature,[25]

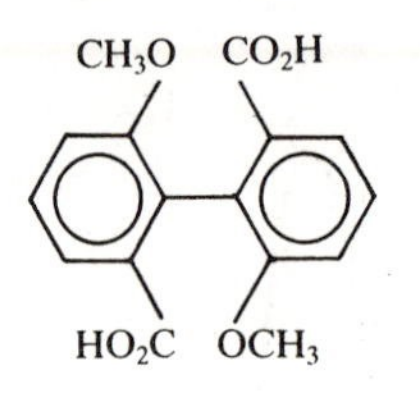

resolvable at room temperature, but easily racemized above 120 °C,[25]

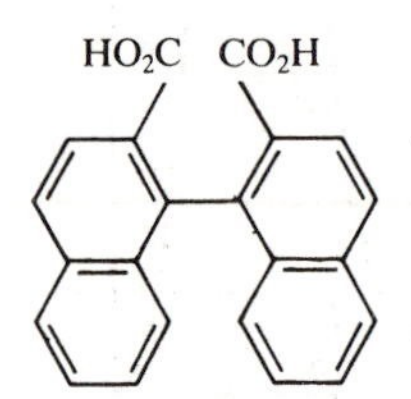

resolvable and configurationally stable even above 140 °C.[26]

1.3.7 Ring puckering and pseudo-rotation

So far we have dealt with pyramidal inversion, interconversion of acyclic molecules by rotations about carbon–carbon single bonds, and *cis-trans* isomerizations of a simple alkene by rotation about a carbon–carbon double bond. Cyclic molecules have very interesting conformational properties and the facility of internal bond rotation and, therefore, conformational mobility is determined to a large extent by the ring size.

1.3.7.1 Four-membered rings[27,28]

The *ring-puckering* vibration shown in eqn (1.10) interconverts different conformers of four-membered ring compounds [8] and, for cyclobutane ([8], $X = CH_2$) and silacyclobutane ([8], $X = SiH_2$),

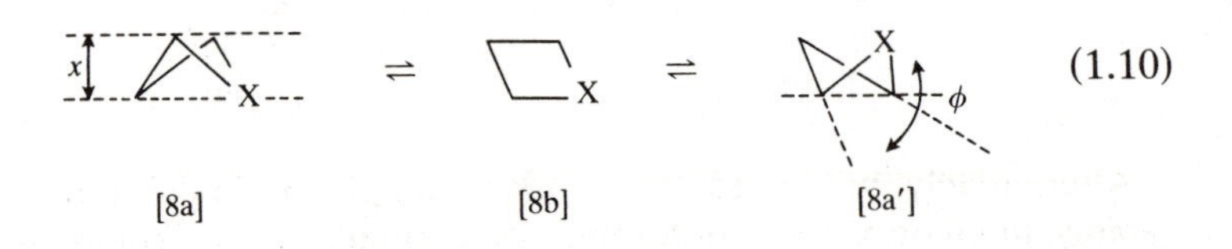

the potential-energy curve (Fig. 1.11) is superficially similar to that for the molecular inversion shown in Fig. 1.6 (p. 20). But whereas Fig. 1.6 (p. 20) is satisfactorily regarded as the interpenetration of two enantiomeric harmonic curves (modified perhaps for anharmonicity), the change in potential energy due to a puckering vibration is generally described by a more complex function such as eqn (1.11)

$$V(x) = Ax^4 - Bx^2, \tag{1.11}$$

where A and B are constants for the particular molecule and x is an

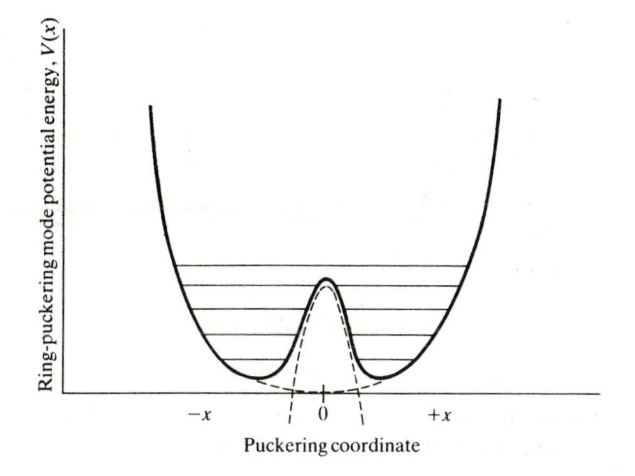

Fig. 1.11. Potential-energy curve for the ring-puckering vibration of cyclobutane and silacyclobutane.

internal coordinate which describes the pucker and is zero when the ring is planar. This equation comprises a quartic term which accounts for the steep-sided U-shaped curve with the broken base-line in Fig. 1.11 *minus* a squared term for the central inverted parabolic broken line. The algebraic sum of these components gives the full line and the size of the central maximum is described by the magnitude of the constant B. Although the conformations [8a], [8b], and [8a′] interconvert by partial rotations about single bonds, the most convenient internal coordinate, x, is not angular. It is the distance between the two diagonals of the four-membered ring shown in [8a].

In the case of cyclobutane[29] and silacyclobutane,[30] the infra-red and microwave spectra are best interpreted by potential-energy curves with central barriers (6.02 and 5.29 kJ mol^{-1} respectively) which are higher than the first few vibrational levels as indicated in Fig. 1.11. Molecules in these lower vibrational levels oscillate about a mean puckered shape ($\phi = 35$ and 28° respectively) where the angle of pucker ϕ is defined in [8a′].

The microwave spectra of cyclobutanone[31] and oxetane[32] ([8], X = C=O and O respectively) are most satisfactorily accommodated by potential energy curves with very small barriers (91 and 183 J mol^{-1} respectively) which are lower even than the ground-state vibrational levels (Fig. 1.12, full line). These molecules, therefore, execute a complete puckering vibration in all levels and the time-averaged configuration is planar even though the potential energy curve has the two minima.

3,3-Difluoro-oxetane is also a planar molecule which undergoes a puckering vibration. But in this case, the spectral results do not require a

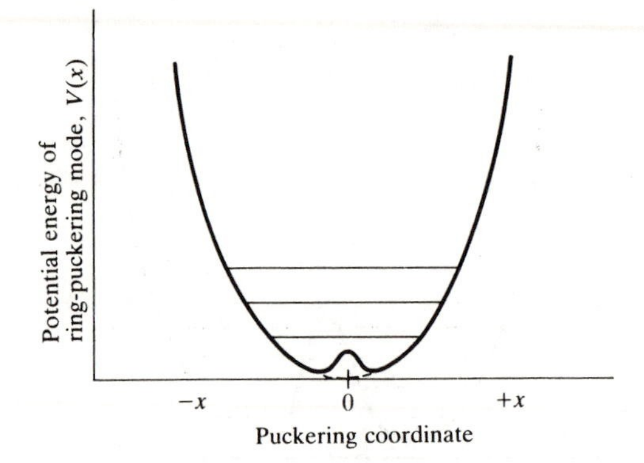

Fig. 1.12. Potential-energy curve for the ring-puckering vibrations of cyclobutanone, oxetane, and 3,3-difluoro-oxetane.

F CH_2
C O
F CH_2

central maximum at all,[33] so the potential-energy curve follows the dotted line in Fig. 1.12.

1.3.7.2 Cyclopentane[27]

A five-membered ring may also undergo a ring-puckering vibration as indicated by diametrical equations in Fig. 1.13. But whereas a four-membered ring has only a single puckering mode, the five-membered analogue has more. Each of the five methylene groups of cyclopentane can oscillate about a mean position in the plane of the ring, and these five vibrations share a common potential energy maximum corresponding to the planar conformation, [9].

However, the *envelope*-shaped conformers can also interconvert by a route which bypasses the planar conformation. This process is represented by the peripheral equations of Fig. 1.13. It can be considered as a ripple moving around the five-membered ring with each carbon atom moving up and down in turn. The molecule as a whole does not rotate but, because the ripple moves round, this process is called *pseudo-rotation.* The envelope conformations [10a–e] and their mirror images [10a′–e′], all of which involve four co-planar carbon atoms, correspond to the maxima in the potential-energy profile for pseudo-rotation shown in Fig. 1.14. The minima are the *twist* conformations [11a–e] and their mirror images [11a′–e′]; none of these involve more than three carbon atoms in any one plane.

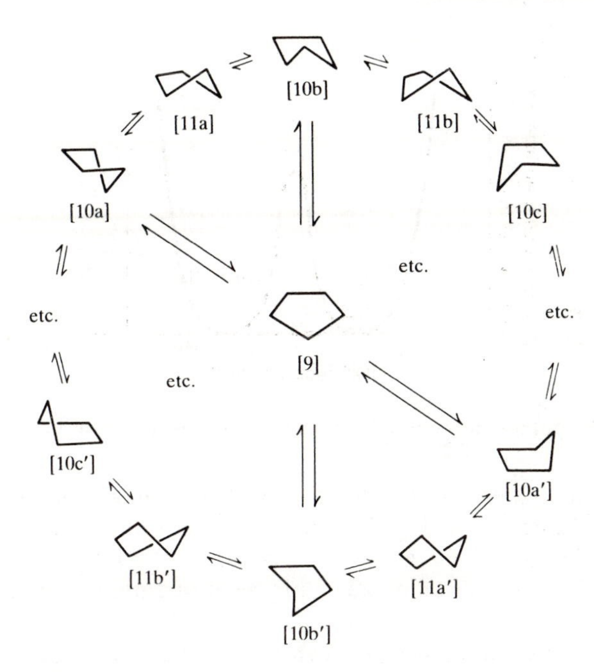

Fig. 1.13. Pseudo-rotation and ring-puckering modes of cyclopentane.

The energy profile in Fig. 1.14 is periodic and involves ten maxima and ten minima in one complete cycle. The potential energy difference between envelope and twist conformers has been calculated[34,35(a)] to be very low (about 1 kJ mol^{-1}) so pseudo-rotation is fast. (The value for tetrahydrofuran, a hetero-analogue of cyclopentane has been measured by microwave spectroscopy and is 682 J mol^{-1}.)[36] Although the time-averaged structure of cyclopentane is planar, there is no planar conformation in the pseudo-rotation itinerary.

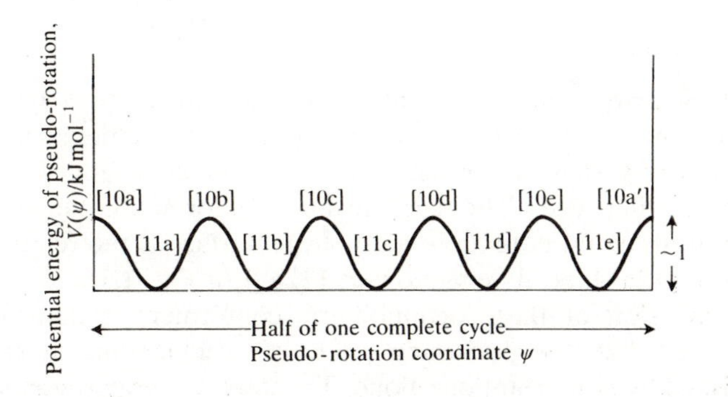

Fig. 1.14. Potential-energy profile for pseudo-rotation of cyclopentane.

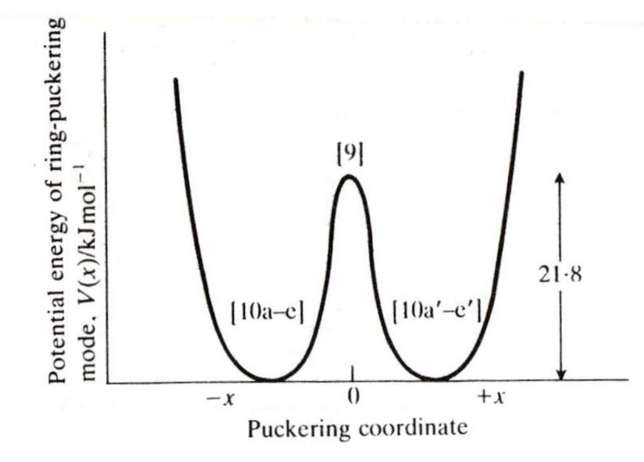

Fig. 1.15. Potential-energy profile of ring-puckering vibrations of cyclopentane.

The potential-energy curve for the puckering mode of the envelope conformers illustrated in Fig. 1.15 has a higher central potential-energy barrier (21.8 kJ mol^{-1} by gas-phase Raman spectroscopy[37]), and steep outer sides. It is not periodic, but reminiscent of earlier ones such as Figs 1.6 and 1.11.

The quantitative difference between Figs 1.14 and 1.15 can be ascribed to the different natures of the two vibrations. Whereas pseudo-rotation involves internal rotations with bond lengths and angles remaining constant, the puckering mode requires bond angle deformation and the force constants for bond angle bending are considerably larger than those for internal rotation.

The conformational processes described by Figs 1.14 and 1.15 (the peripheral and diametrical equations in Fig. 1.13) are conceptually distinguishable and mathematically separable. But if we wish to represent the change in potential energy due to concurrent pseudo-rotation and ring-puckering, we could construct a potential-energy surface of the sort encountered previously (p. 17).

1.3.7.3 Cyclohexane

There was a sharp increase in conformational complexity as we progressed from four- to five-membered rings (the conformational analysis of three-membered rings is trivial) and it increases again with further increases in ring size. The cyclohexane conformation at the lowest minimum in the total potential-energy hypersurface is the relatively rigid *chair* form, [12a] (and its enantiomer, [12a′]) in Fig. 1.16.

Interconversion of these enantiomeric conformers is known as *ring inversion* and, like the puckering mode of cyclopentane, it cannot be accomplished by bond rotations alone. The lowest energy route occurs via a family of flexible conformers which are themselves interconverted by

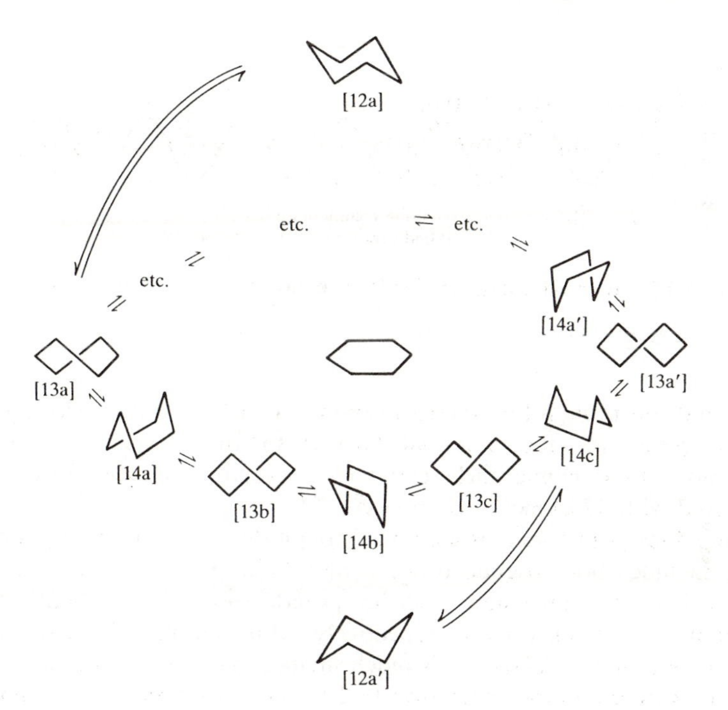

Fig. 1.16. Pseudo-rotation and ring-inversion of cyclohexane.

pseudo-rotation.[38] The equatorial equation in Fig. 1.16 describes this pseudo-rotation which is exactly analogous to the cyclopentane process described earlier. We have here not a planar diagram as in Fig. 1.13, but a spherical one with pseudo-rotation around the equator and [12a] and [12a′] at polar positions. The centre of the sphere represents the extremely high energy cyclohexane with six coplanar carbon atoms.

The local energy minima in the cyclohexane pseudo-rotation correspond to *twist* or *twist-boat* conformations, [13], and the maxima to *boat* conformations, [14]. The pseudo-rotation energy profile is again periodic (Fig. 1.17) but this time there are 6 maxima and 6 minima in one complete cycle. The potential-energy barrier to pseudo-rotation has been calculated to be larger (about 5 kJ mol^{-1}) than in the cyclopentane system,[34,35(a)] but the process is still very fast compared with ring inversion. However, because the potential-energy difference between the chair and twist conformations is about 20–25 kJ mol^{-1} (based upon calorimetric results using perhydroanthracenes[39] and theory[34,35(a)]), cyclohexane is overwhelmingly in the chair conformation at normal temperatures.

The potential-energy curve for cyclohexane ring inversion is shown in Fig. 1.18. The solid line represents the interconversion via a twist-boat

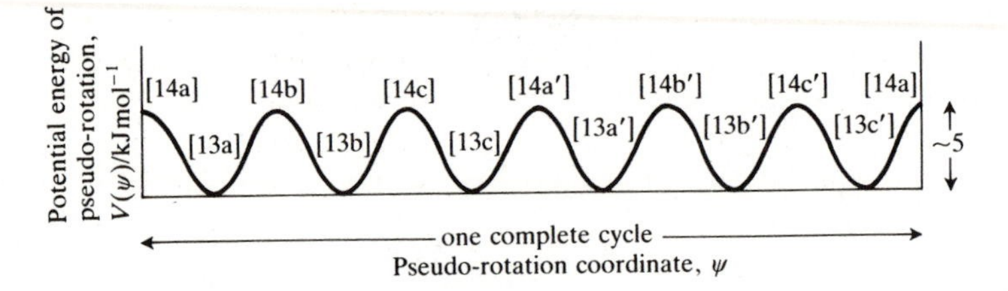

Fig. 1.17. Potential-energy profile for pseudo-rotation in cyclohexane.

form, and the broken line interconversion via a boat form. (The experimental result for the potential barrier to ring inversion is about 45 kJ mol^{-1} by dynamic nuclear magnetic resonance spectroscopy,[40,35(b)] compared with 47 kJ mol^{-1} by theory[34,35(a)]).

These lines can be seen as sections through the potential-energy surface which includes both pseudo-rotation and chair inversion. The full and broken lines for inversion cut the pseudo-rotation coordinate at a minimum and a maximum respectively. But because the barrier to pseudo-rotation (~5 kJ mol^{-1}) is much smaller than the barrier between a twist conformer and the chair form (~20–25 kJ mol^{-1} measured from the twist conformation), it is most improbable that chair inversion passes through only a single twist or boat conformation.

There are obvious similarities between the conformational properties of cyclopentane and cyclohexane. Both can be seen in terms of a family of

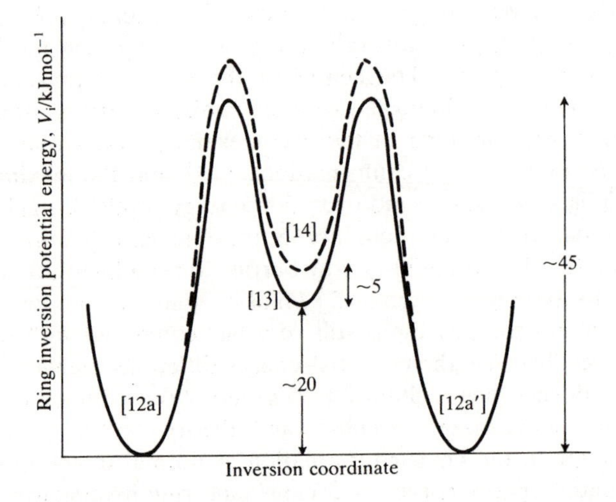

Fig. 1.18. Potential-energy profile for ring-inversion of cyclohexane.

flexible conformers interconverted by periodic pseudo-rotations which involve low potential-energy barriers. But whereas the maxima in the cyclopentane pseudo-rotation are also related by a higher energy route (puckering), the minima and maxima of cyclohexane's pseudo-rotation are a connected series of intermediate states in the inversion process which interconverts the two much more stable enantiomeric chair conformers.

1.4 Problems

1. Calculate (i) the energy per photon and (ii) the energy per mole of photons for the following electromagnetic radiations:
(a) vacuum ultraviolet, $\lambda = 160$ nm;
(b) ultraviolet, $\lambda = 230$ nm;
(c) visible, $\lambda = 500$ nm;
(d) infra-red, $\bar{\nu} = 900\ \text{cm}^{-1}$;
(e) microwave $\bar{\nu} = 1.00\ \text{cm}^{-1}$;
(f) n.m.r. spectrometer radiofrequency, $\nu = 90.0$ MHz.
2. Identify the types of degrees of freedom in the following molecules (translations, rotations, bending and stretching vibrations):
(i) HF
(ii) N_2
(iii) CO_2
(iv) H_2O
(v) HOCl
(vi) COS
(vii) O_3
3. For each of the following compounds, calculate the energy difference between the electronic states corresponding to the ultraviolet spectral absorption band indicated:[1]

Compound (solvent)	λ_m/nm
(i) Benzene (cyclohexane)	256
(ii) Naphthalene (95% aqueous ethanol)	312
(iii) Anthracene (cyclohexane)	378
(iv) Tetracene (benzene)	471

4. Calculate the energy difference between the vibrational states of methanol corresponding to each infra-red absorption band indicated:[4]

Vibration	$\bar{\nu}/\text{cm}^{-1}$
(i) —O—H stretch	3645
(ii) —O—H bend	1346
(iii) >C—O— stretch	1034

5. Frequencies of radiation corresponding to the $v=0$ to $v=1$ vibrational transitions of the following homonuclear diatomic molecules have been determined from Raman spectroscopy.[5,6(a)] Assuming the simple harmonic approximation, calculate (i) the force constants of the bonds and (ii) the molar zero-point energies.

	Molecule	$\bar{\nu}/\text{cm}^{-1}$
(a)	H_2	4160.2
(b)	N_2	2330.7
(c)	O_2	1554.7
(d)	Cl_2	556.9
(e)	I_2	213.4

6. Calculate the equilibrium distribution of molecules at 25 °C between the two energy states referred to in parts (a), (c), and (e) of Problem 5.
7. A weak Raman band due to a puckering transition of cyclobutane has been identified at $197\ \text{cm}^{-1}$.[29] Calculate the energy difference between the two states involved and the equilibrium distribution of molecules between these two states at 25 °C.
8. The unstable compound CF ($\bar{\nu}=1308\ \text{cm}^{-1}$)[8] is isoelectronic with NO ($\bar{\nu}=1876\ \text{cm}^{-1}$, $\kappa=1548\ \text{N m}^{-1}$, and zero-point energy $=11.2\ \text{kJ mol}^{-1}$; see Tables 1.1 and 1.3). Calculate (i) the force constant and (ii) the molar zero-point energy of CF assuming the simple harmonic approximation. Account for the vibrational difference between the two compounds.
9. The magnetic field of a nuclear magnetic resonance spectrometer removes the degeneracy of nuclear spin energy states for hydrogen (and certain other nuclei). Transitions between the energy levels of a hydrogen nucleus may then be induced by radiofrequency electromagnetic radiation. Calculate (i) the molar energy difference between the two spin states when the transition is induced by radiation at 60 MHz, and (ii) the equilibrium distribution of a large assembly of hydrogen nuclei between these two levels at 35 °C.

1.5 References

1. H. H. Jaffe and M. Orchin, *Theory and applications of ultraviolet spectroscopy*, Wiley, New York (1962).
2. N. J. Turro, *Modern molecular photochemistry*, Benjamin/Cummings Publ. Co., USA (1978); R. P. Wayne, *Photochemistry*, Butterworths, London (1970).
3. I. Fleming, *Frontier orbitals and organic chemical reactions*, Wiley, Chichester (1976); R. E. Lehr and A. P. Marchand, *Orbital symmetry*, Academic Press, New York (1972).
4. L. J. Bellamy, *The infra-red spectra of complex molecules*, Vol. 1, 3rd edn, Chapman & Hall, London (1975).
5. G. M. Barrow, *Introduction to molecular spectroscopy*, McGraw-Hill (1962).

6. G. Herzberg, *Molecular spectra and molecular structure*, (*a*) Vol. 1: *Spectra of diatomic molecules*, 2nd edn, Van Nostrand-Reinhold (1950); (*b*) Vol. 2: *Infrared and Raman spectroscopy of polyatomic molecules*, Van Nostrand (1945). These advanced texts, and their companion, Vol. 3: *Electronic spectra and electronic structure of polyatomic molecules*, published in 1966, include rigorous theoretical treatments of molecular spectroscopy and definitive compilations of experimental results available up to the dates of their publications.
7. W. V. F. Brooks and B. Crawford, *J. chem. Phys.* **23,** 363 (1955).
8. *JANEF Thermochemical tables*, 2nd edn (eds. D. R. Stull and H. Prophet) NSRDS-NBS37, National Bureau of Standards, Washington DC (1970).
9. A. G. Gaydon, *Dissociation energies and spectra of diatomic molecules*, 3rd edn, Chapman & Hall, London (1968).
10. R. N. Dixon, *Spectroscopy and structure*, Methuen, London (1965).
11. H. C. Allen and E. K. Plyler, *J. chem. Phys.* **26,** 972 (1957).
12. A. Rauk, L. C. Allen, and K. Mislow, 'Pyramidal inversion', *Angew. Chem. int. edn* **9,** 400 (1970); J. B. Lambert, 'Pyramidal atomic inversion' in *Topics in stereochemistry*, (eds. N. L. Allinger and E. L. Eliel), Vol. 6, p. 19, Wiley-Interscience (1971).
13. H. D. Munro and H. Horner, *Tetrahedron* **26,** 4621 (1970).
14. N. L. Owen (ref. 15(*a*)) gives an account of the spectroscopic determination of low potential-energy barriers including those for pyramidal inversion.
15. *Internal rotation in molecules* (ed. W. J. Orville-Thomas), Wiley (1974): (*a*) Chapt. 6, 'Studies of internal rotation by microwave spectroscopy' by N. L. Owen; (*b*) Chapt. 8, 'Torsional vibrations and rotational isomerism' by G. Allen and S. Fewster; (*c*) Chapt. 4, 'Infra-red and Raman band intensities and conformational change' by P. J. D. Park, R. A. Pethrick, and B. H. Thomas; (*d*) Chapt. 9, 'Molecular acoustics and conformational behaviour' by S. M. Walker; (*e*) Chapt. 11, 'Ab initio calculations of barrier heights' by A. Veillard. See also E. B. Wilson, 'Conformational studies on small molecules', *Chem. Soc. Rev.* **1,** 293 (1972).
16. J. D. Swalen and J. A. Ibers, *J. chem. Phys.* **36,** 1914 (1962).
17. G. Binsch, 'The study of intramolecular rate processes by dynamic nuclear magntic resonance', in *Topics in stereochemistry* (eds. E. L. Eliel and N. Allinger), Vol. 3, p. 97, Wiley-Interscience (1968).
18. D. Höfner, I. Tamir, and G. Binsch, *Org. mag. Res.* **11,** 172 (1978).
19. E. Hirota, S. Saito, and Y. Endo, *J. chem. Phys.* **71,** 1183 (1979).
20. G. Allen, P. N. Brier, and G. Lane, *Trans. Faraday Soc.* **63,** 824 (1967).
21. P. B. Woller and E. W. Garbisch, *J. Am. chem. Soc.* **94,** 5310 (1972); see also S. Kint, J. R. Scherer and R. G. Snyder, *J. chem. Phys.* **73,** 2599 (1980), and reference 15(*c*). Slightly different results for the gas phase of about 2.6 and 4.0 kJ mol^{-1} have been deduced from electron diffraction (R. A. Bonham and L. S. Bartell, *J. Am. chem. Soc.* **81,** 3491 (1959), K. Kuchitsu, *Bull. chem. Soc. Japan* **32,** 748 (1959)) and Raman experiments (A. L. Varma, W. F. Murphy, and H. J. Bernstein, *J. chem. Phys.* **60,** 1540 (1974)), respectively.
22. M. R. Peterson and I. G. Csizmadia, *J. Am. chem. Soc.* **100,** 6911 (1978); see also L. Radom and J. A. Pople, ibid. **92,** 4786 (1970).
23. J. E. Piercy and M. G. S. Roa, *J. chem. Phys.* **46,** 3951 (1967); see also ref. 15(*d*), and E. Wyn-Jones and R. A. Pethrick, 'The use of ultrasonic absorption and vibrational spectroscopy to determine energies associated with conformational changes' in *Topics in Stereochemistry* (eds. E. L. Eliel and N. L. Allinger), Vol. 5, p. 205. Wiley-Interscience (1970).

24. B. S. Rabinovitch and K. W. Michel, *J. Am. chem. Soc.* **81,** 5065 (1959). The potential-energy difference between the minima corresponding to *cis*- and *trans*-isomers has been estimated from their standard enthalpies of formation at 298 K in the gas phase. See p. 51 in the following chapter, and J. D. Cox and G. Pilcher, *Thermochemistry of organic and organometallic compounds*', Academic Press, London (1970).
25. A general account of the stereochemistry of biphenyls is given by E. L. Eliel in *The stereochemistry of carbon compounds*, McGraw-Hill, New York (1962).
26. D. M. Hall and E. E. Turner, *J. chem. Soc.* 1242 (1955).
27. A. C. Legon, 'Equilibrium conformations of four- and five-membered cyclic molecules in the gas phase', *Chem. Rev.* **80,** 231 (1980).
28. R. M. Moriarty, 'Stereochemistry of cyclobutane and heterocyclic analogs' in *Topics in Stereochemistry* (eds. E. L. Eliel and N. L. Allinger), Vol. 8, p. 271, Wiley-Interscience (1974).
29. J. M. R. Stone and I. M. Mills, *Molec. Phys.* **18,** 631 (1970).
30. W. C. Pringle, *J. chem. Phys.* **54,** 4978 (1971).
31. L. H. Scharpen and V. W. Laurie, *J. chem. Phys.* **49,** 221 (1968).
32. S. I. Chan, T. R. Borgers, J. W. Russell, H. L. Strauss, and W. D. Gwinn, *J. chem. Phys.* **44,** 1103 (1966).
33. G. L. McKown and R. A. Beaudet, *J. chem. Phys.* **55,** 3105 (1971).
34. J. R. Hoyland, *J. chem. Phys.* **50,** 2775 (1969).
35. *Conformational analysis: Scope and present limitations* (ed. G. Chiurdoglu), Academic Press, New York, (1971): (*a*) p. 129, 'Theoretical conformational analysis' by J. M. Lehn: see also ref. 15(*e*); (*b*) p. 15, 'Nuclear magnetic resonance studies of the conformations and conformational barriers in cyclic molecules' by F. A. L. Anet.
36. G. G. Engerholm, A. C. Luntz, W. D. Gwinn, and D. O. Harris, *J. chem. Phys.* **50,** 2446 (1969).
37. L. A. Carreira, G. J. Jiang, W. B. Person, and J. N. Willis, *J. chem. Phys.* **56,** 1440 (1972).
38. G. M. Kellie and F. G. Riddell, 'Non-chair conformations of six-membered rings' in *Topics in Stereochemistry*, (eds. E. L. Eliel and N. L. Allinger), Vol. 8, p. 225, Wiley-Interscience (1974).
39. J. L. Margrave, M. A. Frisch, R. G. Bautista, R. L. Clarke, and W. S. Johnson, *J. Am. chem. Soc.* **85,** 546 (1963).
40. F. A. L. Anet and A. J. R. Bourn, *J. Am. chem. Soc.* **89,** 760 (1967).

Supplementary references

G. M. Barrow, *The structure of molecules. An introduction to molecular spectroscopy*, W. A. Benjamin, New York (1964).
J. E. Crooks, *The spectrum in chemistry*, Academic Press, London (1978).
C. N. Banwell, *Fundamentals of molecular spectroscopy*, 2nd edn, McGraw-Hill, (1972).
E. L. Eliel, N. L. Allinger, S. J. Angyal, and G. A. Morrison, *Conformational analysis*, Interscience, New York (1965).
M. Hanack, *Conformation theory*, Academic Press, New York (1965).
J. A. Barltrop and J. D. Coyle, *Principles of photochemistry*, Wiley, Chichester (1978).

2

Thermochemistry

2.1 Introduction

In the previous chapter we focused our attention on some aspects of the structures and properties of discrete molecules. Experimentally, however, we cannot deal with individual molecules; we can deal only with huge assemblies of them. If an organic compound of low molecular weight is investigated in a mass spectrometer or some other instrument operating at low pressure, there will still be about 10^{10} molecules per cm^3 at 10^{-5} Torr; and, in dilute solution, there are more than 10^{15} molecules of solute in $1\,cm^3$ of a solution which is 10^{-5} molar.

In this and the next two chapters we look at the relationship between initial and final states of chemical reactions carried out on a macroscopic scale and see how the difference between the two states can be described other than in terms of molecular structures. This is part of the subject of *chemical thermodynamics*. In a sense, classical thermodynamics (as opposed to statistical thermodynamics) bypasses the atomic and molecular nature of matter; it is concerned with pressure, concentration, volume, temperature, and other familiar bulk (in contrast to molecular) properties, and with energy. Consequently we must, at some stage, reconcile these two quite different descriptions of the same phenomena so that we are able to interpret the thermodynamic account of compounds and reactions in terms of the properties and structures of individual molecules.

Some organic reactions, for example those between alkyl halides and a metal M such as lithium which generate synthetically useful metal alkyls,

$$\mathrm{RHal + 2M \rightarrow RM + MHal},$$

have to be cooled when they are done on a large scale to prevent them from getting out of control.

Cooling is necessary when the early stages of a reaction produce heat which, if not dissipated quickly enough, is sufficient to raise the temperature of the system sharply and, in turn, cause the reaction to become faster with an even more rapid evolution of further heat; and so on until at least one of the reactants has been used up or, in the most extreme cases, the reaction explodes. A chemical reaction which produces heat is called *exothermic.*

A few organic reactions such as the hydrolysis of n-pentanamide:[1]

$$\mathrm{n\text{-}C_4H_9CONH_2 + H_2O \xrightarrow{H_2O} n\text{-}C_4H_9CO_2H + NH_3},$$

proceed with the absorption of heat from their surroundings and are called *endothermic.* If heat is not supplied to an endothermic reaction, it cools down and consequently becomes slower. A reaction which neither evolves nor absorbs heat is *thermoneutral.*

Reactions may be carried out under *isothermal* conditions (at constant temperature) by the use of some thermostatting arrangement, for example a constant temperature bath. In keeping the temperature constant, the bath supplies heat to an endothermic reaction and accepts heat generated by the exothermic reaction.

Alternatively, if the chemical system is thermally insulated so that there can be no heat transfer to or from its environment, any heat change can only cause the temperature of the reaction to rise or fall. Such reaction conditions are called *adiabatic.* A reaction can be carried out isothermally, adiabatically, or more casually with no concern at all for either temperature stability or thermal insulation, depending upon the purpose of the particular reaction.

In the rest of this chapter we consider quantitatively heat changes associated with chemical reactions; this is what we mean by *thermochemistry.* We shall not, however, cover in any detail the experimental side which is called *calorimetry*, and most of the reactions we shall deal with will be isothermal. But as a preliminary, we need to consider the molecular basis of the heat changes which accompany chemical reactions.

2.2 Internal energy

We saw in the previous chapter how the energy of a single molecule at a particular temperature is affected by its electronic configuration and, if it is polyatomic, by its vibrational and rotational state. A molecule also has translational thermal energy of $\frac{1}{2}mv^2$ where m is the molecular mass and v its linear velocity. This translational thermal energy is directly proportional to the absolute temperature.

At normal temperatures, most molecules are in their ground electronic states, but otherwise identical members of a large assembly of polyatomic molecules will be distributed among many vibrational and rotational levels. Consequently, the *total* potential and kinetic energy of a large assembly of molecules, treated as an ideal gas, is determined by

(i) the number of molecules,
(ii) the natures of the vibrations and rotations, and
(iii) the absolute temperature.

To a good level of approximation, this energy can be calculated for simple molecules, but not for large organic molecules.

If we are dealing not with an ideal gas but with a collection of real molecules, then we also have to take intermolecular interactions into account. Currently, this is theoretically practicable for small numbers of only very simple molecules and impossible for large assemblies of complex organic molecules. But when the total energy due to all intermolecular interactions (whatever it might be) is added to the potential and kinetic energy of the molecules, we have the total *intrinsic* or *internal energy* U of the system, (sometimes the symbol E is used).

At a given temperature we can expect the internal energy of one mole (6.022×10^{23} molecules) of one compound to be different from that of one mole of another compound. This is because each type of molecule has its own mass and set of force constants, its own moments of inertia, and, for example, its own dipole moment; consequently, the two types of molecule have different velocities, rotations, vibrations, and intermolecular interactions. An isothermal chemical reaction, therefore, such as eqn (2.1) involves the conversion of one mole of A with internal energy $U(\text{A})$ into B with internal energy $U(\text{B})$ and $U(\text{B})$ is usually unequal to $U(\text{A})$.

$$\text{A} \rightarrow \text{B}. \tag{2.1}$$

Energy (strictly, mass-energy) cannot, of course, be created or destroyed but only converted from one form into another. So a chemical reaction at constant temperature such as eqn (2.1) must involve a conversion between internal energy and some other form of energy. If the reaction is carried out at constant volume so that no mechanical work can be done by or on the system, this other form of energy is heat, and heat changes can be measured.

Although we cannot calculate or measure the molar internal energy U of a compound on an absolute scale, we can measure, sometimes with great precision, internal energy *changes* ΔU. The overall change in internal energy associated with the isothermal formation of products from reactants at constant volume in one mole of a chemical reaction as described by a chemical equation is called simply the *molar internal energy of the reaction*, $\Delta_r U$.

$$\text{Reactants} \xrightarrow[\text{constant temperature, } T]{\text{constant volume, } V} \text{Products;}$$

$$\Delta_r U_{V,T} = \sum U(\text{products})_{V,T} - \sum U(\text{reactants})_{V,T}.$$

If $\sum U(\text{products})_{V,T} < \sum U(\text{reactants})_{V,T}$, internal energy is converted into heat energy, and the evolution of heat (exothermic process) corresponds to $\Delta_r U_{V,T}$ with a *negative* sign. Correspondingly, a positive $\Delta_r U_{V,T}$ corresponds to an endothermic reaction with $\sum U(\text{products})_{V,T} > \sum U(\text{reactants})_{V,T}$.

If a reaction at constant volume is adiabatic and not thermoneutral, the temperature must change so that the total internal energy of the products at the *new* temperature is equal to the total internal energy of reactants at the original temperature.

$$\text{Reactants, } T_1 \xrightarrow[\text{constant volume, } V]{\text{adiabatically and at}} \text{Products, } T_2;$$

$$\sum U(\text{products})_V, T_2 = \sum U(\text{reactants})_V, T_1.$$

2.3 Enthalpy

When an isothermal exothermic reaction takes place not at constant volume, but at constant pressure, and involves an increase in volume as, for example, in the combustion of butane above 100 °C,

$$C_4H_{10} + 6.5O_2 \xrightarrow[\text{constant pressure}]{T > 100\,°C} 4CO_2 + 5H_2O$$

$$1 \text{ vol.} + 6.5 \text{ vol.} \rightarrow 4 \text{ vol.} + 5 \text{ vol.},$$

some of the energy liberated by the reaction is used to expand the system against the constant external pressure and is not evolved as heat. In contrast, the isothermal oxidation of methane at constant pressure P below 100 °C involves a decrease in volume,

$$CH_4 + 2O_2 \xrightarrow[P]{T < 100°C} CO_2 + 2H_2O$$

$$1 \text{ vol.} + 2 \text{ vol.} \rightarrow 1 \text{ vol.} + \text{liquid (negligible volume)},$$

therefore work is done *on* this system by the external pressure compressing the reacting system down to about one third of its original volume. This puts energy into the system additional to any energy change due to the chemical reaction. The *net* heat change which accompanies one mole of a reaction at constant pressure – that which, when the reaction is carried out isothermally, has to be removed (if the reaction is exothermic) or supplied (if it is endothermic) – is called the *molar enthalpy of the reaction*, $\Delta_r H$. It is the overall change in heat energy associated with the isothermal conversion of reactants into products at constant pressure in one mole of chemical reaction as described in a chemical equation.

$$\text{Reactants} \xrightarrow[\text{constant temperature, } T]{\text{constant pressure, } P} \text{Products}$$

$$\Delta_r H_{P,T} = \sum H(\text{products})_{P,T} - \sum H(\text{reactants})_{P,T}.$$

The *molar enthalpy* or *heat content* H of a compound is a property with units of energy per mole which, like internal energy, U, cannot be calculated or determined in an absolute sense. We can however measure *changes* in enthalpy, ΔH, and, just as in the case of internal energy, an evolution of heat corresponds to ΔH = negative and an endothermic process has ΔH = positive.

The difference between $\Delta_r H$ of a reaction carried out at constant pressure and $\Delta_r U$ of the same reaction carried out at constant volume is due entirely to the energy of the pressure-volume change $\Delta(PV)$ between reactants and products. The relationship may be expressed mathematically:

$$\Delta_r H = \Delta_r U + \Delta(PV).$$

The absolute but, in practice, indeterminate enthalpy of an element or compound is given by

$$H = U + PV,$$

so H is the sum of the internal energy (the total potential and kinetic energy of all the molecules plus the energy due to intermolecular interactions), and energy which the material has simply by virtue of being at a finite pressure and occupying space. When an ideal gas undergoes a pressure–volume change at constant temperature, $P_1V_1 = P_2V_2$ consequently $\Delta(PV) = 0$, so the enthalpy of an ideal gas is independent of its pressure.

Reactions which do not involve gases usually have a negligible volume change, if any, and they are generally carried out at atmospheric pressure. In such cases, the change in enthalpy, which is the directly measured heat change, is virtually identical with the change in internal energy, $\Delta_r H \sim \Delta_r U$. But combustions and other reactions involving gases are usually carried out at constant volume for reasons of experimental convenience in which cases $\Delta(PV) = V\,\Delta P$. Consequently, the measured heat change of these reactions gives $\Delta_r U$ and, to calculate $\Delta_r H$ from the heat change of a reaction actually carried out at constant volume, we use

$$\Delta_r H = \Delta_r U + V\,\Delta P. \tag{2.2}$$

Correspondingly, to calculate $\Delta_r U$ from the heat change of a reaction measured at constant pressure, P,

$$\Delta_r U = \Delta_r H - P\,\Delta V.$$

The partial pressure P_i of n_i moles of an ideal gas in a total volume V at absolute temperature T is given by

$$P_i = \frac{n_i}{V} \cdot RT.$$

A change Δn in the total number of moles of gas present due to the conversion of reactants into products at constant volume as in

$$\nu_A A + \nu_B B + \ldots \rightarrow \nu_X X + \nu_Y Y + \ldots$$

is given by

$$\Delta n = \sum \nu_i = (\nu_X + \nu_Y + \ldots) - (\nu_A + \nu_B + \ldots)$$

and this will cause a change in total pressure ΔP where

$$\Delta P = \frac{\Delta n}{V} \cdot RT.$$

Therefore, assuming ideal gas behaviour, eqn (2.2) for one mole of reaction becomes

$$\Delta_r H = \Delta_r U + \Delta n \,.\, RT. \tag{2.3}$$

Example. The heat evolved in the combustion of 1.000 g of crystalline bicyclo[2.2.1]heptane (C_7H_{12}) at constant volume and 25 °C is 45.336 kJ.[2] Calculate the molar enthalpy of combustion.

$$C_7H_{12}(c) + 10O_2(g) \xrightarrow{25\,°C} 7CO_2(g) + 6H_2O(l),$$

where (c), (l), and (g) denote crystalline solid, liquid, and gas respectively.

Heat evolved $= 45.336$ kJ g^{-1}
$= 45.336 \times 96.174$ kJ mol^{-1}
$= 4360.1$ kJ mol^{-1}.

As heat is *evolved*, ΣU(products) $<$ ΣU(reactants), so $\Delta_r U$ is negative,

$$\text{therefore } \Delta_r U = -4360.1 \text{ kJ mol}^{-1}.$$

The complete thermochemical equation can now be written:

$$C_7H_{12}(c) + 10O_2(g) \rightarrow 7CO_2(g) + 6H_2O(l);$$
$$\Delta_r U_{298K} = -4360.1 \text{ kJ mol}^{-1}.$$

$$\Delta_r H = \Delta_r U + \Delta n \,.\, RT.$$

We can ignore the change in volume due to the loss of the crystalline reactant and the gain of the liquid product so

$$\Delta n = -3,$$

therefore $\Delta_r H_{298\text{K}} = -4360.1 - 3 \times 8.314 \times 10^{-3} \times 298.15 \text{ kJ mol}^{-1}$
$= \underline{-4367.5 \text{ kJ mol}^{-1}}$.

Note that the molar heat change was calculated from initial information relating to 1 g of one reactant. Since 1 mole of reaction involves 1 mole of that reactant, the result corresponds to the heat change for 1 mole of reaction.

2.4 Heat capacity

The molar *heat capacity*, *C*, of a compound or element is the temperature coefficient of its molar internal energy or enthalpy (according to whether the temperature change occurs at constant volume or constant pressure)

$$C_P = \left(\frac{\mathrm{d}H}{\mathrm{d}T}\right)_P \quad \text{and} \quad C_V = \left(\frac{\mathrm{d}U}{\mathrm{d}T}\right)_V$$

The difference between C_P and C_V is significant only for gases, in which case (assuming ideal behaviour)

$$C_P - C_V = R. \tag{2.4}$$

Because the heat capacities of different materials, for example reactants and products in a reaction, are not identical, the heat change in an isothermal reaction is not the same at two different temperatures.

$$\begin{array}{ccc} \text{Reactants } (T_2) & \xrightarrow{\hspace{3cm}} & \text{Products } (T_2);\ \Delta_r H(T_2) \\ \uparrow & \textit{all processes at} & \uparrow \\ & \textit{constant pressure} & \\ \text{Reactants } (T_1) & \xrightarrow{\hspace{3cm}} & \text{Products } (T_1);\ \Delta_r H(T_1) \end{array}$$

Fig. 2.1. A thermochemical cycle for the process Reactants $(T_1) \rightarrow$ Products (T_2).

The heat required to raise the temperature of the reactants in 1 mole of the reaction in Fig. 2.1 by ΔT, where $\Delta T = T_2 - T_1$, is given by

$$\text{heat}(r) = \sum C_P(\text{reactants}) \,.\, \Delta T;$$

similarly the heat required to raise the temperature of the products by ΔT is

$$\text{heat}(pr) = \sum C_P(\text{products}) \,.\, \Delta T.$$

The total heat change in going from reactants at T_1 in Fig. 2.1 to products at T_2 at constant pressure must be the same regardless of which route is followed, therefore

$$\sum C_P(\text{reactants})(T_2 - T_1) + \Delta_r H(T_2) = \Delta_r H(T_1) + \sum C_P(\text{products})(T_2 - T_1)$$

so

$$\Delta_r H(T_2) = \Delta_r H(T_1) + \Delta C_P(T_2 - T_1), \tag{2.5}$$

where ΔC_P = the *molar heat capacity of the reaction at constant pressure*

$$= \sum C_P(\text{products}) - \sum C_P(\text{reactants})$$

or, expressing eqn (2.5) in abbreviated differential form,

$$\left(\frac{\delta\, \Delta H}{\delta T}\right)_P = \Delta C_P.$$

Similarly, it can be shown that

$$\Delta_r U(T_2) = \Delta_r U(T_1) + \Delta C_V(T_2 - T_1),$$

or

$$\left(\frac{\delta\, \Delta U}{\delta T}\right)_V = \Delta C_V.$$

Example. The enthalpy of hydrogenation of ethene at 25 °C is -136.8 kJ mol^{-1}.[1] Calculate the value at 127 °C given that the mean heat capacities of ethene, hydrogen, and ethane over this temperature range are 49.52, 29.01, and 57.94 J K^{-1} mol^{-1}.[3]

$$C_2H_4 + H_2 \rightarrow C_2H_6; \qquad \Delta_r H_{298\,\mathrm{K}} = -136.8 \text{ kJ mol}^{-1}.$$

By eqn (2.5)

$$\begin{aligned}\Delta_r H_{400\,\mathrm{K}} &= \Delta_r H_{298\,\mathrm{K}} + \Delta C_P(400 - 298)\\ &= -136.8 + (57.94 - 29.01 - 49.52) \times 10^{-3} \times 102\\ &= -136.8 - 2.1\\ &= \underline{-138.9 \text{ kJ mol}^{-1}}.\end{aligned}$$

2.5 Standard states

We have seen above that the heat change associated with an isothermal chemical reaction depends upon the temperature. To a much lesser extent, it also depends upon the pressure at which the reaction takes place. In order to facilitate comparisons, heat changes for reactions are usually documented for standard conditions. In thermochemical work these are 1 standard atmosphere (101 325 Pa) and, usually, 25 °C (298.15 K).

The isothermal interconversion of gas and liquid, liquid and solid, and gas and solid, or the conversion of one crystalline form of a solid to another, involves a heat change; usually, so do the dissolution of a solid or liquid in a solvent and the dilution of a solution. Consequently, the enthalpy change (or internal energy change) of a chemical reaction also depends upon the physical states of the reactants and products. Heat changes of chemical reactions, therefore, are usually referred not only to standard physical conditions (temperature and pressure) but also to specified *standard states* of all reactants and products. The choice of standard states is arbitrary and is made purely for convenience in the particular context.

In thermochemistry, the preferred standard state of a gaseous material is the hypothetical ideal gas at 1 standard atmosphere pressure and, usually, 298.15 K (25 °C); it has the same enthalpy as the real gas at the same temperature but at zero pressure. For a liquid or a solid, the preferred standard states are the pure liquid or pure crystalline solid under a total pressure of one standard atmosphere and, usually, 298.15 K (25 °C). The enthalpy (or internal energy) change for the conversion of reactants into products in one mole of reaction, all species being in standard states, is known as the *standard molar enthalpy* (or *standard molar internal energy*) of the reaction. When the standard states are the pure materials under 1 standard atmosphere pressure the symbols $\Delta_r H^\circ$ (or $\Delta_r U^\circ$) are used although the subscript may be made more specific to indicate the reaction type e.g. we use $\Delta_c H^\circ$ (or $\Delta_c U^\circ$) for combustion.

We shall see later that for other purposes a different choice of standard state is more convenient – for example, a hypothetical solution of unit molarity ($1\ \text{mol dm}^{-3}$) or unit molality ($1\ \text{mol kg}^{-1}$) in which there are no solute-solute interactions. In such cases, the symbol $\Delta_r H^\ominus$ is used where the superscript $^\ominus$ indicates a specified standard state other than the pure material at 1 standard atmosphere pressure.

We have mentioned above that the standard pressure is 1 standard atmosphere but the standard temperature is only *usually* 25 °C (298.15 K). The convention is that if the pressure is not 1 standard atmosphere, then we are not dealing with a thermochemical standard state at all and the superscript ° must not be used. If temperature is not indicated, it is presumed to be 25 °C and this is the common and preferred choice. But another standard temperature may be used in which case it must be specified as a subscript in kelvin (K).

Thus,

ΔH° refers to a process involving pure materials at 1 standard atmosphere pressure and 25 °C;

$\Delta H^\circ_{300\,\text{K}}$ refers to a process involving pure materials at 1 standard atmosphere pressure and 300 K (26.85 °C); and

$\Delta H_{500\,\text{K}}$ (15 Torr) refers to a process involving materials at 500 K and 15 Torr pressure.

In proper thermochemical equations, the physical states of all species are indicated. Some examples are shown:

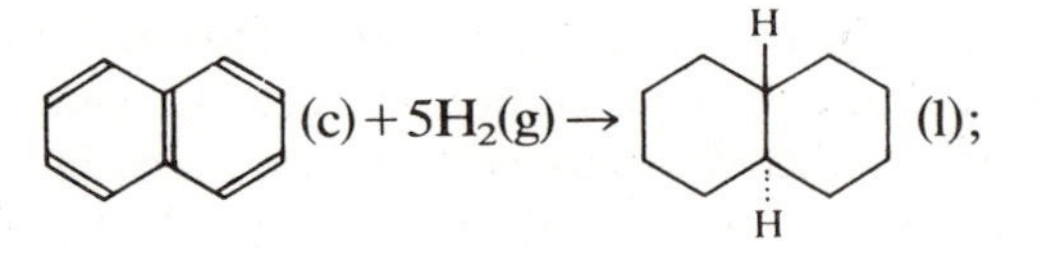

$$\Delta_r H^\circ = -308.6 \text{ kJ mol}^{-1}.$$

This tells us that the reaction between 1 mole of pure crystalline naphthalene under 1 atmosphere pressure at 25 °C and 5 moles of *ideal* hydrogen at 25 °C and 1 atmosphere pressure (or the real gas at zero pressure) to give 1 mole of pure liquid *trans*-decalin under 1 atmosphere pressure at 25 °C evolves 308.6 kJ of heat.[1]

The following can be interpreted similarly:[1]

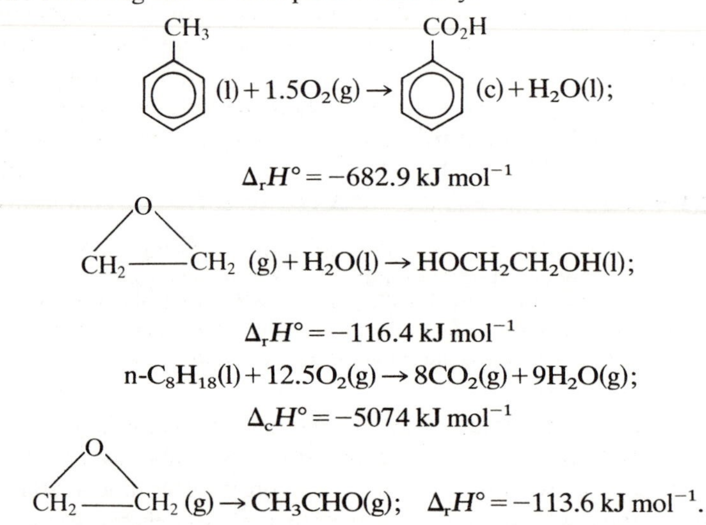

These five thermochemical equations include compounds at 1 standard atmosphere pressure because of the standard state superscripts, and 25 °C because no other temperatures are indicated. In every reaction the standard state of each compound is indicated by (c), (l), or (g).

It may seem inconsistent that, in one of the above equations, water in the gaseous standard state at 25 °C is indicated, and in another reaction, water is in its liquid standard state also at 25 °C. Analogously, other compounds which are normally liquids or even solids at 25 °C may appear in a thermochemical equation in the gas standard state. In such cases, the equilibrium vapour pressures of the materials at 25 °C may be very much less than 1 atmosphere. How, therefore, can we refer to a gaseous standard state at 1 atmosphere pressure for such compounds at 25 °C?

TABLE 2.1
Standard molar enthalpies of vaporization and sublimation at 298.15 K

Substance	$\Delta_v H^\circ_{298\,K}$/kJ mol^{-1}	Substance	$\Delta_s H^\circ_{298\,K}$/kJ mol^{-1}
H_2O[a]	44.016	C(graphite)[a]	716.67
Br_2[a]	30.91	I_2[a]	62.421
Benzene[b]	33.85	Si[a]	450
Cyclohexane[b]	33.1	Naphthalene[b]	72.89

[a] Ref. 4. [b] Ref. 1.

The gaseous standard state is the *hypothetical ideal* gas at 1 atmosphere pressure which has the same enthalpy as the *real* gas at the same temperature but at zero pressure. At zero pressure the real gas behaves ideally, and the enthalpy of an ideal gas is independent of pressure. There is, therefore, no inconsistency. And the enthalpy change involved in the conversion of one mole of substance from the solid or liquid standard state to the gas standard state is simply its *standard molar enthalpy of sublimation* $\Delta_s H^\circ_T$, or *evaporation*, $\Delta_v H^\circ_T$, at the specified temperature, T. A few values for $\Delta_v H^\circ_{298\,K}$ and $\Delta_s H^\circ_{298\,K}$ are shown in Table 2.1. $\Delta_s H^\circ_{298\,K}$ (or $\Delta_v H^\circ_{298\,K}$) may be determined experimentally from the variation of the vapour pressure of the solid (or liquid) with temperature in the region of 298 K. If the vapour pressure measurements are made over a temperature range far from 298 K, then a correction has to be applied to the result which requires knowledge of the heat capacities of the solid (or liquid) and the gas between 298 K and the temperature range of the vapour pressure measurements.

Example. From the information in thermochemical equation (i) for the combustion of crystalline bicyclo[2.2.1]heptane,[2] calculate the standard enthalpy of the corresponding gas-phase reaction in eqn (ii).

(i) $C_7H_{12}(c) + 10O_2(g) \rightarrow 7CO_2(g) + 6H_2O(l)$; $\Delta_c H^\circ = -4367.5$ kJ mol^{-1};

(ii) $C_7H_{12}(g) + 10O_2(g) \rightarrow 7CO_2(g) + 6H_2O(g)$; $\Delta_r H^\circ$.

The standard enthalpy of sublimation of bicyclo[2.2.1]heptane at 25 °C is 40.4 kJ mol^{-1} and the value for the vaporization of water is given in Table 2.1.[2,4]

(iii) $C_7H_{12}(c) \rightarrow C_7H_{12}(g)$; $\Delta_s H^\circ = 40.4$ kJ mol^{-1}

(iv) $H_2O(l) \rightarrow H_2O(g)$; $\Delta_v H^\circ = 44.0$ kJ mol^{-1}.

Reversal of (iv) gives (v)

(v) $H_2O(g) \rightarrow H_2O(l)$; $\Delta H^\circ = -\Delta_v H^\circ = -44.0$ kJ mol^{-1}.

If we add chemical equations (iii), (ii), and $6\times$(v) we get the chemical equation (i), therefore

$$\Delta_c H^\circ = \Delta_s H^\circ + \Delta_r H^\circ - 6\times\Delta_v H^\circ$$

so

$$-4367.5 = 40.4 + \Delta_r H^\circ - 6\times 44.0$$

therefore

$$\underline{\Delta_r H^\circ = -4143.9\ \text{kJ mol}^{-1}}\,.$$

This example illustrates how important it is to specify the physical states of reactants and products in thermochemical equations.

There are some very stringent conditions which must be satisfied in order that the heat change of a reaction can be measured accurately and interpreted meaningfully.

(1) The complete chemical equation for the reaction must be known; there must, therefore, be a complete analysis of reaction products.

(2) The reactant(s) must be pure; this means not less than 99.95 per cent for combustion calorimetry and 99 per cent for reaction calorimetry (but sometimes better depending upon what the impurities are).

(3) The reaction must be complete in a reasonably short time (minutes or hours rather than days).

However, results for reactions which are not amenable to direct calorimetric investigation can be obtained by indirect methods because complete thermochemical equations can be manipulated algebraically just as simple chemical equations can. This has been implicit in some of what we have already covered and is explicitly recognized in Hess's Law: *the overall heat change of a chemical reaction is determined only by the difference between final and initial states, and not at all by the route of the conversion.*

Example. The standard enthalpies of combustion of graphite and diamond are 393.51 and 395.37 kJ mol^{-1} respectively at 25 °C.[1] Calculate the standard enthalpy of conversion of graphite to diamond.

(i) $C(graphite) + O_2(g) \rightarrow CO_2(g)$; $\Delta_{(i)}H° = -393.51$ kJ mol^{-1}

(ii) $C(diamond) + O_2(g) \rightarrow CO_2(g)$; $\Delta_{(ii)}H° = -395.37$ kJ mol^{-1}

Reversal of (ii) gives (iii)

(iii) $CO_2(g) \rightarrow C(diamond) + O_2(g)$; $\Delta_{(iii)}H° = +395.37$ kJ mol^{-1}

Addition of (i) and (iii) gives:

$$C(graphite) \rightarrow C(diamond);\ \Delta_t H° = \Delta_{(i)}H° + \Delta_{(iii)}H°$$

$$\underline{\Delta_t H° = +1.86 \text{ kJ mol}^{-1}.}$$

(The symbol $\Delta_t H°$ is generally used when the reaction is a *phase transition* from one crystal form of a material to another.)

This example also illustrates a particular difficulty in calorimetry. Errors of 0.1 per cent in both $\Delta_{(i)}H°$ and $\Delta_{(ii)}H°$, which in most circumstances would be seen as acceptably small, lead to an accumulated error of 0.78 kJ mol^{-1}. This represents an error of 42 per cent in the standard enthalpy of conversion of graphite into diamond! A calorimetric result is very frequently a small difference between two very large numbers, and achieving a small experimental uncertainty in the outcome requires experimental precision unattained in most other aspects of chemical science.

2.6 Standard enthalpy of formation

It follows from Hess's Law that the standard molar enthalpy change for any chemical reaction is the algebraic sum of the standard molar enthalpies of formation of products and reactants:

$$\Delta_r H^\circ_T = \sum \Delta_f H^\circ_T(\text{products}) - \sum \Delta_f H^\circ_T(\text{reactants}) \qquad (2.6)$$

where $\Delta_f H^\circ_T$ = the *standard molar enthalpy of formation* of a compound at temperature T.

It is the heat change associated with the formation of one mole of the material from the stable forms of its elements, all species being in their thermochemical standard states (pure at 1 standard atmosphere pressure) at the specified temperature. Consequently, the (hypothetical) standard molar enthalpy of formation of the stable form of any element at any temperature is zero and in the previous example,

$$\Delta_f H^\circ(\text{carbon, graphite}) = 0, \qquad \Delta_f H^\circ(\text{carbon, diamond}) = 1.86 \text{ kJ mol}^{-1}.$$

Hess's Law, as embodied in eqn (2.6) is commonly used to calculate standard enthalpies of formation of organic compounds from the much more readily determined standard enthalpies of combustion. This requires knowledge of the standard enthalpies of formation of products of combustion. For the overwhelming majority of organic compounds which contain only carbon, hydrogen and oxygen, these are simply carbon dioxide and water – two of the small number of compounds whose enthalpies of formation have been determined directly and are known to a high degree of precision:[4]

$$C(\text{c, graphite}) + O_2(g) \rightarrow CO_2(g);$$
$$\Delta_r H^\circ_{298\text{ K}} = \Delta_f H^\circ_{298\text{ K}}(CO_2, g) = -393.51 \text{ kJ mol}^{-1}$$
$$H_2(g) + \tfrac{1}{2}O_2(g) \rightarrow H_2O(l);$$
$$\Delta_r H^\circ_{298\text{ K}} = \Delta_f H^\circ_{298\text{ K}}(H_2O, l) = -285.830 \text{ kJ mol}^{-1}.$$

Then when the standard enthalpies of formation of all compounds in some chemical reaction are known, the standard enthalpy of that reaction can be calculated even though for some reason direct calorimetric investigation is not practicable.

Example. The standard enthalpy of combustion of liquid cyclohexane to give gaseous water and carbon dioxide is -3655.4 kJ mol^{-1}.[1] The standard enthalpies of vaporization of cyclohexane and water are given in Table 2.1 (p. 49). From these data and the standard enthalpies of formation of carbon dioxide and liquid water given above, calculate $\Delta_f H^\circ(C_6H_{12}, l)$ and $\Delta_f H^\circ(C_6H_{12}, g)$.

$$C_6H_{12}(l) + 9O_2(g) \rightarrow 6CO_2(g) + 6H_2O(g); \qquad \Delta_c H^\circ = -3655.4 \text{ kJ mol}^{-1}$$
$$H_2(g) + 0.5O_2(g) \rightarrow H_2O(l); \qquad \Delta_f H^\circ(H_2O, l) = -285.83 \text{ kJ mol}^{-1}$$
$$H_2O(l) \rightarrow H_2O(g); \qquad \Delta_v H^\circ = 44.02 \text{ kJ mol}^{-1},$$

therefore

$$\Delta_f H(H_2O, g) = -241.81 \text{ kJ mol}^{-1}.$$

In the combustion equation,

$$\sum \Delta_f H°(\text{products}) = -6 \times (393.51 + 241.81) \text{ kJ mol}^{-1}$$
$$= -3811.9 \text{ kJ mol}^{-1}.$$

By eqn (2.6),

$$-3655.4 = -3811.9 - \Delta_f H°(C_6H_{12}, l)$$

therefore

$$\underline{\Delta_f H°(C_6H_{12}, l) = -156.5 \text{ kJ mol}^{-1}.}$$

$$C_6H_{12}(l) \rightarrow C_6H_{12}(g); \Delta_v H° = +33.1 \text{ kJ mol}^{-1}$$
$$\Delta_f H°(C_6H_{12}, g) = \Delta_f H°(C_6H_{12}, l) + \Delta_v H°$$
$$= -156.5 + 33.1$$
$$= \underline{-123.4 \text{ kJ mol}^{-1}}.$$

Again we see that the result, a standard enthalpy of formation, is a small difference between two much larger terms – the heat of combustion of the compound and the heats of formation of several moles of carbon dioxide and water.

TABLE 2.2
Standard molar enthalpies of formation, 298.15 K

Compound[a]	$\Delta_f H°(g)$/kJ mol^{-1}	Compound	$\Delta_f H°(g)$/kJ mol^{-1}
HF	−273.30	NO[b]	+90.29
HCl	−92.31	ICl[b]	+17.51
HBr	−36.38	CH_4[b]	−74.873
HI	+26.36	CH_3F[b]	−234
CO	−110.53	CH_3Cl[b]	−86.44
H_2O[d]	−241.814	CH_3Br[c]	−38
CO_2	−393.51	CH_3I[c]	+14
SO_2	−296.81	CH_3OH[c]	−201.1
NH_3	−45.94	CF_4[c]	−934.3
BF_3	−1135.95	$HC\equiv CH$[b]	+226.7
SiF_4	−1614.95	$H_2C{=}CH_2$[b]	+52.47
		C_2H_6[c]	−84.68
		C_3H_8[c]	−103.9
		C_6H_6[c,d]	+82.89
		CH_3NH_2[c]	−23.0
		Cyclopropane[c]	+53.3
		Cyclobutane[c]	+28.4
		Cyclopentane[c]	−77.15
		Cyclohexane[c,d]	−123.4

[a] Ref. 4. [b] Ref. 5. [c] Ref. 1. [d] $\Delta_f H°(l)$ may be obtained from $\Delta_f H°(g)$ given here and $\Delta_v H^\circ_{298\,K}$ from Table 2.1, p. 49.

It will be evident from the preceding that standard enthalpies of formation are exceedingly important quantities in chemistry. A few values are given in Table 2.2; more comprehensive lists are available elsewhere.[1,3–5]

2.7 Thermochemical bond dissociation enthalpy

A simple equation such as

$$A + BC \rightarrow AB + C; \Delta_r H°,$$

can be dissected as follows:

$$BC \rightarrow B + C; \qquad \Delta_{(i)}H°,$$

$$AB \rightarrow A + B; \qquad \Delta_{(ii)}H°.$$

These both represent homolytic dissociations (or homolyses); each fragment retains one of the two electrons which initially constitute the bond which suffers fission. If AB has no unpaired electrons and is singly bonded, the dissociated fragments A and B are *free radicals*. Each is one electron short of a valence complement and must contain an unpaired electron.

By Hess's Law,

$$\Delta_r H° = \Delta_{(i)}H° - \Delta_{(ii)}H° \qquad (2.7)$$

where $\Delta_{(i)}H°$ is the *standard bond dissociation enthalpy* of BC which is given a special symbol, $DH°(BC)$. Similarly, $\Delta_{(ii)}H° = DH°(AB)$.

Frequently, $DH°$, which always relates to the gas standard state, is called the bond dissociation energy or even more simply, bond energy.[6] It is, however, a molar enthalpy term and loose names might be taken mistakenly to refer to the change in internal energy or molecular potential energy. To avoid ambiguity, the more precisely descriptive name is preferable.

Equation (2.7) can be rewritten as

$$\Delta_r H° = DH°(BC) - DH°(AB)$$

which, in turn, may be generalized to give eqn (2.8), a very useful expression of Hess's Law, and an alternative to eqn (2.6),

$$\Delta_r H° = \sum DH°(\text{bonds broken}) - \sum DH°(\text{bonds formed}). \qquad (2.8)$$

2.7.1 Diatomic molecules

The standard (thermochemical) bond dissociation enthalpy of a diatomic molecular compound $DH°$ is related to, but not the same as, the (spectroscopic) bond dissociation energy D_0 encountered in the previous chapter. The anharmonic potential-energy curve for a diatomic molecule

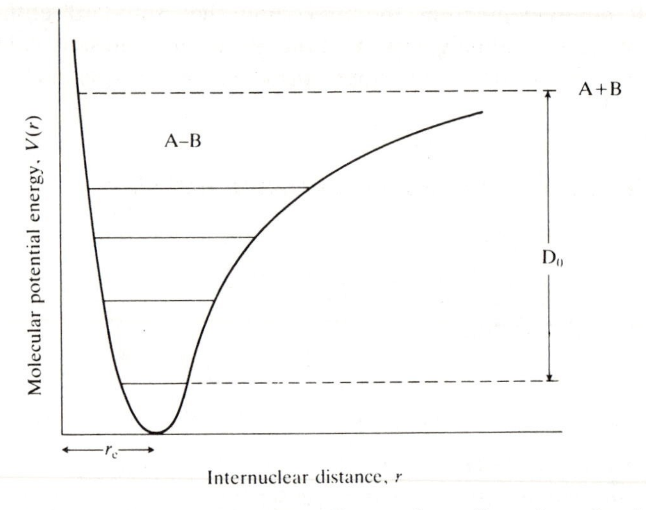

Fig. 2.2. Anharmonic potential-energy diagram for a diatomic molecule A–B.

AB is reproduced in Fig. 2.2. This includes D_0 as defined previously in energy units per molecule. D_0 is the molecular energy difference between the ground state vibrational level of the molecule (with no rotational or translational energy) and the just-dissociated pair of atoms, all species being in their ground electronic states. It may be multiplied by the Avogadro constant to put the result on a molar basis in which case the result is the notional change in internal energy due to hypothetical homolysis of one mole of AB as an ideal gas at absolute zero, eqn (2.9).

$$AB \rightarrow A+B; \qquad \Delta U^{\circ}_{0\,K} = N_A \,.\, D_0. \tag{2.9}$$

The superscript on ΔU eqn (2.9) indicates the gaseous standard state and the subscript indicates the absolute temperature.

The absolute zero temperature specification ensures that the reactant molecules are all in their ground electronic, vibrational, and rotational states with no thermal energy at all, and that the hypothetical dissociation does not impart any kinetic energy to the newly-formed atoms. The species are taken as hypothetical ideal gases since we have not introduced or considered any intermolecular interactions; we have simply taken a molecular property D_0 and multiplied by the Avogadro constant to get onto a molar scale.

Since, by eqn (2.3), ΔU and ΔH are identical at absolute zero,

$$\Delta H^{\circ}_{0\,K} = \Delta U^{\circ}_{0\,K} = N_A \,.\, D_0. \tag{2.10}$$

It follows that the difference between the spectroscopic bond dissociation energy expressed on a molar scale, $N_A \,.\, D_0$ in eqn (2.10), and the standard (thermochemical) bond dissociation enthalpy of the diatomic compound, DH°, is the difference between the enthalpy of a hypothetical reaction at absolute zero, and that of the reaction with molecular reactant and atomic products in the gas standard state, 1 standard atmosphere pressure and 25 °C. Having identified the relationship between D_0 and DH°, which have quite different origins, we are now in a position to show in principle how one can be calculated from the other.

At 25 °C a large assembly of molecules will be partitioned among vibrational and rotational levels so we cannot indicate a *single* energy difference in Fig. 2.2 which, when multiplied by the Avogadro constant, gives the standard internal energy change for the dissociation. (This is an illustration that ΔU°, unlike D_0, is a molar quantity and not a molecular property.)

Upon dissociation, all of this vibrational and rotational thermal energy is lost. Conversely, both molecules and atoms above absolute zero have translational thermal energy, and since one molecule yields two atoms, dissociation at 25 °C takes place with a gain in total translational thermal energy. Thirdly, since DH° relates to *standard* gas-phase reaction conditions, which in thermochemistry means a pressure of 1 atmosphere, the dissociation of one mole of AB into two moles of atoms involves a doubling of the volume, and energy is required to accomplish this.

These three aspects are combined in the formal relationship between D_0 (in molar terms) and DH°

$$DH^\circ = N_A \,.\, D_0 + \int_{0\ \mathrm{K}}^{298.15\ \mathrm{K}} \Delta C_P^\circ \,.\, \mathrm{d}T \qquad (2.11)$$

where ΔC_P° = the standard heat capacity of the dissociation
$= C_P^\circ(\mathrm{A}) + C_P^\circ(\mathrm{B}) - C_P^\circ(\mathrm{AB})$.

To give some idea of the relative size of the second term on the right-hand side of eqn (2.11), we can evaluate the integral for a simplified case.

If the energy levels for the molecular vibration of compound AB in eqn (2.9) are very widely spaced, then even at 25 °C virtually all the molecules will be in the ground vibrational state and there will be no vibrational contribution to $C_P^\circ(\mathrm{AB})$.

If A, B, and AB may be treated classically as ideal gases and the thermal energy is evenly distributed among the translational and rotational degrees of freedom according to the principle of the equipartition of energy, the molar heat capacity for each species at constant volume is $\frac{1}{2}R$ per degree of freedom.

Consequently, $C_V^\circ(\mathrm{AB}) = 2.5R$ and, from eqn (2.4), $C_P^\circ = 3.5R$
and $C_V^\circ(\mathrm{A}) = C_V^\circ(\mathrm{B}) = 1.5R$ and $C_P^\circ(\mathrm{A}) = C_P^\circ(\mathrm{B}) = 2.5R$.

In eqn (2.11), therefore, ΔC_P° becomes $1.5R$ and, in this simple illustration,

$$DH^\circ_{298\,K} = N_A \,.\, D_0 + 1.5R \times 298.15\ K$$

or

$$DH^\circ_{298\,K} = N_A \,.\, D_0 + 3.72\ kJ\ mol^{-1}.$$

Some of the higher vibrational levels of real diatomic molecules will, of course, be occupied at 298.15 K but the rotational contribution to the heat capacity will be less than the classical value R because of the quantization of rotational levels, so there is some cancellation of errors. Numerically, DH° and $N_A \,.\, D_0$ are not likely to differ by more than about 5 kJ mol^{-1}.

However, in principle, D_0 is capable of being determined spectroscopically to a very high degree of precision, for example $D_0(H_2) = 36113.0 \mp 0.3\ cm^{-1}$,[7] (432.007 kJ mol^{-1}) and heat capacity data for atoms and diatomic molecules can be evaluated with low errors, so this spectroscopic method via D_0 can lead to very precise DH° results for diatomic molecules. Unfortunately not all diatomic molecules readily yield spectra of the required quality.

Spectroscopic results for chlorine do allow calculation of DH°.[4]

$$Cl_2(g) \rightarrow 2Cl(g); \qquad \Delta_r H^\circ = DH^\circ(Cl_2) = 242.6\ kJ\ mol^{-1}.$$

The result could also be called the enthalpy of atomization of chlorine, or it could be expressed as

$$\Delta_f H^\circ(Cl, g) = 121.3\ kJ\ mol^{-1}.$$

TABLE 2.3

Standard bond dissociation enthalpies of some gaseous diatomic elements and compounds, 298.15 K[a]

Element(g)	DH°/kJ mol^{-1}	Compound(g)	DH°/kJ mol^{-1}
H_2	435.99	HF	570.7
N_2	945.36	HCl	431.6
O_2	498.34	HBr	366.2
F_2	158.8	HI	298.4
Cl_2	242.60	CO	1076.4
Br_2	192.8	NO	631.6
I_2	151.10	ICl	210.6

[a] The values for the elements were obtained from spectroscopic results.[4] Those for the heteronuclear diatomics were calculated from

$$DH^\circ(AB) = \Delta_f H^\circ(A, g) + \Delta_f H^\circ(B, g) - \Delta_f H^\circ(AB, g);$$

$\Delta_f H^\circ$ data for compounds are given in Table 2.2 (p. 52) and for atomic elements in Table 2.4 (p. 59).

The DH° result for hydrogen chloride, however,

$$HCl(g) \rightarrow H(g) + Cl(g); \qquad \Delta_r H^\circ = DH^\circ(HCl) = 431.6\ \text{kJ mol}^{-1}$$

was obtained from $\Delta_f H^\circ$(HCl, g), which was obtained calorimetrically, and $\Delta_f H^\circ$ values for H(g) and Cl(g) determined spectroscopically.[4] In this case, therefore, a thermochemically determined DH° result led back to a D_0 value which could not be obtained spectroscopically.

Some results for DH° of diatomic elements and compounds are shown in Table 2.3. These may be compared with the four molar values for D_0 in Table 1.3 in the previous chapter (p. 14). More comprehensive lists are available.[6,8]

2.7.2 Polyatomic molecules

The standard molar enthalpy change for the dissociation of a bond in a polyatomic molecular compound can also be identified as a standard bond dissociation enthalpy.

$$R—X \rightarrow R + X; \qquad \Delta_r H^\circ = DH^\circ(R—X).$$

Mass spectrometry[9] is a physical method by which such bond enthalpies can be determined. A very important chemical method involves rate measurements of a reaction which includes homolysis of the bond whose dissociation enthalpy is sought.[8,10] We can illustrate this general method by anticipating a little of what is covered in later chapters on kinetics. The reaction of bromine and methane in the gas phase is complex but one reversible step in the overall sequence is

$$Br + CH_4 \underset{k_{-1}}{\overset{k_1}{\rightleftharpoons}} CH_3 + HBr; \qquad \Delta_r H$$

where k_1 and k_{-1} = the rate constants of the forward and reverse reactions and $\Delta_r H$ = the molar enthalpy of the elementary reaction under the conditions of the experiment.

By eqn (2.8),

$$\Delta_r H^\circ = DH^\circ(CH_3—H) - DH^\circ(HBr) \qquad (2.12)$$

and DH°(HBr) is already known (Table 2.3).

$\Delta_r H$ may be obtained from Arrhenius plots of the temperature dependence of k_1 and k_{-1} (see p. 231) and a correction can be applied to convert $\Delta_r H$ (the enthalpy of the reaction under the actual conditions) to $\Delta_r H^\circ$ (the *standard* enthalpy of the reaction). From eqn (2.12), therefore, the experimental value for DH°(CH_3—H) may be calculated. The result by this method, DH°(CH_3—H) = 435 kJ mol^{-1}, is in excellent agreement with that calculated from photo-ionization mass spectrometric data (438.6 kJ mol^{-1}).[4,5]

We shall not deal here with the frequently difficult problems of obtaining the appropriate elementary rate constants of the various steps of the gas-phase reactions. A wide range of reactions has been investigated by this method to yield $DH°$ results for bonds in many organic compounds.[8,10]

Values for $\Delta_f H°(CH_4, g)$[5], $\Delta_f H°(H, g)$,[4] and $DH°(CH_3—H)$ allow the standard enthalpy of formation of methyl radical to be calculated.

$$CH_4(g) \rightarrow CH_3(g) + H(g); \qquad \Delta_r H° = DH°(CH_3—H) \qquad (2.13)$$

For eqn (2.13),

$$DH°(CH_3—H) = \Delta_f H°(CH_3, g) + \Delta_f H°(H, g) - \Delta_f H°(CH_4, g)$$

$$438.6 = \Delta_f H°(CH_3, g) + 218.0 - (-74.9)$$

therefore

$$\underline{\Delta_f H°(CH_3, g) = 145.7 \text{ kJ mol}^{-1}.}$$

Once the standard enthalpy of formation of methyl radical is known, the way is open to calculate $DH°(CH_3—X)$ from $\Delta_f H°(CH_3X)$ and $\Delta_f H°(X)$. We can illustrate this for methyl chloride.

$$\Delta_f H°(CH_3Cl, g) = -86.44 \text{ kJ mol}^{-1},^{5}$$

$$\Delta_f H°(Cl, g) = +121.30 \text{ kJ mol}^{-1},^{4}$$

$$CH_3—Cl(g) \rightarrow CH_3(g) + Cl(g); \qquad \Delta_r H° = DH°(CH_3—Cl)$$

$$DH°(CH_3—Cl) = \Delta_f H°(CH_3) + \Delta_f H°(Cl) - \Delta_f H°(CH_3Cl)$$

$$= 145.7 + 121.30 - (-86.44)$$

$$= \underline{353.4 \text{ kJ mol}^{-1}.}$$

The dissociation need not simply be of a bond to a single atom. The method can also be used to calculate $DH°(CH_3—CH_3)$ from $\Delta_f H°$(ethane):

$$CH_3—CH_3(g) \rightarrow 2CH_3(g); \qquad \Delta_r H° = DH°(CH_3—CH_3)$$

$$\Delta_f H°(C_2H_6, g) = -84.68 \text{ kJ mol}^{-1},^{1}$$

therefore

$$DH°(CH_3—CH_3) = 2 \times \Delta_f H°(CH_3) - \Delta_f H°(C_2H_6, g)$$

$$= 291.4 - (-84.68)$$

$$= \underline{376.1 \text{ kJ mol}^{-1}.}$$

We can see from the above, that from a relatively small number of standard enthalpies of formation of atoms and free radicals, standard dissociation enthalpies of bonds in many compounds can be calculated from their respective standard enthalpies of formation. Some useful data are collected in Table 2.4, and Table 2.5 shows a few results. Many more are to be found in compilations by Benson[6,10] and Kerr.[8]

TABLE 2.4
Standard molar enthalpies of formation of atoms and free radicals, 298.15 K

Atom[a]	$\Delta_f H°(g)$/kJ mol^{-1}	Radical	$\Delta_f H°(g)$/kJ mol^{-1}
H	217.997	CH_3	145.7[b]
B	560	CH_2	385.2[b]
C	716.67	CH	594.1[b]
N	472.68	C_2H_5	105[c]
O	249.17	n-C_3H_7	95[c]
S	276.98	$(CH_3)_2CH$	74[c]
Si	450	$(CH_3)_3C$	34[c]
F	79.39	C_6H_5	325[d]
Cl	121.302	OH	39.46[b]
Br	111.86	NH	339[b]
I	106.762	NH_2	168[b]

[a] Ref. 4. [b] Ref. 5. [c] Ref. 10. [d] G. A. Chamberlain and E. Whittle, *Trans. Faraday Soc.* **67,** 2077 (1971).

TABLE 2.5
Standard dissociation enthalpies ($DH°$) *for bonds in simple polyatomic compounds,* 298.15 K

Compound–bond	$DH°$/kJ mol^{-1}	Compound–bond	$DH°$/kJ mol^{-1}
CH_3—F	459[a]	CH_3—H	438.6[a]
CH_3—Cl	353.4[a]	C_2H_5—H	408[d]
CH_3—Br	296[a]	$(CH_3)_2CH$—H	396[d]
CH_3—I	238[a]	$(CH_3)_3C$—H	387[d]
CH_3—NH_2	331[b]	$PhCH_2$—H	355[d]
CH_3—OH	381[b]	C_6H_5—C_6H_5	418[c]
CH_3—OCH_3	335[b]	C_6H_5—CH_3	389[c]
HO—H	499.3[a]	CH_3—CH_3	376.1[e]
CH_3O—H	427[b]	$(CH_3)_3C$—CH_3	335[c]
C_6H_5O—H	356[c]	$(CH_3)_3C$—$C(CH_3)_3$	282[c]

[a] From data in Tables 2.2 and 2.4; see p. 58 for the example of CH_3Cl.
[b] Ref. 8. [c] Ref. 6. [d] Ref. 10. [e] Calculated on p. 58.

2.7.3 Mean bond dissociation enthalpy

The enthalpy changes for consecutive bond dissociations of water are not equal nor should we expect them to be. Hydrogen–oxygen single bonds in quite different chemical entities are being dissociated.

$$H_2O(g) \rightarrow HO(g) + H(g); \qquad DH°(HO\text{—}H) = 499 \text{ kJ mol}^{-1}$$

$$HO(g) \rightarrow O(g) + H(g); \qquad DH°(O\text{—}H) = 428 \text{ kJ mol}^{-1}.$$

(Both values calculated from $\Delta_f H°$ data in Tables 2.2 and 2.4). A useful thermochemical quantity, but one which does not relate directly to a real chemical process, is the *mean* standard bond dissociation enthalpy $\overline{DH°}$. In the case of water, this is half of the enthalpy change associated with complete atomization and can be calculated from the standard enthalpies of formation of reactant and products (Tables 2.2 and 2.4).

$$H_2O(g) \rightarrow O(g) + 2H(g); \qquad \Delta_r H° = 927 \text{ kJ mol}^{-1},$$

therefore $\overline{DH°} = 463.5 \text{ kJ mol}^{-1}$.

Similarly for methane, the standard enthalpies for the consecutive dissociation steps are unequal and we can calculate the following results from data in Tables 2.2 and 2.4.

$$CH_4(g) \rightarrow CH_3(g) + H(g); \qquad DH°(CH_3\text{—}H) = 439 \text{ kJ mol}^{-1}$$

$$CH_3(g) \rightarrow CH_2(g) + H(g); \qquad DH°(CH_2\text{—}H) = 457 \text{ kJ mol}^{-1}$$

$$CH_2(g) \rightarrow CH(g) + H(g); \qquad DH°(CH\text{—}H) = 427 \text{ kJ mol}^{-1}$$

$$CH(g) \rightarrow C(g) + H(g); \qquad DH°(C\text{—}H) = 341 \text{ kJ mol}^{-1}.$$

However, we can again define a mean standard bond dissociation

TABLE 2.6

Mean standard bond dissociation enthalpies, 298.15 K

Compound	Bond dissociated	$\overline{DH°}$/kJ mol^{-1}
H_2O	O—H	463[a]
NH_3	N—H	391[b]
CH_4	C—H	416[a]
BF_3	B—F	645[b]
CF_4	C—F	492[b]
SiF_4	Si—F	596[b]
NPh_3	N—C	374[c]
PPh_3	P—C	321[c]
$AsPh_3$	As—C	285[c]
$SbPh_3$	Sb—C	268[c]
$BiPh_3$	Bi—C	194[c]
CPh_4	C—C	404[d]
$SiPh_4$	Si—C	354[d]
$GePh_4$	Ge—C	303[d]
$SnPh_4$	Sn—C	253[d]
$PbPh_4$	Pb—C	203[d]

[a] Calculated on pp. 60, 61.
[b] From data in Tables 2.2 and 2.4.
[c] Given by W. V. Steele, *J. chem. Thermodynamics* **11,** 187 (1979).
[d] Given by W. V. Steele, *J. chem. Thermodynamics* **10,** 445 (1978).

enthalpy which is one quarter of the enthalpy of complete atomization of methane.

Example. Calculate $\overline{DH}^\circ(CH_4)$ from data in Tables 2.2 and 2.4.

$$CH_4(g) \rightarrow C(g) + 4H(g); \qquad \Delta_r H^\circ = 4\overline{DH}^\circ(CH_4)$$

$$4\overline{DH}^\circ(CH_4) = \Delta_f H^\circ(C, g) + 4\,\Delta_f H^\circ(H, g) - \Delta_f H^\circ(CH_4, g)$$
$$= 716.67 + 872.00 - (-74.87)$$
$$= 1664 \text{ kJ mol}^{-1}$$

therefore

$$\underline{\overline{DH}^\circ(CH_4) = 416 \text{ kJ mol}^{-1}.}$$

In general, $\overline{DH}^\circ$ values are easily calculated for compounds such as those above which contain bonds to equivalent atoms or groups. The concept is helpful in discerning trends in reactions such as

$$XR_n \rightarrow X + nR; \qquad \Delta_r H^\circ = n \times \overline{DH}^\circ(XR_n)$$

where R = an atom or polyatomic radical which is kept constant whilst X is varied or, less usually, where X is kept constant and R is varied. Some results are shown in Table 2.6.

2.8 Calculation of standard enthalpy of formation

A long-standing aim of chemists has been to calculate standard enthalpies of formation because their experimental determination is time-consuming in the most propitious cases or, too frequently, not practicable at all. Such calculations cannot be exclusively theoretical because there is no adequate theory yet. But just as an enthalpy of a reaction can be calculated from the standard enthalpies of formation of reactants and products, the standard enthalpy of formation of a polyatomic compound can be regarded as the sum of *standard bond enthalpy of formation terms,*

$$\Delta_f H^\circ = \sum BH^\circ(\text{P–Q})$$

where BH°(P–Q) = the incremental standard enthalpy of formation due to one type of bond in the compound.

However, reactants and products in a chemical reaction have independent existence and their $\Delta_f H^\circ$ values are rigorously independent of the nature of any reaction or calculation in which they are involved. Whilst a $\Delta_f H^\circ$ can *in principle* be regarded as a sum of standard bond enthalpy of formation terms (or, more simply from now on, bond enthalpy terms) with no loss of rigour, in practice the complete dissection of a $\Delta_f H^\circ$ into component bond enthalpy terms is problematical. Furthermore, the expectation that these notional bond enthalpy terms when identified are transferable, and can be used to calculate $\Delta_f H^\circ$ results for other compounds, needs justification. Some justification can be given. If compounds which are used to determine bond enthalpy terms are chemically very similar to those whose standard enthalpies of formation are subsequently calculated, for example if they are all alkanes, satisfactory results are usually obtained. And $\Delta_f H^\circ$ values of some quite complex hydrocarbons calculated by this procedure are in good agreement with independently obtained experimental results. Little more can be required of any empirical method than that it is reasonable in principle and works in practice!

For saturated hydrocarbons, the simplest postulate is that $\Delta_f H^\circ$ comprises only $BH^\circ(\text{C—C})$ and $BH^\circ(\text{C—H})$ terms and by using known $\Delta_f H^\circ$ results for a few members of the series, values of $BH^\circ(\text{C—C})$ and $BH^\circ(\text{C—H})$ can be obtained. Mathematically this corresponds to the statistical fitting of some experimental $\Delta_f H^\circ$ results to the two-parameter equation

$$\Delta_f H^\circ(\text{hydrocarbon}) = \sum BH^\circ(\text{C—C}) + \sum BH^\circ(\text{C—H}). \qquad (2.14)$$

This yields $BH^\circ(\text{C—C})$ and $BH^\circ(\text{C—H})$ values which give the best empirical fit for the experimental results used. Then from these two parameters, $\Delta_f H^\circ$ can be calculated from the molecular formula of any further member of the family where an experimental result is not available.

The incremental standard bond enthalpy of formation term $BH^\circ(\text{P–Q})$ and the bond dissociation enthalpy $DH^\circ(\text{P–Q})$ are obviously related. For the atomization reaction

$$\text{Compound} \rightarrow \text{Atoms},$$

$\Delta H^\circ(\text{atomization}) = \sum \Delta_f H^\circ(\text{atoms}) - \Delta_f H^\circ(\text{compound})$.
But $\Delta H^\circ(\text{atomization}) = \sum DH^\circ(\text{bonds})$
and $\Delta_f H^\circ(\text{compound}) = \sum BH^\circ(\text{bonds})$
therefore $\sum DH^\circ(\text{bonds}) = \sum \Delta_f H^\circ(\text{atoms}) - \sum BH^\circ(\text{bonds})$.
So for any compound,
$\Delta_f H^\circ = \sum BH^\circ(\text{bonds}) = \sum \Delta_f H^\circ(\text{atoms}) - \sum DH^\circ(\text{bonds})$.

This rudimentary approach works tolerably for simple saturated hydrocarbons but not for those with small rings, with highly branched alkyl groups, or for polycyclic compounds. It is obviously doomed to give identical results for structural and stereoisomers.

A beginning towards improving the method is to recognize that a given bond enthalpy term will be affected not only by the identity of the two atoms directly bonded, but also by the next nearest atoms. Thus, for [1], R^1, R^2, and R^3 affect $BH^\circ(\text{C—H})$. Similarly, substituents on carbon affect

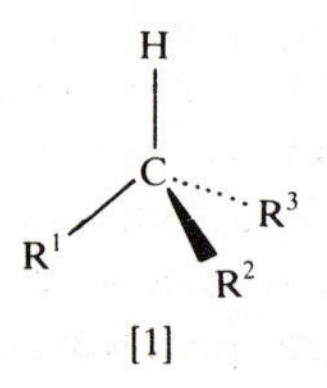

[1]

BH°(C—C). It would, however, be very cumbersome to have different terms for bonds with only slightly different molecular environments, and a successful compromise is to fit the body of experimental results to a four-parameter equation,[1,11]

$$\Delta_f H^\circ(\text{hydrocarbon}) = \sum \mathrm{B}H^\circ(\text{C—C}) + \sum \mathrm{B}H^\circ(\text{C—H})_p + \sum \mathrm{B}H^\circ(\text{C—H})_s + \sum \mathrm{B}H^\circ(\text{C—H})_t \quad (2.15)$$

where p, s, and t refer to the primary, secondary, or tertiary nature of the carbon of each C–H bond.

An alternative (but ultimately equivalent) analytical approach has been to resolve the standard enthalpies of formation not into *bond* standard enthalpy of formation terms, but into *group* standard enthalpy of formation terms.[1,12]

$$\Delta_f H^\circ(\text{hydrocarbon}) = \sum \Delta_f H^\circ(\text{CH}_3) + \sum \Delta_f H^\circ(\text{CH}_2) + \sum \Delta_f H^\circ(\text{CH}) + \sum \Delta_f H^\circ(\text{C}) \quad (2.16)$$

Equation (2.16) is for saturated hydrocarbons and the group terms are determined by fitting it to existing experimental $\Delta_f H^\circ$ results. (It may also be shown that

$$\Delta_f H^\circ = \sum \Delta_f H^\circ(\text{groups}) = \sum \Delta_f H^\circ(\text{atoms}) - \sum \mathrm{D}H^\circ(\text{bonds})$$

for any compound.)

Both eqn (2.15) and (2.16) usually give better agreement with experiment than the simpler two-parameter analogue, eqn (2.14), but of course they require a greater input of experimental $\Delta_f H^\circ$ results for the initial determination of the incremental enthalpy terms—the so-called *parameterization* of the equations. Some results are shown later in Table 2.9 (p. 71).

However, even eqns (2.15) and (2.16) do not work well for highly-branched hydrocarbons, those with three- and four-membered rings, and many bridged hydrocarbons. Such compounds invariably have more positive (or less negative) experimental $\Delta_f H^\circ$ results than are calculated, so molecules of these compounds must be destabilized by features which are not recognized in whichever equation is used.

One way forward to allow more reliable calculations of $\Delta_f H^\circ$ values is to introduce further terms specifically for cyclobutane and cyclopropane rings, and destabilizing features such as gauche dihedral interactions for C–C–C–C groupings. This method is now highly developed due largely to the work of Benson and his colleagues[12,13] and incremental terms are available for a wide variety of organic functional groups and structural features including aromatic and heterocyclic systems. Bridged alicyclic compounds, however, are a major class of organic compounds with which this method still cannot adequately cope.

An alternative strategy is not to modify the basic equation (2.15) or (2.16) but to acknowledge that the bond (or group) incremental terms are derived from simple molecular compounds relatively free of serious bond length and bond angle distortions and other destabilizing features. Consequently, a result calculated from only these terms using either eqn (2.15) or (2.16) will correspond to a value for a hypothetical strain-free compound, $\Delta_f H°$(strain free). The difference between this value and the *real* $\Delta_f H°$ of the compound is the *strain enthalpy* of the compound. Since this difference, according to the hypothesis, is due entirely to molecular strain and will not include a (PV) component, the more usual name, *strain energy*, is justifiable and we shall use it:

$$\text{strain enthalpy} = \text{strain energy} = E(\text{strain})$$

and

$$\text{real } \Delta_f H° = \Delta_f H°(\text{strain free}) + E(\text{strain}). \qquad (2.17)$$

If a calorimetric result for $\Delta_f H°$ of a compound is available, then an experimental value for the strain energy of the compound can be obtained from eqn (2.17), $\Delta_f H°$(strain free) being calculated from eqn (2.15) or (2.16).

It is important to appreciate that the strain energy in eqn (2.17) is not an absolute measure of anything. It is simply the difference between the *real* $\Delta_f H°$ of a compound and a value built up from incremental standard enthalpy of formation terms derived from very simple acyclic alkanes, and although eqn (2.17) may help us to identify a problem, it in itself brings us no nearer to the main objective, which is to *calculate* the *real* $\Delta_f H°$ of a compound. To do this via eqn (2.17) we must be able to *calculate* E(strain).

Largely due to the development of high-speed computers, it is now possible to calculate not only E(strain) from the molecular structure of a compound, but also the structure itself which has the minimum strain for a given atom connectivity. At this stage, it is necessary to return to the molecular level.

2.9 Molecular shape, strain, and steric energy

Methane, ethane, propane, and butane have very similar bond lengths and bond angles. Saturated hydrocarbons such as cyclopropane or cyclobutane which are constrained by their atom connectivities to have bond angles different from those of, for example, ethane and propane are regarded as having *bond angle strain* and some associated *bond length strain*.

Similarly, the *trans* coplanar conformation of n-butane shown in [2] is the most stable arrangement of a C–C–C–C grouping and any molecule with such fragments which are constrained to be other than *trans* and coplanar, for example cyclohexane [3], has *dihedral angle* or *torsional strain.*

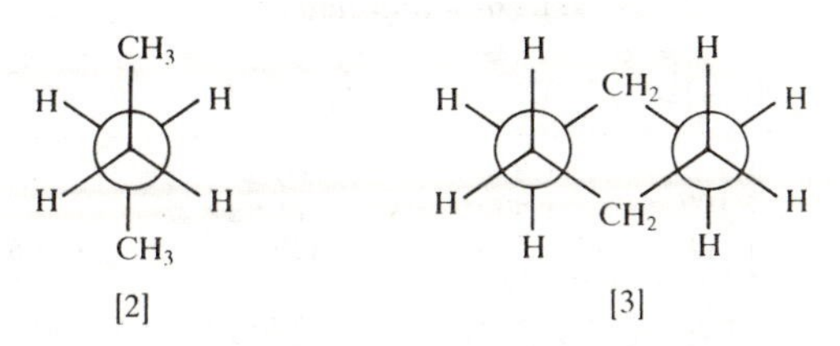

Adverse *non-bonded interactions* also increase strain in a molecule. These occur when atoms which are not directly bonded together are forced by the atom connectivity into too close a proximity as, for

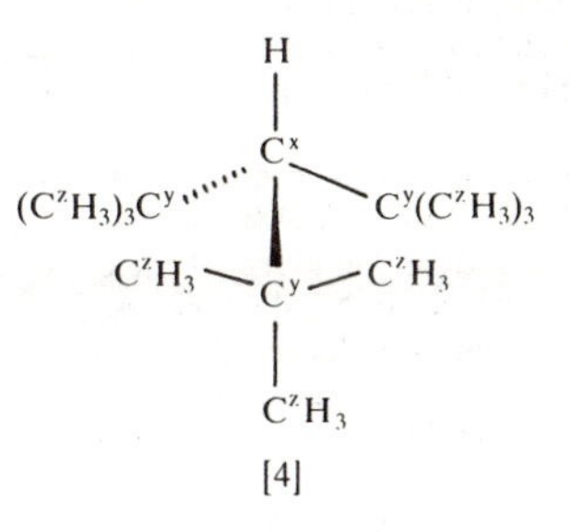

example, in the highly congested tri(t-butyl)methane, [4]. Table 2.7 shows that bond angles and lengths can be significantly different in strained molecules from values typically found in many compounds regarded as unstrained.

The increased molecular potential energy due to these four destabilizing features is known as the *steric energy* E(steric). It can be regarded as the *difference* between the *actual* potential energy of a molecule and that of a *hypothetical* configuration which has the same formal C–C and C–H bonds but each, and all associated bond angles and other intramolecular interactions, contributing only the potential energy which they contribute to ethane, propane, or butane. (Propane has to be considered because ethane does not have a C–C–C bond angle, and n-butane is required as the other two do not have a C–C–C–C torsion.)

In principle, it is not possible for a complex organic molecule to have only one sort of molecular strain. A simple analogy may be helpful. If several different springs are connected end to end, and then the series is stretched between two fixed points, the total elongation will not be provided by a single spring. The potential energy of the whole system is increased by the stretching and it is distributed among the components, the elongation in each spring being large or small according to its 'stiffness'.

Analogously, the atom connectivity of a molecule such as cyclopropane

TABLE 2.7
Structural parameters of some unstrained and strained molecules determined by electron diffraction[a]

	r(C—H)	r(C—C)	$\widehat{C—C—C}$	$\widehat{H—C—C}$	$\widehat{H—C—H}$
Ethane[b]	1.112	1.534	—	110	—
Propane[c]	1.107	1.532	112	—	107[h]
n-Butane[d]	1.108	1.533	112.4	110.5	—
Cyclopropane[e]	1.089	1.510	60	117.7	115.1
Cyclobutane[f]	1.092	1.548	~90	—	~114[i]
Tri(t-butyl)-methane[g], $HC^x(C^y(C^zH_3)_3)_3$	1.111	$1.611C^xC^y$ $1.548C^yC^z$	$116.0C^yC^xC^y$ $113.0C^xC^yC^z$ $105.7C^zC^yC^z$	$101.6HC^xC^y$ $114.2HC^zC^y$	104.3

[a] Bond lengths, r, in ångströms, angles in degrees.
[b] L. S. Bartell and H. K. Higginbotham, *J. chem. Phys.* **42,** 851 (1965).
[c] It was assumed that all C–H bond lengths are the same; T. Iijima, *Bull. chem. Soc. Japan* **45,** 1291 (1972).
[d] It was assumed that all C–C bond lengths are the same, and all C–H bond lengths are the same; R. A. Bonham and L. S. Bartell, *J. Am. chem. Soc.* **81,** 3491 (1959).
[e] O. Bastiansen, F. N. Fritsch, and K. Hedberg, *Acta Cryst.* **17,** 538 (1964).
[f] A. Almenningen, O. Bastiansen, and P. N. Skancke, *Acta chem. Scand.* **15,** 711 (1961).
[g] It was assumed that all C–H bond lengths are the same; H. B. Burgi and L. S. Bartell, *J. Am. chem. Soc.* **94,** 5236 (1972); *see structure* [4].
[h] This angle refers to the methyl groups, the H—C—H of the methylene was assumed to be 106.1°.
[i] J. D. Dunitz and V. Schomaker, *J. chem. Phys.* **20,** 1703 (1952).

requires bond lengths and angles to be distorted from the values in the simpler acyclic alkanes. This gives rise to a strained molecule and the total strain is distributed among the various bond lengths and angles according to their respective force constants. And just as the stretched linear system of springs vibrates if it is displaced and released, the strained molecule undergoes molecular vibrations. The equilibrium configuration of cyclopropane, which corresponds to the lowest minimum in its potential-energy hypersurface, has interatomic distances and angles which are different from the equilibrium values in ethane and propane. And the potential energy at the minimum is higher than the potential energy of three hypothetically bonded motionless unstrained CH_2 groups by an amount equal to the steric energy of cyclopropane.

2.10 Force-field calculations

We have now seen what is meant at the molecular level, by saying that some molecules are more strained than others, and a reasonable next stage in the overall quest to calculate the *molar* $\Delta_f H°$ of a compound is the calculation of the *molecular* steric energy, E(steric).[14–18,20]

$\delta V(r)$ = the contribution towards the steric energy due to the difference δr in one particular bond length from its value in an unstrained molecule.

Similarly, a particular bond angle which is different by $\delta\theta$ from its unstrained value will contribute $\delta V(\theta)$, and a torsion angle which is not at its minimum potential energy value will add a further term $\delta V(\phi)$. If we sum these three contributions for *all* bond lengths, bond angles, and dihedral interactions –

$$E(r) = \sum \delta V(r),$$

$$E(\theta) = \sum \delta V(\theta),$$

and

$$E(\phi) = \sum \delta V(\phi),$$

and then add a fourth composite term $E(l)$ which is the sum of all non-bonded interaction potential-energy terms $E(l) = \sum V(l)$, we obtain the important equation,[14]

$$E(\text{steric}) = E(r) + E(\theta) + E(\phi) + E(l). \tag{2.18}$$

$E(l)$ is different from the others in this equation in that they are all *differences* in potential energy due to distortions from arbitrarily chosen standard bond lengths and angles. The non-bonded interaction term is, in a sense, absolute since it really is zero when all the atoms concerned are very far apart. There is, therefore, no need in this case to relate $E(l)$ to any standard finite but indeterminate interaction which is found in some other standard molecule.

We now need analytical expressions for each of the terms $\delta V(r)$, $\delta V(\theta)$, $\delta V(\phi)$, and $V(l)$. Different authors have used different equations and, in empirical work of this sort, the justification for one rather than another is purely in terms of its success.

Examples for the first three terms are[19]

(i) $$\delta V(r) = \tfrac{1}{2}\kappa_r (r - r_0)^2$$

where κ_r = a bond-stretching force constant,
r_0 = the idealized bond length in a 'strain-free' molecule, and
r = the actual bond length in the configuration of the molecule under consideration;

(ii) $$\delta V(\theta) = \tfrac{1}{2}\kappa_\theta (\theta - \theta_0)^2$$

where κ_θ = a bond-angle bending force constant,
θ_0 = the idealized bond angle in a 'strain-free' molecule, and
θ = the angle in the configuration of the molecule under consideration; and

(iii) $$\delta V(\phi) = \tfrac{1}{2}V_0(1 + \cos 3\,\phi)$$

where V_0 = a parameter in energy units which characterizes a particular internal bond rotation, and
ϕ = the dihedral angle of the particular configuration of the molecule under consideration.

We encountered expressions such as these previously in our considerations of molecular vibrations in Chapter 1. Together they constitute part of the force field of the molecule (see p. 18). Consequently, empirical calculations of the molecular steric energy by this method are called *force-field calculations.*[20] An alternative name is *molecular mechanics.*[15–18]

The first two terms above include the harmonic approximation for bond stretching and angle bending vibrations (see p. 6). More complex expressions can be used which include anharmonicity terms. The total force-field expression for the molecule includes one or other of the above terms for *every* bond length, bond angle, and dihedral angle (H–C–C–H, H–C–C–C, and C–C–C–C).

The expression for non-bonded interactions is more difficult but is generally of the form[15–20]

$$V(l) = A + B\,\mathrm{e}^{-l} + C\,l^{-6},$$

where A, B, and C are constants for a particular type of interaction between non-bonded atoms at distance l apart. For example, one set of A, B, and C will cater for all non-bonded intramolecular H...H interactions, and there will be second and third sets for analogous C...H and C...C interactions.

In the total force-field expression, therefore, there are parameters which are taken to be common to all members of a family of compounds:

(i) a κ_r and an r_0 for both C—C and C—H;

(ii) a κ_θ and a θ_0 for each of

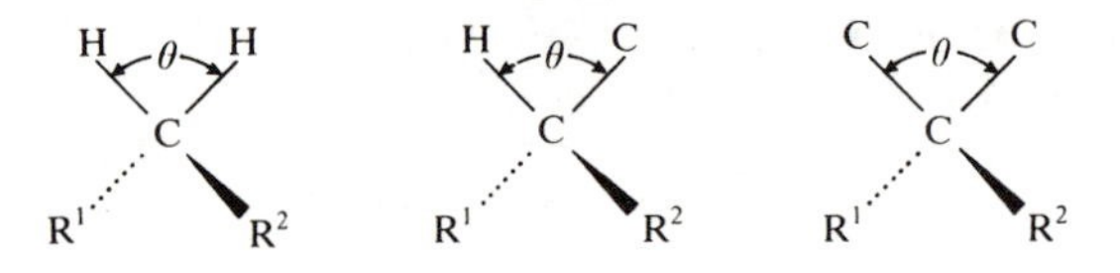

where R^1 and R^2 may be H or alkyl;

(iii) a V_0 value for each of H—C—C—H, H—C—C—C, and C—C—C—C: and

(iv) a set of A, B, and C for each of H...H, H...C, and C...C.

The total number of individual terms in the right-hand side of eqn (2.18) for E(steric) depends upon the number of atoms in one molecule of the particular compound.

Once the above parameters are known for a given family of compounds, E(steric) for any configuration of a molecule of one compound can be calculated using values for the *structure variables* r, θ, ϕ, and l for all bonds, angles, and distances which together describe the particular configuration.

Tentative values for the force constants κ_r and κ_θ may be obtained from infra-red and microwave vibrational spectroscopy and values for r_0 and θ_0 are those corresponding to bond lengths and angles in ethane or propane. V_0 values can also be obtained from spectroscopic results or calculated from *ab initio* theory. Investigations of *intermolecular* interactions lead to provisional A, B, and C values. All these are then used for a series of compounds of *known* molecular structures and, after many iterative calculations by computer, values for the structure variables are obtained which give the lowest E(steric) result for a molecule of each compound. If these calculated minimum energy structures are in good agreement with the molecular configurations found experimentally, then the force-field expression and the initial parameters are satisfactory. If not, then the parameters are systematically modified and the procedure is repeated.

By repetition, this process yields a refined set of parameters for the particular force-field expression and for a particular family of compounds which satisfactorily calculates minimum energy molecular structures, and, for each molecule's minimum energy configuration, the steric energy, E(steric). If no set of parameters can be found which allows accurate calculation of experimental structures, the analytical expression of the force field is inadequate and requires modification. With the best current methods, calculated bond lengths are accurate to ~0.01 Å and bond angles are within about 2° of experimental values. Some results are shown in Table 2.8.

It remains now to establish the relationship between E(steric) of eqn

TABLE 2.8
Calculated molecular structures compared with experimental results[a]

Compound	Structural parameter	Calculated	Experimental
CH_3–CH_2–CH_3	r(C—C)	1.531	1.533
	θ(C—C—C)	112	112
(cyclohexane)	r(C—C)	1.533	1.528
	θ(C—C—C)	111	111
	ϕ(C—C—C—C)	55	55
(norbornane; atoms 1, 2, 3, 4)	r(C1—C2)	1.536	1.538
	r(C2—C3)	1.540	1.552
	θ(C1C2C3)	110	109.7
	l(C1 . . . C4)	2.595	2.592

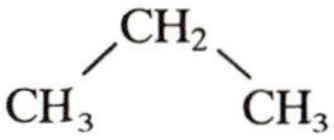

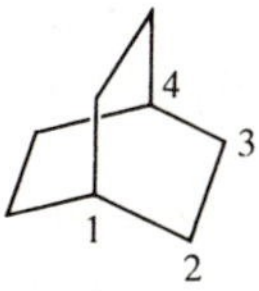

[a] Results taken from ref. 15, bond lengths in ångströms, angles in degrees.

(2.18), a molecular potential-energy term, and the standard molar enthalpy quantity, E(strain), in eqn (2.17). The minimum energy configuration corresponds to the lowest minimum in the potential-energy hypersurface of the molecule and corresponds to a hypothetical motionless state. It is the multidimensional equivalent of the minimum in a potential-energy curve for a diatomic molecule (see p. 13). The lowest energy state of the real molecule, as opposed to the energy minimum of the potential-energy hypersurface, is the lowest vibrational state, so zero-point energy has to be added to E(steric) to obtain a molecular potential-energy term at absolute zero. Multiplication by the Avogadro constant then puts this onto a molar scale to give the strain energy (enthalpy) at absolute zero.

$$\text{Strain energy(absolute zero)} = E(\text{strain})_{0\,\text{K}} = \{E(\text{steric}) + (\text{Zero-Point Energy Correction})\} \times N_{\text{A}}$$

Secondly, heat capacity information is required to calculate the molar strain energy (enthalpy) at 298.15 K, $E(\text{strain})_{298.15\,\text{K}}$, or, taking the 298.15 K subscript as understood,

$$E(\text{strain}) = \{E(\text{steric}) + (\text{Zero-Point Energy Correction})\} \times N_{\text{A}} + (\text{Heat Capacity Correction}) \qquad (2.19)$$

Assuming that zero-point energy and heat capacity corrections for a series of closely related compounds are also incremental and additive, they may be determined by fitting eqn (2.19) (including E(steric) values from eqn (2.18)) to experimental E(strain) results obtained from eq (2.17)

$$E(\text{strain}) = \text{real } \Delta_f H° - \Delta_f H°(\text{strain free}), \qquad (2.17)$$

using experimental results for real $\Delta_f H°$ and $\Delta_f H°$(strain free) values from eqn (2.15) or (2.16). This gives incremental group (or bond) zero-point energy and heat capacity correction parameters and the final fully parameterized eqn (2.19) is now ready to calculate E(strain) at 25 °C for any further member of the family of compounds.

Standard enthalpies of formation, *real* $\Delta_f H°$ values, can now be *calculated* using eqn (2.17) from the calculated E(strain) and $\Delta_f H°$(strain free) obtained from the standard bond (or group) enthalpy of formation terms (eqns (2.15) or (2.16)). In cases where an experimental value is available, the calculated results are usually within 5 kJ mol^{-1} which is only about twice the typical uncertainty in a calorimetric standard enthalpy of formation.

Table 2.9 shows some calculated and experimental results for comounds which were used in the parameterization. The agreement for a wide range of structural types of hydrocarbons is much better than for results calculated simply from eqns (2.14) or (2.15). In Table 2.10, some

TABLE 2.9
Experimental standard enthalpies of formation compared with calculated values[a]

Compound	$\Delta_f H°(g)$, expt.[b]	$\Delta_f H°(g)$, calc.		
		A[c]	B[d]	C[e]
$(CH_3)_2CH_2$	−103.9	−107	−105	−107.9
$(CH_3)_3CH$	−135.6	−127	−134	−138.9
$(CH_3)_4C$	−168.5	−147	−170	−175.4
$(CH_3)_2CHCH_2C_2H_5$	−174.8	−167	−175	−178.3
$(CH_3)_3CC(CH_3)_3$	−225.2	−208	−255	−222.5
Cyclohexane	−123.4	−121	−123	−122.8
Cyclopentane	−77.15	−101	−103	−76.86
Cyclobutane	+28.4	−80	−82	+24.18
Bicyclo[2.2.1]heptane	−61.6	−94	−117	−54.35
Adamantane	−128.8	−108	−152	−136.0
Cubane	+622.2	+26	−58	+621.7

[a] All values in kJ mol^{-1} at 298.15 K.
[b] Experimental results taken from ref. 1 except values for bicyclo[2.2.1]heptane (ref. 2) and adamantane (R. S. Butler, A. S. Carson, P. G. Laye and W. V. Steele, *J. chem. Thermodynamics* **3**, 277 (1971)).
[c] Using eqn (2.14) parameterized using experimental results for n-butane and n-hexane;[1] $BH°(C—H) = -16.7$ and $BH°(C—C) = +13.3$ kJ mol^{-1}.
[d] Using eqn (2.15) and values for $BH°(C—C) = -0.13$, $BH°(C—H)_p = -14.1$, $BH°(C—H)_s = -10.2$, and $BH°(C—H)_t = -7.11$ kJ mol^{-1} recommended in ref. 1.
[e] Results taken from ref. 15.

results *predicted* by different empirical force-field expressions are compared with experimental values *subsequently* obtained. Again, the agreement is such that we can be reasonably confident in other calculated values as long as the compounds concerned are not too dissimilar in structural type to those used in the parameterization.

We have described the initial parameterization using compounds of known structures to get the force-field parameters, and the second stage parameterization using the calculated E(steric) values (based upon the initial parameterization) and some experimental strain energy results (based upon experimental $\Delta_f H°$ values) to obtain group zero-point energy and heat capacity corrections as separate processes. In practice, the experimental structures and $\Delta_f H°$ values may be fed in at the same time and, with skilful computer programming, the whole parameterization is accomplished in a single complex and reiterative process. This complete parameterization does not need to be repeated every time a new structure

TABLE 2.10
Experimental standard enthalpies of formation of hydrocarbons compared with values predicted by different empirical force-field calculations[a]

Compound	$-\Delta_f H°(g)$, expt.	$-\Delta_f H°(g)$, calc.	
		Allinger[b]	Schleyer[c]
	128[d]	126.7	128.2
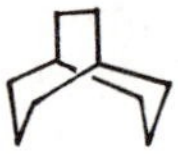	106[d]	105.4[e]	109.5
	89[f]	95.4[e]	105.5
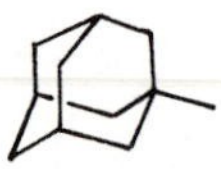	175[g]	179.5[g]	174.9
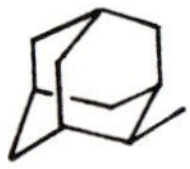	156[g]	163.2[g]	158.6[g]

[a] All values in kJ mol^{-1} at 298.15 K.
[b] Ref. 17.
[c] Ref. 15.
[d] W. Parker, W. V. Steele, and I. Watt, *J. chem. Thermodynamics* **9,** 307 (1977).
[e] Reported in ref. 15.
[f] W. Parker, W. V. Steele, W. Stirling, and I. Watt, *J. chem. Thermodynamics* **7,** 795 (1975).
[g] W. V. Steele and I. Watt, *J. chem. Thermodynamics* **9,** 843 (1977).

or $\Delta_f H°$ is calculated, although, from time to time, the force-field expression should be updated, refined, or re-parameterized as new experimental results become available.

The stage in the calculation of $\Delta_f H°$ which involves eqn (2.15) or (2.16) to obtain $\Delta_f H°$(strain free) is, of course, quite separate from the calculation of E(steric). But here again, the complete programme usually contains a subroutine and stored strainless group or bond standard enthalpy of formation data, so the final output is the desired $\Delta_f H°$ and, if required, the strain energy.

2.11 Problems

1. The enthalpy of combustion of crystalline naphthalene is given by[1]

$$C_{10}H_8(c) + 12O_2(g) \rightarrow 10CO_2(g) + 4H_2O(l);$$
$$\Delta_c H° = -5156.2 \text{ kJ mol}^{-1}.$$

Calculate the enthalpy of the corresponding gas-phase reaction,

$$C_{10}H_8(g) + 12O_2(g) \rightarrow 10CO_2(g) + 4H_2O(g),$$

given[1,4] that $\Delta_s H°(C_{10}H_8) = 72.89 \text{ kJ mol}^{-1}$ and

$$\Delta_v H°(H_2O) = 44.02 \text{ kJ mol}^{-1}.$$

2. Calculate the heat produced when benzene (0.5000 g) is burned at 25 °C and constant volume given[1]

$$C_6H_6(l) + 7.5O_2(g) \xrightarrow{25\,°C} 6CO_2(g) + 3H_2O(l);$$

$$\Delta_c H°_{298\,K} = -3267.6 \text{ kJ mol}^{-1}.$$

3. The enthalpy of hydrogenation of 1-hexene is given by[1]

$$C_6H_{12}(l) + H_2(g) \xrightarrow{25\,°C} C_6H_{14}(l); \quad \Delta_r H°_{298\,K} = -198.6 \text{ kJ mol}^{-1}.$$

From the information given below,[1,3] calculate the enthalpy for the reaction:

$$C_6H_{12}(g) + H_2(g) \xrightarrow{500\,K} C_6H_{14}(g); \quad \Delta_r H°_{500\,K}$$

$$\Delta_v H°_{298\,K}(C_6H_{12}) = 30.7 \text{ kJ mol}^{-1}$$
$$\Delta_v H°_{298\,K}(C_6H_{14}) = 31.5 \text{ kJ mol}^{-1}$$
$$\bar{C}°_P(C_6H_{12}, g), 300\text{–}500 \text{ K} = 167.5 \text{ J K}^{-1} \text{ mol}^{-1}$$
$$\bar{C}°_P(H_2, g), 300\text{–}500 \text{ K} = 29.18 \text{ J K}^{-1} \text{ mol}^{-1}$$
$$\bar{C}°_P(C_6H_{14}, g), 300\text{–}500 \text{ K} = 181.9 \text{ J K}^{-1} \text{ mol}^{-1}.$$

4. The heat evolved in the combustion of 1.000 g of each of the following compounds at 25 °C (constant volume) and their standard molar enthalpies of sublimation are shown.[2]

Compound	*Heat of combustion* (25 °C, const.vol)/kJ g^{-1}	$\Delta_s H°$/kJ mol^{-1}
Norborn-2-ene [1]	−44.936	37.7
Tricyclene [2]	−44.828	38.7
Norborna-2,5-diene [3]	−44.570	33.8
Quadricyclane [4]	−45.528	37.0

[1] [2] [3] [4]

Calculate (i) the standard molar enthalpy of combustion, $\Delta_c H^\circ_{298\,K}$(c), for each compound, (ii) $\Delta_f H^\circ_{298\,K}$(g) for each compound given $\Delta_f H^\circ_{298\,K}(H_2O, l) = -285.83$ kJ mol^{-1} and $\Delta_f H^\circ_{298\,K}(CO_2, g) = -393.51$ kJ mol^{-1}, and (iii) $\Delta_r H^\circ_{298\,K}$(g) for the isomerizations:

(a) [2] → [1]
(b) [4] → [3],

and comment upon their respective values.

5. Calculate $\Delta_r H°$ for the following gas-phase reactions from the data in Table 2.2 (p. 52).
   (i) $CO_2 + 4HF \rightarrow CF_4 + 2H_2O$
   (ii) $CH_4 + F_2 \rightarrow CH_3F + HF$
   (iii) $C_2H_2 + 2H_2 \rightarrow C_2H_6$
   (iv) $3C_2H_2 \rightarrow C_6H_6$
   (v) $C_2H_2 + C_2H_6 \rightarrow 2C_2H_4$.

6. Calculate $\Delta_r H°$ for the following gas-phase reactions from the data in Table 2.3 (p. 56).
   (i) $H_2 + F_2 \rightarrow 2HF$
   (ii) $H_2 + I_2 \rightarrow 2HI$
   (iii) $2ICl \rightarrow I_2 + Cl_2$
   (iv) $ICl + H_2 \rightarrow HI + HCl$
   (v) $2NO \rightarrow N_2 + O_2$.

7. Calculate $\Delta_r H°$ for the following gas-phase reactions from the data in Tables 2.3 and 2.5 (p. 56 and 59).
   (i) $CH_3OH + HI \rightarrow CH_3I + H_2O$
   (ii) $CH_3OH + HF \rightarrow CH_3F + H_2O$
   (iii) $CH_4 + H_2O \rightarrow CH_3OH + H_2$
   (iv) $C_2H_6 + F_2 \rightarrow 2CH_3F$
   (v) $C_2H_6 + I_2 \rightarrow 2CH_3I$
   (vi) $(CH_3)_2O + HF \rightarrow CH_3OH + CH_3F$.

2.12 References

1. J. D. Cox and G. Pilcher, *Thermochemistry of organic and organometallic compounds*, Academic Press (1970). This book contains compilations of standard enthalpies of formation and combustion of several thousand organic and organo-metallic compounds and authoritative notes on the results. It also contains information on practical aspects of calorimetry and is indispensable to anyone interested in thermochemistry of organic compounds.
2. W. V. Steele, *J. chem. Thermodynamics* **10,** 919 (1978).
3. D. R. Stull, E. F. Westrum, and G. C. Sinke, *The chemical thermodynamics of organic compounds*, Wiley, New York, (1969). Besides standard enthalpies of formation, this book also includes heat capacity, entropy, and free energy data, for some inorganic and many organic compounds.
4. *J. chem. Thermodynamics* **10,** 903 (1978). This is one of a series of authoritative publications of recommended values for important thermochemical quantities.
5. *JANEF Thermochemical tables*, 2nd edn (ed. by D. R. Stull and H. Prophet), National Bureau of Standards, USA (1970).
6. S. W. Benson 'Bond energies', *J. chem. Educ.* **42,** 502 (1965). See also L. Pauling, *The nature of the chemical bond*, 3rd edn, Cornell University Press, New York (1960).
7. G. Herzberg and A. Monfils, *J. molec. Spectrosc.* **5,** 482 (1960).
8. J. A. Kerr, 'Bond dissociation energies by kinetic methods', *Chem. Rev.* **66,** 465 (1966).
9. J. H. Beynon, *Mass spectrometry and its applications to organic chemistry*, p. 459, Elsevier (1960). A. G. Harrison, 'Correlations of fragment ion structures with energetics of formation', p. 121, in *Topics in organic mass spectrometry*, Vol. 8 (ed. A. L. Burlingame), Wiley-Interscience (1970).
10. D. M. Golden and S. W. Benson, 'Free radical and molecule thermochemistry from studies of gas phase iodine–atom reactions', *Chem. Rev.* **69,** 125 (1969).
11. K. J. Laidler, *Canad. J. Chem.* **34,** 626 (1956).
12. S. W. Benson and J. H. Buss, *J. chem. Phys.* **29,** 546 (1958).
13. S. W. Benson, F. R. Cruickshank, D. M. Golden, G. R. Haugen, H. E. O'Neal, A. S. Rodgers, R. Shaw, and R. Walsh, 'Additivity rules for the estimation of thermochemical properties', *Chem. Rev.* **69,** 279 (1969). See also S. W. Benson, *Thermochemical Kinetics*, 2nd edn, Wiley, New York (1976), and H. K. Eigenmann, D. M. Golden, and S. W. Benson, *J. phys. Chem.* **77,** 1687 (1973).
14. F. H. Westheimer, 'Calculation of the magnitude of steric effects', Chapt. 12 in *Steric effects in organic chemistry* (ed. M. S. Newman), Wiley (1956).
15. E. M. Engler, J. D. Andose, and P. von R. Schleyer, 'Critical evaluation of molecular mechanics', *J. Am. chem. Soc.* **95,** 8005 (1973).
16. J. E. Williams, P. J. Stang, and P. von R. Schleyer, *A. Rev. phys. Chem.* **19,** 531 (1968).
17. N. L. Allinger, M. T. Tribble, M. A. Miller, and D. H. Herz, *J. Am. chem. Soc.* **93,** 1637 (1971).
18. N. L. Allinger, *J. Am. chem. Soc.* **99,** 8127 (1977); *Adv. phys. org. Chem.* **13,** 1 (1976).
19. S.-J. Chang, D. McNally, S. Shary-Tehrany, M. J. Hickey, and R. H. Boyd, *J. Am. chem. Soc.* **92,** 3109 (1970).
20. O. Ermer, *Structure and Bonding* **27,** 161 (1976); D. B. Boyd and K. B. Lipkowitz, *J. chem. Educ.* **59,** 269 (1982).

Supplementary references

U. Burkert and N. L. Allinger, '*Molecular Mechanics*', ACS Monograph No. 177, American Chemical Society, Washington DC (1982).

T. L. Cottrell, *The strengths of chemical bonds*, Butterworths, London (1954). This important volume gives an account of earlier work on bond energies and related aspects. Unfortunately many of the results are now out of date.

J. M. Sturtevant, 'Calorimetry', Chapt. 7 in *Techniques of chemistry* Vol. 1, Pt V (eds. A. Weissberger and B. W. Rossiter), Wiley-Interscience (1971).

M. L. McGlashan, 'Internationally recommended names and symbols for physico-chemical quantities and units', *A. Rev. phys. Chem.* **24,** 51 (1973).

3

Free energy and entropy

3.1 Introduction

The minimization of enthalpy is the evident driving force for the combustion processes considered in the previous chapter. In a simple exothermic reaction shown in eqn (3.1),

$$\text{(methylenecyclopropane)}\ (\mathrm{l}) \longrightarrow \text{(1-methylcyclobutene)}\ (\mathrm{l});\quad \Delta H^\circ_{298\,\mathrm{K}} = -86.9\ \mathrm{kJ\ mol^{-1}}, \qquad (3.1)$$

a reactant of high enthalpy of formation ($\Delta_f H^\circ = 49.0\ \mathrm{kJ\ mol^{-1}}$)[1] isomerizes into a product of lower enthalpy of formation ($\Delta_f H^\circ = -37.9\ \mathrm{kJ\ mol^{-1}}$)[1] and heat is evolved.

In contrast, a process such as the dissolution of potassium chloride in water (eqn (3.2)) is endothermic;[2]

$$\mathrm{KCl(c)} \xrightarrow{\mathrm{H_2O}} \mathrm{KCl(aq,\ molality = 0.0500\ mol\ kg^{-1})}; \qquad (3.2)$$
$$\Delta H_{298\,\mathrm{K}} = 17.5\ \mathrm{kJ\ mol^{-1}},$$

but because the potassium chloride dissolves spontaneously, the system as a whole must become more stable in some sense.

The reactions in eqns (3.1) and (3.2) proceed from left to right as written (although the former is slow at 25 °C); there is no tendency for anhydrous potassium chloride to precipitate from a 0.05 molal aqueous solution at 25 °C, and 1-methylcyclopentene shows no detectable inclination to revert to 2-cyclopropylpropene. In other words, the state represented by the left-hand side of each equation is thermodynamically unstable with respect to that represented by the right-hand side.

3.2 Free energy, entropy, and standard states

The processes in eqns (3.1) and (3.2), one being exothermic and the other endothermic, establish that the direction of spontaneous reaction is determined by more than just the minimization of the total enthalpy. Indeed, it is the minimization of the total *free energy* or *Gibbs function*, G, of the system which is the driving force at constant temperature and pressure, and eqn (3.3), which relates the corresponding changes in free energy and enthalpy,[3] is one of the most important in chemistry.

$$\text{Reactants} \xrightarrow{P,\,T} \text{Products}; \quad \Delta G, \Delta H, \Delta S;$$
$$\Delta G = \Delta H - T\,.\,\Delta S, \tag{3.3}$$

where ΔG, ΔH, and ΔS = the changes in *free energy*, *enthalpy*, and *entropy* associated with 1 mole of reaction at constant temperature and pressure.

When $\sum G(\text{products}) < \sum G(\text{reactants})$, ΔG is negative and the reaction is thermodynamically feasible or favourable. If ΔG is positive, then the *reverse* reaction as written is thermodynamically favourable.

But a negative ΔG does not necessarily mean that the reaction will take place at an appreciable rate. Chemical thermodynamics deals only with the relative stability of one state and another. Concern with the *rate* at which one state becomes another is the subject of chemical *kinetics*; and *how*, at the molecular level, this transformation takes place is the *mechanism* of the reaction. Thermodynamics tells us little about kinetics and mechanism; and, as we shall see, there are many materials which are *thermodynamically unstable*, but which do not decompose to any perceptible extent even in a very long time. They are *kinetically stable*.

3.2.1 Standard free energy of formation

We saw in the previous chapter that $\Delta_r H_T$ is replaced by $\Delta_r H^\circ_T$ when reactants and products are in thermochemical standard states (pure materials at 1 standard atmosphere pressure). In just the same way, $\Delta_r G^\circ_T$ and $\Delta_r S^\circ_T$ are used.

$$\text{Reactants(g, l, or c)} \xrightarrow[1\ \text{atm}]{T} \text{Products(g, l, or c)}; \quad \Delta_r H^\circ_T, \Delta_r S^\circ_T, \Delta_r G^\circ_T.$$

By analogy with eqn (3.4), which we encountered in the previous chapter,

$$\Delta_r H^\circ_T = \sum \Delta_f H^\circ_T(\text{products}) - \sum \Delta_f H^\circ_T(\text{reactants}), \qquad (3.4)$$

we can write

$$\Delta_r G^\circ_T = \sum \Delta_f G^\circ_T(\text{products}) - \sum \Delta_f G^\circ_T(\text{reactants})$$

where $\Delta_r G^\circ_T$ = the *standard molar free energy of the reaction*, and

$\Delta_f G^\circ_T$ = the *standard molar free energy of formation* of a substance in its thermochemical standard state (g, l, or c) at temperature T.

By convention, $\Delta_f G^\circ_T$ is the free-energy change associated with the formation of 1 mole of a material at temperature T from the stable forms of its elements, all species being in thermochemical standard states; the hypothetical standard free energy of formation of the pure stable form of an element at 1 standard atmosphere pressure and any temperature is, therefore, zero. Free energies of chemical reactions can be manipulated in just the same way as enthalpies using Hess's Law. For the sake of brevity, we shall continue occasionally to use the convention of using ΔG etc. rather than $\Delta_r G$ and of omitting or truncating the temperature subscript when it is understood to be 298.15 K. Thus ΔG° and $\Delta G^\circ_{298\,\text{K}}$ both mean $\Delta G^\circ_{298.15\,\text{K}}$. Otherwise a temperature will be specified e.g. $\Delta G^\circ_{600\,\text{K}}$.

3.2.2 Absolute standard entropy

In contrast to enthalpy and free energy, an *absolute* standard molar entropy can be attributed to compounds *and* elements, so there is no need for tabulated conventional $\Delta_f S^\circ$ terms. As we shall see later, this arises because *the entropy of a pure regular crystal is zero at zero* K – an expression of the Third Law of Thermodynamics.[3] Consequently, we can write

$$\Delta_r S^\circ_T = \sum S^\circ_T(\text{products}) - \sum S^\circ_T(\text{reactants}) \qquad (3.5)$$

where $\Delta_r S^\circ_T$ = *the standard molar entropy of the reaction, and*

S°_T = the *absolute standard molar entropy* of the substance in its thermochemical standard state (the pure material under 1 standard atmosphere pressure) at temperature T.

Some values are given in Table 3.1.

TABLE 3.1
Standard entropies of some elements and compounds at 298.15 K

Material[a] (state)	$S°$/J K^{-1} mol^{-1}	Material (state)	$S°$/J K^{-1} mol^{-1}
C (c)	5.74	CH_4 (g)	186.1[b]
C (g)	157.988	CH_3F (g)	222.8[b]
H_2 (g)	130.570	CH_3Cl (g)	234.3[b]
N_2 (g)	191.502	CH_3Br (g)	246.3[c]
O_2 (g)	205.037	CH_3I (g)	254.0[c]
F_2 (g)	202.685	CF_4 (g)	261.3[b]
Cl_2 (g)	222.965	CH_3NH_2 (g)	243.3[c]
Br_2 (g)	245.350	CH_3OH (g)	239.7[c]
Br_2 (l)	152.210	C_2H_2 (g)	200.85[b]
I_2 (g)	260.567	C_2H_4 (g)	219.22[b]
I_2 (c)	116.139	C_2H_6 (g)	229.5[c]
HF (g)	173.665	C_3H_8 (g)	269.9[d]
HCl (g)	186.786	Cyclopropane (g)	237.4[d]
HBr (g)	198.585	Cyclobutane (g)	265.4[d]
HI (g)	206.480	Cyclopentane (g)	292.9[d]
H_2O (g)	188.724	Cyclohexane (g)	298.2[d]
H_2O (l)	69.950	Cyclohexane (l)	204.3[d]
NH_3 (g)	192.67	Benzene (g)	269.2[d]
CO_2 (g)	213.677	Benzene (l)	173.3[d]

[a] Ref. 4.
[b] Ref. 5.
[c] Ref. 6.
[d] Ref. 1.

Example. Calculate the standard molar entropy change in the following gas-phase reactions from data given in Table 3.1.

(i) $C_2H_2 + 2H_2 \xrightarrow[1\text{ atm}]{25\,°C} C_2H_6$

(ii) $CH_4 + F_2 \xrightarrow[1\text{ atm}]{25\,°C} CH_3F + HF$

(iii) $\text{Cyclobutane} \xrightarrow[1\text{ atm}]{25\,°C} 2C_2H_4$

(i) From eqn (3.5)

$$\begin{aligned}\Delta_r S° &= S°(C_2H_6) - \{S°(C_2H_2) + 2 \times S°(H_2)\}\\ &= 229.5 - (200.85 + 2 \times 130.57)\\ &= 229.5 - 461.99\\ &= \underline{-232.5\text{ J K}^{-1}\text{ mol}^{-1}}.\end{aligned}$$

(ii)
$$\begin{aligned}\Delta_r S° &= \{S°(CH_3F) + S°(HF)\} - \{S°(CH_4) + S°(F_2)\}\\ &= 222.8 + 173.665 - 186.1 - 202.685\\ &= \underline{7.68\text{ J K}^{-1}\text{ mol}^{-1}}.\end{aligned}$$

(iii) $\Delta_r S° = 2 \times S°(C_2H_4) - S°(\text{cyclobutane})$
$= 2 \times 219.2 - 265.4$
$= \underline{173\ \text{J K}^{-1}\ \text{mol}^{-1}}$.

Note the units of entropy; a change in entropy has to be multiplied by an absolute temperature to be directly comparable with an energy change and there is usually a factor of 10^{-3} as energies are more conveniently expressed in kJ rather than J. For example, at 25 °C (298 K) an entropy change of 40 J K^{-1} corresponds to 11920 J or about 12 kJ; and at 100 °C (373 K) this same entropy change is equivalent to about 15 kJ.

3.3 Free energy and chemical reactions

When ΔH and ΔS for a reaction are of opposite sign, then by eqn (3.3) they reinforce each other and the two aspects of free energy are pushing the reaction in the same direction. For example, if ΔH is negative and ΔS is positive for the reaction written from left to right,

$$\text{Reactants} \rightarrow \text{Products}; \quad \Delta G, \Delta H, \Delta S,$$

then ΔG must always be negative and the reaction is thermodynamically favourable in the direction shown at all temperatures.

When ΔH and ΔS are of the same sign, then by eqn (3.3), they are in opposition. The minimization of enthalpy is achieved by progress of the reaction in one direction, but the maximization of entropy requires the reaction to proceed in the opposite direction. The *overall* thermodynamically favourable direction of the reaction (ΔG negative) will be determined by the relative magnitudes of ΔH and $T.\Delta S$.

Example. From the enthalpy data given in Tables 2.1 and 2.2 of Chapter 2 (p. 49 and p. 52) and the entropy data in Table 3.1 above, calculate $\Delta_f G°$ for (i) methylamine (g), (ii) benzene (l), (iii) ethane (g), and (iv) carbon dioxide (g).

(i) $C(c) + 2.5H_2(g) + 0.5N_2(g) \xrightarrow[1\ \text{atm}]{25\ °\text{C}} CH_3NH_2(g)$

$$\Delta_f H°(CH_3NH_2, g) = \underline{-23.0\ \text{kJ mol}^{-1}}$$

By eqn (3.5),

$$\begin{aligned}\Delta_f S°(CH_3NH_2, g) &= 243.3 - (5.74 + 2.5 \times 130.57 + 0.5 \times 191.5)\\ &= 243.3 - 427.9\\ &= \underline{-184.6\ \text{J K}^{-1}\ \text{mol}^{-1}}\end{aligned}$$

By eqn (3.3),

$$\begin{aligned}\Delta_f G°(CH_3NH_2, g) &= -23.0 + 298 \times 184.6 \times 10^{-3}\\ &= -23.0 + 55.0\\ &= \underline{+32.0\ \text{kJ mol}^{-1}}.\end{aligned}$$

This is a relatively uncommon example of a compound whose formation is exothermic yet which has a positive standard free energy of formation at normal temperatures. It is, therefore, thermodynamically unstable at 25 °C with respect to decomposition into the elements of which it is composed. Such a reaction does not take place at any perceptible rate, so the compound is kinetically stable.

(ii) $6C(c) + 3H_2(g) \xrightarrow[1\text{ atm}]{25\,°C} C_6H_6(l)$

$$\Delta_f H°(C_6H_6, g) = 82.89 \text{ kJ mol}^{-1}, \qquad \Delta_v H°(C_6H_6) = 33.85 \text{ kJ mol}^{-1}$$

therefore

$$\Delta_f H°(C_6H_6, l) = \underline{49.04 \text{ kJ mol}^{-1}}$$

$$\begin{aligned}\Delta_f S°(C_6H_6, l) &= 173.3 - (6 \times 5.74 + 3 \times 130.57)\\ &= 173.3 - 426\\ &= \underline{-253 \text{ J K}^{-1} \text{ mol}^{-1}}\end{aligned}$$

$$\begin{aligned}\Delta_f G°(C_6H_6, l) &= 49.04 + 298 \times 253 \times 10^{-3}\\ &= 49.04 + 75.4\\ &= \underline{124 \text{ kJ mol}^{-1}}.\end{aligned}$$

In this case, the standard free energy of formation is positive like the (endothermic) standard enthalpy of formation, but numerically it is even larger. Like $CH_3NH_2(g)$, therefore, benzene is thermodyamically unstable with respect to decomposition into its elements at room temperature, but kinetically stable.

(iii) $2C(c) + 3H_2(g) \xrightarrow[1\text{ atm}]{25\,°C} C_2H_6(g)$

$$\Delta_f H°(C_2H_6, g) = \underline{-84.68 \text{ kJ mol}^{-1}}$$

$$\begin{aligned}\Delta_f S°(C_2H_6, g) &= 229.5 - (2 \times 5.74 + 3 \times 130.57)\\ &= 229.5 - 403.2\\ &= \underline{-173.7 \text{ J K}^{-1} \text{ mol}^{-1}}\end{aligned}$$

$$\begin{aligned}\Delta_f G°(C_2H_6, g) &= -84.68 + 298 \times 173.7 \times 10^{-3}\\ &= -84.68 + 51.76\\ &= \underline{-32.92 \text{ kJ mol}^{-1}}.\end{aligned}$$

The formation of ethane from its elements is both exothermic and has a negative standard free energy.

(iv) $C(c) + O_2(g) \xrightarrow[1\text{ atm}]{25\,°C} CO_2(g)$

$$\Delta_f H°(CO_2, g) = \underline{-393.51 \text{ kJ mol}^{-1}}$$

$$\begin{aligned}\Delta_f S°(CO_2, g) &= 213.68 - (5.74 + 205.04)\\ &= 213.68 - 210.78\\ &= \underline{2.9 \text{ J K}^{-1} \text{ mol}^{-1}}\end{aligned}$$

$$\Delta_f G^\circ(CO_2, g) = -393.51 - 298 \times 2.9 \times 10^{-3}$$
$$= -393.51 - 0.86$$
$$= \underline{-394.37 \text{ kJ mol}^{-1}}.$$

In the first three of these examples, the standard entropies of formation of the compounds are negative. In the case of benzene it reinforces an unfavourable enthalpy of formation; for methylamine it more than counterbalances a small exothermic enthalpy so that methylamine is overall thermodynamically unstable; and in the case of ethane, the entropy term merely reduces a very substantial negative enthalpy of formation.

In general, we can expect organic compounds to have unfavourable (negative) entropies of formation, and consequently $\Delta_f G^\circ$ will be more positive (or less negative) than $\Delta_f H^\circ$.

In the fourth example above, the formation of carbon dioxide is thermodynamically very favourable due to a huge exothermicity and a very small but nevertheless favourable (positive) entropy contribution.

Example. From the enthalpy results in Table 2.2 of Chapter 2 (p. 52) and the entropy data in Table 3.1 above, calculate $\Delta_r G^\circ$ for the following reactions:

(i) $CH_4(g) + Cl_2(g) \xrightarrow[1 \text{ atm}]{25\,^\circ C} CH_3Cl(g) + HCl(g)$

By eqn (3.4),

$$\Delta_r H^\circ = (-86.44 - 92.31) - (-74.87)$$
$$= \underline{-103.9 \text{ kJ mol}^{-1}}$$

By eqn (3.5)

$$\Delta_r S^\circ = (234.3 + 186.79) - (186.1 + 222.97)$$
$$= \underline{12.0 \text{ J K}^{-1} \text{ mol}^{-1}}$$

By eqn (3.3),

$$\Delta_r G^\circ = -103.9 - 298 \times 12.0 \times 10^{-3}$$
$$= -103.9 - 3.6$$
$$= \underline{-107.5 \text{ kJ mol}^{-1}}.$$

In this reaction, a small positive entropy term reinforces a large exothermicity and the reaction is thermodynamically very favourable.

(ii) $CH{\equiv}CH(g) + 2H_2(g) \xrightarrow[1 \text{ atm}]{25\,^\circ C} C_2H_6(g)$

$$\Delta_r H^\circ = (-84.68) - (222.7)$$
$$= \underline{-307.4 \text{ kJ mol}^{-1}}$$

$$\Delta_r S^\circ = 229.5 - (200.85 + 2 \times 130.57)$$
$$= 229.5 - 462.0$$
$$= \underline{-232.5 \text{ J K}^{-1} \text{ mol}^{-1}}$$

$$
\begin{aligned}
\Delta_r G^\circ &= -307.4 + 298 \times 232.5 \times 10^{-3} \\
&= -307.4 + 69.3 \\
&= \underline{-238.1 \text{ kJ mol}^{-1}}.
\end{aligned}
$$

The hydrogenation of ethyne is strongly exothermic but entropically unfavourable. At 25 °C, the net effect is that the reaction is *overall* thermodynamically favourable. This is a very common situation at normal temperatures – the free energy dominated by the enthalpy term. A change in temperature can however alter matters as the following reaction shows.

Example. Determine the thermodynamic feasibility of the following gas-phase reaction at 400 and 600 K using the data provided.[1]

$$C_6H_5CH{=}CH_2 + CH_4 \xrightarrow[T]{1\text{ atm}} C_6H_5{-}C_3H_7$$

$$\Delta_r H^\circ_{400\text{ K}} = -64.98 \text{ kJ mol}^{-1}; \qquad \Delta_r H^\circ_{600\text{ K}} = -64.52 \text{ kJ mol}^{-1};$$

$$\Delta_r S^\circ_{400\text{ K}} = -131.7 \text{ J K}^{-1}\text{ mol}^{-1}; \qquad \Delta_r S^\circ_{600\text{ K}} = -130.8 \text{ J K}^{-1}\text{ mol}^{-1}.$$

The forward reaction is exothermic but associated with a loss of entropy; note that $\Delta_r H^\circ$ and $\Delta_r S^\circ$ are not significantly different at the two temperatures.

$$
\begin{aligned}
\Delta_r G^\circ &= \Delta_r H^\circ - T\,.\,\Delta_r S^\circ \\
\Delta_r G^\circ_{400\text{ K}} &= -64.98 + 400 \times 131.7 \times 10^{-3} \\
&= -64.98 + 52.68 \\
&= \underline{-12.3 \text{ kJ mol}^{-1}}
\end{aligned}
$$

$$
\begin{aligned}
\Delta_r G^\circ_{600\text{ K}} &= -64.52 + 600 \times 130.8 \times 10^{-3} \\
&= -64.52 + 78.48 \\
&= \underline{+13.96 \text{ kJ mol}^{-1}}.
\end{aligned}
$$

The reaction as written, therefore, is thermodynamically feasible at 400 K, but not at 600 K. At the lower temperature, the reaction is directed by the enthalpy; but because $T\,.\,\Delta S$ becomes numerically larger as T increases, whereas ΔH hardly changes, the entropy term dominates at high temperatures.

In this section, we have seen that the thermodynamic feasibility of a reaction can be investigated by simple calculations if $\Delta_f G^\circ$ (or $\Delta_f H^\circ$ and S°) data are available for reactants and products. We shall leave until the next chapter consideration of those processes for which $\Delta_r G^\circ$, on paper, is equal to or close to zero, and of what this means as far as the real reactions are concerned. We shall also leave until the next chapter further discussion of those interesting reactions in which $\Delta_r G^\circ$ changes sign when the temperature is altered.

3.4 Entropy, molecular properties, and physical change

In the previous chapter we saw that $\Delta_f H^\circ$ is a molar energy term appropriate to constant pressure conditions and we were able to relate it directly to the molecular level. Analogously, $\Delta_r H^\circ$ is the net molar energy change due to the breaking and making of bonds in molecules and any change in volume when reactants become products as described by the chemical equation under conditions of constant temperature and pressure.

In order to develop a molecular understanding of free energy, we now need to consider entropy in molecular terms. This is not so easy because entropy is related to more than just the properties of individual molecules. Although we were able to start with the energetics of a single molecule and progress directly towards molar internal energy and then to enthalpy, entropy can relate *only* to large assemblies of molecules; there cannot be an entropy of a single molecule.

There are two common approaches to the treatment of entropy itself which necessarily converge at some stage. The most illuminating approach lies in statistical or molecular thermodynamics. This combines mathematical rigour and the possibility of a physical interpretation in terms of the statistical probability associated with an assembly of molecules under particular conditions of pressure, volume, and temperature.[3,7] Without such a treatment, an understanding of entropy would be incomplete.

We shall not, however, begin from this point of view. Rather, we shall consider the *change* in entropy due to some familiar physical processes from the standpoint of classical thermodynamics, and then proceed to illustrate qualitatively in the particular contexts how entropy *per se* relates to what many authors have called the randomness or disorder of an assembly of molecules.

Our basis in classical thermodynamics is provided by the equation[3]

$$\delta S \geqslant \frac{\delta q}{T},$$

where δS = the change in entropy of the system,
δq = the heat transfer to or from the system, and
T = its absolute temperature.

In accordance with our earlier convention, an endothermic process (heat supplied *to* the system) corresponds to δq *positive* and an increase in entropy (δS = positive), and an exothermic process (heat removed *from* the system) has δq *negative* and corresponds to a decrease in the entropy of the system (δS = negative).

Heat transfer is *reversible* in the thermodynamic sense when the temperature of the system remains identical with that of its surroundings, and under these conditions

$$\delta S = \frac{\delta q}{T}. \tag{3.6}$$

When the temperature of the system T is higher or lower than that of its surroundings, heat δq flows *irreversibly* from the hot to the cold. Under these conditions, the entropy change of the *system* is given by

$$\delta S > \frac{\delta q}{T}.$$

This equation by itself is of no use if we wish to calculate δS for an irreversible process. But entropy, like enthalpy, is a state function; it is independent of *how* a given state is attained. Therefore, we devise an alternative reversible route between the same initial and final states. This may involve several steps but we shall be able to apply eqn (3.6) to each of these and thereby calculate the entropy change for the overall process.

3.4.1 Entropy change due to a change in temperature

A system initially with entropy S_1 at temperature T_1 is transformed by some means at constant pressure (the condition of most chemical reactions in condensed phases) to temperature T_2 with entropy S_2. As mentioned above, in order that we may calculate $(S_2 - S_1)$ in terms of recognizable thermal properties of the system, we must effect the temperature change by transferring heat reversibly.

The incremental temperature change δT brought about in one mole of material at constant pressure by a (reversible) heat transfer δq is given by

$$\delta q = C_P \, . \, \delta T$$

where C_P = the molar heat capacity of the system at constant pressure. From eqn (3.6), $\delta S = \delta q/T$ since δq is transferred reversibly therefore

$$\delta S_P = \frac{C_P \, . \, \delta T}{T}$$

and the change in entropy $(S_2 - S_1)$ caused by the temperature change from T_1 to T_2 at constant pressure is given by

$$\Delta S_P = \int_{T_1}^{T_2} \frac{C_P}{T} \cdot \mathrm{d}T. \tag{3.7}$$

We used the device of transferring heat reversibly to change the system from T_1 to T_2 in order to use eqn (3.6) to derive eqn (3.7). But eqn (3.7) is valid regardless of how the system is changed from T_1 to T_2 at constant pressure because, as mentioned, S_1 and S_2 are state functions.

If an analytical expression is available for C_P in terms of T, then eqn

(3.7) can be evaluated. For an ideal monatomic gas this is particularly easy as $C_P = 2.5R$,[3] therefore

$$\Delta S_P = 2.5R \ln\left\{\frac{T_2}{T_1}\right\}.$$

By eqn (3.7), the *absolute* standard entropy of a crystalline material at temperature T and 1 standard atmosphere pressure is given by

$$S^\circ_T = S^\circ_{0\,\mathrm{K}} + \int_{0\,\mathrm{K}}^{T} \frac{C^\circ_P}{T} \,.\, \mathrm{d}T. \qquad (3.8)$$

Assuming the validity of the Third Law of Thermodynamics (the entropy of a perfect crystalline solid at zero K is zero), we can set $S^\circ_{0\,\mathrm{K}}$ in eqn (3.8) equal to zero and the absolute standard entropy at any other temperature T, S°_T, is then given by the second term.

The standard heat capacity C°_P may be determined calorimetrically down to about 4 K. Between zero and 4 K (or even higher), the heat capacity can be evaluated by the Debye relationship between heat capacity and temperature.[3]

If the material has residual entropy at zero K, then a calculated value for $S^\circ_{0\,\mathrm{K}}$ has to be included in eqn (3.8) (see p. 96).

If one mole of the material undergoes a temperature change from T_1 to T_2 at constant volume (the condition of many gas-phase reactions) we can equally well show that the change in entropy is given by

$$\Delta S_V = \int_{T_1}^{T_2} \frac{C_V}{T} \,.\, \mathrm{d}T,$$

where C_V = the molar heat capacity at constant volume. If the material is a monatomic ideal gas,[3] $C_V = 1.5R$ therefore $\Delta S_V = 1.5R \ln\{T_2/T_1\}$.

3.4.2 Entropy change of a phase transition

3.4.2.1 Fusion

A familiar example of a thermodynamically reversible process is the fusion of a pure crystalline material at its normal melting point, T_{mp}. At this temperature the pure liquid and crystalline modifications are in equilibrium and exert equal vapour pressures.

$$\mathrm{A(c)} \xrightarrow[1\ \mathrm{atm}]{T_{\mathrm{mp}}} \mathrm{A(l)}; \quad \Delta_{\mathrm{m}}H^\circ_{T_{\mathrm{mp}}}, \Delta_{\mathrm{m}}S^\circ_{T_{\mathrm{mp}}}.$$

By eqn (3.6),

$$\Delta_{\mathrm{m}}S^\circ_{T_{\mathrm{mp}}} = \frac{\Delta_{\mathrm{m}}H^\circ_{T_{\mathrm{mp}}}}{T_{\mathrm{mp}}}$$

TABLE 3.2
Standard molar enthalpy and entropy changes of reversible fusion

Material	T_{mp}/K	$\Delta_m H^{\circ}_{T_{mp}}$/kJ mol^{-1}	$\Delta_m S^{\circ}_{T_{mp}}$/J K^{-1} mol^{-1}
HCl[a]	158.9	1.99	12.5
H_2O[a]	273.15	6.009	22.00
Benzene[b]	278.68	9.837	35.30
Naphthalene[b]	353.4	18.98	53.7
CBr_4[a]	363.3	4.1	11.3
Benzoic acid[b]	395.52	18.1	45.7
CHI_3[a]	398	16	41
Urea[a]	405.8	15.1	37

[a] Ref. 8. [b] Ref. 1.

where $\Delta_m H^{\circ}_{T_{mp}}$ and $\Delta_m S^{\circ}_{T_{mp}}$ = the standard molar enthalpy and entropy of fusion at T_{mp}.

The standard state superscripts indicate that these values refer to the pure material under 1 standard atmosphere total pressure.

Heat has to be supplied to melt the crystals so the entropy of the material is higher in the liquid (molten) state than in the crystal. Some results are given in Table 3.2; there appears to be no pattern or trend in the $\Delta_m S^{\circ}_{T_{mp}}$ values over quite a wide range of temperature and type of material.

3.4.2.2 *Evaporation*

A liquid boils at the temperature when its vapour pressure becomes equal to the external pressure, and vaporization at this temperature is thermodynamically reversible. If the external pressure is 1 standard atmosphere,

$$A(l) \xrightarrow[1\ \text{atm}]{T_{bp}} A(g); \quad \Delta_v H^{\circ}_{T_{bp}}, \Delta_v S^{\circ}_{T_{bp}};$$

and by eqn (3.6),

$$\Delta_v S^{\circ}_{T_{bp}} = \frac{\Delta_v H^{\circ}_{T_{bp}}}{T_{bp}}$$

where $\Delta_v H^{\circ}_{T_{bp}}$ and $\Delta_v S^{\circ}_{T_{bp}}$ = the standard molar enthalpy and entropy of vaporization at the boiling point, T_{bp}, under 1 standard atmosphere pressure.

The standard state is still the pure material at 1 standard atmosphere pressure. For the gas, this means the *hypothetical* ideal gas at 1 standard atmosphere pressure which has the same enthalpy as the *real* gas at the

TABLE 3.3
Standard molar enthalpy and entropy of reversible evaporation

Material	T_{bp}/K	$\Delta_v H^\circ_{T_{bp}}$/kJ mol^{-1}	$\Delta_v S^\circ_{T_{bp}}$/J K^{-1} mol^{-1}
F_2[a]	85.24	6.32	74.1
Cl_2[b]	239.10	20.41	85.4
HCN[b]	298.85	25.22	84.4
HBr[a]	206.43	17.61	85.3
NH_3[a]	239.73	23.35	97.4
CH_3NH_2[a]	266.84	25.82	96.7
H_2O[a]	373.15	40.66	109.0
CH_3OH[a]	337.9	35.3	104.4
$(CH_3)_3N$[b]	276.02	22.94	83.1
CH_3CHO[b]	293.3	25.71	87.7
$(C_2H_5)_2O$[b]	307.7	26.69	86.8
$CHCl_3$[a]	334.4	29.4	87.8
Ethyl acetate[b]	349.15	32.30	92.5
Benzene[b]	353.25	30.76	87.1
Cyclohexane[b]	353.88	30.08	85.0
HCO_2H[a]	373.7	22.3	59.6
$C_2H_5NO_2$[b]	387.22	35.1	90.8
Pyridine[b]	388.38	35.11	90.4
CH_3CO_2H[b]	391.25	23.69	60.6

[a] Ref. 8. [b] Ref. 1.

same temperature but at zero pressure; see p. 49. Vaporization is endothermic so the entropy of A is higher in the gaseous state than as the liquid.

As may be seen from Table 3.3, $\Delta_v S^\circ_{T_{bp}}$ is remarkably constant at about 87 J K^{-1} mol^{-1} for quite a wide range of materials (the thermodynamic expression of Trouton's Rule). The anomolously high values for liquids such as water and methanol implicate a higher liquid structure than normal due, no doubt, to intermolecular hydrogen bonding. This relates to the liquid at its boiling point, and we can expect such hydrogen-bonding to be considerably more extensive at lower temperatures. The unusually low values for simple carboxylic acids are probably due to hydrogen-bonded dimers in the gas state.[9]

3.4.2.3 *Crystal transition*

Below the melting point of a pure material, reversible transitions between different crystal forms may occur, indeed, they are very common.

$$A(c_1) \xrightarrow[1\text{ atm}]{T_t} A(c_2); \quad \Delta_t H^\circ_{T_t}, \Delta_t S^\circ_{T_t}.$$

TABLE 3.4
Standard molar enthalpy and entropy changes of some reversible phase transitions in crystalline materials[a]

Material	T_t/K	$\Delta_t H^\circ_{T_t}$/kJ mol^{-1}	$\Delta_t S^\circ_{T_t}$/J K^{-1} mol^{-1}
F_2	45.55	0.728	16.0
HCl	98.38	1.19	12.1
Hexamethylbenzene	116.48	1.13	9.7
CH_3CN	216.9	0.900	4.15
Cyclohexanol	263.5	8.20	31.1
1,4-Dioxan	272.9	2.35	8.62
2-Methylnaphthalene	288.5	5.61	19.4
C_2Cl_6	344.4	7.95	23.1
Sulphur	368.46	0.402	1.1

[a] All data taken from ref. 1.

By eqn (3.6),

$$\Delta_t S^\circ_{T_t} = \frac{\Delta_t H^\circ_{T_t}}{T_t}$$

where $\Delta_t H^\circ_{T_t}$ and $\Delta_t S^\circ_{T_t}$ = the standard molar enthalpy and entropy of the phase transition at T_t, the transition temperature at which both crystalline modifications are stable under 1 standard atmosphere total pressure. Some examples are given in Table 3.4.

It is evident from the data in Tables 3.2–3.4 that, in general,

$$\Delta_v S^\circ_{T_{bp}} > \Delta_m S^\circ_{T_{mp}} > \Delta_t S^\circ_{T_t}.$$

This is in accordance with current views on the nature of intermolecular interactions and structure in the three states of matter. Crystalline solids are very highly ordered compared with liquids, and gases usually have no intermolecular structure at all.

We can now give a comprehensive expression for the thermochemical standard absolute molar entropy S°_T of a material in the gas phase at temperature T which has for example a single crystal transition at temperature T_t:

$$S^\circ_T(g) = S^\circ_{0\,K} + \int_{0\,K}^{T_t} \frac{C^\circ_P(c_1)}{T}\,.\,dT + \frac{\Delta_t H^\circ_{T_t}}{T_t} + \int_{T_t}^{T_{mp}} \frac{C^\circ_P(c_2)}{T}\,.\,dT$$

$$+\frac{\Delta_m H^\circ_{T_{mp}}}{T_{mp}} + \int_{T_{mp}}^{T_{bp}} \frac{C^\circ_P(l)}{T}\,.\,dT + \frac{\Delta_v H^\circ_{T_{bp}}}{T_{bp}} + \int_{T_{bp}}^{T} \frac{C^\circ_P(g)}{T}\,.\,dT \quad (3.9)$$

If the material is a liquid at T, the last two terms are absent and the integration in the third last term is between T_{mp} and T. If it is a solid, then the expression comprises only the first four terms and the integral in the fourth is between T_t and T. Results obtained by this type of expression using calorimetry and a calculated value for S°_{0K} (see p. 96) are called thermochemical or Third Law entropies.

TABLE 3.5
Standard molar enthalpy and entropy changes for reversible sublimations[a]

Material	T_s/K	$\Delta_s H^\circ_{T_s}$/kJ mol^{-1}	$\Delta_s S^\circ_{T_s}$/J K^{-1} mol^{-1}
CO_2	194.68	25.23	129.6
SF_6	209.5	22.8	109.0
POF_3	233.7	38	161
IF_7	276.6	30.8	111.5
ICN	413.0	59.41	144

[a] All data taken from ref. 8.

3.4.2.4 *Sublimation*

As the temperature of some solids is raised, their vapours reach the external pressure before the crystals melt. At this temperature, T_s, a further supply of heat causes such a solid to sublime reversibly and isothermally. If the sublimation takes place at atmospheric pressure,

$$A(c) \xrightarrow[1\ \text{atm}]{T_s} A(g); \quad \Delta_s H^\circ_{T_s}, \Delta_s S^\circ_{T_s},$$

and by eqn (3.6),

$$\Delta_s S^\circ_{T_s} = \frac{\Delta_s H^\circ_{T_s}}{T_s},$$

where $\Delta_s H^\circ_{T_s}$ and $\Delta_s S^\circ_{T_s}$ = the standard molar enthalpy and entropy of sublimation at T_s, the temperature at which both gas and solid are stable at 1 standard atmosphere pressure. Some results are shown in Table 3.5.

3.4.2.5 *Non-reversible phase transition*

If a phase change takes place at a temperature T and pressure P which represent a combination other than one at which both modifications are stable, then the transition will not be thermodynamically reversible. For example, in the hypothetical case of a sublimation at 298.15 K and 1 standard atmosphere pressure,

$$A(c) \xrightarrow[1\ \text{atm}]{298.15\ \text{K}} A(g); \quad \Delta_s H^\circ_{298\ \text{K}}, \Delta_s S^\circ_{298\ \text{K}},$$

$$\Delta_s H^\circ_{298\ \text{K}} = \Delta_f H^\circ_{298\ \text{K}}(g) - \Delta_f H^\circ_{298\ \text{K}}(c), \tag{3.10}$$

and

$$\Delta_s S^\circ_{298\ \text{K}} = S^\circ_{298\ \text{K}}(g) - S^\circ_{298\ \text{K}}(c).$$

TABLE 3.6
Standard molar enthalpy and entropy changes of non-reversible phase transitions at 1 standard atmosphere pressure and 25 °C

Material	Transition[a]	$\Delta H^\circ_{298\,K}$/kJ mol^{-1}	$\Delta S^\circ_{298\,K}$/J K^{-1} mol^{-1}
C[b]	s	716.67	152.25
Si[b]	s	450	149.1
Br_2[b]	v	30.91	93.14
I_2[b]	s	62.421	144.43
H_2O[b]	v	44.016	118.77
ICN[c]	s	59.4	160
Dimethylsulphoxide[c]	s	79.9	168
Benzene	v	33.85[d]	95.9[e]
Cyclohexane[e]	v	33.10	93.9
t-Butanol[e]	s	40.1	155.4
Phenol[e]	s	68.66	171.6
Benzoic acid[e]	s	94.99	201.4
Naphthalene[e]	s	72.89	168.7
$(CH_3)_3C—C(CH_3)_3$[f]	s	43.37	115.6

[a] s = sublimation; v = vaporization.
[b] Ref. 4.
[c] Ref. 6.
[d] J. D. Cox and G. Pilcher, *Thermochemistry of organic and organometallic compounds*, Academic Press, London (1970).
[e] Ref. 1.
[f] D. W. Scott, D. R. Douslin, M. E. Gross, G. D. Oliver, and H. M. Huffman, *J. Am. chem. Soc.* **74,** 883 (1952).

But $\Delta_s S^\circ_{298\,K}$ is *not* normally equal to $\Delta_s H^\circ_{298\,K}/298$ K because at 1 atmosphere pressure and 298 K, only one of the two modifications is likely to be stable, so the interconversion of material between the phases is not thermodynamically reversible.

We encountered eqn (3.10) in the previous chapter. It is used to determine $\Delta_f H^\circ_{298\,K}(g)$ from experimental results for $\Delta_f H^\circ_{298\,K}(c)$ (by combustion calorimetry) and $\Delta_s H^\circ_{298\,K}$ (by vapour pressure–temperature measurements). $S^\circ_{298\,K}(g)$ may be calculated from statistical thermodynamics as we shall see later, and $S^\circ_{298\,K}(c)$ can be obtained calorimetrically from the truncated version of eqn (3.9) for solids.

Some results are given in Table 3.6. Comparison of values for ICN in Tables 3.5 and 3.6 illustrates that although $\Delta_s H^\circ_{298\,K}$ and $\Delta_s H^\circ_{T_s}$ may be very similar, $\Delta_s S^\circ_{T_s}$ and $\Delta_s S^\circ_{298\,K}$ for reversible and non-reversible processes are significantly different, the latter being the larger. Correspondingly, data in Tables 3.3 and 3.6 show that the entropy of non-reversible evaporation at 298 K is higher than for the reversible process at T_{bp}; now, however, $\Delta_v H^\circ_{298\,K}$ values for the non-reversible process are appreciably higher than $\Delta_v H^\circ_{T_{bp}}$ for reversible evaporation.

3.4.3 Entropy change due to an isothermal change in the volume of an ideal gas

We may apply the First Law of Thermodynamics to a system and write

$$\delta U = \delta q + w$$

where δU = the change in the internal energy,
δq = heat transfer, and
w = work.

We now specify the system as an ideal gas initially with entropy S_1, at pressure P_1 and volume V_1, and allow it to undergo a thermodynamically reversible isothermal change to a final state with entropy S_2 at pressure P_2 and volume V_2. For such a process to be reversible, the temperature and pressure of the surroundings which contain the system must always be identical with the temperature and pressure of the system. Under these conditions, the work involved in an incremental volume change δV will be maximal and given by

$$w = -P\,.\,\delta V,$$

where P = the external pressure which is equal to the pressure of the gas. When δV is positive (gas expansion), the system does work; and when δV is negative (a compression), work is done on the system.

The internal energy of an ideal gas is independent of volume at constant temperature so for an isothermal volume change,

$$\delta U = 0,$$

and the First Law expression becomes

$$0 = \delta q - P\,.\,\delta V,$$

or

$$\delta q = P\,.\,\delta V.$$

If we are dealing with n moles of an ideal gas,

$$PV = nRT,$$

therefore

$$\delta q = \frac{nRT}{V}\,.\,\delta V,$$

and the change in entropy δS associated with this reversible isothermal change in volume is given by

$$\delta S = \frac{\delta q}{T} = \frac{nR\,.\,\delta V}{V}.$$

The overall change is from P_1V_1 to P_2V_2 and $(S_2-S_1)=\Delta S_T$ at constant temperature T. Therefore

$$\Delta S_T = nR \,.\int_{V_1}^{V_2}\frac{dV}{V}$$

or

$$\Delta S_T = nR \,.\ln\left(\frac{V_2}{V_1}\right). \tag{3.11}$$

If it is more convenient, this equation can be written as

$$\Delta S_T = nR \,.\ln\left(\frac{P_1}{P_2}\right) \tag{3.12}$$

since, for the ideal gas, $P_1V_1 \overset{T}{=} P_2V_2$. When the gas expands, $V_2 > V_1 (P_2 < P_1)$ so its entropy increases. Because entropy is a state function, eqns (3.11) and (3.12) may also be used to calculate the entropy change of an ideal gas which undergoes an *irreversible* PV change as long as the initial and final temperatures are identical.

Generally, standard state entropies of gases in thermochemical reference tables refer to 298 K and a standard state of 1 atmosphere pressure. For comparison with quantities derived from kinetics, a standard state of 1 mol dm^{-3} is frequently more appropriate as we shall see later. It is, therefore, necessary to be able to calculate the standard entropy of a gas at any temperature T and the standard state of 1 mol dm^{-3} ($S_T^{\ominus}$) from a value at 298 K and a standard state of 1 atmosphere ($S^{\circ}_{298\,K}$). This is straightforward using equations which we have now derived and the assumption that the gas can be treated as ideal.

We illustrate the method by calculating the molar entropy of ethane at 1 mol dm^{-3} and 858 K from $S^{\circ}_{298\,K}$ and C°_P data.

(i) The temperature change at constant pressure. From eqn (3.7),

$$S^{\circ}_{T_2} = S^{\circ}_{T_1} + \int_{T_1}^{T_2}\frac{C^{\circ}_P}{T}\,dT$$

$$= S^{\circ}_{T_1} + \bar{C}^{\circ}_P \ln\left(\frac{T_2}{T_1}\right)$$

where $\bar{C}^{\circ}_P$ is the mean standard molar heat capacity of the gas at constant pressure over the temperature range T_2-T_1. For ethane,[12] $S^{\circ}_{298\,K}(1\ \text{atm}) = 229\ \text{J K}^{-1}\ \text{mol}^{-1}$ and $\bar{C}^{\circ}_P(298\text{–}858\ \text{K}) = 84\ \text{J K}^{-1}\ \text{mol}^{-1}$, therefore

$$S^{\circ}_{858\,K}(1\ \text{atm}) = \left(229 + 84\ln\left(\frac{858}{298}\right)\right)\ \text{J K}^{-1}\ \text{mol}^{-1}$$

$$= 318\ \text{J K}^{-1}\ \text{mol}^{-1}.$$

(ii) The change in standard state from 1 atm to 1 mol dm^{-3} at constant temperature. This corresponds to one mole of ethane as an ideal gas undergoing an isothermal compression at 858 K from its volume at 1 atm pressure to 1 dm^3.

By equation (3.11),

$$\Delta S_T = R\,.\,\ln\left(\frac{V_2}{V_1}\right)$$

therefore

$$S_T^{\ominus}(1\ \text{mol dm}^{-3}) = S_T^{\circ}(1\ \text{atm}) + R\,.\,\ln\left(\frac{V_2}{V_1}\right).$$

Since $P_1V_1 = RT$,

$$S_T^{\ominus}(1\ \text{mol dm}^{-3}) = S_T^{\circ}(1\ \text{atm}) + R\,.\,\ln\left(\frac{P_1V_2}{RT}\right)$$

where $P_1 = 1$ atm, $V_2 = 1\ \text{dm}^3$, and R *within* the logarithmic term must have the proper units i.e. $0.0821\ \text{dm}^3\ \text{atm K}^{-1}$ for 1 mole of gas. Therefore

$$S_T^{\ominus}(1\ \text{mol dm}^{-3}) = S_T^{\circ}(1\ \text{atm}) - R\,.\,\ln(0.0821T). \qquad (3.13)$$

This is a very important and generally useful expression which we may apply here to give

$$S_{858\ \text{K}}^{\ominus}(\text{ethane, } 1\ \text{mol dm}^{-3}) = 318 - 8.314\ln(0.0821\times 858)$$
$$= \underline{283\ \text{J K}^{-1}\ \text{mol}^{-1}}.$$

3.4.4 The entropy of mixing

If valves connecting different ideal gases at the same pressure P are opened, the gases will diffuse into each other to give a homogeneous mixture at an unchanged total pressure P. The volume available to each gas becomes the total volume of the system V, and the final partial pressure P_i of component i initially occupying V_i (at P) is given by

$$P_i = X_i\,.\,P$$

where X_i = the mole fraction of component i in the final mixture.[3]

From eqn (3.12), the change in entropy of gas i upon mixing with the others $\Delta_{mx}S_i$ is given by

$$\Delta_{mx}S_i = n_iR\,.\,\ln\left\{\frac{P}{P_i}\right\} = -n_iR\,.\,\ln\left\{\frac{P_i}{P}\right\},$$

therefore

$$\Delta_{mx}S_i = -n_iR\,.\,\ln X_i.$$

If the total change in entropy due to mixing the gases $= \Delta_{mx}S$ then

$$\underline{\Delta_{mx}S = \sum \Delta_{mx}S_i = -R\sum n_i \ln X_i.} \qquad (3.14)$$

Other proofs show that eqn (3.14) applies to the entropy of mixing of any ideal solutions, not just gases.[3]

Equation (3.14) allows us to calculate the residual standard entropy at zero K, $S^\circ_{0\,\mathrm{K}}$ in eqns (3.8) and (3.9), when it arises from random mixing of more than one form of the crystalline material. The difference between the forms could be due to the direction of alignment of molecules in the lattice as in the case of carbon monoxide (OC... OC and OC... CO). Or, in the case of a racemic modification of a chiral compound, it is due to randomly mixed enantiomers.

In both of these examples, the molar entropy at zero K, $S^\circ_{0\,\mathrm{K}}$, is the entropy increase upon the hypothetical mixing of 0.5 mol of one pure form with zero entropy with 0.5 mol of the other.

$$\begin{aligned} S^\circ_{0\,\mathrm{K}} &= -R(0.5 \ln 0.5 + 0.5 \ln 0.5) \\ &= -R \ln 0.5 \\ &= R \ln 2 \\ &= 2.90\ \mathrm{J\,K^{-1}\,mol^{-1}}. \end{aligned}$$

3.5 The resolution and calculation of entropy

Calorimetric results for the entropies of many elements and compounds using eqn (3.9) are available to a satisfactory degree of precision. However, as can be imagined, the necessary experimental work is not accomplished very rapidly. It would be desirable to have a quicker method for obtaining entropy results; and if, at the same time, this could provide an independent check on the accuracy of those calorimetric values which have already been determined, so much the better. Such a method exists; absolute entropies of ideal gases can be calculated from spectroscopic and structural results using rigorous theory based in statistical thermodynamics.[3,7]

To a good level of approximation, the energy of the electronic ground state of a molecule can be resolved into components due to vibration, rotation, and translation; this was implicit in our treatment of molecular vibrations in Chapter 1. Analogously, the molar entropy of the electronic ground state of a material can also be resolved:

$$S = S(\mathrm{vib}) + S(\mathrm{rot}) + S(\mathrm{trans}),$$

and, within the validity of the Third Law of Thermodynamics, each of these terms is absolute since, at zero K, their total is equal to zero.

In Fig. 1.1 of Chapter 1 (p. 2), we saw a generalized molecular-energy diagram which included electronic, vibrational, and rotational levels. In a similar way, the first few energy levels of two different molecules X and Y are represented diagrammatically in Fig. 3.1 below.

It does not matter at the moment whether these refer to vibrational, rotational, or any other sort of energy.

If the total number of molecules of X is N, and the numbers in the

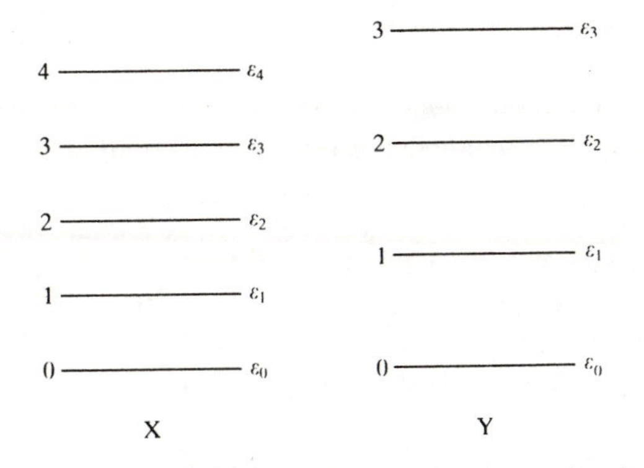

Fig. 3.1. Energy levels of high and low entropy for molecules X and Y.

molecular energy levels with quantum numbers 0, 1, 2, . . . i . . . and energies $\varepsilon_0, \varepsilon_1, \varepsilon_2, \ldots \varepsilon_i \ldots$ are $n_0, n_1, n_2 \ldots n_i \ldots$, then, by the Boltzmann Distribution Law,[3]

$$\frac{n_i}{n_0} = \mathrm{e}^{-(\varepsilon_i - \varepsilon_0)/k_B T} \tag{3.15}$$

$$N = \sum n_i = n_0 \sum \mathrm{e}^{-(\varepsilon_i - \varepsilon_0)/k_B T} \tag{3.16}$$

therefore

$$n_i = \frac{N}{\sum \mathrm{e}^{-(\varepsilon_i - \varepsilon_0)/k_B T}} \cdot \mathrm{e}^{-(\varepsilon_i - \varepsilon_0)/k_B T}$$

and the term $\sum \mathrm{e}^{-(\varepsilon_i - \varepsilon_0)/k_B T}$ is known as the *molecular partition function, Q*.

It is seen from eqn (3.16) that

$$Q = \frac{\text{total number of molecules}}{\text{number in the zero energy level}} = \frac{N}{n_0}.$$

When all the molecules are in the zero level, $n_0 = N$ so $Q = 1$; and, as more molecules occupy higher energy levels for example as the temperature is raised, Q increases and in principle it has no upper limit.

In the examples of Fig. 3.1, the occupation of the higher levels for X will be greater than for Y at the same temperature because the levels in X are closer together and consequently are energetically more accessible. Suppose $Q(\mathrm{X}) = 1.25$, then 80 per cent of the molecules are in the ground state and the other 20 per cent are distributed among the upper levels according to eqn (3.15). If, at the same temperature, $Q(\mathrm{Y}) = 1.01$ then 99 per cent of these molecules are in their ground state and only 1 per cent are in the upper levels. In neither case does it matter which 20 per cent or 1 per cent are out of the ground state; indeed the molecules may undergo rapid intermolecular energy transfer and be constantly exchanging their states. But at any instant the Boltzmann Distribution Law defines the statistical partitioning of any large assembly of molecules among the available energy levels.

There are many more ways of distributing any 20 per cent of the molecules of X among its available upper levels than of distributing any 1 per cent of Y among its more widely spaced upper levels, abiding in each case by eqn (3.15). In the language of statistical thermodynamics, molecules X with a partition function of 1.25 have a higher probability than molecules Y at the same temperature with their partition function of 1.01, and consequently a large assembly of molecules X has a higher entropy than an equal number of molecules Y.

If the temperature is altered, $Q(\mathrm{X})$ and $Q(\mathrm{Y})$ will change but at all temperatures $Q(\mathrm{X}) > Q(\mathrm{Y})$ simply because of the relative energy level spacings. Consequently, it is sometimes stated that a set of closely spaced energy levels has a higher entropy than a more widely-spaced set.

For one mole of non-localized molecules such as a gas, the relationship between entropy S and the total molecular partition function Q is[7]

$$S = k_{\mathrm{B}}N_{\mathrm{A}} \ln Q + k_{\mathrm{B}}N_{\mathrm{A}}T\frac{\partial(\ln Q)}{\partial T} - k_{\mathrm{B}} \ln(N_{\mathrm{A}}!)$$

where N_{A} = the Avogadro constant and k_{B} = the Boltzmann constant. In the case of a mole of localized particles as in a crystal,[7]

$$S = k_{\mathrm{B}}N_{\mathrm{A}} \ln Q + k_{\mathrm{B}}N_{\mathrm{A}}T\frac{\partial(\ln Q)}{\partial T}.$$

In order to use either of these equations, Q must be known as a function of the absolute temperature and this in turn requires knowledge of ε_i as a function of temperature and molecular parameters.

As the molar internal energy U can also be calculated directly from the partition function,[7]

$$U = k_{\mathrm{B}}N_{\mathrm{A}}T^2\frac{\partial(\ln Q)}{\partial T},$$

all other thermodynamic functions can in turn be determined. Obviously, the molecular partition function is an exceedingly important quantity.

We can now consider what the molecular energy levels in Fig. 3.1 above might correspond to, and see how the resolved entropy terms can be calculated.

Molecules of the overwhelming majority of stable covalent compounds are in their electronic ground states at normal temperatures. Consequently, the electronic molecular partition function, $Q(\mathrm{el})$, for such a compound is 1 and there is no electronic contribution towards the molar entropy.

3.5.1 Translational entropy

A monatomic molecule has no internal degrees of freedom and consequently a large assembly of such molecules in their ground electronic state has only translational entropy. Using the particle-in-a-box approach, an equation can be derived for molecular kinetic energy of an ideal gas as a function of temperature, and this leads to the following equation for its translational molecular partition function, Q(trans):[3,7]

$$Q(\text{trans}) = \left(\frac{2\pi m k_B T}{h^2}\right)^{3/2} . V, \tag{3.17}$$

where m = the molecular mass,
T = the absolute temperature,
V = volume,
k_B = the Boltzmann constant, and
h = the Planck constant.

In the derivation of eqn (3.17), the summation in the definition of the partition function has been replaced by an integration. This is effectively regarding the translational energy as a continuum comprising infinitesimally close quantized levels. Consequently, levels higher than the ground state will be easily attained even at low temperatures and we may anticipate that translation has a high entropy.

From Q(trans), an expression for the translational entropy of n moles of an ideal gas can be obtained (the Sackur–Tetrode equation):[3,7]

$$S(\text{trans}) = nR . \ln\left\{\frac{e^{5/2} . V}{N_A}\left(\frac{2\pi m k_B T}{h^2}\right)^{3/2}\right\} \tag{3.18}$$

where N_A = the Avogadro constant, e = the base of natural logarithms, and the other symbols are as in eqn (3.17).

Equation (3.18) shows that the absolute molar translational entropy is a function of the molecular mass, absolute temperature, volume, and various fundamental constants.

Example. Calculate (a) the absolute standard molar entropy of neon at 298.15 K, and (b) the translational component of the absolute standard molar entropy of ammonia at 298.15 K. In both cases assume ideal gas behaviour.

(a) Neon at 298.15 K and 1 standard atmosphere pressure.

$$\text{Mass of 1 neon atom, } m = 20.179 \text{ amu}$$
$$= 20.179 \times 1.6606 \times 10^{-27} \text{ kg}$$
$$\underline{m = 3.3509 \times 10^{-26} \text{ kg.}}$$

Volume of 1 mole (6.02205×10^{23} atoms) of neon at 1 atmosphere pressure and

298 K,

$$V = 22.414 \times \frac{298.15}{273.15}\ \mathrm{dm^3}$$

or

$$\underline{V = 2.4465 \times 10^{-2}\ \mathrm{m^3}.}$$

$$\left(\frac{2\pi m k_B T}{h^2}\right)^{3/2} = \left(\frac{2\pi \times 3.3509 \times 10^{-26}\ \mathrm{kg} \times 1.3807 \times 10^{-23}\ \mathrm{J\,K^{-1}} \times 298.15\ \mathrm{K}}{6.6262^2 \times 10^{-68}\ \mathrm{J^2\,s^2}}\right)^{3/2}$$

$$= (1.9740 \times 10^{21}\ \mathrm{kg\,J^{-1}\,s^{-2}})^{3/2}$$

$$= (1.9740 \times 10^{21}\ \mathrm{m^{-2}})^{3/2}$$

$$\underline{= 8.7704 \times 10^{31}\ \mathrm{m^{-3}}.}$$

$$S(\mathrm{trans})_{298\,\mathrm{K}} = 8.314\ \mathrm{J\,K^{-1}\,mol^{-1}}\,.\,\ln\left(\frac{e^{5/2} \times 2.4465 \times 10^{-2}\ \mathrm{m^3}}{6.0220 \times 10^{23}} \times 8.7704 \times 10^{31}\ \mathrm{m^{-3}}\right)$$

$$= 8.314\ \mathrm{J\,K^{-1}\,mol^{-1}}\,.\,\ln(4.3407 \times 10^7)$$

$$= 8.314 \times 17.586\ \mathrm{J\,K^{-1}\,mol^{-1}}$$

$$\underline{= 146.2\ \mathrm{J\,K^{-1}\,mol^{-1}}.}$$

Neon has no rotational or vibrational degrees of freedom so it only has translational entropy; and since the gas is at 1 standard atmosphere pressure, its translational entropy constitutes its (absolute) standard entropy i.e.

$$S(\mathrm{trans})_{298.15\,\mathrm{K}}(1\ \mathrm{atm}) = S^\circ_{298.15\,\mathrm{K}} = 146.2\ \mathrm{J\,K^{-1}\,mol^{-1}}.$$

The experimental calorimetric value using eqn (3.9), assuming $S^\circ_{0\,\mathrm{K}} = 0$, is $146.5\ \mathrm{J\,K^{-1}\,mol^{-1}}$.[10] Note from the above that $Q(\mathrm{trans}) = (2\pi m k_B T/h^2)^{3/2}\,.\,V \approx 2 \times 10^{30}$ for neon at 1 atmosphere pressure and 298 K. Such huge numbers are typical of translational molecular partition functions of gases.

(b) Translational entropy of ammonia at 298.15 K and 1 standard atmosphere pressure.

$$\text{Mass of 1 } NH_3 \text{ molecule, } m = 17.031\ \text{amu}$$

$$= 17.031 \times 1.6606 \times 10^{-27}\ \mathrm{kg}$$

$$\underline{= 2.8282 \times 10^{-26}\ \mathrm{kg}.}$$

As in the previous calculation, $V = \underline{2.4465 \times 10^{-2}\ \mathrm{m^3}.}$

$$\left(\frac{2\pi m k_B T}{h^2}\right)^{3/2} = \left(\frac{2\pi \times 2.8282 \times 10^{-26}\ \mathrm{kg} \times 1.3807 \times 10^{-23}\ \mathrm{J\,K^{-1}} \times 298.15\ \mathrm{K}}{6.6262^2 \times 10^{-68}\ \mathrm{J^2\,s^2}}\right)^{3/2}$$

$$= (1.6661 \times 10^{21}\ \mathrm{m^{-2}})^{3/2}$$

$$= \underline{6.8005 \times 10^{31}\ \mathrm{m^{-3}}.}$$

$$S^\circ(\mathrm{trans}) = 8.314\ \mathrm{J\,K^{-1}\,mol^{-1}}\,.\,\ln\left(\frac{e^{5/2} \times 2.4465 \times 10^{-2}\ \mathrm{m^3}}{6.0220 \times 10^{23}} \times 6.8005 \times 10^{31}\ \mathrm{m^{-3}}\right)$$

$$= 8.314\ \mathrm{J\,K^{-1}\,mol^{-1}}\,.\,\ln(3.3657 \times 10^7)$$

$$S^\circ(\mathrm{trans}) = \underline{144.1\ \mathrm{J\,K^{-1}\,mol^{-1}}.}$$

If we compare this result with $S^{\circ}_{298\,K} = 192.67$ for $NH_3(g)$ in Table 3.1 (p. 80), we see that the translational contribution is a major proportion of the total entropy.

This is generally true of gaseous materials, consequently a chemical reaction in the gas phase which gives rise to a change in the number of molecules will proceed with a substantial change in entropy. And the greater the change in the number of molecules per mole of reaction, the greater the entropy change. For example:[11]

$$cis\text{-}CH_3CH{=}CHCO_2CH_3(g) \rightarrow trans\text{-}CH_3CH{=}CHCO_2CH_3(g); \quad \Delta S^{\circ} = 2\ \mathrm{J\,K^{-1}\,mol^{-1}}$$

$$(CH_3)_3COH(g) \rightarrow (CH_3)_2C{=}CH_2(g) + H_2O(g); \quad \Delta S^{\circ} = 156\ \mathrm{J\,K^{-1}\,mol^{-1}}$$

[1,3,5-trioxane] (g) $\rightarrow$ $3CH_2O(g)$; $\quad \Delta S^{\circ} = 350\ \mathrm{J\,K^{-1}\,mol^{-1}}$

2 [cyclopentadiene] (g) $\rightarrow$ [dicyclopentadiene] (g); $\quad \Delta S^{\circ} = -200\ \mathrm{J\,K^{-1}\,mol^{-1}}$.

3.5.2 Vibrational entropy

The number of vibrational modes of a molecule is determined by its symmetry and the number of atoms. Consequently, because all vibrations may contribute, the total vibrational entropy increases with molecular size and complexity. But the nature of each individual mode is also important.

The total vibrational molecular partition function $Q(\text{vib})$ can be factorized:

$$\begin{aligned} Q(\text{vib}) &= Q(\text{vib}\,.\,1)\,.\,Q(\text{vib}\,.2)\,\ldots .\,Q(\text{vib}\,.\,i)\ldots \\ &= \prod Q(\text{vib}\,.\,i), \end{aligned} \tag{3.19}$$

where $\prod$ means 'the product of' a set of terms and $Q(\text{vib}\,.\,i)$, the value for a single harmonic mode i of frequency ν_i, is given by[3,7]

$$Q(\text{vib}\,.\,i) = \frac{1}{1 - e^{-h\nu_i/k_BT}}. \tag{3.20}$$

Equation (3.20), unlike eqn (3.18), is not restricted to gases. It expresses quantitatively and in a particular context what we have already seen qualitatively and in general. The vibrational partition function is higher for a low frequency vibration with closely-spaced quantum levels than for a high frequency vibration with widely-spaced levels.

Two vibrational potential-energy curves are shown in Fig. 3.2. One is broad and shallow with closely-spaced energy levels. This corresponds to

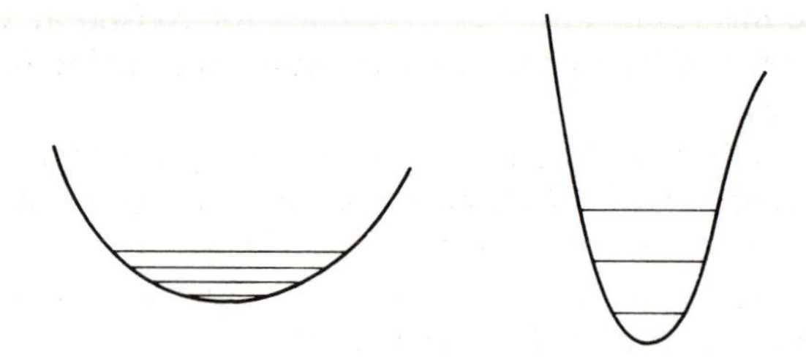

Fig. 3.2. Molecular potential-energy curves for vibrations of high and low entropy.

a group bending or wagging mode with a low frequency; or it could be a molecular vibration of a species with considerable conformational flexibility, or one which is very loosely bonded.

The other vibration has a higher fundamental frequency and more widely spaced energy levels. This is typical of a bond stretching mode or a molecular vibration of a species with very little conformational freedom, or of one which is very tightly bonded.

To put this into perspective with translational partition functions, solutions to eqn (3.20) for vibrations of different frequencies at 298 K are shown below:

$\bar{\nu}_i/\text{cm}^{-1}$	$10^{-12}\nu_i/\text{s}^{-1}$	$Q_{298\,\text{K}}(\text{vib}\,.\,i)$
4000	120	1.0000001
400	12	1.17
40	1.2	5.70
4	0.12	52.3

These results indicate that the vibrational contribution to a compound's entropy will be significant only if there are very low frequency vibrations and, by eqn (3.19), when there are many of them. The vibrational molecular partition functions of compounds such as HCl or N_2 with single high-frequency vibrations are virtually unity at normal temperatures and, like the electronic partition function, contribute almost nothing to the molar entropy at 25 °C. In contrast, macromolecular compounds such as enzymic proteins have very high molecular weights and loose tertiary structures; consequently, they have substantial vibrational entropy at 25 °C.

An isomerization involves no change in translational entropy. If the product has much higher conformational freedom than the reactant, the isomerization will proceed with an increase in vibrational entropy. For

example[11]

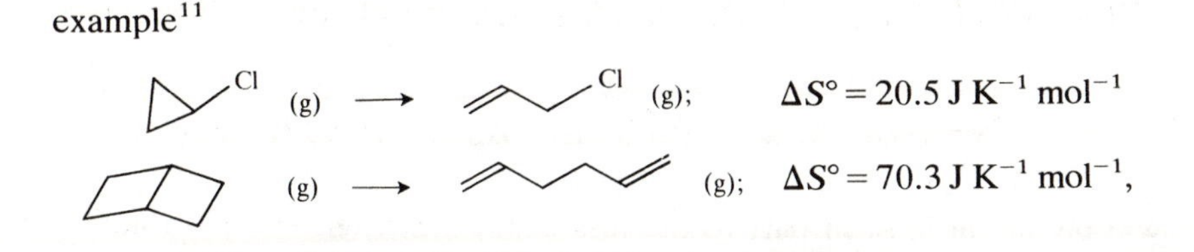

but the following reaction between two very similar structures involves almost no change in entropy:

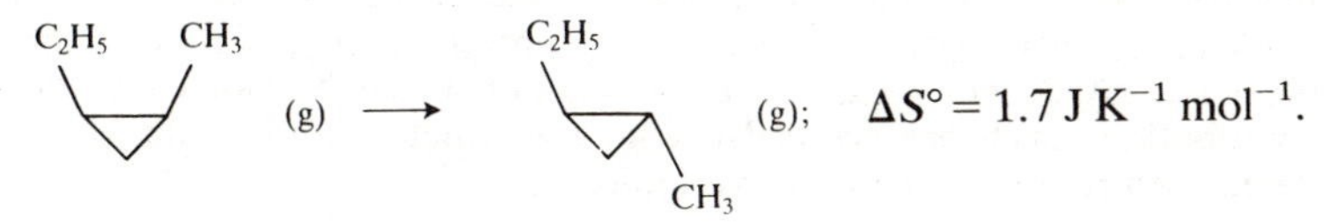

The following thermolysis is particularly interesting.[11]

N N (g) ⟶ (g) + N_2 (g); $\Delta S° = 219\ \text{J K}^{-1}\ \text{mol}^{-1}$.

There is an increase in the number of molecules, but also a compound comprising fairly rigid bicyclic molecules yields an acyclic organic product. Consequently, this reaction proceeds with gains in both translational and vibrational entropy (and probably rotational as well), and $\Delta S°$ is large.

An *internal rotation* of one part of a molecule with respect to the rest about a single bond which is inhibited in some way may be regarded as a type of molecular vibration that cannot be treated adequately by the simple harmonic approximation (see p. 23). The molecular potential energy $V(\omega)$ of such a *restricted rotation* is given by

$$V(\omega) = 0.5V_0(1 - \cos n\omega)$$

where ω = the angle of rotation,
V_0 = the potential-energy barrier to rotation, and
n = the number of times the molecular rotation potential-energy profile repeats itself in 360° of rotation.

If the potential barrier V_0 is low, then rotation is relatively free and will be associated with a high entropy – much higher than that of most simple harmonic vibrational modes. 1,1,1-Trifluoroethane is a compound which has the high entropy due to a restricted internal rotation (see Table 3.7, p. 104).[1] But as V_0 increases, the restricted rotation becomes more like a torsional vibration (libration), and its entropy decreases.

The loss of entropy[11] in the gas-phase reaction of eqn (3.1) (p. 77)

(g) ⟶ (g); $\Delta S° = -15.5\ \mathrm{J\ K^{-1}\ mol^{-1}}$,

will be due at least in part to the loss of the restricted internal rotation between the cyclopropyl and the 2-propenyl groups.

The partition function for an internal rotation and the associated entropy are given by expressions which include moments of inertia of parts of the molecule about the axis of rotation. Possibly because of this formal mathematical similarity, and because their entropies are generally comparable in magnitude with molecular rotational rather than harmonic vibrational contributions, such degrees of freedom are sometimes misleadingly considered with rotations of the whole molecule.

3.5.3 Rotational entropy

The number of rotational degrees of freedom of a polyatomic molecule is determined by its symmetry and is either 2 (linear molecule) or 3 (non-linear). Molecular energy levels of a rotational mode are much closer together than vibrational levels, consequently the absolute rotational entropy of one mole of a compound is invariably higher than its vibrational entropy.

TABLE 3.7

Relative contributions to the calculated absolute standard molar entropies of some elements and compounds, and comparision with experimental thermochemical results[a]

Material	S°(trans)	S°(rot)	S°(vib)	S°(calc)	S°(exp)
Ne[b]	146.2	—	—	146.2	146.5
Ar[c]	154.7	—	—	154.7	154.6
N_2[d]	150.42	41.20	~0	191.62	192.2
F_2[c]	154.7	48.1	0.6	203.4	203.2
HCl[d]	153.71	33.08	~0	186.79	186.4
CS_2[e]	162.06	65.93	9.71	237.70	237.8
CH_3Br[e]	164.81	77.01	2.94	244.76	245.2
C_2H_4[d]	150.45	66.54	2.60	219.6	220
CH_3OH[d]	152.10	79.61	7.93	239.64	241
CH_3CF_3[e]	163.29	96.26	26.53[f]	286.08	287.2

[a] All results in $\mathrm{J\ K^{-1}\ mol^{-1}}$; gas standard state = pure gas at 1 standard atmosphere pressure and 298.15 K.
[b] See p. 99.
[c] Ref. 3, p. 195.
[d] Ref. 7; 300 K.
[e] Ref. 1.
[f] The high vibrational contribution includes a restricted internal rotation, see p. 103.

The expression for the rotational molecular partition function includes the moments of inertia of the molecule, its symmetry number (the number of configurations of the molecule which can be generated by rotation of the whole molecule and which are indistinguishable in the absence of isotopic labelling), the absolute temperature, and various fundamental constants.[3,7] In order that moments of inertia can be calculated, bond lengths and angles need to be known and these are usually obtained from electron diffraction and molecular spectroscopy. Even in the case of fairly small organic molecules, accurate calculations of rotational molecular partition functions, and hence the rotational contribution to the molar entropy, are not easy. For large flexible organic molecules, such calculations are virtually impossible.

Some results for the different contributions towards the total entropy of a few elements and compounds are shown in Table 3.7. The order is always the same:

$$S(\text{trans}) > S(\text{rot}) > S(\text{vib})(> S(\text{el}) \sim 0).$$

3.6 Empirical determination of entropy

In order to assess the thermodynamic feasibility of a reaction, we need to know $\Delta_f G^\circ$ values for reactants and products. These can be calculated from $\Delta_f H^\circ$ and S° data. The enthalpies of formation are obtained either by experiment (usually combustion calorimetry) or by empirical calculations as described in the previous chapter. In sections of this chapter so far, we have seen that entropy may be determined *either* calorimetrically from heat capacity measurements *or* calculated from spectroscopic and structural data using theoretical relationships derived by the methods of statistical thermodynamics. (No counterpart exists for determining $\Delta_f H^\circ$ values by such calculations.)

The experimental measurement of C_P at different temperatures is in principle applicable to any material but it is neither easy nor quick. The statistical thermodynamic calculation of S° requires accurate structural data and interpretable spectra. These requirements are stringent and in practice restrict the method to relatively simple compounds in the gas phase.

A third method of determining S° of a compound is available which involves using known calorimetric or statistical thermodynamic S° values of other related compounds. This empirical approach is very closely related methodologically to that which was described in the previous chapter for estimating $\Delta_f H^\circ$ using group or bond increments.[13]

The simplest approximation is

$$S^\circ = \sum S^\circ(\text{bond})$$

and an average bond entropy term is allocated to C—H, C—C, C—Cl, etc., from known results. This method inevitably gives identical values for compounds which are structural or stereoisomers.

If a group increment scheme is used,

$$S° = \sum S°(\text{group})$$

where CH_3, CH_2, OH, etc., have group average values, it may be possible to distinguish between structural but not stereoisomers. This scheme can be improved by incorporating corrections for particular structural features, e.g. rings, non-bonded interactions, and symmetry elements both within groups and of the molecule as a whole. The method has now become sufficiently refined, due almost entirely to the efforts of Benson[13] and his colleagues, that computed values are in excellent agreement with experimental results, where these exist, for a wide range of structural types (Table 3.8), and we can be confident in purely estimated values.

TABLE 3.8
Comparison of estimated and measured absolute standard molar gas phase entropies, $S^\circ_{298\,K}$

Compound	$S^\circ_{298\,K}$/J K^{-1} mol^{-1} Measured[a]	Estimated[b]
But-1-ene	305.6	307.6
2-Methylpropene	293.6	294.4
cis-Pent-2-ene	346.3	343.7
trans-Pent-2-ene	340.4	343.7
Pent-1-yne	329.8	330.5
Benzene	269.2	268.9
n-Propylbenzene	400.7	399.9
Styrene	345.1	345.1
Cyclohexane	298.2	299.5
Ethanol	282.6	282.8
Propan-2-ol	309.9	311.7
Phenol	315.6	314.6
Ethanethiol	296.1	296.4;
Acetaldehyde	264.2	264.4
Pentan-3-one	370.0	373.6
Acetic acid	282.5	285.8
t-Butyl chloride	322.2	322.2
1-Bromobutane	330.9	328.0
Fluorobenzene	302.6	302.1

[a] Data from ref. 1.
[b] Data from S. W. Benson, F. R. Cruickshank, D. M. Golden, G. R. Haugen, H. E. O'Neal, A. S. Rodgers, R. Shaw, and R. Walsh, *Chem. Rev.* **69,** 279 (1969).

3.7 Enthalpy and entropy of solution

In this chapter so far, we have considered phase transitions and some reactions of gases and liquids. For such processes, the pure materials are a convenient choice of standard state. Most reactions of interest to the organic chemist, however, take place in solution and a more convenient standard state than the pure material is a solution of unit molality (1 mole of solute in 1 kg of solvent) or unit molarity (1 mole of solute in 1 dm^3 of solution).

These are not ordinary or real solutions however, but hypothetical ones in which the solute has the same enthalpy as in a *real* solution at infinite dilution. In the hypothetical solution at unit molality (or molarity) each solute molecule behaves *as though it were* at infinite dilution.

Before we consider solution reactions in later chapters, we need to look first at the changes in enthalpy and entropy due simply to the dissolution of a material in a solvent[14] – the change from the thermochemical to a solution standard state. This involves the use of different symbols (see p. 47). For example

$$\mathrm{A(c, l, or\ g)} \xrightarrow{\text{solvent}} \mathrm{A(1\ mol\ dm^{-3})}; \qquad \Delta_{\mathrm{sol}}H^{\ominus}, \Delta_{\mathrm{sol}}S^{\ominus};$$

$$\Delta_{\mathrm{sol}}H^{\ominus} = \Delta_{\mathrm{f}}H^{\ominus}(1\ \mathrm{mol\ dm^{-3}}) - \Delta_{\mathrm{f}}H^{\circ}(\mathrm{c, l, or\ g}),$$

and

$$\Delta_{\mathrm{sol}}S^{\ominus} = S^{\ominus}(1\ \mathrm{mol\ dm^{-3}}) - S^{\circ}(\mathrm{c, l, or\ g}),$$

where $\Delta_{\mathrm{sol}}H^{\ominus}$ and $\Delta_{\mathrm{sol}}S^{\ominus}$ = the standard molar enthalpy and entropy of solution, $\Delta_{\mathrm{f}}H^{\ominus}(1\ \mathrm{mol\ dm^{-3}})$ and $S^{\ominus}(1\ \mathrm{mol\ dm^{-3}})$ = the standard molar enthalpy of formation and the absolute standard molar entropy of the solute in the specified solution standard state, and $\Delta_{\mathrm{f}}H^{\circ}(\mathrm{c, l, or\ g})$ and $S^{\circ}(\mathrm{c, l, or\ g})$ have their usual meanings and relate to 1 mole of the pure solute in the crystal, liquid, or gas standard state at 1 standard atmosphere pressure.

We have been careful not to refer to the hypothetical solution standard state as an *ideal* solution. This term is reserved for systems which obey Raoult's Law. In an ideal solution, intermolecular interactions between like and unlike molecules are identical consequently there can be no enthalpy change upon dissolution, or dilution. The decrease in free energy associated with these processes is due exclusively to the gain in entropy.

In the gas phase, there are only weak intermolecular attractive forces (none if the gas is ideal) but upon dissolution, the solute molecules become solvated. If the decrease in enthalpy brought about by the association of solvent molecules around each solute molecule is greater than the enthalpy required for the concomitant disruption of the solvent's liquid structure, the dissolution will be exothermic.

When one mole of an ideal gas at 1 atmosphere pressure and 298 K dissolves in 1 kg of solvent, the volume occupied by the gas decreases from 24.5 dm^3 to about 1 dm^3 (depending upon the density of the unit molal solution). There will, therefore, be a loss of translational entropy. Furthermore, the solvation of the solute molecules by solvent restricts not only the mobility of the solute but also its freedom of molecular rotation. This causes the solute's rotational energy levels to become further apart so the rotational molecular partition function decreases. Solvation, therefore, causes a loss of rotational entropy of the solute. Solvation may also restrict translational and rotational freedom of *solvent* molecules which will also contribute towards a negative entropy of solution. And the more strongly solvent molecules are bound, the greater the overall loss of entropy (but in compensation, this should also correspond to a more exothermic dissolution). Qualitatively, therefore, we expect dissolution of a gas to be mildly exothermic but with a loss of entropy. Some results are shown in Table 3.9.

1 mol dm^{-3} may be considered a more appropriate standard state for the gas than 1 atmosphere pressure. In this case, there will be a smaller decrease in the translational component of $\Delta_{sol}S^\ominus$, and the entropy results in Table 3.9 will be less negative by 26.6 J K^{-1} mol^{-1} (calculated using eqn (3.13), p. 95). In all the examples shown, however, $\Delta_{sol}S^\ominus$ is still negative. $\Delta_{sol}H^\ominus$ is, of course, unaffected by this change in gas standard state as the enthalpy of an ideal gas is independent of volume.

The dissolution of a crystalline or liquid solute could in principle be exothermic, endothermic, or thermoneutral depending upon the particular system. In the case of a crystal, it depends upon how the decrease in enthalpy due to solvation compares with the enthalpy required to disrupt the solvent structure and the crystal lattice, and disperse the solute. For

TABLE 3.9
Standard molar enthalpy and entropy of solution of gases in water at 298.15 K[a]

Gas	$\Delta_{sol}H^\ominus$/kJ mol^{-1}	$\Delta_{sol}S^\ominus$/J K^{-1} mol^{-1}
Cl_2	−23	−102
I_2[b]	−40 (−37)	−123 (−97.4)
NH_3[c]	−31.18	−81.0
CO_2	−20.29	−96.0
HCN[c]	−28.0	−77.0
HNO_2[c]	−39.7	−101
CH_3Cl	−20.8	−89.7
CH_3NH_2	−47.2	−119.9

[a] Ref. 6; standard states are the ideal gas at 1 atmosphere pressure and the hypothetical unit molal solution.
[b] Results in parentheses are for solution in CCl_4.
[c] Un-ionized in solution.

TABLE 3.10
Standard molar enthalpy and entropy of solution of liquids in water at 298.15 K[a]

Liquid	$\Delta_{sol}H^{\ominus}$/kJ mol^{-1}	$\Delta_{sol}S^{\ominus}$/J K^{-1} mol^{-1}
H_2O_2	−3.39	34.3
CH_3OH	−7.28	6.3
NH_2NH_2	−16.3	16.9
HCO_2H[b]	0.71	34
CH_3CO_2H[b]	−1.26	18.8
C_2H_5OH	−10.6	−12.1
Br_2	−2.6	−21.7
$(CH_3)_2NH$	−26.7	−49.3
$(CH_3)_3N$	−30.0	−75.0

[a] Ref. 6; standard states are the pure liquid and the hypothetical unit molal solution.
[b] Un-ionized in solution.

the solution of a liquid, it is a matter of how solute–solvent interactions in solution compare with solute–solute interactions in the pure liquid, and solvent–solvent interactions in the pure solvent.

As can be seen from Table 3.10, $\Delta_{sol}H^{\ominus}$ for liquids in water tend to be negative but rather small; results for solids in Table 3.11 include negative

TABLE 3.11
Standard molar enthalpy and entropy of solution of crystalline solids in water at 298.15 K[a]

Solid	Note	$\Delta_{sol}H^{\ominus}$/kJ mol^{-1}	$\Delta_{sol}S^{\ominus}$/J K^{-1} mol^{-1}
I_2	*b*	23	21.1
ICN	*b*	11.67	28.9
$NH_2CH_2CO_2H$	*b*	14.1	54.8
NH_4Cl	*b*	14.77	75.3
CH_3NH_3Cl	*b*	5.77	60.5
$(CH_3)_4NBH_4$	*b*	23.0	131
NaCl	*c*	3.89	43.1
KNO_3	*c*	34.9	158
CsBr	*c*	25.9	92
$Ca(NO_3)_2$	*d*	−19.2	46.4
$BaBr_2$	*d*	−23.4	28.0
$CaCl_2$	*d*	−81.3	−44.8
$ZnSO_4$	*b*	−80.3	−212
$Al_2(SO_4)_3$	*b*	−350	−823

Notes

[a] Standard states are the pure crystalline solid and the hypothetical unit molal solution.
[b] Ref. 6.
[c] Ref. 8.
[d] V. B. Parker, D. D. Wagman, and W. H. Evans, NBS Technical Note 270-6, Washington DC (1971).

TABLE 3.12
Standard molar enthalpy and entropy of solution in non-aqueous solvents, 298.15 K[a]

Solute	Solvent	$\Delta_{sol}H^{\ominus}$/kJ mol^{-1}	$\Delta_{sol}S^{\ominus}$/J K^{-1} mol^{-1} [d]
H_2O[b]	CH_3NO_2	14.5	48 (24)
	DMF	−3.8	11 (21)
	CH_3CN	6.15	29 (25)
	DMSO	−5.36	14 (22)
CH_3OH[b]	CH_3CN	4.60	31 (25)
	DMSO	−1.1	25 (22)
$(CH_3)_2CO$[b]	CH_3CN	−0.3	25 (25)
	DMSO	1.5	21 (22)
	Formamide	−1.8	17 (27)
CH_3CN[b]	DMF	−0.92	19 (21)
	CH_3NO_2	−0.04	27 (24)
LiCl[c]	CH_3OH	−50.6	−94 (27)
	DMF	−47.7	−101 (21)
NaBr[c]	CH_3OH	−17.2	−66 (27)
	DMF	−30.1	−110 (21)
$CsBPh_4$[c]	CH_3OH	36.8	10 (27)
	DMF	−3.8	−22 (21)
$(C_2H_5)_4NI$[c]	CH_3OH	34.7	76 (27)
	DMF	14.6	−4 (21)
	CH_3CN	19.7	15 (25)

[a] DMF = dimethylformamide, DMSO = dimethyl sulphoxide; standard states are the pure solute and the hypothetical solution of unit molarity.
[b] Data from B. G. Cox, A. J. Parker, and W. E. Waghorne, *J. Am. chem. Soc.* **95,** 1010 (1973); $\Delta_{sol}H^{\ominus}$ values at 23 °C and entropy results calculated for unit molarity from free energy results at unit mole fraction actually reported.
[c] Data from B. G. Cox, *Annual Reports*, Part A, The Chemical Society 249 (1973).
[d] Numbers in parentheses are values calculated using eqn (3.14), p. 95, assuming that the density of the 1 molar solution is the same as the density of the pure solvent. Clearly, this gives reasonable results for some of the liquid–liquid solutions, but not for solutions of the ionic solids.

and positive values for $\Delta_{sol}H^{\ominus}$ over quite a wide numerical range (+34.9 to −350 kJ mol^{-1}). Some results for non-aqueous solvents are given in Table 3.12.

Using the simplest model, we would expect the dissolution of a condensed phase in a solvent to be accompanied by an increase in entropy, the increase being larger for a crystalline solute than for a liquid. Few data are available, but this appears to be found for covalent materials in non-aqueous solvents as shown in Table 3.12; results for electrolytes in the same solvents are seen to be positive or negative and may be fairly substantial.

Widely divergent results are found for $\Delta_{sol}S^{\ominus}$ of liquids and solids in water ranging from large and positive through approximately zero, to very large and negative (Tables 3.10 and 3.11).

Water is exceptional in being very effective both as a Lewis acid and a Lewis base. We have already seen one consequence of this: the pure solvent has an extended hydrogen-bonded structure which causes $\Delta_v S^\circ_{373\,K}$ to be anomolously high. Another is that water is able to solvate cations, anions, and dipolar covalent solutes very effectively; this causes an appreciable loss of translational and rotational freedom of solvent molecules and corresponds to a decrease in entropy. The large negative $\Delta_{sol}S^\ominus$ values for compounds such as $CaCl_2$, $ZnSO_4$, and $Al_2(SO_4)_3$ in water (Table 3.11) can only be explained by the contribution from the solvent through the dissociated ions being very extensively solvated. And it is only because of the strongly exothermic nature of this process that these compounds are able to dissolve at all.

Negative entropies of solution of organic compounds in water are more intriguing. Amines are strongly solvated through exothermic hydrogen-bonding, and are analogous to ionic compounds which dissolve with a loss of entropy through solvent-ordering. But higher alcohols and many other types of organic compounds which are not especially good at hydrogen-bonding also dissolve in water with a decrease in entropy.[15] This is thought to be due to an enhanced solvent structure by hydrogen-bonding between water molecules in the solvent shell immediately around these covalent non-polar solute molecules.

Although the physical chemistry of liquid and solution structure is still not well understood, it is becoming clearer that many unusual effects and unexpected trends in the solution reactions of organic compounds are due largely to solvent and solvation entropy effects.

3.8 Problems

1. Calculate $\Delta_r S^\circ_{298.15\,K}$ for the following gas-phase reactions from the data given in Table 3.1 (p. 80).

(i) $H_2 + I_2 \rightarrow 2HI$

(ii) $H_2 + F_2 \rightarrow 2HF$

(iii) $CH_3OH + HI \rightarrow CH_3I + H_2O$

(iv) $CH_3OH + HF \rightarrow CH_3F + H_2O$

(v) Cyclobutane $\rightarrow 2C_2H_4$

(vi) Benzene $\rightarrow 3C_2H_2$

(vii) Cyclohexane $\rightarrow$ Benzene $+ 3H_2$

(viii) Cyclopropane $+ H_2 \rightarrow C_3H_8$.

2. Calculate (i) $\Delta_f S^\circ_{298.15\,K}(g)$ and (ii) $\Delta_f G^\circ_{298.15\,K}(g)$ for the following gaseous compounds from the data given in Table 3.1 (p. 80) and Table 2.2 of Chapter 2 (p. 52).

(a) C_2H_2 (d) H_2O

(b) CF_4 (e) CH_3I

(c) CH_4 (f) NH_3

3. Calculate $\Delta_r G^\circ_{298.15\,K}$ for the following gas-phase reactions from the data given in Table 3.1 (p. 80) and Table 2.2 of Chapter 2 (p. 52).

(i) $CH_3OH + HI \rightarrow CH_3I + H_2O$

(ii) $CH_3OH + HF \rightarrow CH_3F + H_2O$

(iii) $C_2H_6 + C_2H_2 \rightarrow 2C_2H_4$

(iv) $3C_2H_2 \rightarrow C_6H_6$.

4. Calculate $\Delta_r S^\ominus_{298\,K}$ (standard state = 1 mol dm^{-3}) for the gas-phase reactions (i), (v), (vi), (vii) and (viii) in question 1 from your results for $\Delta_r S^\circ_{298\,K}$ (standard state = pure gas at 1 standard atmosphere pressure). Assume ideal gas behaviour.

5. Calculate $\Delta_r S^\ominus_{298\,K}$ (standard state = 1 mol dm^{-3}) for the following gas-phase reactions from the results given.[11] Assume ideal gas behaviour.

Reaction	$\Delta_r S^\circ_{298\,K}$/J K^{-1} mol^{-1}
(i) *cyclo*-$C_3H_5Cl \rightarrow CH_2{=}CHCH_2Cl$;	20.5
(ii) $(CH_3)_3COH \rightarrow (CH_3)_2C{=}CH_2 + H_2O$;	156
(iii) 2 Cyclopentadiene → Cyclopentadiene dimer;	-200
(iv) 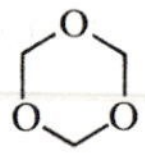 $\rightarrow 3CH_2O$;	350

6. From the data in Table 3.9 (p. 108), calculate the standard molar free energy of solution in water at 298.15 K (solution standard state = 1 mol kg^{-1}) for the following gases initially at (a) 1 standard atmosphere pressure, and (b) 1 mol dm^{-3}.

(i) CH_3Cl (ii) CH_3NH_2.

7. Calculate the standard molar free energy of solution in water at 298.15 K for the following substances (solution standard state = 1 mol kg^{-1}) from the data in Tables 3.10 and 3.11 (p. 109).

(i) $HCO_2H(l)$ (iv) $ICN(c)$

(ii) $Br_2(l)$ (v) $Ca(NO_3)_2(c)$

(iii) $(CH_3)_2NH(l)$ (vi) $ZnSO_4(c)$.

8. From the results in Table 3.12 (p. 110), calculate the standard free energy of solution at 298.15 K (standard state = 1 mol dm^{-3}) for the following compounds in the solvents indicated.

(i) $H_2O(l)$ in DMSO

(ii) $H_2O(l)$ in CH_3NO_2

(iii) NaBr(c) in DMF

(iv) LiCl(c) in CH_3OH

(v) $CsBPh_4(c)$ in DMF

(vi) $(C_2H_5)_4NI(c)$ in DMF.

3.9 References

1. D. R. Stull, E. F. Westrum, and G. C. Sinke, *The chemical thermodynamics of organic compounds*, Wiley, New York (1969). This advanced monograph contains chapters on principles and applications of chemical thermodynamics. It includes very useful tables of thermodynamic properties of many organic and some inorganic compounds with expert critical comments.
2. R. L. Montgomery, R. A. Melaugh, C-C. Lau, G. H. Meier, H. H. Chan, and F. D. Rossini, *J. chem. Thermodynamics* **9,** 915 (1977).
3. W. J. Moore, *Physical chemistry*, 5th edn, Longman (1972).
4. *J. chem. Thermodynamics* **10,** 903 (1978). See references to Chapter 2.
5. *JANEF Thermochemical tables*, 2nd edn. NSRDS-NBS37 (eds. D. R. Stull and H. Prophet), National Bureau of Standards, Washington, DC, USA (1970).
6. D. D. Wagman, W. H. Evans, V. B. Parker, I. Halow, S. M. Bailey, and R. H. Schumm, Technical Note 270-3, *Selected values of chemical thermodynamic properties*, National Bureau of Standards, Washington, DC, USA (1968).
7. J. H. Knox, *Molecular thermodynamics*, rev. edn, Wiley, Chichester (1978).
8. F. D. Rossini, D. D. Wagman, W. H. Evans, S. Levine, and I. Jaffe, *Selected values of chemical thermodynamic properties*, National Bureau of Standards Circular 500, Washington, DC, USA (1952).
9. G. Allen and E. F. Caldin, *Q. Rev.* (*Chem. Soc.*) **7,** 255 (1953); W. Weltner, *J. Am. chem. Soc.* **77,** 3941 (1955).
10. G. N. Lewis and M. Randall, *Thermodynamics*, 2nd rev. edn (eds. K. S. Pitzer and L. Brewer), p. 421, McGraw-Hill, New York (1961).
11. *Kinetic data on gas phase unimolecular reactions* (eds. S. W. Benson and H. E. O'Neal). National Standard Reference Data Series, National Bureau of Standards, (NSRDS-NBS21), Washington, DC, USA (1970).
12. US Dept. of the Interior, Bureau of Mines, Bulletin 666 (1974).
13. S. W. Benson, *Thermochemical kinetics*, 2nd edn, Wiley, New York (1976). See also S. W. Benson and J. H. Buss, *J. chem. Phys.* **29,** 546 (1958), and S. W. Benson, F. R. Cruickshank, D. M. Golden, G. R. Haugen, H. E. O'Neal, A. S. Rodgers, R. Shaw, and R. Walsh, *Chem. Rev.* **69,** 279 (1969).
14. There has been fairly extensive work to determine enthalpies of solution – e.g. C. V. Krishnan and H. L. Friedman, Chapter 9 in *Solute–solvent interactions*, Vol. 2 (eds. J. F. Coetzee and C. D. Ritchie), Marcel Dekker, New York (1976) – but much less on entropies. See, however, D. H. Wertz, *J. Am. chem. Soc.* **102,** 5316 (1980).
15. H. S. Frank and M. W. Evans, *J. chem. Phys.* **13,** 507 (1945). Some aspects of the thermodynamics of aqueous solutions of electrolytes and non-electrolytes are discussed by R. E. Robertson, *Progr. phys. org. Chem.* **4,** 213 (1967), and M. J. Blandamer, *Adv. phys. org. Chem.* **14,** 203 (1977).

4

Chemical equilibrium

4.1 Introduction

We saw in the previous chapter that the free energy of a chemical reaction, $\Delta_r G$, is the change in total free energy of the system due to one mole of reaction as described by the stoichiometric equation. It is equal to the algebraic sum of the free energies of formation of reactants and products, and, if all species are in standard states, appropriate superscripts $^\circ$ or $^\ominus$ are included:

e.g. $$\Delta_r G^\circ = \sum \Delta_f G^\circ(\text{products}) - \sum \Delta_f G^\circ(\text{reactants}).$$

The implication in the two previous chapters was that the reactions considered were unidirectional.

The standard free-energy change for the reaction of any balanced chemical equation can in principle be calculated from compilations of $\Delta_f H^\circ$ and S° data (since $\Delta_f G^\circ = \Delta_f H^\circ - T\,\Delta_f S^\circ$) regardless of the practicality of the process, consequently $\Delta_f G^\circ$ may be only a notional quantity.

Reactions such as those represented in eqns (4.1)–(4.3) ($\Delta_r G_T^\circ$ results calculated from published enthalpy and entropy data[1]) proceed from left to right and, in practical terms, the systems achieve equilibrium only when no reactants can be detected.

$$\text{cyclopropane (g)} \longrightarrow \text{propene (g)}; \; \Delta_r G^\circ_{500\,K} = -47.9 \text{ kJ mol}^{-1}. \quad (4.1)$$

$$\text{cycloheptatriene (g)} \longrightarrow \text{toluene } (CH_3) \text{ (g)}; \; \Delta_r G^\circ_{400\,K} = -133.7 \text{ kJ mol}^{-1}. \quad (4.2)$$

$$CH_4(g) + 2O_2(g) \rightarrow CO_2(g) + 2H_2O(1);$$
$$\Delta_c G^\circ_{298\,K} = -818 \text{ kJ mol}^{-1}. \quad (4.3)$$

This is not to say that the states represented by the right-hand sides of these equations are completely stable in all respects. For example, the standard free energies of formation[1] of propene and toluene at 500 and 400 K respectively are +93.93 and +147.7 kJ mol^{-1} so both of these compounds are thermodynamically unstable (but kinetically stable) with respect to decomposition to elemental carbon and hydrogen. However, propene and toluene show no discernible tendency to isomerize to cyclopropane and cyclohepta-1,3,5-triene at 500 and 400 K respectively, and are thermodynamically perfectly stable as far as eqns (4.1) and (4.2) are concerned. So we describe equilibrium for the moment as the state in which the concentrations of reactants and products cease to change with time, it being understood that change may be only by the reaction specified in the chemical equation.

4.2 Reversible reactions, equilibrium, and free energy

There are many chemical reactions which are not unidirectional. They are reversible and yield mixtures which contain appreciable quantities of reactants and products at equilibrium. For example, if we start with a sample of the 7,7-disubstituted bicycloheptadiene (1) (the stereochemistry is unimportant) in deuteriochloroform at 67 °C (eqn (4.4)), it will isomerize to the 7,7-disubstituted cycloheptatriene (2), and this reaction can be monitored by proton magnetic resonance spectroscopy.[2] But the conversion of (1) into (2) never becomes complete; the rate of decrease in concentration of (1) becomes slower and slower until (1) and (2) are present as a 0.40:0.60 mixture, then no further change takes place at 67 °C.

Correspondingly, if we had started with pure (2) in $CDCl_3$, it would isomerize into (1) but not completely; the same stable 0.40:0.60 mixture of (1):(2) would ultimately be obtained at 67 °C. In this reaction, the designation of (1) or (2) as reactant is arbitrary, and eqn (4.4) could equally well have been written the other way round.

$$\underset{(1)}{\text{7-Ph-7-}CO_2CH_3\text{-bicycloheptadiene}} \xrightleftharpoons[-9 \text{ to } 113\,^\circ C]{CDCl_3} \underset{(2)}{\text{7-Ph-7-}CO_2CH_3\text{-cycloheptatriene}} \quad (4.4)$$

TABLE 4.1
Variation of the equilibrium constant K with temperature for the reaction of eqn (4.4)[2]

Temperature/°C	K
−9	0.151
32.5	0.616
67	1.52
75	1.83
113	3.98

We saw in the previous chapter that the driving force for any reaction is the minimization of the total free energy of the system, G. In reactions such as those represented by eqns (4.1)–(4.3), G is minimized by what is virtually the complete conversion of reactants into products. But for the reversible reaction in eqn (4.4), the free energy of the system is at a minimum when the mixture contains a significant amount of reactant in equilibrium with product. None the less, $\Delta_r G^\ominus$, $\Delta_r H^\ominus$, and $\Delta_r S^\ominus$ for eqn (4.4) retain their earlier meanings: they correspond to the changes in standard free energy, enthalpy, and entropy associated with the conversion of 1 mole of (1) into 1 mole of (2) even though a solution of pure (1) will not in practice yield a solution of pure (2).

For our present purposes, we can define an *equilibrium constant* K for the reaction of eqn (4.4) (as written) by

$$K = \frac{[2]_e}{[1]_e},$$

where $[1]_e$ and $[2]_e$ = the molar concentrations of (1) and (2) at equilibrium, and Table 4.1 shows how K varies with temperature. There will obviously be a temperature between 32.5 and 67 °C (actually 50 °C) at which $K = 1$ and $[1]_e = [2]_e$. We may say, perhaps rather loosely, that the position of equilibrium changes from the left-hand side to the right-hand side of eqn (4.4) as the temperature is raised, being on neither side completely in the range considered. But it would be wrong to interpret this as meaning simply that the direction of spontaneous reaction changes at 50 °C, being from right to left below 50 °C and from left to right above this temperature. A solution of pure (1) in deuteriochloroform at −9 °C will undergo some extent of spontaneous reaction from left to right to give the 0.87 : 0.13 mixture corresponding to K (−9 °C) = 0.151. During this equilibration, the conversion of δn moles of (1) into (2) causes the total free energy of the system G to *decrease* by an amount δG, therefore $\delta G/\delta n$ = negative.

Analogously, a solution of initially pure (2) at 113 °C will undergo some spontaneous reaction from right to left to give the 0.20 : 0.80 equilibrium mixture, $K(113\,^{\circ}\mathrm{C}) = 3.98$. And again, as equilibrium is approached, this time from the right-hand side, the free energy of the system G *decreases* by δG when δn moles of (2) change into (1), i.e.

$$\frac{\delta G}{\delta n} = \text{negative.}$$

We can now see that regardless of which side equilibrium is approached from, and regardless of the temperature and composition of the equilibrium mixture, $\delta G/\delta n$ = *negative* as the system moves *towards* equilibrium, and for an infinitesimal change in either direction *at* equilibrium $\delta G/\delta n = 0$.

4.3 Extent and degree of reaction

The transformation of a system from reactants to products corresponds to a linear change in some function and this function is the *extent of reaction*, ξ,[3,4] defined by

$$\xi = \frac{n_i(\xi) - n_i(0)}{\nu_i}$$

or

$$n_i(\xi) = n_i(0) + \nu_i \xi,$$

where $n_i(0)$ = the number of moles of component i actually present when the system is represented by the left-hand side of the equation ($\xi = 0$),
$n_i(\xi)$ = the number of moles of component i actually present at extent of reaction ξ, and
ν_i = the stoichiometric number of moles of i in the balanced chemical equation (positive if i is a product, negative if i is a reactant).

The unit of ξ is the mole.

As seen from the definition, the maximum value that ξ can reach for a given reaction depends not only upon the conditions under which it is carried out, but also upon the scale.

A useful related parameter which is independent of the scale of the process is the dimensionless *degree of reaction* α (degree of dissociation, association, isomerization, etc.):[3]

$$\alpha = \frac{\xi}{\xi_{max}}$$

where ξ_{max} = the maximum possible extent of reaction – that achieved when the components on the left-hand side of the equation have reacted completely.

It follows from this that

$$\alpha = \frac{n_i(\alpha) - n_i(0)}{n_i(1) - n_i(0)}$$

where $n_i(0)$ is as defined above,

$n_i(1)$ = the number of moles of i present when the reaction is represented by the right-hand side of the equation, and

$n_i(\alpha)$ = the number of moles of i present at degree of rèaction α.

Complete conversion of the state represented by the left hand side of any balanced chemical equation into that represented by the right corresponds to a change in α from 0 to 1 regardless of the stoichiometry or of the scale of the reaction.

It also follows that the number of moles of the *limiting reactant* l remaining at degree of reaction α is given by

$$n_l(\alpha) = (1-\alpha) \,.\, n_l(0)$$

where $n_l(0)$ = the number of moles of l present at $\alpha = 0$. And the number of moles of any product j formed at degree of reaction α from n_l moles of the limiting reagent l is given by

$$n_j(\alpha) = -\alpha \,.\, \frac{\nu_j}{\nu_l} \,.\, n_l(0)$$

(ν_l is negative since l is a reactant, consequently $n_j(\alpha)$ is always positive).

Example. Calculate the degree of reaction at equilibrium for the isomerization in eqn (4.4) at 75 °C from the results in Table 4.1.

$$(1) \rightleftharpoons (2); \qquad K = 1.83 \text{ at } 75\,°\text{C}.$$

Let Z mol dm^{-3} = the concentration of (1) at degree of reaction = 0.

Degree of reaction	[1]/mol dm^{-3}	[2]/mol dm^{-3}
0	Z	0
1	0	Z
α	$(1-\alpha)Z$	αZ

At equilibrium $\alpha = \alpha_e$ and $K = [2]_e/[1]_e = 1.83$ at 75 °C therefore

$$\frac{\alpha_e Z}{(1-\alpha_e)Z} = \frac{\alpha_e}{1-\alpha_e} = 1.83 \text{ at } 75\,°\text{C}$$

$$2.83\alpha_e = 1.83$$

$$\alpha_e = 0.647 \text{ at } 75\,°\text{C}.$$

In this example, there is the same number of moles on both sides of the equation, ($\sum \nu_i = 0$) so although α_e depends upon K, it is independent of Z, the reactant concentation at $\xi = 0$ ($\alpha = 0$). As we shall see later, this is not true of reactions in which the total number of moles changes as the reaction proceeds.

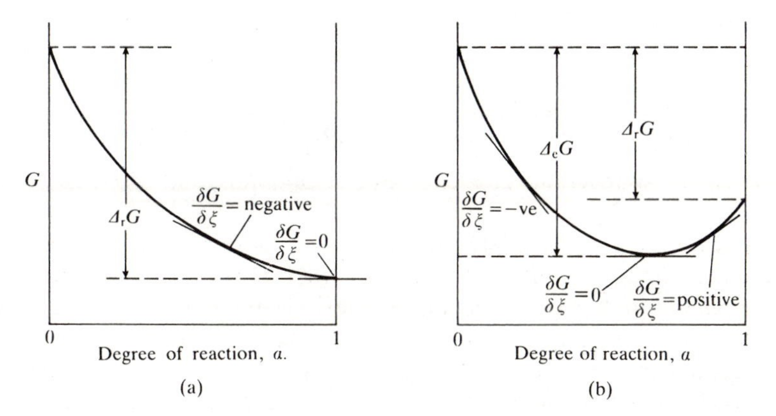

Fig. 4.1. Free-energy profiles for unidirectional and reversible reactions.

Figure 4.1(a) represents the free-energy profile of a unidirectional reaction such as the combustion process in eqn (4.3) above. As α changes from 0 to 1, the free energy of the system decreases non-linearly and the differential of the free energy with respect to the extent of the reaction $\delta G/\delta\xi$ becomes progressively less negative as α approaches 1 and is equal to 0 at $\alpha = 1$. This differential function under specified conditions (including constant temperature) can be thought of as the instantaneous free-energy gradient of the process – the thermodynamic driving force of the reaction.

The free-energy profile in Fig. 4.1(b) for a reversible process is characteristically different. It could correspond to the reaction of eqn (4.4) in the region of 60 °C. The free-energy minimum is not at $\alpha = 1$ but at a value between 0.5 and 1. As ξ increases from $\xi = 0$, $\delta G/\delta\xi$ is negative but becomes progressively less so as G decreases.

At equilibrium G is minimal ($\delta G/\delta\xi = 0$). As ξ increases beyond the equilibrium state, $\delta G/\delta\xi$ becomes positive and G increases again; $\delta G/\delta\xi$ becomes increasingly positive as α approaches 1.

As shown in Fig. 4.1(b), the decrease in free energy associated with the complete conversion of reactants into products ($\alpha = 0$ to $\alpha = 1$) is less than that for the conversion of reactants into the equilibrium mixture ($\alpha = 0$ to $\alpha = \alpha_e$): $\Delta_e G < \Delta_r G$. The precise shape of a free-energy profile is a characteristic of the particular reaction at the temperature and other conditions under which it is carried out.

So far, our considerations of reversible reactions have been qualitative. In order to gain further insight into the nature of chemical equilibria, we shall derive quantitative relationships between the equilibrium constant, free-energy change, and temperature. This is not offered purely as a formal exercise in chemical mathematics, but rather as a means of establishing the origins of familiar equations, and of identifying sources of potential difficulty in our subsequent uses of these important equations.

4.4 Chemical potential and the equilibrium condition

In a chemical reaction, elements and compounds undergo changes in concentration; the concentrations of products increase as those of reactants decrease, and the relationships between these changes are given by the stoichiometry of the reaction. In order to describe precisely the instantaneous total free energy of the system as the reaction proceeds, we need to be able to express the change in the total free energy due to changes in concentration of all components in the system. In other words, we need the *partial molar free energy* of every component under the particular reaction conditions. This term is usually called the *chemical potential*,[4] μ_i, of component i, the system being at total pressure P, at temperature T, and includes other components j, k, etc.

$$\mu_i = \left(\frac{\delta G}{\delta n_i}\right)_{P,T,j,k,\text{ etc.}}, \tag{4.5}$$

where δG = the change in the total free energy of the system caused by a change in the number of moles δn_i of component i. The dimensions of μ_i are those of energy per mole. The chemical potential μ_i of a pure material i is its absolute molar free energy G_i which is, in practice, indeterminate.

A system comprising a general reversible chemical reaction with reactants A, B, etc. and products X, Y, etc., at total pressure P and temperature T is represented by eqn (4.6).

$$\nu_A\,.\,A + \nu_B\,.\,B + \ldots \overset{P,T}{\rightleftharpoons} \nu_X\,.\,X + \nu_Y\,.\,Y + \ldots. \tag{4.6}$$

At any stage of the reaction, a small change $\delta\xi$ in the extent of reaction causes, for example, the number of moles of A and of X to change by δn_A and δn_X respectively:

$$\delta n_A = -\nu_A\,.\,\delta\xi,$$

and

$$\delta n_X = \nu_X\,.\,\delta\xi.$$

If $\delta\xi$ is an increase, δn_A is negative and δn_X is positive and in general,

$$\delta n_i = \nu_i\,.\,\delta\xi, \tag{4.7}$$

where ν_i is positive for products and negative for reactants. In fact, eqn (4.7) is the differential version of the earlier equation which defined ξ (p. 117). By eqn (4.5), the change in free energy δG_i of a system due to a small change δn_i in the number of moles of i at constant pressure and temperature is given by

$$\delta G_i = \mu_i \, \delta n_i,$$

and the total change $(\delta G)_{P,T}$ in a reacting system due to a small change $\delta\xi$ in the extent of reaction is

$$(\delta G)_{P,T} = \sum \delta G_i,$$

$$= \sum \mu_i \, \delta n_i,$$

so, by eqn (4.7),

$$(\delta G)_{P,T} = \sum \mu_i \nu_i \, . \, \delta\xi,$$

or

$$\left(\frac{\delta G}{\delta \xi}\right)_{P,T} = \sum \nu_i \mu_i.$$

We have already seen that the equilibrium condition is

$$\left(\frac{\delta G}{\delta \xi}\right)_{P,T} = 0,$$

so we can now write:

$$\underline{\text{at equilibrium, } \sum \nu_i \mu_i^{e} = 0,} \qquad (4.8)$$

where μ_i^{e} = the chemical potential of component i in the equilibrium mixture.

In order to proceed further, we now need to be able to determine μ_i^{e}.

4.5 Chemical potential, activity, and standard states

A general equation[4] for the chemical potential of a material i which defines its *activity* a_i is

$$\mu_i = \mu_i^{\ominus} + RT \ln \left(\frac{a_i}{a_i^{\ominus}}\right), \qquad (4.9)$$

where $\mu_i^{\ominus}$ = the chemical potential of i in a chosen standard state and is usually called the *standard chemical potential* of i,

$a_i^{\ominus}$ = the activity of i in this standard state, and

μ_i = the chemical potential of i in any other state specified by activity a_i.

The standard chemical potential $\mu_i^{\ominus}$ depends only upon the temperature, the scale we choose for activity, and what value on this scale we select as the standard state activity. Equation (4.9) is usually replaced by

$$\mu_i = \mu_i^{\ominus} + RT \ln a_i^{r}, \qquad (4.10)$$

where $a_i^{r} = a_i/a_i^{\ominus}$ = the *relative* activity of i and is dimensionless. Since $a_i^{\ominus}$ is invariably defined as unity on some chosen scale, a_i^{r} is equal to the *numerical value* of the absolute activity, a_i.

And in eqn (4.10), $\mu_i^\ominus$, the standard chemical potential of i, is the chemical potential of i at unit (absolute) activity on the same scale.

4.5.1 Gases

If the reacting system involves only ideal gases, then pressure in atmospheres is a suitable activity scale and the pure ideal gas at 1 standard atmosphere pressure is an appropriate standard state.

For this, the thermochemical standard state which we have already encountered in previous chapters,

(i) the superscript $^\ominus$ is usually replaced by $^\circ$,
(ii) $a_i^\circ = 1$ standard atmosphere pressure, and
(iii) $a_i =$ the partial pressure P_i (in atmospheres) of the ideal gas i.

Equation (4.9) may now be written

$$\mu_i = \mu_i^\circ + RT \ln\left(\frac{P_i \text{ atm}}{1 \text{ atm}}\right), \tag{4.11}$$

and eqn (4.10) becomes

$$\mu_i = \mu_i^\circ + RT \ln P_i^r, \tag{4.12}$$

where P_i^r is the relative activity of i which is equal to the numerical value of the (absolute) activity (partial pressure) of component i in atmospheres, and μ_i° is the chemical potential of the pure gas i behaving ideally at 1 standard atmosphere pressure (its standard chemical potential on this scale).

An alternative standard state is the pure ideal gas at activity 1 mol dm^{-3} in which case, by eqn (4.9),

$$\mu_i = \mu_i^\ominus + RT \ln\left(\frac{c_i \text{ mol dm}^{-3}}{1 \text{ mol dm}^{-3}}\right),$$

or

$$\mu_i = \mu_i^\ominus + RT \ln c_i^r, \tag{4.13}$$

where $\mu_i^\ominus =$ the chemical potential of the pure ideal gas at unit molar activity (its standard chemical potential on this scale),
$\mu_i =$ the chemical potential of the gas at activity $c_i \text{ mol dm}^{-3}$,

and c_i^r in eqn (4.13) is the relative molar activity of i (equal to the numerical value of the absolute molar activity).

It is important to appreciate that, for a given gas i, μ_i° in eqn (4.12) is not the same as $\mu_i^\ominus$ in eqn (4.13) since they relate to different activity scales and different standard states. So before eqns (4.9) or (4.10) can ever be used, the standard state and its activity must be specified. But of course, the *actual* chemical potential of any gas μ_i (in energy units per mole) depends only upon its physical conditions and is independent of how these physical conditions are described.

4.5.2 Liquids and solvents

Mole fraction (X) is the usual activity scale for components in liquid mixtures and for the solvent if a solution is dilute even if the solute, when pure, is a solid or a gas. Using this scale, the pure liquid (mole fraction activity = 1) is normally the standard state.

Equation (4.9) then becomes

$$\mu_i = \mu_i^{\ominus} + RT \ln\left(\frac{X_i}{X_i^{\ominus}}\right),$$

or

$$\mu_i = \mu_i^{\ominus} + RT \ln X_i, \quad \text{since} \quad X_i^{\ominus} = 1,$$

where μ_i = the chemical potential of i when present at mole fraction activity X_i, and

$\mu_i^{\ominus}$ = the chemical potential of i at unit mole fraction activity (its standard chemical potential on this scale).

With the mole fraction scale, $\mu_i^{\ominus}$ is sometimes replaced by μ_i° since unit mole fraction activity (the pure liquid) corresponds to the material in its thermochemical standard state.

Note that mole fraction has no units, and that, strictly, the pressure above the liquid should be specified. This would normally be taken to be 1 standard atmosphere but the properties of liquids are not very sensitive to the external pressure within broad limits and this aspect is commonly ignored.

4.5.3 Solutes in dilute solution

It will usually be convenient for our purposes to use molarity as the activity scale for compounds in dilute solution. Thus,

$$a_i^{\ominus} = 1 \text{ mol dm}^{-3},$$

$\mu_i^{\ominus}$ = the chemical potential of the pure solute i at activity 1 mol dm^{-3} in solution in the particular solvent (the defined standard chemical potential of i),

and, from eqn (4.9),

$$\mu_i = \mu_i^{\ominus} + RT \ln\left(\frac{c_i \text{ mol dm}^{-3}}{1 \text{ mol dm}^{-3}}\right),$$

or, as it is commonly expressed,

$$\mu_i = \mu_i^{\ominus} + RT \ln c_i^{\mathrm{r}}, \tag{4.13}$$

where μ_i = the chemical potential of i in this solvent at activity $c_i \text{ mol dm}^{-3}$, and

c_i^{r} = the relative molar activity of i (equal to the numerical value of the absolute molar activity).

4.6 Standard free-energy change and the thermodynamic equilibrium constant

We now return to the general reversible chemical reaction of eqn (4.6) at temperature T and total pressure P:

$$\nu_A \,.\, \mathrm{A} + \nu_B \,.\, \mathrm{B} + \ldots \underset{}{\overset{P,T}{\rightleftharpoons}} \nu_X \,.\, \mathrm{X} + \nu_Y \,.\, \mathrm{Y} + \ldots . \tag{4.6}$$

The change in total free energy due to the formation of products from reactants in one mole of reaction under these arbitrary but constant conditions is given by

$$\Delta_r G = \sum \nu_i \mu_i$$

and so from eqn (4.9)

$$\Delta_r G = \sum \nu_i \left[\mu_i^{\ominus} + RT \ln\left(\frac{a_i}{a_i^{\ominus}}\right)\right]$$

or, from eqn (4.10),

$$\Delta_r G = \sum \nu_i [\mu_i^{\ominus} + RT \ln a_i^{r}].$$

Since $\sum \nu_i \mu_i^{\ominus} = \Delta_r G^{\ominus}$, the standard molar free energy of the reaction as defined in the previous chapter, we may write

$$\Delta_r G = \Delta_r G^{\ominus} + RT \sum \nu_i \ln a_i^{r},$$

or

$$\Delta_r G = \Delta_r G^{\ominus} + RT \ln\left(\prod (a_i^{r})^{\nu_i}\right),$$

where $\prod$ means 'the product of' just as $\sum$ means 'the sum of'. This general equation relates $\Delta_r G$, the free-energy change for one mole of reaction under the particular conditions specified by P, T, and relative activities a_i^{r}, to the *standard* free energy of the reaction, $\Delta_r G^{\ominus}$, at the same temperature.

We shall now represent the relative activity and chemical potential of i when the system in eqn (4.6) is at equilibrium by $a_i^{r,e}$ and μ_i^{e}. From eqn (4.8), (p. 121), the equilibrium condition,

$$\sum \nu_i \mu_i^{e} = \Delta_r G^{e} = 0,$$

may now be written

$$0 = \Delta_r G^{\ominus} + RT \ln\left(\prod (a_i^{r,e})^{\nu_i}\right)$$

or

$$\Delta_r G^{\ominus} = -RT \ln K_a^{\ominus}, \tag{4.14}$$

where $K_a^{\ominus}$, the *thermodynamic equilibrium constant*, is defined by

$$K_a^{\ominus} = \prod (a_i^{r,e})^{\nu_i} = \frac{\prod (a_i^{e})^{\nu_i}}{\prod (a_i^{\ominus})^{\nu_i}}. \tag{4.15}$$

Equation (4.15) expresses the relationship between the activities of reactants and products when the general reversible reaction of eqn (4.6) is at equilibrium under particular experimental conditions of pressure and temperature. It may be written out more fully as follows:

$$K_a^{\ominus} = \left(\frac{(a_X^{r,e})^{\nu_X} . (a_Y^{r,e})^{\nu_Y} \dots}{(a_A^{r,e})^{\nu_A} . (a_B^{r,e})^{\nu_B} \dots}\right)$$

or

$$K_a^{\ominus} = \left(\frac{(a_X^{e})^{\nu_X} . (a_Y^{e})^{\nu_Y} \dots}{(a_A^{e})^{\nu_A} . (a_B^{e})^{\nu_B} \dots}\right) . \left(\frac{(a_A^{\ominus})^{\nu_A} . (a_B^{\ominus})^{\nu_B} \dots}{(a_X^{\ominus})^{\nu_X} . (a_Y^{\ominus})^{\nu_Y} \dots}\right).$$

If the reaction is in a single phase, a common activity scale and unit standard state activity for all species may conveniently be used. The second quotient in this last expression is then *numerically* unity, and the thermodynamic equilibrium constant may be expressed as

$$K_a^{\ominus} = \left(\frac{(a_X^{e})^{\nu_X} . (a_Y^{e})^{\nu_Y} \dots}{(a_A^{e})^{\nu_A} . (a_B^{e})^{\nu_B} \dots}\right) . (a^{\ominus})^{-\Sigma\nu_i}, \tag{4.16}$$

where $a^{\ominus}$ is simply the common unit standard state activity, including units, and a_i^e the absolute activity of component i on the same scale (also including units) when the system is at equilibrium. The factor $(a^{\ominus})^{-\Sigma\nu_i}$ in eqn (4.16) must not be omitted otherwise it may be inferred, mistakenly, that $K_a^{\ominus}$ has units when $\sum \nu_i \neq 0$. Thermodynamic equilibrium constants are dimensionless numbers.

The more specific symbol K_P° rather than $K_a^{\ominus}$ is used for the thermodynamic equilibrium constant if the reaction is between ideal gases and pressure in standard atmospheres is used as the activity scale, the standard state being the pure ideal gas at 1 standard atmosphere pressure.

We shall use other specific symbols for chemical equilibria between gases, between liquids, and between solutes in solution:

$K_c^{\ominus}$: molar scale, standard state = the pure material at an activity of 1 mol dm^{-3} (in a specified solvent for solutions); and

$K_X^{\ominus}$: mole fraction scale, standard state = an activity of unit mole fraction.

But $K_a^{\ominus}$ will still be used as the general symbol and could, according to context, mean any of the above.

Because $\Delta_r G^{\ominus}$ in eqn (4.14) ($= \sum \Delta_f G^{\ominus}$(products) $- \sum \Delta_f G^{\ominus}$(reactants)) is affected only by the temperature and the choice of standard states for reactants and products, $K_a^{\ominus}$ is also a function of only the chosen standard states and temperature. Consequently, K_P° is unaffected by the total pressure of the system when an ideal gas reaction achieves equilibrium and $K_c^{\ominus}$ is unaffected by the overall concentration regardless of whether the equilibrium is between ideal gases or in solution. But the value of K_P° for an equilibrium between ideal gases (obtained experimentally or calculated from thermochemical $\Delta_f G^\circ$ data) will not necessarily be identical with the value of $K_c^\ominus$ for the same gas-phase equilibrium.

Analogously, $K_c^\ominus$ for a given equilibrium in solution, measured experimentally or derived from $\Delta_f G^\ominus$ data relating to a solution standard state activity of 1 mol dm^{-3}, will not normally be identical with the equilibrium's $K_X^\ominus$ value which relates to the unit mole fraction solution standard state.

Equations (4.15) and (4.16) apply to systems *at equilibrium* regardless of the relative or absolute amounts of the materials from which the equilibrium mixtures are derived and regardless of how equilibration is achieved. In other words, a $K_a^\ominus$ describes an equilibrium *state* whose previous history is irrelevant. This is in contrast to the $\Delta_r H$ (or $\Delta_r U$), $\Delta_r S$, and $\Delta_r G$ values of a *transformation* all of which relate to one mole of reaction and depend not only upon the final stage, but also upon the initial state.

4.6.1 The relationship between K_P°, $K_c^\ominus$, and $K_X^\ominus$ for gas-phase reactions

It is occasionally necessary to convert a K_P° result for an ideal gas reaction (standard state activity = 1 atmosphere pressure) into $K_c^\ominus$ or $K_X^\ominus$ values (unit molarity or mole fraction standard state activities, respectively). In terms of first principles, such a calculation requires $\Delta_r G^\ominus$ from $\Delta_r H^\ominus$ and $S^\ominus$ data for the components in the appropriate standard states, and then the use of eqn (4.14). However, such data are sparse so it is more usual (and simpler) to carry out a conversion on K_P° directly, based upon a knowledge of the properties of ideal gases.[5]

4.6.1.1. K_P° and $K_c^\ominus$

$P_i V = n_i \,.\, RT$, therefore

$$P_i = \frac{n_i}{V} \,.\, RT,$$

or

$$P_i = c_i \,.\, RT,$$

where P_i and c_i are the partial pressure and molar concentration of component i in the ideal gas mixture which occupies a total volume V at absolute temperature T. Equation (4.16) may be put explicitly into pressure activities:

$$K_P^\circ = \left(\frac{(P_X^e)^{\nu_X} \,.\, (P_Y^e)^{\nu_Y} \ldots}{(P_A^e)^{\nu_A} \,.\, (P_B^e)^{\nu_B} \ldots}\right)(1\ \text{atm})^{-\Sigma\nu_i} \tag{4.17}$$

$$= \left(\frac{(c_X^e)^{\nu_X} \,.\, (c_Y^e)^{\nu_Y} \ldots}{(c_A^e)^{\nu_A} \,.\, (c_B^e)^{\nu_B} \ldots}\right)(RT)^{\Sigma\nu_i} \,.\, (1\ \text{atm})^{-\Sigma\nu_i}$$

and also into molar activities (assuming for the ideal gas system that molar activity = molar concentration):

$$K_c^{\ominus} = \left(\frac{(c_X^e)^{\nu_X} \cdot (c_Y^e)^{\nu_Y} \ldots}{(c_A^e)^{\nu_A} \cdot (c_B^e)^{\nu_B} \ldots}\right)(1\ \text{mol dm}^{-3})^{-\Sigma\nu_i}.$$

From these expressions for K_P° and $K_c^{\ominus}$ we can see that

$$\underline{K_P^{\circ} \cdot (1\ \text{atm})^{\Sigma\nu_i} = K_c^{\ominus} \cdot (1\ \text{mol dm}^{-3})^{\Sigma\nu_i} \cdot (RT)^{\Sigma\nu_i},}$$

and $K_P^{\circ} = K_c^{\ominus}$ only when $\sum \nu_i = 0$. This equation must include R in the proper units i.e. $R = 0.082056\ \text{dm}^3\ \text{atm K}^{-1}\ \text{mol}^{-1}$ so that all dimensions cancel.

4.6.1.2 K_P° and $K_X^{\ominus}$

We can substitute for P_i^e in eqn (4.17) by using the following expression for the partial pressure of a component in a mixture of ideal gases:

$$P_i^e = X_i^e \cdot P_T$$

where X_i^e = the mole fraction of component i at equilibrium, and P_T = the total pressure of the equilibrium mixture of ideal gases in atmospheres.

$$K_P^{\circ} = \left(\frac{(X_X^e)^{\nu_X} \cdot (X_Y^e)^{\nu_Y} \ldots}{(X_A^e)^{\nu_A} \cdot (X_B^e)^{\nu_B} \ldots}\right)(P_T\ \text{atm})^{\Sigma\nu_i} \cdot (1\ \text{atm})^{-\Sigma\nu_i}.$$

But $K_X^{\ominus}$, which relates to the unit mole fraction standard state, may also be defined from eqn (4.16):

$$K_X^{\ominus} = \left(\frac{(X_X^e)^{\nu_X} \cdot (X_Y^e)^{\nu_Y} \ldots}{(X_A^e)^{\nu_A} \cdot (X_B^e)^{\nu_B} \ldots}\right)(1)^{-\Sigma\nu_i}.$$

Therefore

$$K_P^{\circ} = K_X^{\ominus} \cdot (P_T\ \text{atm})^{\Sigma\nu_i} \cdot (1\ \text{atm})^{-\Sigma\nu_i}$$

hence

$$\underline{K_P^{\circ} = K_X^{\ominus} \cdot (P_T)^{\Sigma\nu_i},}$$

where P_T is now the *numerical* value of the total equilibrium pressure in atmospheres, the units having cancelled exactly.

For a gas-phase reaction at equilibrium at temperature T, therefore, it follows that

(i) unlike K_P° and $K_c^{\ominus}$, $K_X^{\ominus}$ is not independent of the total pressure of the system except when $\sum \nu_i = 0$, consequently $K_X^{\ominus}$ is not a very useful reaction parameter for a gas-phase equilibrium, and

(ii) $K_P^{\circ} = K_c^{\ominus} = K_X^{\ominus}$ only when $\sum \nu_i = 0$.

Example. The thermodynamic equilibrium constant K_P° for the gas-phase dehydration of t-butanol is 1 at 428 K (estimated from thermochemical data[1]). Calculate the degree of reaction when the total pressure at equilibrium is (a) 1 and (b) 10 atmospheres at 428 K. Assume ideal gas behaviour.

$$P_i = X_i \,.\, P_\alpha$$

where P_i atm = partial pressure of i,
X_i = mole fraction of i, and
P_α atm = the total pressure of the system at degree of reaction α.

Let
Z = the number of moles of t-butanol when $\alpha = 0$.

	$(CH_3)_3COH(g)$ ⇌	$(CH_3)_2C{=}CH_2(g)$ +	$H_2O(g)$
Number of moles at $\alpha = 0$,	Z	0	0
Number of moles at $\alpha = 1$,	0	Z	Z
Number of moles at α,	$Z(1-\alpha)$	$Z\alpha$	$Z\alpha$
Partial pressure/atm at degree of reaction α,	$\left(\frac{1-\alpha}{1+\alpha}\right)P_\alpha$	$\left(\frac{\alpha}{1+\alpha}\right)P_\alpha$	$\left(\frac{\alpha}{1+\alpha}\right)P_\alpha$

Let $\alpha = \alpha_e$ and $P_\alpha = P_e$ at equilibrium then, by eqn (4.16),

$$K_P^\circ = \frac{\left(\dfrac{\alpha_e}{1+\alpha_e}\right)^2 . (P_e\,\text{atm})^2}{\left(\dfrac{1-\alpha_e}{1+\alpha_e}\right) . P_e\,\text{atm}} \times (1\,\text{atm})^{-1} = 1 \text{ at } 428\text{ K},$$

therefore

$$\frac{\alpha_e^2 . P_e}{1-\alpha_e^2} = 1 \text{ at } 428\text{ K}.$$

Notice that we have not had to specify whether the reaction is at constant volume (in which case P_α would double if α were to increase from 0 to 1), or at constant pressure (in which case the volume would double).

(a) When $P_e = 1$ at 428 K

$$\frac{\alpha_e^2}{1-\alpha_e^2} = 1$$

$$2\alpha_e^2 = 1$$

therefore

$$\underline{\alpha_e = 0.707.}$$

(b) When $P_e = 10$ at 428 K

$$\frac{10\alpha_e^2}{1-\alpha_e^2} = 1$$

$$11\alpha_e^2 = 1$$

and

$$\underline{\alpha_e = 0.302.}$$

Example. Calculate the equilibrium degree of dimerization of ethene to give cyclobutane and $K_X^\ominus$ (the mole fraction thermodynamic equilibrium constant) at (a) 1 and (b) 10 atmospheres total pressure at 447 K, the temperature at which $K_P^\circ = 1$ (see p. 144). Assume ideal gas behaviour.

Let $P_\alpha$atm = the total pressure at degree of reaction α and Z = the number of moles of ethene present at $\alpha = 0$.

$$2C_2H_4(g) \underset{}{\overset{447\ K}{\rightleftharpoons}} C_4H_8(g);$$

$$K_P^\circ = 1.$$

Since $K_P^\circ = K_X^\ominus . P_\alpha^{\sum \nu_i}$ and $\sum \nu_i = -1$,

$$K_X^\ominus = P_\alpha.$$

	C_2H_4	C_4H_8
Number of moles present at degree of reaction α,	$Z(1-\alpha)$	$0.5Z\alpha$
Partial pressure/atm at degree of reaction α,	$\left(\frac{1-\alpha}{1-0.5\alpha}\right)P_\alpha$	$\left(\frac{0.5\alpha}{1-0.5\alpha}\right)P_\alpha$

Let $\alpha = \alpha_e$ and $P_\alpha = P_e$ at equilibrium then, by eqn (4.16),

$$K_P^\circ = \frac{\left(\frac{0.5\alpha_e}{1-0.5\alpha_e}\right)P_e\ \text{atm}}{\left(\frac{1-\alpha_e}{1-0.5\alpha_e}\right)^2 (P_e\ \text{atm})^2} \times (1\ \text{atm}) = 1 \text{ at } 447\ \text{K},$$

therefore

$$\frac{0.5\alpha_e(1-0.5\alpha_e)}{(1-\alpha_e)^2 P_e} = 1 \text{ at } 447\ \text{K}.$$

(a) When $P_e = 1$, $K_X^\ominus = 1$ and $1.25\alpha_e^2 - 2.5\alpha_e + 1 = 0$

$$\alpha_e = \frac{2.5 \mp \sqrt{1.25}}{2.5}$$

$$\underline{\alpha_e = 0.553.}$$

(b) When $P_e = 10$, $K_X^\ominus = 10$ and $10.25\alpha_e^2 - 20.5\alpha_e + 10 = 0$

$$\alpha_e = \frac{20.5 \mp \sqrt{10.25}}{20.5}$$

$$\underline{\alpha_e = 0.844.}$$

In contrast to the elimination of water from t-butanol which is inhibited by high pressures, a higher total pressure causes an increase in the equilibrium extent of this dimerization. These opposing tendencies are in accord with Le Chatelier's Principle.[4]

Example. Calculate the degree of reaction at equilibrium for the following reaction in solution when $K_c^\ominus = 1$ starting with A at activity (a) 1 and (b) 0.1 mol dm^{-3}.

$$A \underset{\text{solvent}}{\overset{P,\ T}{\rightleftharpoons}} B + C.$$

Let Z mol dm^{-3} = activity of A at degree of reaction = 0.

	Activity/mol dm^{-3}				
Degree of reaction	A	$\rightleftharpoons$	B	+	C
0	Z		0		0
1	0		Z		Z
α	$(1-\alpha).Z$		$\alpha.Z$		$\alpha.Z$

At equilibrium $\alpha=\alpha_e$ and, by eqn (4.16),

$$\frac{\alpha_e^2(Z\ \text{mol dm}^{-3})^2}{(1-\alpha_e)Z\ \text{mol dm}^{-3}}\times(1\ \text{mol dm}^{-3})^{-1}=1$$

$$Z\alpha_e^2+\alpha_e-1=0.$$

We see again in this reaction for which $\sum\nu_i\neq 0$, that α_e is dependent not only upon the magnitude of $K_c^{\ominus}$, but also upon Z – the activity of the reactant when $\xi=0$.

(a) When $Z=1$, $\alpha_e^2+\alpha_e-1=0$

$$\alpha_e=\frac{-1\mp\sqrt{5}}{2}$$

$$\underline{\alpha_e=0.618.}$$

(b) When $Z=0.1$, $\alpha_e^2+10\alpha_e-10=0$

$$\alpha_e=\frac{-10\mp\sqrt{140}}{2}$$

$$\underline{\alpha_e=0.916.}$$

In this solution reaction with $K_a^{\ominus}=1$ and reactant at unit activity when $\xi=0$, equilibrium is at a different degree of reaction from in the gas-phase dehydration of t-butanol considered above which has the same stoichiometry, with $K_a^{\ominus}=1$, and unit reactant activity at $\xi=0$. This is because the gas-phase reaction at constant volume undergoes a pressure change as it attains equilibrium whereas the reaction in dilute solution takes place at constant pressure with no change in volume.

Above, we have calculated α_e from equilibrium constants for different types of reactions using eqn (4.16). However, by using eqn (4.14), we are also able to carry out the prior calculation of the equilibrium constant from $\Delta_rG^{\ominus}$. The standard free energy of the reaction can be computed from $\Delta_fH^{\ominus}$ and $S^{\ominus}$ values of all reactants and products. So in principle we are able to start with experimental calorimetric and spectroscopic results for individual components of a reaction, and finish with a calculated value for the degree of reaction at equilibrium. This is one of the major achievements of chemical science. On the theoretical side, we have presumed that 'activity' relates to some real property of a material, and indeed we have equated it with pressure for ideal gases. (We shall discuss the activity of materials in solution more fully in the next section.) We have also assumed that the actual reacting system, to which our calculated results apply, attains equilibrium.

Example. Calculate the thermodynamic equilibrium constant and the degree of reaction at equilibrium in the following gas-phase reaction at 400 K from the thermodynamic data provided.[1] Assume ideal gas behaviour.

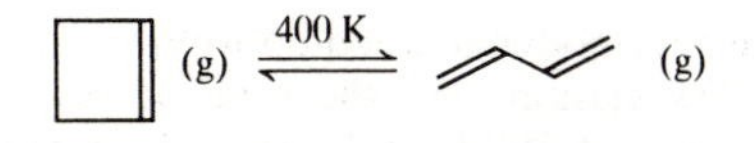

$$\Delta_f G^\circ_{400\,K}(\text{cyclobutene, g}) = 190.9 \text{ kJ mol}^{-1}$$
$$\Delta_f G^\circ_{400\,K}(\text{buta-1,3-diene, g}) = 165.1 \text{ kJ mol}^{-1}$$

therefore

$$\Delta_r G^\circ_{400\,K} = (165.1 - 190.9) \text{ kJ mol}^{-1}$$
$$= -25.8 \text{ kJ mol}^{-1}.$$

By eqn (4.14),

$$-25.8 = -8.314 \times 10^{-3} \times 400 \times \ln K^\circ_P$$
$$\ln K^\circ_P = 7.758$$
$$K^\circ_P = 2340 \text{ at } 400 \text{ K}.$$

The method and results of the first parts of this example apply generally, regardless of the stoichiometry of the reaction and of the choice of the standard states. But of course, what a particular numerical result for $K^\ominus_a$ means in terms of the extent of the reaction at equilibrium can only be deduced from the stoichiometry and standard state specification.

Let P atm = total pressure of the system at degree of reaction α. Because $\sum \nu_i = 0$ in this reaction, there is no pressure change as equilibrium is attained at constant volume.

Degree of reaction	Activity/atm	
	Cyclobutene ⇌	Buta-1,3-diene
0	P	0
1	0	P
α	$(1-\alpha) . P$	$\alpha . P$

At equilibrium, $\alpha = \alpha_e$ and, by eqn (4.16),

$$\frac{\alpha_e}{(1-\alpha_e)} = 2340 \text{ at } 400 \text{ K}$$
$$2341\alpha_e = 2340$$
$$\alpha_e > 0{\cdot}999.$$

This reaction is very far over to the right-hand side at 400 K but is still an equilibrium comprising a reversible process. In contrast to the dehydration of t-butanol and the dimerization of ethene, α_e of this reaction is determined only by K_P°, being independent of the total pressure of the system. (In this respect, it is the gas-phase equivalent of the isomerization in solution of eqn (4.4) (p. 115) where α_e is independent of the reactant activity at $\xi = 0$.) And also, because $\sum \nu_i = 0$,

$$K_P^\circ = K_c^\ominus = K_X^\ominus.$$

Example. Calculate the thermodynamic equilibrium constants, K_P° and $K_c^\ominus$, the equilibrium degree of dissociation, and the total pressure at equilibrium in the following gas-phase reaction at 500 K and constant volume starting from t-butyl chloride at activity (a) 1 and (b) 10 atmospheres. Use the thermodynamic data provided[1] and assume ideal gas behaviour.

$$\Delta_f G^\circ_{500\,K}((CH_3)_3CCl, g) = 20.0 \text{ kJ mol}^{-1}$$

$$\Delta_f G^\circ_{500\,K}((CH_3)_2C{=}CH_2, g) = 112.0 \text{ kJ mol}^{-1}$$

$$\Delta_f G^\circ_{500\,K}(HCl, g) = -97.15 \text{ kJ mol}^{-1}$$

$$(CH_3)_3CCl(g) \underset{}{\overset{500\,K}{\rightleftharpoons}} (CH_3)_2C{=}CH_2(g) + HCl(g)$$

$$\Delta_r G^\circ_{500\,K} = (112.0 - 97.15 - 20.0) \text{ kJ mol}^{-1}$$
$$= -5.15 \text{ kJ mol}^{-1}.$$

By eqn (4.14),

$$-5.15 = -8.314 \times 10^{-3} \times 500 \times \ln K_P^\circ$$
$$\ln K_P^\circ = 1.239$$
$$K_P^\circ = 3.45 \text{ at 500 K.}$$

In this gas reaction $\sum \nu_i = 1$ therefore as the reaction proceeds from pure gaseous $(CH_3)_3CCl$ at constant volume, the pressure increases to its equilibrium value.

Let P_I atm = the initial pressure of the system (at degree of reaction = 0)

$$(CH_3)_3CCl \overset{V,\,T}{\rightleftharpoons} (CH_3)_2C{=}CH_2 + HCl$$

	$(CH_3)_3CCl$	$(CH_3)_2C{=}CH_2$	HCl
Partial pressure/atm at degree of reaction α	$(1-\alpha)P_I$	αP_I	αP_I

Let $\alpha = \alpha_e$ at equilibrium then, by eqn (4.16),

$$K_P^\circ = \frac{\alpha_e^2 (P_I \text{ atm})^2}{(1-\alpha_e) P_I \text{ atm}} \times (1 \text{ atm})^{-1} = 3.45 \text{ at 500 K,}$$

therefore

$$\frac{\alpha_e^2 P_I}{(1-\alpha_e)} = 3.45 \text{ at 500 K.}$$

(a) When $P_I = 1$,

$$\frac{\alpha_e^2}{(1-\alpha_e)} = 3.45$$

$$\alpha_e^2 + 3.45\alpha_e - 3.45 = 0$$

$$\alpha_e = \frac{-3.45 \mp \sqrt{25.7}}{2}$$

$$\underline{\alpha_e = 0.81.}$$

The total pressure P_T at equilibrium is the sum of the partial pressures of reactants and products present

$$P_T = (1+\alpha_e)P_I,$$

$$\underline{P_T = 1.81 \text{ atm.}}$$

(b) When $P_I = 10$,

$$\frac{10\alpha_e^2}{(1-\alpha_e)} = 3.45$$

$$10\alpha_e^2 + 3.45\alpha_e - 3.45 = 0$$

$$\alpha_e = \frac{-3.45 \mp \sqrt{149.9}}{20}$$

$$\underline{\alpha_e = 0.44.}$$

And the total pressure P_T at equilibrium in this case is given by

$$P_T = (1+\alpha_e)10 \text{ atm}$$

$$\underline{P_T = 14.4 \text{ atm.}}$$

As seen on p. 127, the relationship between K_P° and $K_c^\ominus$ is

$$K_P^\circ(1 \text{ atm})^{\Sigma\nu_i} = K_c^\ominus(1 \text{ mol dm}^{-3})^{\Sigma\nu_i}(RT)^{\Sigma\nu_i}.$$

In this example, $K_P^\circ = 3.45$,

$$R = 0.082057 \text{ dm}^3 \text{ atm K}^{-1} \text{ mol}^{-1},$$

$$T = 500 \text{ K},$$

and

$$\sum \nu_i = +1,$$

therefore $3.45 \times (1 \text{ atm}) = K_c^\ominus(1 \text{ mol dm}^{-3}) \times (0.082057 \text{ dm}^3 \text{ atm K}^{-1} \text{ mol}^{-1} \times 500 \text{ K})$

$$\underline{K_c^\ominus = 0.0841 \text{ at } 500 \text{ K.}}$$

Another important use of eqn (4.14), besides calculating equilibria from thermochemical data, is to determine $\Delta_f G^\ominus$ results from equilibrium measurements. If $\Delta_f G^\ominus$ values for all components except one in an equilibrium are known, then an experimental value for this one can be obtained if $\Delta_r G^\ominus$ can be determined since

$$\Delta_r G^\ominus = \sum \Delta_f G^\ominus(\text{products}) - \sum \Delta_f G^\ominus(\text{reactants}).$$

An experimental value for $\Delta_r G^\ominus$ may be obtained from $K_a^\ominus$ using eqn (4.14) if the equilibrium constant is available from measured activities of all components in the equilibrium.

4.7 Activity coefficients and intermolecular interactions in solution

4.7.1 Solute–solute interactions

If we wish to determine $K_a^\ominus$ for an equilibrium in solution rather than calculate it from standard thermochemical data, we need to be able to measure activities. Sometimes, the activity of a component in solution can be measured directly, for example, by vapour pressure or solubility measurements or, if it is ionic, by an electrochemical method. But, in general, it is easier to measure concentrations, and it is always easier to make up a solution of known concentration than one of known activity.

At infinite dilution, each molecule of a single, pure solute is surrounded only by solvent molecules. In this *reference state*, the solvent's properties and solute–solvent interactions are unperturbed by solute–solute effects, and molar concentration and activity of the solute on the molar scale are identical. At any other concentration, they are not usually identical. We can conceive, however, of a *hypothetical* solution at finite dilution in which each solute molecule behaves *as though it were* at infinite dilution – unperturbed by other solute molecules. In such a hypothetical solution, solute–solute interactions are zero and molar concentration is equal to activity on the molar scale; when this is equal to 1 mol dm^{-3} we have the standard state of the solute in the particular solvent.

For activity on the mole fraction scale, the reference state is the pure material (activity = mole fraction = 1). This is satisfactory for a gas or liquid when the pure material is also a convenient standard state (the thermochemical standard state). But for solutes in dilute solution, unit mole fraction is neither a satisfactory reference state nor standard state. Consequently, we shall restrict ourselves in the following to the molar (or sometimes the molal) scale for activity of solutes in dilute solution and an infinite dilution reference state.

In a *real* solution of a single, pure solute i at finite dilution, the relationship between molar activity and concentration is given by

$$c_i = [i]\gamma_i,$$

where c_i = activity of i on the molarity scale,
$[i]$ = the actual concentration of i in mol dm^{-3}, and
γ_i = the activity coefficient of i (related to the molarity scale) under the particular experimental conditions.[6]

Note that the reference state when using the molarity scale (real solution at infinite dilution) is not the same as the standard state (hypothetical solution at $c_i = [i] = 1\ \text{mol dm}^{-3}$) even though $\gamma_i = 1$ in both.

We shall now consider the chemical potential μ_i of solute i in a real solution using the molar scale for both activity and concentration.

From eqn (4.13) (p. 122),

$$\mu_i = \mu_i^{\ominus} + RT \ln c_i^{\mathrm{r}}$$

where c_i^{r} = the relative activity of i in solution; we can substitute for c_i^{r} to give

$$\mu_i = \mu_i^{\ominus} + RT \ln[i]^{\mathrm{r}}\gamma_i$$

or

$$\mu_i = \mu_i^{\ominus} + RT \ln[i]^{\mathrm{r}} + RT \ln \gamma_i,$$

and it is here understood that $[i]^{\mathrm{r}}$ means the numerical value of the molar concentration of the solute (the *relative* molar concentration.)

We now see that the difference between the chemical potential of i in a *real* solution with solute–solute interactions and that of i at the same concentration in a hypothetical solution in the same solvent but with no effects due to solute–solute interactions is the term $RT \ln \gamma_i$. In other words, transfer of solute i from the hypothetical (h) to the real solution (r) at the same concentration in the same solvent at the same temperature involves a change in its chemical potential given by:

$$\mu_i(\mathrm{r}) - \mu_i(\mathrm{h}) = RT \ln \gamma_i \qquad (4.18)$$

If $\gamma_i > 1$, then the free energy of i is higher in the real than in the corresponding hypothetical solution and the solute–solute interactions constitute a de-stabilizing effect. Conversely, when $\gamma_i < 1$, i has a lower free energy in the real solution than in the hypothetical, and the solute–solute interactions stabilize the system. But either way, as the real solution becomes more dilute and closer to the reference state, solute–solute interactions diminish, γ_i tends towards unity, and $RT \ln \gamma_i$ towards zero.

Some values of activity coefficients (actually related to the molal scale but the molar values would be very similar) are given in Table 4.2.

As can be seen, values are appreciably different from unity only for relatively concentrated solutions and electrolytes with highly charged ions.

4.7.2 The practical equilibrium constant

If the substitution

$$a_i^{\mathrm{e}} = c_i^{\mathrm{e}} = [i]_{\mathrm{e}}\gamma_i$$

TABLE 4.2
Molal activity coefficients, aqueous solution at 25 °C

Solute	Molality	γ_i
HBr[a]	0.001	0.966
	0.01	0.906
	0.1	0.805
	1.0	0.871
	3.0	1.67
NaCl[a]	0.001	0.966
	0.01	0.904
	0.1	0.778
	1.0	0.656
	3.0	0.719
$ZnSO_4$[a]	0.001	0.734
	0.01	0.387
	0.1	0.148
	1.0	0.044
	3.0	0.041
$Ce_2(SO_4)_3$[a]	0.01	0.171
	0.05	0.063
	0.1	0.041
1-Butanol[b]	0.001	0.9990
	0.01	0.9906
	0.1	0.9433
	1.0	0.8227
Sucrose[c]	0.1	1.017
	0.5	1.085
	1.0	1.188
	2.0	1.442
	3.0	1.751

[a] S. Glasstone, *Thermodynamics for chemists*, p. 402, Van Nostrand, New York (1947).
[b] 0 °C; W. D. Harkins and R. W. Wampler, *J. Am. chem. Soc.* **53,** 850 (1931).
[c] R. A. Robinson and R. H. Stokes, *Electrolyte solutions*, 2nd edn rev., p. 244, Butterworths, London (1965).

is made in eqn (4.16) (p. 125), it may be rewritten

$$K_c^{\ominus} = \left(\frac{[X]_e^{\nu_X} \,.\, [Y]_e^{\nu_Y} \ldots}{[A]_e^{\nu_A} \,.\, [B]_e^{\nu_B} \ldots}\right) \cdot \left(\frac{\gamma_X^{\nu_X} \,.\, \gamma_Y^{\nu_Y} \ldots}{\gamma_A^{\nu_A} \,.\, \gamma_B^{\nu_B} \ldots}\right)_e \,.\, (a^{\ominus})^{-\Sigma\nu_i}$$

$$= K_c \,.\, K_\gamma \,.\, (a^{\ominus})^{-\Sigma\nu_i}$$

where

$$K_c(\text{the } \textit{practical equilibrium constant}) = \left(\frac{[X]_e^{\nu_X} \,.\, [Y]_e^{\nu_Y} \ldots}{[A]_e^{\nu_A} \,.\, [B]_e^{\nu_B} \ldots}\right), \quad (4.19)$$

$$K_\gamma = \left(\frac{\gamma_X^{\nu_X} \cdot \gamma_Y^{\nu_Y} \cdots}{\gamma_A^{\nu_A} \cdot \gamma_B^{\nu_B} \cdots}\right)_e, \tag{4.20}$$

and the e subscript indicates the equilibrium condition. Unlike $K_c^\ominus$, K_c is not a true equilibrium constant. Because γ_i depends upon the dilution of component i, K_c for an equilibrium in a particular solvent at a given temperature is concentration dependent.

However, for many purposes, K_c (defined by eqn (4.19) in terms of the molar concentrations of components at equilibrium) is an adequate measure of the degree of the reaction at equilibrium and is relatively easily determined. Indeed, in some cases K_γ in eqn (4.20) will be close to unity even though there may be reasons for believing that individual activity coefficients are not. In such cases, numerically, $K_c^\ominus \sim K_c$, but of course K_c has units (except when $\sum \nu_i = 0$) whereas $K_c^\ominus$ does not.

Example. Calculate the practical equilibrium constant for the dissociation of dinitrogen tetroxide in chloroform solution at 8.2 °C from the following results (obtained by J. T. Cundall and quoted by S. Glastone, *Thermodynamics for chemists*, p. 280, Van Nostrand, New York (1947)).

$$N_2O_4 \underset{8.2\,°C}{\overset{CHCl_3}{\rightleftharpoons}} 2NO_2$$

$[N_2O_4]_e$/mol dm^{-3}	0.129	0.227	0.324	0.405	0.778
$10^3[NO_2]_e$/mol dm^{-3}	1.17	1.61	1.85	2.13	2.84

From $K_c = [NO_2]_e^2/[N_2O_4]_e$, the following values for K_c are obtained:

$10^5 K_c$/mol dm^{-3}	1.06	1.14	1.06	1.12	1.04

Mean value:

$$K_c = 1.08 \times 10^{-5} \text{ mol dm}^{-3} \text{ at } 8.2\,°C.$$

Since K_c shows no systematic variation with $[N_2O_4]$ in the range investigated, K_γ must be constant which suggests that it is in fact close to unity. Consequently, $K_c^\ominus \sim 10^{-5}$ at this temperature.

A practical equilibrium constant K_X calculated from concentrations expressed as mole fractions (instead of mole fraction activities which require knowledge of the appropriate activity coefficients) is sometimes used as an approximate $K_X^\ominus$ for reactions between liquids as the following illustrates.

Example. A mixture initially comprising propyl formate (42.67 mmol), water (45.09 mmol), and a trace of HCl is equilibrated at 100 °C in a sealed vessel. At equilibrium, the mixture comprises propyl formate (28.01 mmol), water, formic acid, and propanol (plus the catalytic acid). Calculate the practical equilibrium constant and the degree of hydrolysis at equilibrium. (Results taken from R. F. Schultz, *J. Am. chem. Soc.* **61,** 1443 (1939).) Assume that the only reaction is the one described in the equation:

$$HCO_2C_3H_7(l) + H_2O(l) \underset{100\,°C}{\overset{H^+}{\rightleftharpoons}} HCO_2H(l) + C_3H_7OH(l)$$

	$HCO_2C_3H_7(l)$	$H_2O(l)$	$HCO_2H(l)$	$C_3H_7OH(l)$
mmol present at degree of reaction = 0,	42.67	45.09	0	0
mmol present at degree of reaction = 1	0	2.42	42.67	42.67
mmol present at equilibrium,	28.01	(28.01 + 2.42) = 30.43	(42.67 − 28.01) = 14.66	(42.67 − 28.01) = 14.66

It follows from the definition of the degree of reaction, α (p. 118) that, at equilibrium,

$$\alpha_e = \frac{(28.01 - 42.67)}{-42.67}$$

$$\alpha_e = 0.344.$$

And

$$K_X = \frac{X_e(HCO_2H)\,.\,X_e(C_3H_7OH)}{X_e(HCO_2C_3H_7)\,.\,X_e(H_2O)}$$

$$= \frac{14.66^2}{28.01 \times 30.43}$$

$$K_X = 0.252 \text{ at } 100\,°C.$$

4.7.3 Solute–solvent interactions and free energy of transfer

An activity coefficient different from unity for a solute in a given solvent, regardless of the activity scale to which it relates (mole fraction, molality, or molarity), is due to the difference between solute–solvent interactions in the solution and those in the reference state in the same solvent; this difference is caused by solute–solute interactions in the non-reference state.

By expanding our earlier concept of activity coefficient, we can derive a parameter which measures the change in chemical potential of a solute i when it is transferred from a solution in one solvent to a solution in another at the same concentration, both being hypothetical in the sense of having no solute–solute effects upon the respective chemical potentials.

Consider first a solution of a single, pure solute i in solvent 1 separated by a hypothetical membrane from another solution of the same solute in solvent 2 (Fig. 4.2). 1 and 2 may be either two pure solvents or different mixed solvents with or without a common component.

The hypothetical system in Fig. 4.2 must not be confused with osmosis in which solvent, but not solute, passes through a real semi-permeable membrane from a dilute to a concentrated solution.

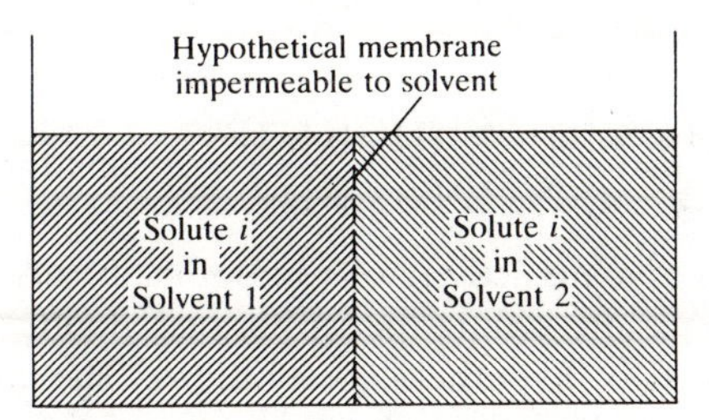

Fig. 4.2. Different solutions separated by a membrane permeable only to the common solute.

The solute in our hypothetical system of Fig. 4.2 freely partitions itself between the two solvents,

$$i(1) \rightleftharpoons i(2),$$

and at equilibrium

$$\sum \nu_i \mu_i^{\mathrm{e}} = 0 \quad \text{(eqn (4.8), p. 121)},$$

or

$$\mu_i^{\mathrm{e}}(1) = \mu_i^{\mathrm{e}}(2),$$

where $\mu_i^{\mathrm{e}}(1)$ and $\mu_i^{\mathrm{e}}(2)$ are the chemical potentials of i in solvents 1 and 2, respectively, at equilibrium. From eqn (4.10),

$$\mu_i^{\mathrm{e}}(1) = \mu_i^{\ominus}(1) + RT \ln a_i^{\mathrm{r,e}}(1),$$

and

$$\mu_i^{\mathrm{e}}(2) = \mu_i^{\ominus}(2) + RT \ln a_i^{\mathrm{r,e}}(2),$$

where $a_i^{\mathrm{r,e}}(1)$ and $a_i^{\mathrm{r,e}}(2)$ are the relative activities of i in solvents 1 and 2 at equilibrium, and $\mu_i^{\ominus}(1)$ and $\mu_i^{\ominus}(2)$ are the standard chemical potentials of i in solvents 1 and 2, respectively, *each related to its own standard state.* So at equilibrium,

$$\mu_i^{\ominus}(1) + RT \ln a_i^{\mathrm{r,e}}(1) = \mu_i^{\ominus}(2) + RT \ln a_i^{\mathrm{r,e}}(2)$$

or

$$\mu_i^{\ominus}(2) - \mu_i^{\ominus}(1) = RT \ln\left(\frac{a_i^{\mathrm{r,e}}(1)}{a_i^{\mathrm{r,e}}(2)}\right) = \Delta_{\mathrm{tr}} G^{\ominus}(i, 1 \rightarrow 2) \qquad (4.21)$$

where $\Delta_{\mathrm{tr}} G^{\ominus}(i, 1 \rightarrow 2) =$ the *standard molar free energy of transfer* of i from solvent 1 to solvent 2.

The difference between the standard chemical potential of i in solvents 1 and 2 on the left-hand side of eqn (4.21) is constant and, like $\mu_i^{\ominus}(1)$ and $\mu_i^{\ominus}(2)$ separately, depends only upon the solute, the solvents, the temperature, and the scales which are used for activity. Consequently, the central term in this equation must be constant. Again restricting ourselves for the sake of illustration to molarity as the activity scale, we may, therefore, write

$$\frac{a_i^{r,e}(1)}{a_i^{r,e}(2)}=\frac{a_i^{e}(1)}{a_i^{e}(2)}=\frac{c_i^{e}(1)}{c_i^{e}(2)}=\frac{[i]_e(1)}{[i]_e(2)}=\text{constant}={}^1\gamma_i^2, \tag{4.22}$$

since (i) the standard state activity of i in both solvents $= 1\ \text{mol dm}^{-3}$, and (ii) there are no solute–solute perturbations upon the solute–solvent interactions in the two (hypothetical) solutions. The constant ${}^1\gamma_i^2$ in eqn (4.22) is sometimes called the *solvent transfer activity coefficient* of the solute.[6,8] This equation shows that ${}^1\gamma_i^2$ is the ratio of the activities of i in solvents 1 and 2 when i has the same chemical potential in the two.

If solute is transferred from solvent 1 to solvent 2 at the same arbitrary activity (or concentration since $\gamma_i = 1$ in both hypothetical solutions), the change in chemical potential is given by

$$\mu_i(2)-\mu_i(1)=\{\mu_i^{\ominus}(2)+RT\ln c_i^{r}(2)\}-\{\mu_i^{\ominus}(1)+RT\ln c_i^{r}(1)\}$$
$$=\mu_i^{\ominus}(2)-\mu_i^{\ominus}(1) \quad \text{since } c_i^{r}(1)=c_i^{r}(2),$$

therefore

$$\mu_i(2)-\mu_i(1)=RT\ln {}^1\gamma_i^2=\Delta_{tr}G^{\ominus}(i, 1\rightarrow 2). \tag{4.23}$$

Equation (4.23) shows that $RT\ln {}^1\gamma_i^2$ is the change in chemical potential of i upon transfer from solvent 1 to the same *arbitrary* concentration (activity) in solvent 2 due to different solute–solvent interactions in the two solvents, unperturbed in both cases by solute–solute effects. And the molar free-energy change accompanying this transfer, $\Delta_{tr}G^{\ominus}(i, 1\rightarrow 2)$, is independent of the activity (concentration) of i. These equations are important in the investigation of solvent effects upon reaction equilibria and rates.[9]

Results are reported either as $\Delta_{tr}G^{\ominus}$ values[10] or as ${}^1\gamma_i^2$ values.[9] We include both for a few electrolytes and non-electrolytes in Table 4.3 and, in contrast to the activity coefficients which account for solute–solute effects in a single solvent, solvent transfer activity coefficients range over several orders of magnitude.

In general terms, the results are as expected. For an ionic compound, ${}^{H_2O}\gamma_i^{\text{org.solv.}}$ is usually greater than unity which corresponds to a positive (unfavourable) free energy of transfer from water to the organic solvent. However, Ph_4AsI has a negative (favourable) free energy of transfer from water to methanol or acetone (${}^{H_2O}\gamma_i^{\text{org.solv.}}<1$). Organic compounds have ${}^{H_2O}\gamma_i^{\text{org.solv.}}$ less than unity and usually ${}^{CH_3OH}\gamma_i^{S}<1$ where S is an aprotic solvent, exceptions being alkyl halides between methanol and the very polar aprotic solvent dimethyl sulphoxide. In loose terms, organic compounds prefer to be in non-polar solvents; however, with ${}^1\gamma_i^2$ or $\Delta_{tr}G^{\ominus}(i, 1\rightarrow 2)$ parameters, we are able to see this effect quantitatively.

So far in this section on solute–solvent interactions and solvent transfer activity coefficients, we have dealt only with hypothetical solutions uncomplicated

TABLE 4.3

Solvent transfer activity coefficients and the corresponding standard molar free energies of transfer of some electrolytes and non-electrolytes, molar scale, 25 °C[a]

Solute	Solvent 1	Solvent 2	${}^1\gamma^2$	$\Delta_{tr}G^{\ominus}/kJ\ mol^{-1}$
$LiCl$[b]	H_2O	CH_3OH	1.97×10^3	18.8
	H_2O	DMF	1.49×10^4	23.8
$NaBr$[b]	H_2O	CH_3OH	2.84×10^3	19.7
	H_2O	DMF	2.84×10^3	19.7
$(C_2H_5)_4NI$[b]	H_2O	CH_3OH	17.6	7.1
	H_2O	DMF	81.4	10.9
Ph_4AsI[b]	H_2O	CH_3OH	1.63×10^{-3}	−15.9
	H_2O	DMF	1.45×10^{-5}	−27.6
I_2[c]	H_2O	CH_3OH	5.0×10^{-3}	−13.1
	CH_3OH	DMF	1.6×10^{-2}	−10.2
	CH_3OH	CH_3CN	0.63	−1.1
	CH_3OH	DMSO	7.9×10^{-5}	−23.4
CH_3I[c]	H_2O	CH_3OH	4.0×10^{-2}	−8.0
	H_2O	70% DMSO-H_2O	7.9×10^{-2}	−6.3
	CH_3OH	DMF	0.32	−2.9
	CH_3OH	CH_3CN	0.40	−2.3
n-Butyl bromide[c]	CH_3OH	DMF	0.79	−0.57
	CH_3OH	DMSO	1.3	0.57
	CH_3OH	CH_3CN	0.63	−1.1
$(CH_3)_3CCl$[c]	H_2O	CH_3OH	1.0×10^{-3}	−17.1
	CH_3OH	DMF	0.63	−1.1
	CH_3OH	DMSO	1.3	0.57
Ph_4C[c]	CH_3OH	DMF	2.5×10^{-2}	−9.1
	CH_3OH	CH_3CN	0.32	−2.9
$CH_3CO_2C_2H_5$[c]	H_2O	70% DMSO-H_2O	0.27	−3.2
$(CH_3O)_3PO$[c]	CH_3OH	DMF	0.40	−2.3

[a] DMF = dimethylformamide, DMSO = dimethyl sulphoxide.
[b] Ref. 10.
[c] Ref. 9.

by solute–solute effects. If we wish to consider concentrations of real solutions then we have to introduce 'solute' activity coefficients to accommodate solute–solute effects in both solvents.

From eqn (4.22) and assuming $a_i^{\ominus} = c_i^{\ominus} = 1\ mol\ dm^{-3}$ in both solvents,

$$ {}^1\gamma_i^2 = \frac{c_i^e(1)}{c_i^e(2)} = \frac{[i]_e(1)\,.\,\gamma_i(1)}{[i]_e(2)\,.\,\gamma_i(2)}. $$

But it is important not to confuse what have been introduced as two types of parameter: ${}^1\gamma_i^2$ on the one hand, and $\gamma_i(1)$ and $\gamma_i(2)$ on the other. If a real two-solution system related to molar (or molal) scales is extrapolated to infinite dilution (the common reference state for these scales), $\gamma_i(1)$ and $\gamma_i(2)$ become unity, but ${}^1\gamma_i^2$ remains unchanged. This does not mean that there is a fundamental difference between the two sorts of activity coefficient or between the physical processes which they relate to. Both measure the effect of intermolecular interactions upon the chemical potential of the solute and, to generalize,

$$\mu_i(\mathrm{r}) - \mu_i(\mathrm{h}) = RT \ln \gamma_i,$$

where $\mu_i(\mathrm{r})$ = the chemical potential of solute i in a real solution under particular conditions of solvent, temperature, concentration, etc.,

$\mu_i(\mathrm{h})$ = the chemical potential of solute i in a hypothetical solution at the same concentration in some specified solvent, temperature, etc...., but with no solute–solute interactions, and solute–solvent interactions identical with those in the reference state (infinite dilution in the specified solvent for the molar and molal scales),

γ_i = the activity coefficient of the solute in the real solution, and

$RT \ln \gamma_i$ = the *total* contribution towards the chemical potential of the solute i in the real solution due to *all* intermolecular interactions which are different from those in the reference state.

4.8 Temperature dependence of chemical equilibria

4.8.1 Reactions between gases

Equation (4.14), (p. 124), applied to a chemical equilibrium between ideal gases (standard state activity = pure gas at 1 standard atmosphere pressure) at temperature T, may be expanded to give

$$\Delta_r H^\circ - T\,.\,\Delta_r S^\circ = -RT\,.\ln K_P^\circ$$

or

$$\ln K_P^\circ = -\frac{\Delta_r H^\circ}{RT} + \frac{\Delta_r S^\circ}{R},$$

therefore

$$\frac{\mathrm{d}(\ln K_P^\circ)}{\mathrm{d}T} = \frac{\Delta_r H^\circ}{RT^2} - \frac{1}{RT}\cdot\frac{\mathrm{d}(\Delta_r H^\circ)}{\mathrm{d}T} + \frac{1}{R}\cdot\frac{\mathrm{d}(\Delta_r S^\circ)}{\mathrm{d}T}.$$

But

$$\frac{\mathrm{d}(\Delta_r H^\circ)}{\mathrm{d}T} = \Delta C_P^\circ,$$

and

$$\frac{\mathrm{d}(\Delta_r S^\circ)}{\mathrm{d}T} = \frac{\Delta C_P^\circ}{T},$$

where $\Delta C_P^\circ = \sum C_P^\circ(\text{products}) - \sum C_P^\circ(\text{reactants})$,

therefore

$$\frac{1}{RT}\cdot\frac{\mathrm{d}(\Delta_r H^\circ)}{\mathrm{d}T} = \frac{1}{R}\cdot\frac{\mathrm{d}(\Delta_r S^\circ)}{\mathrm{d}T},$$

so

$$\frac{\mathrm{d}(\ln K_P^\circ)}{\mathrm{d}T} = \frac{\Delta_r H^\circ}{RT^2}, \tag{4.24}$$

and, correspondingly,

$$\frac{d(\ln K_P^\circ)}{d(1/T)} = -\frac{\Delta_r H^\circ}{R}. \tag{4.25}$$

Equations (4.24) and (4.25) are exact differential equations and, with others to follow all of which express the relationship between $K_a^\ominus$ and T, are formulations of the van't Hoff equation.[4] Qualitatively, eqn (4.24) shows that if a gas-phase reaction is exothermic then its K_P° decreases with increasing temperature; and if it is endothermic, then its K_P° increases as the temperature increases. Quantitatively, the standard molar enthalpy of the reaction is seen to be the temperature coefficient of K_P°; the larger it is numerically, the more sensitive the equilibrium is towards a change in temperature. And, of course, the equilibrium constant of a thermoneutral reaction ($\Delta_r H^\circ = 0$) is independent of temperature.

In principle, eqn (4.25) allows the determination of $\Delta_r H^\circ$ if K_P° can be measured at different temperatures; or, if $\Delta_r H^\circ$ is known from thermochemical data, K_P° at one temperature may be calculated from a value at another. These applications of the integrated version of eqn (4.25) require that either $\Delta_r H^\circ$ be constant over the temperature range involved or that it be known as a function of temperature. In other words, ΔC_P° should be known. In most simple applications, however, ΔC_P° is assumed to be zero ($\Delta_r H^\circ$ constant).

The results in Table 4.4 below are for the dimerization of ethene (eqn (4.26)) in which K_P° changes from being more than 1 to less than 1 as the temperature is raised.

$$2C_2H_4(g) \overset{T}{\rightleftharpoons} \text{Cyclobutane(g)}. \tag{4.26}$$

The values for $\Delta_r H^\circ$ given in Table 4.4, which have been calculated from $\Delta_f H^\circ$ results,[1] do not vary much—ΔC_P° is small. Consequently, a plot of $\ln K_P^\circ$ against $1/T$ using the integrated version of eqn (4.25) is linear (Fig. 4.3). The $\Delta_r S_T^\circ$ values in Table 4.4 were also calculated (from S_T° values of reactants and products) and are also seen to be relatively

TABLE 4.4
Thermodynamic data for the gas-phase dimerization of ethene[a]

T/K	$\Delta_r H^\circ$/kJ mol^{-1}	$\Delta_r S^\circ$/J K^{-1} mol^{-1}	$T.\Delta_r S^\circ$/kJ mol^{-1}	$\Delta_r G^\circ$/kJ mol^{-1}	K_P°
300	−78.0	−174	−52.2	−25.8	3.1×10^4
400	−79.1	−177	−70.8	−8.3	12
500	−79.5	−178	−89.0	9.5	0.10
600	−79.5	−178	−106.8	27.3	4.2×10^{-3}

[a] Data from ref. 1.

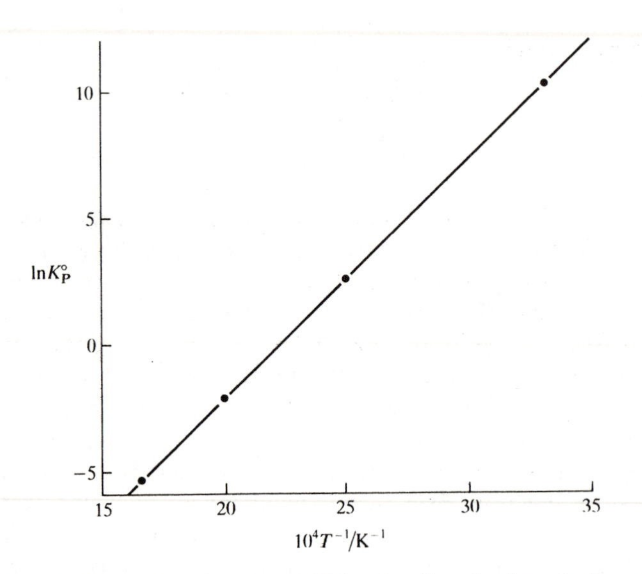

Fig. 4.3. Plot of ln K_P° against $1/T$ for the dimerization of ethene.

constant. However, $T.\Delta_r S_T^\circ$ becomes increasingly negative (less favourable) as the temperature is raised, and, at one particular temperature between 400 and 500 K, it becomes equal to the exothermic (favourable) enthalpy term.

We can calculate the particular temperature T of this equality:

$$\Delta_r H_T^\circ = T.\Delta_r S_T^\circ$$

$$79.3 \times 10^3 \text{ J mol}^{-1} = T \times 177.5 \text{ J K}^{-1} \text{ mol}^{-1}$$

therefore

$$\underline{T = 447 \text{ K.}}$$

And, since $\Delta_r G^\circ = \Delta_r H^\circ - T\,\Delta_r S^\circ = -RT \ln K_P^\circ$, this is the temperature at which $\Delta_r G^\circ = 0$ and $K_P^\circ = 1$, a result used in the example on p. 129. The effect is illustrated in Fig. 4.4.

Example. From the equilibrium results given below for the gas-phase dehydrogenation of propan-2-ol

$$(CH_3)_2CHOH(g) \overset{T}{\rightleftharpoons} (CH_3)_2CO(g) + H_2(g),$$

calculate the standard molar enthalpy of the reaction. Assume ideal gas behaviour (the values of K_P°, which relate to the ideal gas standard state of 1 standard atmosphere pressure, were shown to be independent of the total pressure at equilibrium). Results are taken from H. J. Kolb and R. L. Burwell, *J. Am. chem.*

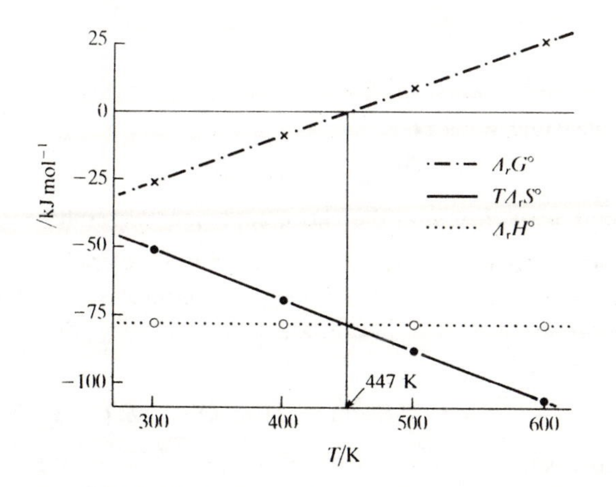

Fig. 4.4. Variation of $\Delta_rH°$, $T.\Delta_rS°$, and $\Delta_rG°$ with temperature for the dimerization of ethene.

Soc. **67,** 1084 (1945), and have already been discussed by Stull, Westrum, and Sinke.[1]

T/K	416.7	436.4	452.2	464.3	491.6
$10^4\ T^{-1}/K^{-1}$	24.00	22.92	22.11	21.54	20.34
K_P°	0.124	0.276	0.435	0.683	1.57
ln K_P°	−2.09	−1.29	−0.83	−0.38	0.45

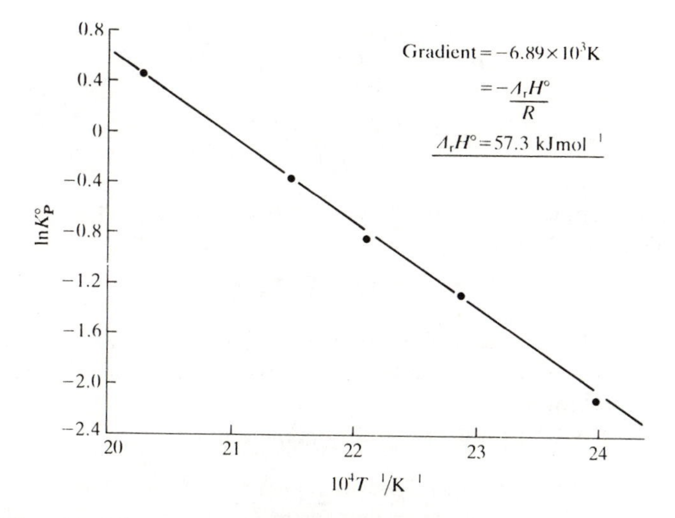

Plot of ln K_P° against $1/T$ for the dehydrogenation of propan-2-ol.

From thermochemical data[1] and using

$$\Delta_r H_T^\circ = \sum \Delta_f H_T^\circ(\text{products}) - \sum \Delta_f H_T^\circ(\text{reactants}),$$

we obtain a calculated value of

$$\Delta_r H_{450\,\text{K}}^\circ = 56.6 \text{ kJ mol}^{-1},$$

which is in good agreement with the above experimental result.

As we have already observed, it is sometimes desirable to describe a gas-phase equilibrium in terms of $K_c^\ominus$ (standard state = ideal gas at 1 mol dm^{-3}) rather than K_P°. The relationship between the two has been shown to be (p. 127):

$$K_P^\circ(1 \text{ atm})^{\sum \nu_i} = K_c^\ominus(1 \text{ mol dm}^{-3})^{\sum \nu_i} . (RT)^{\sum \nu_i}.$$

Since this expression includes the absolute temperature, the two thermodynamic equilibrium constants do not have the same temperature dependence. The temperature variation of $K_c^\ominus$ can be derived and the first step is to rearrange the above equation and express it in logarithmic form:

$$\ln K_P^\circ = \ln K_c^\ominus + \sum \nu_i . \ln\left\{\frac{1 \text{ mol dm}^{-3}}{1 \text{ atm}} . RT\right\} \qquad (4.27)$$

therefore

$$\frac{\mathrm{d}(\ln K_P^\circ)}{\mathrm{d}T} = \frac{\mathrm{d}(\ln K_c^\ominus)}{\mathrm{d}T} + \frac{\sum \nu_i}{T}, \qquad (4.28)$$

where T here is the *numerical* value of the absolute temperature, all dimensions within the bracketed term of eqn (4.27) having cancelled out.

From eqns (4.28) and (4.24) we can now write

$$\frac{\Delta_r H^\circ}{RT^2} = \frac{\mathrm{d}(\ln K_c^\ominus)}{\mathrm{d}T} + \frac{\sum \nu_i}{T},$$

or

$$\frac{\mathrm{d}(\ln K_c^\ominus)}{\mathrm{d}T} = \frac{\Delta_r H^\circ - \sum \nu_i RT}{RT^2}.$$

But we have already seen in Chapter 2 (eqn (2.3), p. 44) that for a reaction between ideal gases,

$$\Delta_r H^\circ = \Delta_r U^\circ + \sum \nu_i RT,$$

where $\Delta_r U^\circ$ = the standard molar internal energy of the reaction (the heat change when 1 mole of reaction is carried out at constant volume)
$= \sum \Delta_f U^\circ(\text{products}) - \sum \Delta_f U^\circ(\text{reactants})$

therefore

$$\frac{\mathrm{d}(\ln K_c^\ominus)}{\mathrm{d}T} = \frac{\Delta_r U^\circ}{RT^2}, \qquad (4.29)$$

or

$$\frac{\mathrm{d}(\ln K_c^{\ominus})}{\mathrm{d}(1/T)} = -\frac{\Delta_r U^\circ}{R}. \tag{4.30}$$

We see, therefore, that it is the standard molar heat change of the gas-phase reaction carried out at constant volume which is the measure of the temperature variation of $K_c^{\ominus}$.

Equations (4.29) and (4.30) include two standard state specifications, and the variation of $K_c^{\ominus}$ with T for an ideal gas appears to be anomalous in being an internal energy rather than an enthalpy term. These observations are not unrelated. We could have reverted to first principles and rewritten eqn (4.14) for a gas-phase reaction as:

$$\Delta_r G^{\ominus} = -RT \ln K_c^{\ominus}$$

the superscript indicating the standard state activity of the ideal gas = 1 mol dm^{-3}. This allows derivation of the expressions

$$\frac{\mathrm{d}(\ln K_c^{\ominus})}{\mathrm{d}T} = \frac{\Delta_r H^{\ominus}}{RT^2}, \tag{4.31}$$

and

$$\frac{\mathrm{d}(\ln K_c^{\ominus})}{\mathrm{d}(1/T)} = -\frac{\Delta_r H^{\ominus}}{R}, \tag{4.32}$$

in exactly the same way that we obtained eqns (4.24) and (4.25). In these equations, $K_c^{\ominus}$, like K_p° before, relates to a system *at equilibrium* and its magnitude for a given reaction is determined only by the temperature. It is not affected by *how* equilibrium is attained. But $\Delta_r H^{\ominus}$ in eqns (4.31) and (4.32) is the molar heat change due to the *transformation*

$$\text{Reactants(g)} \rightarrow \text{Products(g)},$$

and the standard state specification requires that reactants initially and products finally be at 1 mol dm^{-3} (multiplied in each case by ν_i as required by the stoichiometry of the reaction). In other words, we have a *reaction* at 1 mol dm^{-3}.

So although $\Delta_r H^{\ominus}$ may be *calculated* from $\Delta_r H^{\ominus} = \sum \Delta_f H^{\ominus}(\text{products}) - \sum \Delta_f H^{\ominus}(\text{reactants})$ if $\Delta_f H^{\ominus}$ data are available, *experimentally* $\Delta_r H^{\ominus}$ is the molar heat change of the reaction *at constant volume.* And, of course, the molar heat change of an ideal gas reaction at constant volume is $\Delta_r U^\circ$, so for such a reaction

$$\Delta_r H^{\ominus} = \Delta_r U^\circ.$$

This illustrates again just how important a proper understanding of standard states is in chemical thermodynamics.

Example. From the following results for the equilibrium between monomeric and dimeric acetic acid in the vapour phase (gas standard state = 1 mol dm^{-3}), calculate the molar heat change of the reaction at constant volume. Assume ideal gas behaviour (results taken from T. M. Fenton and W. E. Garner, *J. chem. Soc.* 694 (1930)).

$$2CH_3CO_2H(g) \overset{T}{\rightleftharpoons} (CH_3CO_2H)_2(g); \qquad \Delta_r U^\circ$$

T/K	383.6	405.0	428.8	456.9
$10^4\,T^{-1}$/K^{-1}	26.07	24.69	23.32	21.89
$K_c^\ominus$	117.0	44.16	16.86	6.296
$\ln K_c^\ominus$	4.762	3.788	2.825	1.840

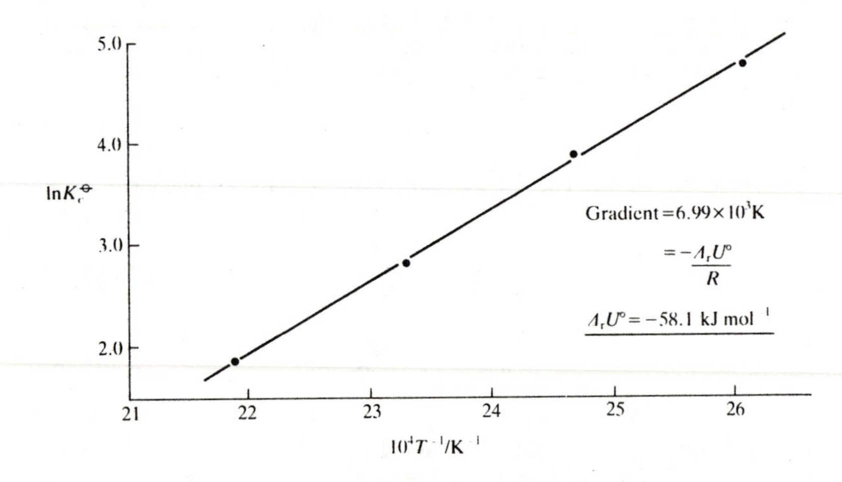

Plot of $\ln K_c^\ominus$ against $1/T$ for the gas-phase dimerization of acetic acid.

If $\Delta_r H^\circ$ were required, it could be obtained either from the slope of a van't Hoff plot of $\ln K_P^\circ$ against $1/T$ or from $\Delta_r U^\circ$ using the equation

$$\Delta_r H^\circ = \Delta_r U^\circ + \sum \nu_i R T_m,$$

where T_m is the mean temperature of the van't Hoff plot:

$$\Delta_r H^\circ = \{-58.1 + (-1) \times 8.314 \times 420 \times 10^{-3}\}\ \text{kJ mol}^{-1}$$
$$= -61.6\ \text{kJ mol}^{-1}.$$

As the enthalpy of formation of a hydrogen bond between oxygen atoms is known to be approximately 30 kJ mol^{-1},[11] this result suggests that the acetic acid dimer is held together by two hydrogen bonds:

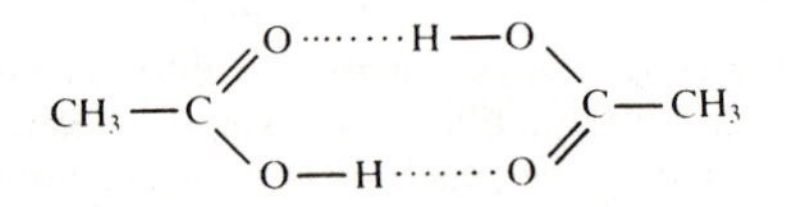

We have seen previously (p. 127) that, for a reaction involving ideal gases,

$$K_P^\circ = K_X^\ominus \,.\, (P_T)^{\sum \nu_i}$$

where P_T = the numerical value of the total pressure (in atmospheres) of the gaseous system at equilibrium and $K_X^\ominus$ = the thermodynamic equilibrium constant relating to the mole fraction scale for activity. If this is put into logarithmic form:

$$\ln K_P^\circ = \ln K_X^\ominus + \sum \nu_i \,.\, \ln P_T$$

and differentiated with respect to temperature *at constant total pressure*, we see that

$$\left(\frac{\mathrm{d}(\ln K_P^\circ)}{\mathrm{d}T}\right)_{P_T} = \left(\frac{\mathrm{d}(\ln K_X^\ominus)}{\mathrm{d}T}\right)_{P_T}$$

consequently

$$\left(\frac{\mathrm{d}(\ln K_X^\ominus)}{\mathrm{d}T}\right)_{P_T} = \frac{\Delta_r H^\circ}{RT^2} \qquad (4.33)$$

or

$$\left(\frac{\mathrm{d}(\ln K_X^\ominus)}{\mathrm{d}(1/T)}\right)_{P_T} = -\frac{\Delta_r H^\circ}{R}. \qquad (4.34)$$

The heat term on the right-hand side of eqns (4.33) and (4.34) is the familiar thermochemical standard enthalpy of the gas-phase reaction:

$$\Delta_r H^\circ = \sum \Delta_f H^\circ(\text{products}) - \sum \Delta_f H^\circ(\text{reactants}).$$

The constant total pressure constraint indicated on the left hand side of these equations is because, as we have already seen, $K_X^\ominus$ (unlike K_P° and $K_c^\ominus$) is not independent of the total pressure at equilibrium. But this does not mean that the equilibrium between ideal gases at any one temperature must be achieved under constant pressure conditions. It means that if a van't Hoff plot is to be constructed from $K_X^\ominus$ and T results, the total pressure of the gaseous system at equilibrium must be the same at all temperatures used in the investigation.

4.8.2 Reactions in dilute solution

We can rewrite eqn (4.14) (p. 124) as

$$\Delta_r G^\ominus = -RT \ln K_c^\ominus,$$

or

$$\ln K_c^\ominus = -\frac{\Delta_r H^\ominus}{RT} + \frac{\Delta_r S^\ominus}{R},$$

and, by proceeding as we did for ideal gas reactions, we obtain as exact differential equations:

$$\left(\frac{\mathrm{d}(\ln K_c^\ominus)}{\mathrm{d}T}\right)_{1\,\text{atm}} = \frac{\Delta_r H^\ominus}{RT^2}, \qquad (4.35)$$

and

$$\left(\frac{\mathrm{d}(\ln K_c^\ominus)}{\mathrm{d}(1/T)}\right)_{1\ \mathrm{atm}} = -\frac{\Delta_r H^\ominus}{R}. \tag{4.36}$$

A significant difference between these and eqns (4.24) and (4.25) is the (constant) 1 atmosphere pressure constraint. It is included in eqns (4.35) and (4.36) simply because of the scale which is used for solute activity – molarity (or molality) – and the standard state specification: a (hypothetical) solution of 1 mol dm^{-3} (or 1 mol kg^{-1}) *under 1 standard atmosphere pressure* with the solute behaving as though it were at infinite dilution (the reference state for dilute solutions at which γ(solute) = 1).

Thermodynamic equilibrium constants are involved in eqns (4.35) and (4.36) and these are not as accessible experimentally as practical equilibrium constants (see p. 135). If the numerical value of the practical equilibrium constant is used in a van't Hoff plot, the presumption is not that K_γ is unity, but that it is independent of temperature. This approximation is probably justifiable if the products and reactants are similar in type, especially if $\sum \nu_i = 0$ as in an isomerization, but is less so when the reaction is, for example, an ionization. Furthermore, the difference between $\Delta_r H^\ominus$ (the heat change due to 1 mole of reaction in solution at activity 1 mol dm^{-3} under 1 standard atmosphere pressure) and $\Delta_r H$ (the heat change due to 1 mole of reaction in some arbitrary volume of solution at ambient atmospheric pressure) is likely to be small compared with the uncertainty in the experimental measurements.

Consequently, eqns (4.35) and (4.36) are in practice approximated to

$$\frac{\mathrm{d}(\ln K_c)}{\mathrm{d}T} = \frac{\Delta_r H}{RT^2},$$

and

$$\frac{\mathrm{d}(\ln K_c)}{\mathrm{d}(1/T)} = -\frac{\Delta_r H}{R},$$

with omission of the standard state superscript.

Example. Calculate the molar enthalpy of the reaction in eqn (4.4) from the results in Table 4.1, (p. 116).

T/K	$10^4\ T^{-1}$/K^{-1}	K_c	ln K_c
264.2	37.9	0.151	−1.89
305.7	32.7	0.616	−0.48
340.2	29.4	1.52	0.42
348.2	28.7	1.83	0.60
386.2	25.9	3.98	1.38

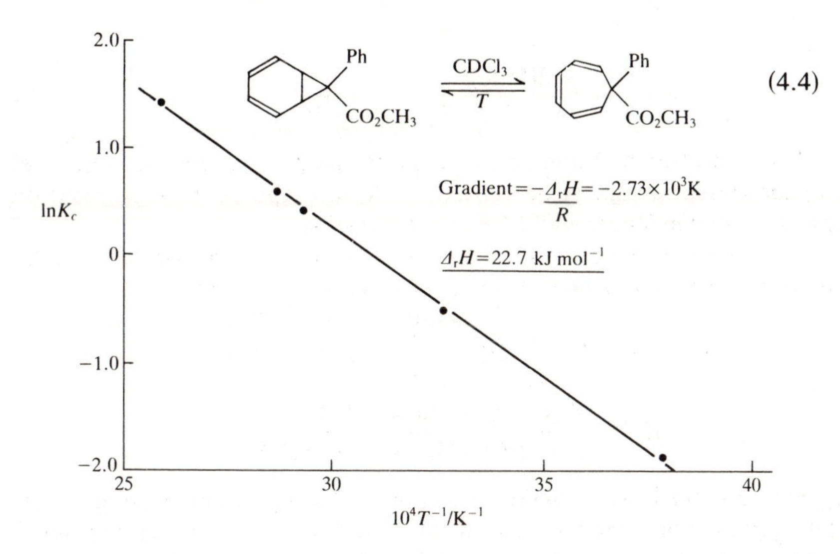

van't Hoff plot of ln K_c against $1/T$ for the equilibrium of eqn (4.4).

These equilibrium constants for the reaction in eqn (4.4), like the ones for the dehydrogenation of propan-2-ol, increase with increasing temperatures, both reactions being endothermic. The opposite trend was observed for the dimerizations of ethene and acetic acid both of which are exothermic.

4.8.3 Reactions between liquids

For a reaction between pure liquids, eqn (4.14) (p. 124) becomes

$$\Delta_r G^{\ominus} = -RT \ln K_x^{\ominus}, \tag{4.37}$$

where the symbolism now indicates the mole fraction activity scale, the reference state ($\gamma = 1$) being the pure liquid, and the standard state is unit mole fraction activity. If this is the liquid under one standard atmosphere total pressure, it is identical with the thermochemical standard state. Consequently, the thermochemical symbols for the components of the resolved free-energy term on the left-hand side of eqn (4.37) may be used:

$$\Delta_r H^{\circ} - T\,\Delta_r S^{\circ} = -RT \ln K_x^{\ominus} \text{ (under 1 atm)},$$

from which it follows that

$$\left(\frac{\mathrm{d}(\ln K_x^{\ominus})}{\mathrm{d}T}\right)_{1\,\mathrm{atm}} = \frac{\Delta_r H^{\circ}}{RT^2}, \tag{4.38}$$

and

$$\left(\frac{\mathrm{d}(\ln K_X^{\ominus})}{\mathrm{d}(1/T)}\right)_{1\,\mathrm{atm}} = -\frac{\Delta_r H^\circ}{R}. \qquad (4.39)$$

The indication of 1 atm pressure is a formality since reactions between liquids would invariably be carried out at atmospheric pressure, and the pressure dependence would be minute anyway.

These equations are convenient for investigating reactions such as the formation of ethyl acetate from ethanol and acetic acid:

$$CH_3CO_2H(l) + C_2H_5OH(l) \underset{1\,\mathrm{atm}}{\overset{K_X^{\ominus}}{\rightleftharpoons}} CH_3CO_2C_2H_5(l) + H_2O(l);$$

$$K_X^{\ominus} = \frac{X_e(CH_3CO_2C_2H_5) \,.\, X_e(H_2O)}{X_e(CH_3CO_2H) \,.\, X_e(C_2H_5OH)}.$$

For this particular reaction, the practical equilibrium constant calculated from mole fraction concentration measurements at equilibrium is rather insensitive to temperature at $K_X \sim 4$.[12] This implies that the heat of reaction is rather small. Experimental confirmation of this is found in the very low standard enthalpy of the reverse process – the hydrolysis of ethyl acetate: $\Delta_r H^\circ_{298\,K} = 3.7$ kJ mol^{-1}.[13]

4.9 Problems

1. Methanol (15.10 mmol) and acetic acid (7.05 mmol) were heated in a sealed vessel to 100 °C. At equilibrium, the solution contained 6.12 mmol of water. Calculate the practical mole fraction equilibrium constant, K_X, for this reaction at 100 °C.

$$CH_3OH(l) + CH_3CO_2H(l) \overset{100\,^\circ C}{\rightleftharpoons} CH_3CO_2CH_3(l) + H_2O(l)$$

(Data taken from R. J. Williams, A. Gabriel, and R. C. Andrews, *J. Am. chem. Soc.* **50,** 1267 (1928).)

2. Calculate the equilibrium constant K_P° for the following reaction from the thermodynamic data provided:[1]

$$CH_3OH(g) + HF(g) \overset{400\,K}{\rightleftharpoons} CH_3F(g) + H_2O(g)$$

	$CH_3OH(g)$	$HF(g)$	$CH_3F(g)$	$H_2O(g)$
$\Delta_f H^\circ_{400\,K}$/kJ mol^{-1}	−204.8	−271.3	−236.9	−242.8
$S^\circ_{400\,K}$/J K^{-1} mol^{-1}	253.6	182.3	234.8	198.7

3. $K_P^\circ = 1$ for the gas-phase dehydration of t-butanol at 428 K (see p. 128):

$$(CH_3)_3COH(g) \overset{428\,K}{\rightleftharpoons} (CH_3)_2C{=}CH_2(g) + H_2O(g).$$

For a reaction starting from 0.20 mol of t-butanol in a 1 dm^3 vessel at 428 K, calculate (i) the equilibrium degree of reaction, and (ii) the final total equilibrium gas pressure.

4. From the results given below for the *cis-trans*-isomerization of 1-ethyl-2-methylcyclopropane, calculate (i) $\Delta_r H^\circ_{700\,K}$ (the mean value over the temperature range of the experiments), and (ii) the equilibrium degree of reaction at 440.0 °C. At what temperature is (iii) $K^\circ_P = 2.800$, and (iv) the degree of isomerization $\alpha = 0.739$.

$$cis\text{-isomer(g)} \overset{T}{\rightleftharpoons} trans\text{-isomer(g)}; \quad \Delta_r H^\circ_T$$

$$K^\circ_P = P^e_{trans}/P^e_{cis}$$

Temp/°C	414.1	419.7	425.6	432.3	440.0
K°_P	2.831	2.817	2.788	2.774	2.745

(Data taken from C. S. Elliott and H. M. Frey, *J. chem. Soc.* 900 (1964).)

5. Calculate ΔH for the dimerization of acetic acid in cyclohexane,

$$2CH_3CO_2H \underset{C_6H_{12}}{\overset{T}{\rightleftharpoons}} (CH_3CO_2H)_2; \qquad \Delta H$$

$$K_c = [(CH_3CO_2H)_2]/[CH_3CO_2H]^2,$$

from the following data (taken from U. Jentschura and E. Lippert, *Ber. Buns. Phys. Chem.* **75,** 556 (1971)).

T/K	302.6	313.9	326.0	337.7	348.9	359.7
$K_c/dm^3\,mol^{-1}$	7515	3511	1698	887	487	274

Compare the result with the value obtained for the corresponding gas-phase reaction (p. 148).

6. Deduce, by a graphical method using the data given below,[1] the temperatures at which $\Delta_r G^\circ$ for the gas-phase dehydration of t-butanol is equal to (i) 7.0 and (ii) $-9.0\ kJ\,mol^{-1}$. Calculate K°_P and $K^\ominus_c$ for this reaction at both temperatures. Assume that $\Delta_r H^\circ$ and $\Delta_r S^\circ$ are constant over this temperature range and (iii) calculate their values from your graph.

	$\Delta_f G^\circ/kJ\,mol^{-1}$		
T/K	$(CH_3)_3COH(g)$ ⇌	$(CH_3)_2C{=}CH_2(g)$ +	$H_2O(g)$
300	−190.2	58.5	−228.5
400	−143.8	84.6	−223.9
500	−95.8	112.0	−219.1
600	−46.7	140.4	−214.1

7. Calculate (i) $\Delta_r U°$ and (ii) $\Delta_r H°$ (the mean values over the range 548–591 K) for the Diels–Alder reaction,

$$\text{C}_6\text{H}_8\,(g) + CH_2{=}CH_2(g) \overset{T}{\rightleftharpoons} \text{C}_8\text{H}_{12}\,(g),$$

from the data given below (calculated from kinetic results provided by G. Huybrechts, D. Rigaux, J. Vankeerberghen, and B. Van Mele, *Int. J. chem. Kin.* **12,** 253 (1980)).

T/K	548	561	574	591
$10^{-3}\,.\,K_c^{\ominus}$	10.8	5.55	2.94	1.33

4.10 References

1. D. R. Stull, E. F. Westrum, and G. C. Sinke, *The chemical thermodynamics of organic compounds*, Wiley, New York (1969).
2. G. E. Hall and J. D. Roberts, *J. Am. chem. Soc.* **93,** 2203 (1971).
3. The usefulness of the *extent of reaction* concept has been discussed by T. Cvitas and N. Kallay, *Chemistry in Britain*, 290 (1978) and, together with the *degree of reaction*, by H. Maskill, *Education in Chemistry* **18,** 146 (1981).
4. W. J. Moore, *Physical chemistry*, 5th edn, Longman, London (1972).
5. An account of equilibrium constants and standard states, especially of how the former depend upon the latter, is given by D. M. Golden, *J. chem. Educ.* **48,** 235 (1971).
6. Most general physical chemistry texts[4,7] deal exhaustively with activity coefficients of electrolytes (ionic solutes) at low concentration in water but hardly at all with organic solutes, or with any solutes in non-aqueous solvents. More widely ranging accounts of activity and activity coefficients are to be found in refs 9 and 10 below and also in: C. D. Ritchie, *Physical organic chemistry: the fundamental concepts*, Marcel Dekker, New York (1975); J. E. Leffler and E. Grunwald, *Rates and equilibria of organic reactions*, Wiley, New York (1973); K. Yates and R. A. McClelland, *Progr. phys. Org. Chem.* **11,** 323 (1974); and E. Buncel and H. Wilson, *Adv. phys. org. Chem.* **14,** 133 (1977).
7. D. H. Everett, *An introduction to the study of chemical thermodynamics*, 2nd edn, Longman, London (1971); K. Denbigh, *The principles of chemical equilibria*, 2nd edn, Cambridge University Press (1966); E. B. Smith, *The principles of chemical thermodynamics*, Oxford University Press (1973); R. A. Robinson and R. H. Stokes, *Electrolyte solutions*, 2nd edn rev., Butterworths, London (1965); P. W. Atkins, *Physical chemistry*, Oxford University Press (1978).
8. Solvent transfer activity coefficients seldom appear in general physical chemistry textbooks at all except in disguise when the effect of one solute upon the activity of another in the same solution is treated by Debye–Hückel theory.
9. A. J. Parker, *Chem. Rev.* **69,** 1 (1969). See also A. J. Parker, *Adv. phys. org. Chem.* **5,** 173 (1967).

10. B. G. Cox, *Annual Reports*, Part A, p. 249, The Chemical Society, London (1973).
11. G. C. Pimental and A. L. McClellan, *The hydrogen bond*, Freeman, San Francisco (1960).
12. S. Glasstone, *Textbook of physical chemistry*, 2nd edn, Macmillan, London (1960).
13. I. Wadsö, *Acta chem. Scand.* **12,** 630 (1958).

5

Acids and bases

5.1 Introduction

In previous chapters, we considered molecular structure and properties, and saw how a chemical equilibrium is described thermodynamically. These matters will be linked towards the end of the present chapter, but we shall start with a qualitative treatment of Lewis acids and bases. We then move on to give a quantitative account of Brönsted acid–base reactions which will be treated in some detail. In part this is because proton-transfer steps have been implicated in many catalysed reactions in solution, so any understanding of catalysis, which we cover in Chapter 8, requires a sound knowledge of Brönsted acid–base behaviour.

5.2 Lewis acids and bases[1]

A *Lewis base* is a molecule or anion which acts as an *electron pair donor*; usually, a non-bonding or lone pair of valence electrons is used to form a covalent bond with another molecule or a cation. The species to which the new bond is formed, the *electron pair acceptor*, is the *Lewis acid*.

Generally, the Lewis acid is initially two electrons short of a closed valence shell. Familiar examples of Lewis acid–base reactions are:[1,2]

base		*acid*	
Et_2O	+	BF_3	$\rightleftharpoons Et_2\overset{+}{O}\text{–}\overset{-}{B}F_3$
Cl^-	+	$AlCl_3$	$\rightleftharpoons AlCl_4^-$
pyridine	+	$AlCl_3$	$\rightleftharpoons C_5H_5\overset{+}{N}\text{—}\overset{-}{Al}Cl_3$
Me_3N	+	BMe_3	$\rightleftharpoons Me_3\overset{+}{N}\text{–}\overset{-}{B}Me_3$

Other reactions which may be regarded in the same way are:[1]

base	*acid*		
O^{2-}	+ SO_3	$\rightleftharpoons$	SO_4^{2-}
$2F^-$	+ SiF_4	$\rightleftharpoons$	SiF_6^{2-}
OH^-	+ CO_2	$\rightleftharpoons$	HCO_3^-
I^-	+ I_2	$\rightleftharpoons$	I_3^-

We can also designate molecular fragments as Lewis acids and bases by dissection of the molecule using heterolytic steps. Such molecular dissections are not intended always to represent feasible chemical reactions, but rather as means of seeing how *in principle* a molecule is built up using steps which do not offend known principles of chemical bonding.

	Lewis acid		*Lewis base*
$(CH_3)_3COH$	$\rightarrow (CH_3)_3C^+$	+	OH^-
$(CH_3)_3COH$	$\rightarrow H^+$	+	$(CH_3)_3CO^-$
$CH_3CO_2C_2H_5$	$\rightarrow C_2H_5^+$	+	$CH_3CO_2^-$
$CH_3CO_2^-$	$\rightarrow CO_2$	+	CH_3^-
$CH_3CO_2C_2H_5$	$\rightarrow CH_3CO^+$	+	$C_2H_5O^-$
$(CH_3)_2\overset{+}{S}\text{–}\overset{-}{B}H_3$	$\rightarrow BH_3$	+	$(CH_3)_2S$
$(CH_3)_2S$	$\rightarrow 2CH_3^+$	+	S^{2-}

Notice that a given molecule may be dissected in more than one way. Some molecular fragments, e.g. alkyl groups and halogens, are either Lewis acids or bases depending upon their electronic complement. This is determined by the direction of the heterolysis involved in their notional generation e.g.

$$(CH_3)_3C\text{—}Cl \rightarrow (CH_3)_3C^+ + Cl^-$$

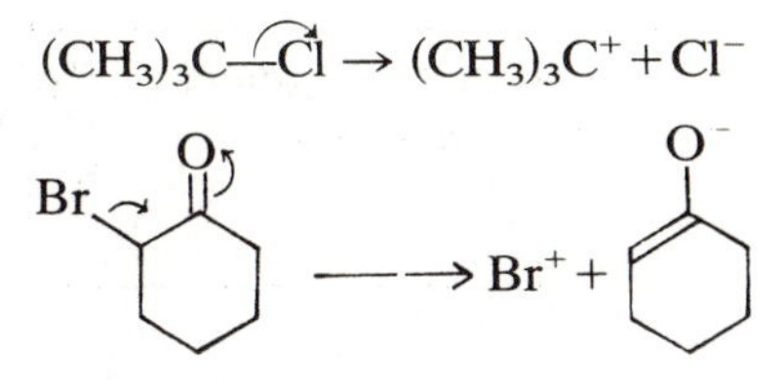

Lewis acid and base products of such retrosynthetic steps need not be stable molecules or ions, or, indeed, known species.

5.2.1 Hard and soft acids and bases[1,3]

A *soft* Lewis base is one in which the donor atom is easily polarizable. Such bases tend to be large and of low electronegativity e.g. S^{2-}, $(Et)_3P$. The donor atom of a *hard* base is not easily polarizable. These bases are small and very electronegative, e.g. F^-, OH^-.

Some bases are intermediate in this hard–soft scale, e.g. Br^-, $C_6H_5NH_2$, and orders or trends are usually more significant than absolute designations. For example, the hardness of a base increases up a group of the periodic table:

$$\text{in hardness, } I^- < Br^- < Cl^- \ll F^-.$$

Hardness also increases for an iso-electronic series across a period from left to right:

$$\text{in hardness, } NH_2^- < OH^- < F^-.$$

Table 5.1 includes a classification of some bases. Given this qualitative hard–soft scale for bases, it was realized from experimental results that some Lewis acids form bonds preferentially with hard bases and are called *hard* acids, whereas others bond preferentially with soft bases and are termed *soft* acids. A hard acid is one in which the lone pair acceptor atom is of low polarizability (high polarizing power). They tend to be small cations in high oxidation states, molecules with incomplete valence shells and electron-withdrawing substituents, and multiply-bonded carbon electrophiles with oxygen or halogen substituents, e.g. H^+, Al^{3+}, BF_3, CO_2.

The acceptor atom of a soft acid is of high polarizability (low polarizing power). These include large cations in low oxidation states, molecules containing incomplete valence shells, and most singly-bonded carbon electrophiles, e.g. Ag^+, 'BH_3', $(CH_3)_3C^+$. Some Lewis acids are intermediate in this hard–soft scale (Zn^{2+}, BMe_3) and, again, trends are usually more important than absolute designations. Thus in hardness,

TABLE 5.1
Some hard and soft Lewis bases[1,3]

Hard	Intermediate	Soft
NH_3, RNH_2, H_2O, ROH; OH^-, O^{2-}, RO^-, F^-, Cl^-, RCO_2^-, SO_4^{2-}, CO_3^{2-}.	$C_6H_5NH_2$, pyridine; N_3^-, SO_3^{2-}, Br^-.	R_3P, $(RO)_3P$, RSH; H^-, R^-, SCN^-, CN^-, S^{2-}, RS^-, $S_2O_3^{2-}$, I^-.

TABLE 5.2
Some hard and soft Lewis acids [1,3]

Hard	Intermediate	Soft
BF_3, $B(OR)_3$, CO_2, SO_3; H^+, Li^+, Mg^{2+}, Al^{3+}, Fe^{3+}; RCO^+.	BR_3, SO_2; Zn^{2+}, Fe^{2+}; NO^+, R_3C^+.	'BH_3', quinones; Ag^+, Hg^+, Hg^{2+}, metal(0); HO^+, RO^+, RS^+, RSe^+, Br^+, I^+, CH_3^+, RCH_2^+.

$Li^+ > Na^+ > K^+$ and $BF_3 > BMe_3 >$ 'BH_3'. Examples are given in Table 5.2.

The basis of these classifications is the experimental generalization that hard acid–hard base and soft acid–soft base interactions or bonds are more favourable than ones involving either soft acid–hard base or hard acid–soft base. This is the so-called *Hard and Soft Acid and Base* (*HSAB*) *Principle*[3] which covers the overwhelming majority of reactions which can be regarded formally (but not always mechanistically) as Lewis acid–base reactions. It has led to a systematization of very diverse chemistry and to predictions which have been extensively verified.

Two simple examples based upon these guidelines illustrate its general use and validity within the thermodynamic context.

Methanol and methyl iodide are both formally the products of a soft acid (CH_3^+) and a base (OH^-, hard; I^-, soft). We 'predict', on the basis of the HSAB principle that, in the following reaction,

$$\underset{\text{s–h}}{CH_3OH(g)} + \underset{\text{h–s}}{H{-}I(g)} \rightleftharpoons \underset{\text{s–s}}{CH_3I(g)} + \underset{\text{h–h}}{H{-}OH(g)}; \quad K$$

$$K = \frac{[H_2O][CH_3I]}{[HI][CH_3OH]},$$

equilibrium lies to the right-hand side since this includes the favourable hard–hard and soft–soft bonds. Indeed, the equilibrium constant is very large indeed ($\sim 10^9$ at 25 °C).[4]

The following reaction includes acyl transfer (CH_3CO^+, hard acid) from the soft base RS^- to the hard one RO^- and equilibrium 'should' lie to the right-hand side:

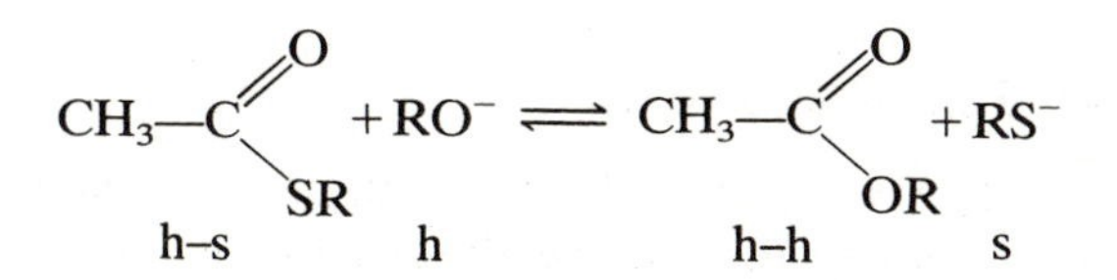

This is the case experimentally, thioesters being very easily solvolysed in slightly basic water or alcoholic media in high yields.[5]

The HSAB principle has been widely applied in organic chemistry, mostly in accounting for relative reactivities of substrates with a common reactant. In particular it has been very useful in describing different reaction modes of ambident nucleophiles such as enolate anions with electrophiles:

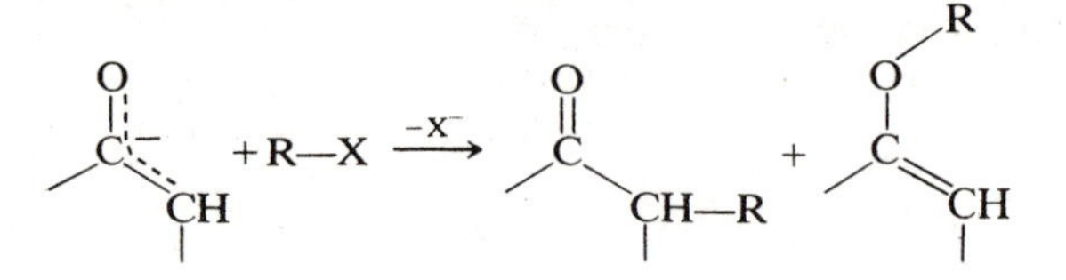

and ambident electrophiles such as α,β-unsaturated carbonyl compounds with nucleophiles:

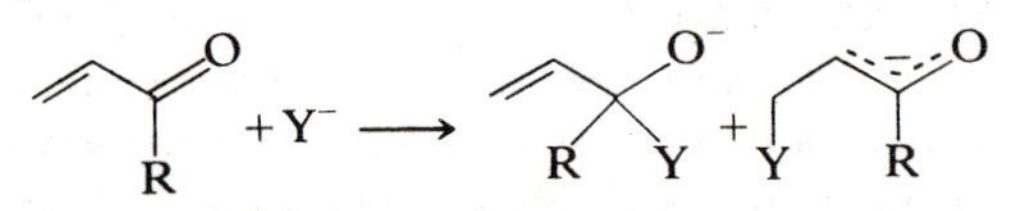

In the former case, the nucleophile has one reaction site softer than the other (carbon softer than oxygen) and the proportion of reaction at carbon increases as the softness of the Lewis acid site of the electrophile R—X increases. In the second reaction, the softer the nucleophile Y^-, the higher its affinity for the softer electrophilic centre of the β-carbon rather than the harder carbonyl. These, however, are matters of substrate reactivity as opposed to product stability (kinetic rather than thermodynamic control, see p. 272) so we shall not consider them further in this chapter which is primarily concerned with equilibria.

In spite of the practical utility of the HSAB principle, it remains at the moment an empirical guide without a single, widely-accepted theoretical basis. Some of the chemistry to which it has been applied may be explained satisfactorily in terms of solvent effects; elementary molecular orbital theory adequately accounts for other results.[3] But the value of the HSAB principle is its wide applicability as a predictive guide even when the mechanism of the reaction is not well understood. Empirical rules of this sort which pre-date sound theoretical explanations have an honourable record in mechanistic organic chemistry. But like all such rules, the HSAB principle occasionally leads to a prediction which subsequently is shown to be wrong, so it must be used with circumspection.[3] The failures occur when some other aspect of a reaction completely dominates the hard and soft matching of Lewis acid–base pairs.

5.3 Brönsted acids and bases[6,7]

A Brönsted acid, AH in eqn (5.1), is a compound which acts as a *proton donor*; the *proton acceptor*, B, is the Brönsted base.

$$AH + B \rightleftharpoons A^- + BH^+ \tag{5.1}$$

In the reverse reaction, A^- is the base and BH^+ the acid, so A^- is called the *conjugate base* of acid AH and BH^+ is the *conjugate acid* of base B.

Acids and bases may be charged or neutral as the following illustrate:

Brönsted acid		*Brönsted base*		*conjugate base*		*conjugate acid*	
CH_3CO_2H	+	CH_3O^-	$\rightleftharpoons$	$CH_3CO_2^-$	+	CH_3OH	(5.2)
H_3O^+	+	NH_3	$\rightleftharpoons$	H_2O	+	NH_4^+	(5.3)
H_2SO_4	+	H_2O	$\rightleftharpoons$	HSO_4^-	+	H_3O^+	(5.4)
$[Fe(H_2O)_6]^{3+}$	+	H_2O	$\rightleftharpoons$	$[Fe(H_2O)_5OH]^{2+}$	+	H_3O^+	
HSO_4^-	+	OH^-	$\rightleftharpoons$	SO_4^{2-}	+	H_2O	

Unifying features of Brönsted acid–base reactions are that they all involve proton transfer and the reactions are, at least in principle, reversible. Consequently, each of the above reactions may be seen as a competition between two Brönsted bases for a single proton; and in each case, the equilibrium extent of reaction is most strongly affected by which base, under the experimental conditions, is the stronger:

in eqn (5.2), CH_3O^- or $CH_3CO_2^-$,
in eqn (5.3), NH_3 or H_2O,
and in eqn (5.4), H_2O or HSO_4^-.

5.3.1 Amphoteric compounds and autoprotolysis

Some compounds may act either as Brönsted acids or bases and are called *amphoteric*. This may arise simply through the presence of independent acidic and basic functionality as in 4-aminophenol:

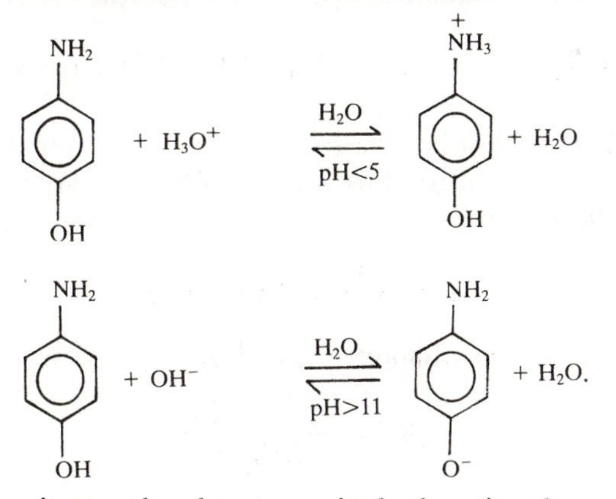

Less commonly, it may be due to a single functional group which has acidic and basic properties and acts one way or the other according to the experimental conditions, e.g. cyanamide:

$$NH_2CN + H_3O^+ \underset{pH<1}{\overset{H_2O}{\rightleftharpoons}} \overset{+}{N}H_3CN + H_2O,$$

and

$$NH_2CN + OH^- \underset{pH>11}{\overset{H_2O}{\rightleftharpoons}} \overset{-}{N}HCN + H_2O.$$

Zwitter-ionic α-amino-acids such as glycine are important amphoteric materials:

$$\overset{+}{N}H_3CH_2CO_2^- + H_3O^+ \underset{pH<2}{\overset{H_2O}{\rightleftharpoons}} \overset{+}{N}H_3CH_2CO_2H + H_2O$$

and

$$\overset{+}{N}H_3CH_2CO_2^- + OH^- \underset{pH>10}{\overset{H_2O}{\rightleftharpoons}} NH_2CH_2CO_2^- + H_2O.$$

Many common solvents have amphoteric properties:

$$H_2O + HNO_3 \overset{H_2O}{\rightleftharpoons} H_3O^+ + NO_3^-$$

$$H_2O + HN{=}C(NH_2)_2 \overset{H_2O}{\rightleftharpoons} OH^- + H_2\overset{+}{N}{=}C(NH_2)_2$$

$$CH_3CO_2H + H_2SO_4 \overset{CH_3CO_2H}{\rightleftharpoons} CH_3C(OH)_2^+ + HSO_4^-$$

$$CH_3CO_2H + C_2H_5NH_2 \overset{CH_3CO_2H}{\rightleftharpoons} CH_3CO_2^- + C_2H_5\overset{+}{N}H_3.$$

Amphoteric compounds may also undergo some observable extent of *autoprotolysis* (self-protonation). It can be intramolecular via the solvent and very extensive as in the case of α-amino-acids in aqueous solution,

$$NH_2CH_2CO_2H \overset{H_2O}{\rightleftharpoons} \overset{+}{N}H_3CH_2CO_2^-; \qquad K > 1,$$

or to a much smaller extent as with 5-aminopentanoic acid,

$$NH_2(CH_2)_4CO_2H \overset{H_2O}{\rightleftharpoons} \overset{+}{N}H_3(CH_2)_4CO_2^-; \qquad K < 1.$$

Autoprotolysis of amphoteric solvents such as water is intermolecular but occurs to only a very low extent. We can write an equilibrium constant for the reaction

$$2H_2O \overset{H_2O}{\rightleftharpoons} H_3O^+ + OH^- \tag{5.5}$$

in the normal way (see Chapter 4, p. 125):

$$K_a^\ominus = \frac{a_{H_3O^+}^{r,e} \cdot a_{OH^-}^{r,e}}{(a_{H_2O}^{r,e})^2}$$

where $a_i^{r,e} = a_i^e / a_i^\ominus$
= the relative activity of i in the equilibrium mixture,
a_i^e = the absolute activity of i in the equilibrium mixture, and
$a_i^\ominus$ = the absolute activity of i in its defined standard state.

However, because the extent of this reaction is so exceedingly small, the relative activity of water itself at equilibrium ($a_{H_2O}^{r,e} = a_{H_2O}^e / a_{H_2O}^\ominus$) is imperceptibly different from unity if we select mole fraction as the activity scale and, as standard state, the hypothetical pure un-ionized liquid. Also on account of the low equilibrium extent of reaction, the activity coefficients of H_3O^+ and OH^- in pure water are very close to unity if we select infinite dilution as the reference state (see p. 134). Consequently, the *relative* activities of H_3O^+ and OH^- are virtually identical with the numerical values of their molar concentrations if we use the molar scale for activity and choose an activity of 1 mol dm^{-3} as the standard state. We can write, therefore,

$$K_a^{\ominus} \sim K_{AP} = [H_3O^+][OH^-]$$

where $[H_3O^+]$ and $[OH^-]$ are the numerical values of the molar concentrations of H_3O^+ and OH^- and K_{AP} is the *autoprotolysis constant* of water (the *ionization constant* K_w in the older literature) which has the value 1.00×10^{-14} at 25 °C.[6]

Equation (5.5) also shows that, in pure water, $[H_3O^+]$ and $[OH^-]$ must be equal; therefore, in pure water,

$$[H_3O^+] = [OH^-] = \sqrt{1.00 \times 10^{-14}} = 1.00 \times 10^{-7} \text{ at } 25\,°\text{C},$$

and $pH = -\log[H_3O^+] = 7.00$.

In acidic solutions, $[H_3O^+] > [OH^-]$ ($pH < 7$) and in alkaline solutions, $[H_3O^+] < [OH^-]$ ($pH > 7$).

When either $[H_3O^+]$ or $[OH^-]$ ceases to be very small, molar concentration and activity may no longer be taken to be identical and activity coefficients should be taken into account.[7]

Autoprotolysis constants for some other liquids at 25 °C are indicated below ($pK_{AP} = -\log K_{AP}$):[8]

System	pK_{AP}
$2H_2SO_4 \rightleftharpoons H_3SO_4^+ + HSO_4^-$;	2.9
$2HCO_2H \rightleftharpoons HC(OH)_2^+ + HCO_2^-$;	6.2
$2HF \rightleftharpoons H_2F^+ + F^-$;	10.7 (0 °C)
$2CH_3OH \rightleftharpoons CH_3OH_2^+ + CH_3O^-$;	16.7
$2CH_3CN \rightleftharpoons CH_3CNH^+ + CH_2CN^-$;	19.5
$2NH_3 \rightleftharpoons NH_4^+ + NH_2^-$;	27.7 (−50 °C)

5.3.2 Dissociation and acidity constants

The bare proton, H^+, is not known except in the gas phase at very low pressures. When Brönsted acids react with water, they give solutions containing the *hydronium* ion H_3O^+,[6,9] for example:

$$CF_3CO_2H + H_2O \xrightleftharpoons{H_2O} CF_3CO_2^- + H_3O^+.$$

Structural studies by diffraction and spectroscopic methods have shown H_3O^+ to be pyramidal. It is hydrated in aqueous solution like other cations and gives reversible proton transfer to many bases under equilibrium control at normal temperatures:

$$H_3O^+ + H_2O \rightleftharpoons H_2O + H_3O^+$$
$$H_3O^+ + C_2H_5OH \rightleftharpoons H_2O + C_2H_5OH_2^+$$
$$H_3O^+ + NH_3 \rightleftharpoons H_2O + NH_4^+$$
$$H_3O^+ + H_2S \rightleftharpoons H_2O + H_3S^+.$$

The dissociation of an acid AH in water may be represented by eqn (5.6).

$$AH + H_2O \overset{H_2O}{\rightleftharpoons} A^- + H_3O^+. \tag{5.6}$$

The thermodynamic equilibrium constant for this reaction expressed fully (see Chapter 4, p. 125) would be

$$K_a^\ominus = \left(\frac{a_{A^-}^e \cdot a_{H_3O^+}^e}{a_{AH}^e \cdot a_{H_2O}^e}\right) \cdot \left(\frac{a_{AH}^\ominus \cdot a_{H_2O}^\ominus}{a_{A^-}^\ominus \cdot a_{H_3O^+}^\ominus}\right)$$

where a_i^e and $a_i^\ominus$ are the absolute activities of i in the equilibrium mixture and the standard state, respectively.

However, the *relative* activity of water at equilibrium, $a_{H_2O}^e / a_{H_2O}^\ominus$, will not vary much from the dilute solution of one acid to another and will become unity as the acid approaches infinite dilution if we take mole fraction as the scale for its activity and pure water as its standard state. For convenience, therefore, a special thermodynamic equilibrium constant, sometimes called the thermodynamic acidity constant, $K_a^\ominus(AH)$, is defined for the dissociation of acid AH at a specified temperature:

$$K_a^\ominus(AH) = \left(\frac{a_{A^-}^e \cdot a_{H_3O^+}^e}{a_{AH}^e}\right) \cdot \left(\frac{a_{AH}^\ominus}{a_{A^-}^\ominus \cdot a_{H_3O^+}^\ominus}\right) \tag{5.7}$$

The molarity scale is normally used for a solute i, and its standard state is an aqueous solution in which the activity of $i = 1$ mol dm^{-3} at 25 °C. Consequently, all dimensions and the standard state terms cancel out in eqn (5.7). Furthermore, if we are dealing only with dilute solutions, the ratio of activity coefficients $\gamma_{A^-} \cdot \gamma_{H_3O^+} / \gamma_{AH}$ may be taken to be much the same from one acid to another and, indeed, close to unity. As a comparative measure of the strength of an acid in dilute aqueous solution at a specified temperature, therefore, it is usual simply to employ an *acidity constant* K_{AH} defined by

$$K_{AH} = \frac{[H_3O^+][A^-]}{[AH]}, \tag{5.8}$$

where $[i]$ is the numerical value of the molar concentration of species i at equilibrium in the dilute solution at the specified temperature, usually 25 °C. In practice, K_{AH} will be concentration-dependent because its definition leaves out activity coefficients. But extrapolation of values to infinite dilution would give the thermodynamic parameter $K_a^{\ominus}(AH)$ of eqn (5.7).

Determination of the acidity (or dissociation) constant of an acid in water by direct measurements of concentrations (or activities) of dissociated and undissociated species is only possible if the acid undergoes perceptible but incomplete dissociation.[7,10–12] When the acid is so weak that dissociation is imperceptible, or so strong that dissociation is virtually complete, the concentrations of the dissociated form in the first case and the undissociated form in the second will not be accurately measurable. For such acids, indirect methods of acidity (or dissociation) constant determination have to be used.[7,10–14]

In order to compare conveniently a wide range of acid strengths, acidity constants are normally expressed logarithmically as pK_{AH} values where

$$pK_{AH} = -\log K_{AH}$$

so

$$K_{AH} = \text{antilog}(-pK_{AH}).$$

Clearly, the more dissociated the acid is at a given finite concentration, the larger its K_{AH} and the *smaller* (or more negative) its pK_{AH}. Some pK_{AH} values are given in Table 5.3. Those which are less than about 0 or more than about 20 may not be accurate. Some of the results, however, are to a high degree of precision corrected to infinite dilution and so are proper thermodynamic $pK_a^{\ominus}(AH)$ values.

Notice that water itself is included in Table 5.3; this follows from regarding its autoprotolysis as the dissociation of a weak acid:

$$H_2O + H_2O \rightleftharpoons H_3O^+ + OH^-.$$

By eqn (5.8),

$$K_{AH} = \frac{[H_3O^+][OH^-]}{[H_2O]} = \frac{K_{AP}}{[H_2O]} = \frac{1.00 \times 10^{-14}}{55.56}$$

$$pK_{AH}(H_2O) = 15.7 \quad \text{at } 25\,°C.$$

It is implicit in this result that the standard state of water *considered as a weak acid* is the same as for all other acidic solutes, i.e. an aqueous solution, hypothetical in the case of water as solute, in which the activity of the solute is 1 mol dm^{-3}. But the activity scale for water *as the solvent* remains mole fraction and its standard state is still the pure material. Only by treating water in the above chemical equation in these two different ways are we able to ascribe a measure of acid strength to water in aqueous

TABLE 5.3
pK_{AH} *values of some acids in water, 25 °C*

Inorganic acids	pK_{AH}	Note	Organic acids	pK_{AH}	Note	Organic acids	pK_{AH}	Note
HI	~−9	*a*	Pentacyanocyclo-			$CH_2(CN)_2$	11.2	*c*
HBr	~−8	*a*	pentadiene	<−11	*d*	2,4,6-Trinitroaniline	12.2	*n*
HCl	~−7	*a*	$HC(CN)_3$	~−5	*e*	CF_3CH_2OH	12.4	*o*
H_2SO_4	~−3	*a*	CH_3SO_3H	0	*f*	$(CH_3)_2C{=}NOH$	12.4	*f*
HNO_3	~−1.4	*a*	$HC(NO_2)_3$	0	*g*	2,4-Dinitroaniline	15.0	*n*
NH_2SO_3H	0.99	*a*	CF_3CO_2H	0.23	*f*	CH_3CONH_2	15.1	*f*
HSO_4^-	2.0	*a*	CCl_3CO_2H	0.66	*f*	CH_3OH	15.5	*f*
H_3PO_4	2.15	*a*	2,4,6-Trinitrophenol			C_2H_5OH	15.9	*p*
HF	3.18	*a*	(picric acid)	0.71	*f*	Cyclopentadiene	16.0	*q*
HNO_2	3.3	*a*	$CHCl_2CO_2H$	1.26	*h*	$(CH_3)_2CHOH$	17.0	*r*
H_2CO_3	3.88	*b*	$CNCH_2CO_2H$	2.47	*h*	$(CH_3)_3COH$	18.0	*r*
H_2S	7.0	*a*	CH_2ClCO_2H	2.86	*f*	4-Nitroaniline	18.4	*n*
$H_2PO_4^-$	7.20	*a*	HCO_2H	3.75	*f*	$C_6H_5COCH_3$	19	*s*
HCN	9.21	*c*	2,4-Dinitrophenol	4.09	*h*	CH_3COCH_3	20	*k*
$B(OH)_3$	9.24	*a*	$C_6H_5CO_2H$	4.20	*h*	$C_6H_5C{\equiv}CH$	21	*t*
NH_4^+	9.25	Table 5						
H_2O_2	11.7	*a*	CH_3CO_2H	4.76	*f*	$CH_3CO_2C_2H_5$	~24	*g*
HPO_4^{2-}	12.4	*a*	$(CH_3)_3CCO_2H$	5.03	*i*	CH_3CN	~25	*g*
HS^-	14.0	*a*	C_6H_5SH	6.5	*f*	$HC{\equiv}CH$	~26	*t*
H_2O	15.7	see text	4-Nitrophenol	7.15	*f*	$C_6H_5NH_2$	~27	*f*
			2-Nitrophenol	7.23	*f*	$(C_6H_5)_3CH$	31.5	*u*
OH^-	~21	*a*	3-Nitrophenol	8.36	*j*	CH_3SOCH_3	33	*v*
NH_3	~33	*a*	$CH_3COCH_2COCH_3$	9	*k*	$C_6H_5CH_3$	~41	*w*
			Hexafluoropropan-2-ol	9.3	*l*	Benzene	~43	*x*
			Phenol	10.0	*h*	Methane	~48	*y*
			CH_3NO_2	10.2	*f*	Cyclohexane	~51	*z*
			C_2H_5SH	10.6	*m*			

Notes

[a] Ref. 10.

[b] Carbon dioxide dissolved in water produces H_2CO_3, carbonic acid: $CO_2 + H_2O \rightleftharpoons H_2CO_3$
The $pK_{AH} = 3.88$ refers to the *subsequent* acid dissociation:[10] $H_2CO_3 + H_2O \rightleftharpoons H_3O^+ + HCO_3^-$
The *apparent* pK_{AH} based upon the total dissolved carbon dioxide $2H_2O + CO_2 \rightleftharpoons H_3O^+ + HCO_3^-$ is 6.35 (ref. 10).

[c] R. H. Boyd and C. H. Wang, *J. Am. chem. Soc.* **87,** 430 (1965).

[d] O. W. Webster, *J. Am. chem. Soc.* **88,** 3046 (1966).

[e] R. H. Boyd, *J. phys. Chem.* **67,** 737 (1963).

[f] Ref. 11.

[g] R. G. Pearson and R. L. Dillon, *J. Am. chem. Soc.* **75,** 2439 (1953).

[h] Ref. 12.

[i] D. H. Everett, D. A. Landsman, and B. R. W. Pinsent, *Proc. R. Soc. A* **215,** 403 (1952).

[j] R. A. Robinson and A. Peiperl, *J. phys. Chem.* **67,** 2860 (1963).

[k] Ref. 6, p. 105.

[l] W. J. Middleton and R. V. Lindsey, *J. Am. chem. Soc.* **86,** 4948 (1964).

[m] R. J. Irving, L. Nelander, and I. Wadsö, *Acta chem. Scand.* **18,** 769 (1964).

[n] Ref. 13.

[o] P. Ballinger and F. A. Long, *J. Am. chem. Soc.* **82,** 795 (1960).

[p] S. Takahashi, L. A. Cohen, H. K. Miller, and E. G. Peake, *J. org. Chem.* **36,** 1205 (1971).

[q] A. Streitwieser and L. L. Nebenzahl, *J. Am. chem. Soc.* **98,** 2188 (1976).

[r] Value based upon several published results given in note *p*.

[s] D. J. Cram, *Fundamentals of carbanion chemistry*, p. 4. Academic Press, New York (1964).

[t] N. S. Wooding and W. C. E. Higginson, *J. chem. Soc.* 774 (1952).

[u] A. Streitwieser, E. Ciuffarin, and J. H. Hammons, *J. Am. chem. Soc* **89,** 63 (1967).

[v] R. Stewart and J. R. Jones, *J. Am. chem. Soc.* **89,** 5069 (1967).

[w] A. Streitwieser, M. R. Granger, F. Mares, and R. A. Wolf, *J. Am. chem. Soc.* **95,** 4257 (1973).

[x] A. Streitwieser, P. J. Scannon, and H. M. Niemeyer, *J. Am. chem. Soc.* **94,** 7936 (1972).

[y] Estimated to be about 10^3 times more acidic than cyclohexane, A. Streitwieser and D. R. Taylor, *Chem. Commun.* 1248 (1970).

[z] Estimated to be about 10^8 times less acidic than benzene, A. Streitwieser, W. R. Young, and R. A. Caldwell, *J. Am. chem. Soc.* **91,** 527 (1969).

solution which allows a reasonable and proper comparison with the pK_{AH} values of other acids.

There are several methods for measuring dissociation constants of acids[7,10–12,14] but all rely upon the determination of either the activity or the concentration of *dissociated* ions. A compound may *ionize* to an *intimate* or *contact ion-pair* which then dissociates to only a limited extent. The dissociation constant of hydrogen fluoride in water is anomalously low ($pK_{AH} = 3.18$; $K_{AH} = 6.61 \times 10^{-4}$) but this does not indicate a low extent of ionization:[9]

$$HF + H_2O \underset{1}{\overset{\text{ionization}}{\rightleftharpoons}} H_3O^+F^- \underset{2}{\overset{\text{dissociation}}{\rightleftharpoons}} H_3O^+ + F^- \qquad (5.9)$$

$$K_{AH} = \frac{[H_3O^+][F^-]}{[HF]} = \frac{[H_3O^+][F^-]}{[H_3O^+F^-]} \times \frac{[H_3O^+F^-]}{[HF]}.$$

Rather, the ion-pair formed in step 1 of eqn (5.9) undergoes only a low extent of subsequent dissociation in step 2. Because of the possible incomplete dissociation of ion-pairs in general, the terms *degree of ionization* or *ionization constant* should not be used for the overall ionization–dissociation process.

Example. The pH of an aqueous 0.035 molar solution of 3-nitrophenol is 4.91; calculate its pK_{AH} and percentage dissociation under these conditions.

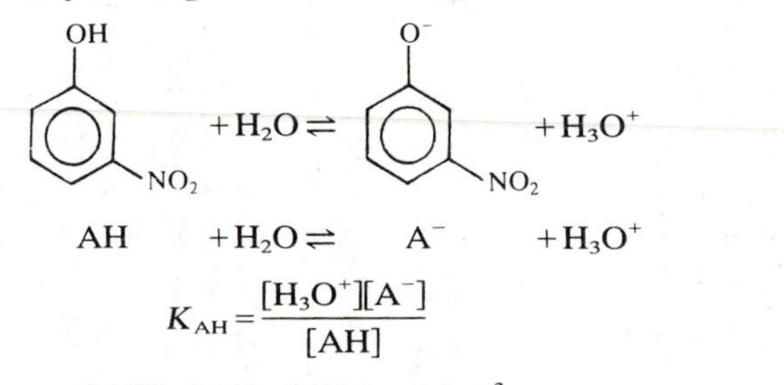

$$K_{AH} = \frac{[H_3O^+][A^-]}{[AH]} \qquad (5.8)$$

$$[AH] + [A^-] = 0.035 \text{ mol dm}^{-3},$$

and, since the solution is electrically neutral,

$$[OH^-] + [A^-] = [H_3O^+].$$

Substituting for $[A^-]$ and $[AH]$ in eqn (5.8) we obtain

$$K_{AH} = \frac{[H_3O^+]([H_3O^+] - [OH^-])}{\{0.035 - ([H_3O^+] - [OH^-])\}}.$$

$$[H_3O^+] = \text{antilog}(-4.91) = 1.23 \times 10^{-5} \text{ mol dm}^{-3},$$

and $[H_3O^+][OH^-] = 10^{-14}$, therefore

$$K_{AH} = \frac{(1.23 \times 10^{-5})^2 - 10^{-14}}{(0.035 - 1.23 \times 10^{-5} + 10^{-14}/1.23 \times 10^{-5})}.$$

We may approximate this to give

$$K_{AH} = \frac{(1.23 \times 10^{-5})^2}{0.035}$$
$$= 4.32 \times 10^{-9}$$

therefore

$$\underline{pK_{AH} = 8.36.}$$

The percentage dissociation under these conditions is given by

$$\frac{[A^-]}{0.035} \times 100 = \frac{([H_3O^+] - [OH^-])}{0.035} \times 100$$
$$\sim \frac{[H_3O^+]}{0.035} \times 100$$
$$= \frac{1.23 \times 10^{-5}}{0.035} \times 100$$
$$\underline{= 3.5 \times 10^{-2} \text{ per cent.}}$$

Table 5.4 gives approximate results for the percentage dissociation of acids of different pK_{AH} values at different total concentrations in water, and the pH of the resulting solutions. It also shows approximate values for the standard molar free-energy change for the acidic dissociation at 298.15 K, $\Delta_{AH}G^{\ominus}_{298\,K}$ (standard state of solutes = 1 mol dm^{-3}).

$$AH + H_2O \underset{H_2O}{\overset{T}{\rightleftharpoons}} H_3O^+ + A^-; \quad \Delta_{AH}G^{\ominus}_T$$

$$\Delta_{AH}G^{\ominus}_T = -RT \ln K^{\ominus}_a$$

Taking $\ln K^{\ominus}_a = 2.303 \log K^{\ominus}_a$ and $K^{\ominus}_a \sim K_{AH}$,

$$\Delta_{AH}G^{\ominus}_T \sim 2.303\, RT\, pK_{AH}$$

or

$$\underline{\Delta_{AH}G^{\ominus}_{298\,K} \sim 5.71\, pK_{AH}\ \text{kJ mol}^{-1}.}$$

If $[AH]_0$ = concentration of AH at extent of reaction = 0, then $[AH]_0 = [AH] + [A^-]$, where [AH] and $[A^-]$ = concentrations of the undissociated acid and its conjugate base at equilibrium. Consequently,

$$\text{percentage dissociation} = \frac{[A^-]}{[AH]_0} \times 100.$$

The results of Table (5.4) demonstrate that dissociation of acids with $pK_{AH} \leqslant \sim -2$ is virtually complete in dilute aqueous solution and, therefore, independent of concentration. At any given (low) concentration, the pH of a solution of one of these so-called *strong* acids is independent of its identity and its pK_{AH}, and it is not possible to have any appreciable concentration of an undissociated strong acid in dilute aqueous solution.

Acids whose dissociation is incomplete at finite dilution and dependent upon concentration are called *weak* acids; at any given concentration, the pH of a solution of a weak acid depends upon its pK_{AH}. Notice that the transition from a weak to a strong acid takes place over a relatively narrow pK_{AH} range (~2–3 pK_{AH} units). The results of Table 5.4 also show that once $pK_{AH} \geqslant \sim 15$, an acid is unable to protonate water to any perceptible extent, and dilute aqueous solutions of such compounds will appear to be non-acidic with pH ~ 7.

TABLE 5.4

Standard molar free-energy change, percentage dissociation, and approximate pH *of the solutions for the dissociation of acids of different* pK_{AH} *in water at* 298.15 K

pK_{AH}	$\Delta_{AH}G^{\ominus}_{298\,K}$/kJ mol^{-1}	% Dissociation (and pH) at total concentrations $[AH]_0$		
		1 mol dm^{-3}	0.1 mol dm^{-3}	0.01 mol dm^{-3}
−3	−17.1	99.9 (0.00)	99.99 (1.00)	99.999 (2.00)
−2	−11.4	99 (0.00)	99.9 (1.00)	99.99 (2.00)
−1	−5.7	92 (0.04)	99 (1.00)	99.9 (2.00)
0	0	62 (0.21)	92 (1.04)	99 (2.00)
1	5.7	27 (0.57)	62 (1.21)	92 (2.04)
2	11.4	9.5 (1.02)	27 (1.57)	62 (2.21)
3	17.1	3.1 (1.51)	9.5 (2.02)	27 (2.57)
5	28.5	0.32 (2.50)	1.0 (3.00)	3.1 (3.51)
8	45.7	0.01 (4.00)	0.03 (4.50)	0.10 (5.00)
11	62.8	3×10^{-4} (5.50)	10^{-3} (6.00)	3×10^{-3} (6.48)
14	79.9	$<10^{-5}$ (6.85)	$<10^{-5}$ (6.98)	10^{-5} (7.00)
15	85.6	$<10^{-6}$ (6.98)	$<10^{-6}$ (7.00)	10^{-6} (7.00)
16	91.3	$<10^{-7}$ (7.00)	$<10^{-7}$ (7.00)	10^{-7} (7.00)

Example. Calculate the percentage degree of dissociation and pH of a 3.88×10^{-2} mol dm^{-3} solution of dichloroacetic acid ($pK_{AH}=1.26$)

$$CHCl_2CO_2H+H_2O\rightleftharpoons CHCl_2CO_2^-+H_3O^+$$

$$AH+H_2O\rightleftharpoons A^- \quad + \quad H_3O^+ \tag{5.6}$$

$$K_{AH}=\frac{[A^-][H_3O^+]}{[AH]}=\text{antilog}(-1.26)=5.50\times10^{-2}$$

and

$$[AH]+[A^-]=3.88\times10^{-2}\ \text{mol dm}^{-3}.$$

If we assume that, compared with the dissociation of dichloroacetic acid, the autoprotolysis of water does not contribute appreciably to $[H_3O^+]$, then by eqn (5.6) $[H_3O^+]\sim[A^-]$, so

$$5.50\times10^{-2}=\frac{[A^-]^2}{(3.88\times10^{-2}-[A^-])}$$

$$[A^-]^2+5.50\times10^{-2}\times[A^-]-2.13\times10^{-3}=0$$

and the positive root of this quadratic equation is

$$\underline{[A^-]=2.62\times10^{-2}\ \text{mol dm}^{-3}.}$$

$$\text{The percentage degree of dissociation}=\frac{[A^-]}{3.88\times10^{-2}}\times100$$

$$=\underline{68\ \text{per cent}}$$

and, since $[A^-]\sim[H_3O^+]$,

$$pH=-\log(2.62\times10^{-2})$$

therefore

$$\underline{pH=1.58.}$$

In this example, we ignored the contribution of the dissociation of water itself towards $[H_3O^+]$. In view of the relative acid strengths of $CHCl_2CO_2H$ and H_2O, this is a good approximation. If, however, we are dealing with a much weaker acid, then the autoprotolysis of water cannot be ignored and the problem is mathematically more complicated. We have to deal with simultaneous equilibria and these lead to a cubic equation for $[H_3O^+]$.

$$AH + H_2O \rightleftharpoons H_3O^+ + A^-;$$

$$K_{AH} = \frac{[H_3O^+][A^-]}{[AH]} \tag{5.8}$$

$$2H_2O \rightleftharpoons H_3O^+ + OH^-;$$

$$K_{AP} = [H_3O^+][OH^-] \tag{5.10}$$

$$[AH]_0 = [AH] + [A^-], \tag{5.11}$$

where $[AH]_0$ = the molar concentration of the weak acid at degree of dissociation = 0, i.e. the total molar concentration of undissociated acid [AH] and conjugate base $[A^-]$ at equilibrium,
and, due to the overall electrical neutrality of the solution,

$$[H_3O^+] = [OH^-] + [A^-]. \tag{5.12}$$

From eqns (5.8) and (5.11)

$$K_{AH} = \frac{[H_3O^+][A^-]}{([AH]_0 - [A^-])}.$$

Incorporating eqn (5.12) we get

$$K_{AH} = \frac{[H_3O^+]([H_3O^+] - [OH^-])}{([AH]_0 - [H_3O^+] + [OH^-])}$$

and using eqn (5.10)

$$K_{AH} = \frac{[H_3O^+]([H_3O^+] - K_{AP}/[H_3O^+])}{([AH]_0 - [H_3O^+] + K_{AP}/[H_3O^+])}. \tag{5.13}$$

Equation (5.13) may be rearranged to give

$$[H_3O^+]^3 + K_{AH}[H_3O^+]^2 - (K_{AP} + K_{AH}[AH]_0)[H_3O^+] - K_{AP} \cdot K_{AH} = 0.$$

Solution of this, which is not easy, gives $[H_3O^+]$ from which the pH is obtained. Then, by eqn (5.12), $[A^-]$ can be calculated and, finally, the degree of dissociation. This procedure was followed to obtain the results for weak acids in Table 5.4.

Example. Calculate (1) the percentage dissociation of 3-nitrophenol (pK_{AH} = 8.36) when it is dissolved in an aqueous solution buffered at pH 9.50, and (2) the pH of the buffered solution which causes 3-nitrophenol to be 50 per cent dissociated.

$$AH + H_2O \rightleftharpoons A^- + H_3O^+$$

$$K_{AH} = \frac{[A^-][H_3O^+]}{[AH]} \tag{5.8}$$

$$= \text{antilog}(-8.36) = 4.37 \times 10^{-9}.$$

At pH 9.50, $[H_3O^+] = \text{antilog}(-9.50) = 3.16 \times 10^{-10}\ \text{mol dm}^{-3}$.

(1) Let x = the equilibrium degree of dissociation, then $[A^-] = xM$ and $[AH] = (1-x)M$ where M = concentration of 3-nitrophenol at degree (or extent) of reaction = 0. From eqn (5.8),

$$4.37 \times 10^{-9} = \frac{x \times 3.16 \times 10^{-10}}{(1-x)}$$

$$\frac{x}{(1-x)} = 13.8$$

$$x = 13.8 - 13.8x$$

$$x = \frac{13.8}{14.8} = 0.93,$$

or degree of dissociation = 93 per cent.

Notice that when the dissociation of a weak acid is established in a buffered solution, $[A^-]$ and $[H_3O^+]$ are usually unequal.

(2) It follows from eqn (5.8) that

$$[H_3O^+] = K_{AH} \cdot \frac{[AH]}{[A^-]},$$

therefore

$$pH = pK_{AH} + \log\left\{\frac{[A^-]}{[AH]}\right\}.$$

In general, this equation (sometimes known as the Henderson equation) allows us to calculate the pH of a solution made up from known amounts of a weak acid and a salt of the weak acid. In the present example, however, we can use it to calculate the pH of the buffer which causes AH to become 50 per cent dissociated. (We assume that the buffer system itself contains no A^- or AH).

At 50 per cent dissociation, $[AH] = [A^-]$ therefore

$$\log\left\{\frac{[A^-]}{[AH]}\right\} = \log 1 = 0$$

and

$$pH = pK_{AH}.$$

So 3-nitrophenol becomes 50 per cent dissociated when added to an aqueous solution buffered at pH = 8.36.

5.3.3 Brönsted bases

The older convention for describing the strength of a base B was by its ability to abstract protons from water and generate hydroxide ions, eqn (5.14).

$$B + H_2O \overset{H_2O}{\rightleftharpoons} BH^+ + OH^-. \qquad (5.14)$$

This has the merit that it indicates quite clearly what happens when a strong neutral base is put into water. An obsolete practical equilibrium constant, the *basicity constant*, K_B, was given by

$$K_B = \frac{[BH^+][OH^-]}{[B]} \tag{5.15}$$

with

$$pK_B = -\log K_B.$$

This is redundant because information about the strength of a base B is included in the acidity constant of its conjugate acid, BH^+.

$$BH^+ + H_2O \overset{H_2O}{\rightleftharpoons} B + H_3O^+ \tag{5.16}$$

$$K_{AH} = \frac{[H_3O^+][B]}{[BH^+]}. \tag{5.17}$$

In other words, a single parameter, the acidity constant K_{AH} of an acid AH (or BH^+), indicates not only the strength of AH (or BH^+) as a proton donor, but also the strength of A^- (or B) as a proton acceptor.

A strong acid AH (or BH^+) corresponds to a weak conjugate base A^- (or B) and its pK_{AH} is small and perhaps even negative. And, expressed the other way round, a strong base A^- (or B) corresponds to a weak conjugate acid AH (or BH^+) whose pK_{AH} is large and positive.

The relationship between the obsolete basicity constant K_B of a base (still to be found in the older literature) and the acidity constant K_{AH} of its conjugate acid is a function of the solvent.

It follows from eqns (5.15) and (5.17) that, for aqueous systems, $K_{AH} \cdot K_B = [H_3O^+][OH^-] = K_{AP}(H_2O)$ so, at 25 °C, $pK_{AH} + pK_B = 14.00$.

The use of pK_{AH} to describe the strength of B (or A^-) as a base as well as the strength of BH^+ (or AH) as an acid is, of course, analogous to using pH to describe not only $[H_3O^+]$ but also $[OH^-]$.

There is, unfortunately, some scope in the literature for confusion over what exactly an acidity constant refers to when a compound is potentially amphoteric. For example, we may read that the pK of *p*-nitroaniline is 0.99 when the author means that the pK_{AH} of the *conjugate acid*, *p*-nitroanilinium cation is 0.99:

$$NO_2-C_6H_4-NH_3^+ + H_2O \rightleftharpoons NO_2-C_6H_4-NH_2 + H_3O^+ \tag{5.18}$$

$$K_{AH} = 0.102,$$

$$pK_{AH} = 0.99.$$

The pK_{AH} of *p*-nitroaniline itself (a very weak acid[13]) refers to the equation

$$NO_2-C_6H_4-NH_2 + H_2O \rightleftharpoons NO_2-C_6H_4-NH^- + H_3O^+$$

$$K_{AH} = 4.266 \times 10^{-19},$$

$$pK_{AH} = 18.37.$$

To avoid potential ambiguities of this sort, we shall refer to the K_{AH} (or pK_{AH}) of an amphoteric compound when the compound *itself* acts as an acid, i.e. donates a proton,

e.g.

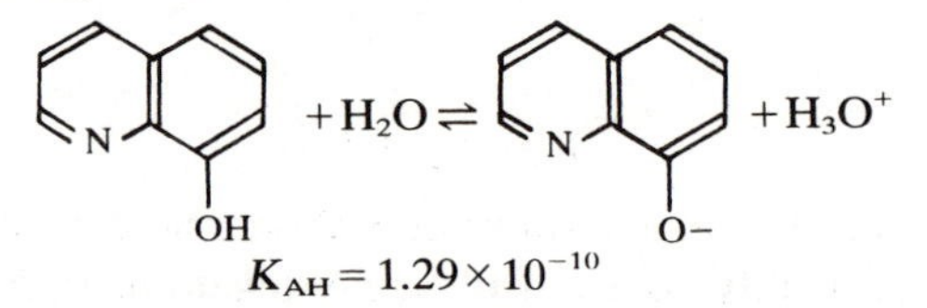

$$K_{AH} = 1.29 \times 10^{-10}$$

therefore the pK_{AH} of 8-hydroxyquinoline[11] = 9.89.

And we shall refer to the K_{BH} (or pK_{BH}) of an amphoteric compound when the compound itself acts as a *base* i.e. receives a proton,

e.g. (8-hydroxyquinolinium cation, N–H, OH) $+ H_2O \rightleftharpoons$ (8-hydroxyquinoline, OH) $+ H_3O^+$

$$BH^+ \quad + H_2O \rightleftharpoons \quad B \quad + H_3O^+$$

$$K_{BH} = \frac{[B][H_3O^+]}{[BH^+]} \tag{5.19}$$

$$K_{BH} = 7.41 \times 10^{-6},$$

or, pK_{BH} of 8-hydroxyquinoline[11] = 5.13.

Either way, the chemical equation is always written with the proton donor and the solvent molecule (H_2O) which receives the proton on the left-hand side, and the base and H_3O^+ on the right. According to this system, the pK_{BH} of a basic compound is the same as the pK_{AH} of its conjugate acid. So from eqn (5.18), for example, the pK_{BH} of *p*-nitroaniline is the same as the pK_{AH} of *p*-nitroanilinium cation = 0.99.

Example. Calculate the pK_{BH} of morpholine given that a 0.050 mol dm^{-3} aqueous solution has pH = 10.69 at 25 °C. Calculate the percentage protonation of morpholine in this solution.

(morpholinium cation, $\overset{+}{N}H_2$) $+ H_2O \underset{25\,°C}{\overset{H_2O}{\rightleftharpoons}}$ (morpholine, N–H) $+ H_3O^+$

$$BH^+ \quad + H_2O \rightleftharpoons \quad B \quad + H_3O^+$$

$$K_{BH} = \frac{[B][H_3O^+]}{[BH^+]}$$

$$[BH^+] + [B] = 0.050 \text{ mol dm}^{-3},$$

therefore

$$K_{BH} = \frac{(0.050 - [BH^+])[H_3O^+]}{[BH^+]}.$$

Since the solution is electrically neutral,

$$[BH^+]+[H_3O^+]=[OH^-],$$

therefore

$$K_{BH}=\frac{(0.050+[H_3O^+]-[OH^-])[H_3O^+]}{([OH^-]-[H_3O^+])}$$

$$[H_3O^+]=\text{antilog}(-10.69)=2.04\times10^{-11}\ \text{mol dm}^{-3}$$

and

$$[OH^-]=\text{antilog}(10.69-14.00)=4.90\times10^{-4}\ \text{mol dm}^{-3}$$

so

$$K_{BH}=\frac{(0.050+2.04\times10^{-11}-4.90\times10^{-4})\times2.04\times10^{-11}}{(4.90\times10^{-4}-2.04\times10^{-11})},$$

or, approximately,

$$K_{BH}=\frac{0.050\times2.04\times10^{-11}}{4.90\times10^{-4}}$$

$$=2.1\times10^{-9},$$

therefore

$$\underline{pK_{BH}=8.7.}$$

$$\text{The percentage protonation}=\frac{[BH^+]}{0.050}\times100$$

$$=\frac{([OH^-]-[H_3O^+])}{0.050}\times100\sim\frac{[OH^-]}{0.050}\times100$$

$$=\frac{4.90\times10^{-4}}{0.050}\times100$$

$$\underline{=1\ \text{per cent.}}$$

Notice that it can be misleading to refer to the *dissociation* of a base. Acids dissociate according to eqns (5.1) or (5.6), but bases become protonated and thereby induce the dissociation of an acid (eqn (5.1)) or the solvent (eqn (5.14)).

Some pK_{BH} values for bases are given in Table 5.5. As was mentioned earlier in the section on acids, pK_{AH} (pK_{BH}) values in only an intermediate and fairly narrow range can be determined by direct measurements of equilibria in aqueous solution.[7,10–12] Determinations of pK_{AH} (pK_{BH}) values outside this range have to be made by indirect methods[13,14,16] and results outside the approximate range $pK_{BH}\sim-2$ to 17 may be inaccurate.

The inclusion of water as a base in Table 5.5 follows by applying eqn (5.17) to the proton-transfer reaction between itself and its own conjugate acid,

TABLE 5.5

pK_{BH} values of some bases B (B^-) *in water*, 25 °C

Inorganic base, B (B^-)	pK_{BH}	Note	Organic base, B	pK_{BH}	Note	Organic base, B	pK_{BH}	Note
NH_2^-	~33	*a*	1,8-Bis(diethylamino)-2,7-			Pyridine oxide	0.8	*d*
O^{2-}	~21	*a*	dimethoxynaphthalene	~16.3	*c*	$C_6H_5NHCOCH_3$	0.4	*d*
OH^-	15.7	Table 4	$NH{=}C(NH_2)_2$	13.6	*d*	Urea	0.2	*d*
S^{2-}	14.0	*a*	$NH{=}C(CH_3)NH_2$	12.4	*d*	CH_3CONH_2	~0	*h*
NH_3	9.25	*b*	1,8-Bis(dimethylamino)-			2-Nitroaniline	−0.25	*g*
NH_2NH_2	8.0	*a*	naphthalene	12.1	*c*	Azulene	−1.7	*g*
HS^-	7.0	*a*	Quinuclidine	11.15	*b*	CH_3SOCH_3	−1.8	*g*
NH_2OH	6.0	*a*	$(C_2H_5)_2NH$	11.0	*b*	CH_3OH	−2.0	*g*
H_2O	−1.74	see	Me_2NH	10.8	*b*	$(CH_3)_3COH$	−3.8	*e*
		text	$(C_2H_5)_3N$	10.7	*b*	2,4-Dinitroaniline	−4.5	*g*
			$C_2H_5NH_2$	10.7	*b*	C_2H_5OH	−5.1	*i*
			CH_3NH_2	10.65	*b*	$(C_2H_5)_2O$	−5.1	*i*
			Me_3N	9.80	*b*	CH_3CO_2H	−6.2	*h*
			Imidazole	7.03	*d*	PhCHO	−7.1	*h*
			Pyridine	5.2	*e*	CH_3COCH_3	−7.2	*e*
			$C_6H_5NHCH_3$	4.85	*d*	Anthraquinone	−8.27	*h*
			$C_6H_5NH_2$	4.60	*f*	2,4,6-Trinitroaniline	−9.41	*h*
			$2\text{-}ClC_6H_4NH_2$	2.64	*d*	CH_3CN	−10.1	*j*
			3-Nitroaniline	2.47	*e*	$C_6H_5NO_2$	−11.3	*h*
			4-Nitroaniline	0.99	*g*	CH_3NO_2	−11.9	*h*
			$(C_6H_5)_2NH$	0.9	*d*			

Notes to Table 5.5

[a] Ref. 10.

[b] pK_{BH} values related to molar activity scale calculated from molal values given in ref. 15.

[c] F. Hibbert and K. P. P. Hunte, *J. chem Soc., Perkin Trans.* 2, 1562 (1981).

[d] Ref. 11.

[e] Ref. 16.

[f] P. D. Bolton and F. M. Hall, *Aust. J. Chem.* **20,** 1797 (1967).

[g] Ref. 2, pp. 145, 188.

[h] E. M. Arnett, *Progr. phys. org. Chem.* **1,** 223 (1963).

[i] E. M. Arnett, R. P. Quirk, and J. J. Burke, *J. Am. chem. Soc.* **92,** 1260 (1970).

[j] N. C. Deno, R. W. Gaugler, and M. J. Wisotsky, *J. org. Chem.* **31,** 1967 (1966).

the hydronium ion:

$$H_3O^+ + H_2O \underset{25\,°C}{\overset{H_2O}{\rightleftharpoons}} H_2O + H_3O^+$$

$$BH^+ + H_2O \rightleftharpoons B + H_3O^+$$

$$K_{BH} = \frac{[H_2O][H_3O^+]}{[H_3O^+]} = [H_2O] = 55.56$$

$$\underline{pK_{BH}(H_2O) = -1.74.}$$

Quantifying the base strength of water to allow legitimate comparisons with other bases in aqueous solution has required treating it in the above chemical equation in two ways. Water *as solute* in aqueous solution relates to the molar activity scale and its (hypothetical) standard state = 1 mol dm^{-3}. Mole fraction is the appropriate activity scale for water *as solvent* and its standard state is the pure liquid. This treatment is exactly analogous to that given on p. 165 for water as an acid.

Example. Calculate (1) the pH of an aqueous 0.0309 molar solution of imidazole ($pK_{BH} = 7.03$), and (2) the percentage protonation of imidazole when it is added to an aqueous buffer at pH = 6.23.

$$\text{(imidazolium) } HN\!\!\;\;NH^+ + H_2O \rightleftharpoons H_3O^+ + HN\;\;N \text{ (imidazole)}$$

$$BH^+ \quad + H_2O \rightleftharpoons H_3O^+ + \quad B$$

$$K_{BH} = \frac{[H_3O^+][B]}{[BH^+]} = \text{antilog}(-7.03) = 9.33 \times 10^{-8}.$$

(1) $[B] + [BH^+] = 0.0309$ mol dm^{-3} therefore

$$9.33 \times 10^{-8} = \frac{[H_3O^+](0.0309 - [BH^+])}{[BH^+]}.$$

But since the solution is electrically neutral, $[BH^+] + [H_3O^+] = [OH^-]$, therefore

$$9.33 \times 10^{-8} = \frac{[H_3O^+](0.0309 - [OH^-] + [H_3O^+])}{([OH^-] - [H_3O^+])}.$$

If $[OH^-]$ is replaced by $10^{-14}/[H_3O^+]$, we have a cubic equation which can be solved with difficulty for $[H_3O^+]$. However, since $[OH^-] \gg [H_3O^+]$ in alkaline solution, and both $[OH^-]$ and $[H_3O^+]$ are much smaller than 0.0309, the expression may be approximated to give

$$9.33 \times 10^{-8} = \frac{0.0309[H_3O^+]}{[OH^-]} = \frac{0.0309[H_3O^+]^2}{10^{-14}},$$

therefore

$$[H_3O^+]^2 = 3.02 \times 10^{-20}$$

or

$$[H_3O^+] = 1.74 \times 10^{-10} \text{ mol dm}^{-3}$$

and

$$\underline{pH = 9.76.}$$

(2) At pH 6.23, $[H_3O^+] = \text{antilog}(-6.23) = 5.89 \times 10^{-7}$ mol dm^{-3} therefore

$$9.33 \times 10^{-8} = \frac{[B] \times 5.89 \times 10^{-7}}{[BH^+]}$$

$$\frac{[BH^+]}{[B]} = 6.31.$$

So, if x = the equilibrium degree of protonation,

$$[BH^+] = xM \quad \text{and} \quad [B] = (1-x)M$$

where M = molar concentration of imidazole at degree (extent) of reaction = 0 therefore

$$\frac{x}{(1-x)} = 6.31$$

$$x = 6.31 - 6.31x$$

$$x = 0.86.$$

So imidazole is 86 per cent protonated in an aqueous solution buffered at pH = 6.23.

Table 5.6 contains pK_{AH} and pK_{BH} values of some amphoteric compounds.

Normally, there is no simple general relationship between an amphoteric compound's pK_{AH} and pK_{BH} values.

TABLE 5.6

pK_{BH} and pK_{AH} values of some amphoteric compounds, H_2O, 25 °C[a]

Compound	pK_{BH}	pK_{AH}
Water	−1.74	15.7
Sulphanilic acid[b,c] (4-aminobenzenesulphonic acid)	0.58	3.23
Cyanamide[c] (NH_2CN)	1.1	10.3
2-Hydroxypyridine[c]	1.25	12.0
Nicotinic acid[b] (pyridine-3-carboxylic acid)	2.07	4.73
Anthranilic acid[b] (2-aminobenzoic acid)	2.11	4.95
Glycine ($\overset{+}{N}H_3CH_2CO_2^-$)[c]	2.35	9.78
4-(4′-Imidazolyl)-butanoic acid[c]	4.26	7.62
5-Aminopentanoic acid[b]	4.27	10.8
8-Hydroxyquinoline[b] (20 °C)	5.13	9.89
4-Aminophenol[b]	5.50	10.3

[a] Other than water, only compounds which can actually exhibit acidic *and* basic properties in aqueous solution are included i.e. compounds with $pK_{BH} > -1$ and $pK_{AH} < 16$.

[b] Ref. 11.

[c] Ref. 16.

Example. Calculate from the data in Table 5.6 the relative proportions of the protonated, neutral, and deprotonated forms of 8-hydroxyquinoline in aqueous solutions buffered at (1) pH = 4.40, (2) pH = 7.50, and (3) pH = 10.90.

Let y and z = the degrees of protonation and deprotonation, respectively, of 8-hydroxyquinoline at equilibrium.

For 8-hydroxyquinoline (XH) acting as a base:

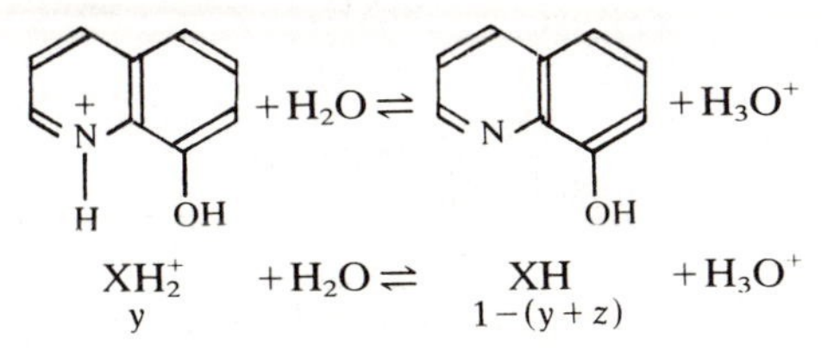

$$\underset{y}{XH_2^+} \quad + H_2O \rightleftharpoons \quad \underset{1-(y+z)}{XH} \quad + H_3O^+$$

For 8-hydroxyquinoline (XH) acting as an acid:

$+H_2O \rightleftharpoons$ $+H_3O^+$

N OH N O^-

$$\underset{1-(y+z)}{XH} \quad + H_2O \rightleftharpoons \quad \underset{z}{X^-} \quad + H_3O^+$$

$$K_{BH} = \frac{[XH][H_3O^+]}{[XH_2^+]} = \text{antilog}(-5.13) = 7.41 \times 10^{-6}$$

$$K_{AH} = \frac{[X^-][H_3O^+]}{[XH]} = \text{antilog}(-9.89) = 1.29 \times 10^{-10}.$$

(1) At pH = 4.40, $[H_3O^+] = \text{antilog}(-4.40) = 3.98 \times 10^{-5}$ mol dm^{-3}, therefore

$$\text{(i)}\ 7.41 \times 10^{-6} = \frac{[XH]}{[XH_2^+]} \times 3.98 \times 10^{-5},$$

so

$$\frac{[XH_2^+]}{[XH]} = 5.37 = \frac{y}{1-(y+z)},$$

and

$$\text{(ii)}\ 1.29 \times 10^{-10} = \frac{[X^-]}{[XH]} \times 3.98 \times 10^{-5},$$

so

$$\frac{[X^-]}{[XH]} = 3.24 \times 10^{-6} = \frac{z}{1-(y+z)}.$$

Clearly, at pH = 4.40, there is a negligible proportion of deprotonated 8-hydroxyquinoline ($z \sim 0$); the approximate proportion of the protonated form, y, is given by

$$\frac{y}{(1-y)} = 5.37$$

$$y = 5.37 - 5.37y$$

$$y = \frac{5.37}{6.37} = 0.84,$$

so 84 per cent of the 8-hydroxyquinoline is protonated in aqueous solution buffered at pH 4.40.

(2) At pH = 7.50, $[H_3O^+]$ = antilog(−7.50) = 3.16×10^{-8} mol dm^{-3}, therefore

$$\text{(i)}\ 7.41 \times 10^{-6} = \frac{[XH]}{[XH_2^+]} \times 3.16 \times 10^{-8},$$

so

$$\frac{[XH_2^+]}{[XH]} = 4.26 \times 10^{-3},$$

and

$$\text{(ii)}\ 1.29 \times 10^{-10} = \frac{[X^-]}{[XH]} \times 3.16 \times 10^{-8},$$

so

$$\frac{[X^-]}{[XH]} = 4.08 \times 10^{-3}.$$

At pH = 7.50, therefore, 8-hydroxyquinoline is neither protonated nor deprotonated to any appreciable extent.

(3) At pH = 10.90, $[H_3O^+]$ = antilog(−10.90) = 1.26×10^{-11} mol dm^{-3}, therefore

$$\text{(i)}\ 7.41 \times 10^{-6} = \frac{[XH]}{[XH_2^+]} \cdot 1.26 \times 10^{-11}$$

so

$$\frac{[XH_2^+]}{[XH]} = 1.70 \times 10^{-6} = \frac{y}{1-(y+z)},$$

and

$$\text{(ii)}\ 1.29 \times 10^{-10} = \frac{[X^-]}{[XH]} \times 1.26 \times 10^{-11}$$

so

$$\frac{[X^-]}{[XH]} = 10.2 = \frac{z}{1-(y+z)}.$$

At pH = 10.90, therefore, there is a negligible proportion of protonated 8-hydroxyquinoline ($y \sim 0$) and the approximate proportion, z, of the deprotonated form is given by

$$\frac{z}{1-z} = 10.2$$

$$z = 10.2 - 10.2z$$

$$z = \frac{10.2}{11.2} = 0.91,$$

so 8-hydroxyquinoline is 91 per cent deprotonated in an aqueous buffer at pH 10.90.

Table 5.7 shows the approximate relationship between the pK_{BH} defined by eqn (5.19), the standard molar free-energy change for the

TABLE 5.7

Standard molar free-energy change, percentage protonation, and approximate pH *of the solutions of bases of different* pK_{BH} *values in water at* 298.15 K

pK_{BH}	pK_B	$\Delta_B G^{\ominus}_{298\,K}$/ kJ mol^{-1}	% Protonation (and pH) at total concentrations $[B]_0$ 1 mol dm^{-3}	0.1 mol dm^{-3}	0.01 mol dm^{-3}
17	−3	−17.1	99.9 (14.00)	99.99 (13.00)	99.999 (12.00)
16	−2	−11.4	99 (14.00)	99.9 (13.00)	99.99 (12.00)
15	−1	−5.7	92 (13.96)	99 (13.00)	99.9 (12.00)
14	0	0	62 (13.79)	92 (12.96)	99 (12.00)
13	1	5.7	27 (13.43)	62 (12.79)	92 (11.96)
12	2	11.4	9.5 (12.98)	27 (12.43)	62 (11.79)
11	3	17.1	3.1 (12.49)	9.5 (11.98)	27 (11.43)
9	5	28.5	0.32 (11.50)	1.0 (11.00)	3.1 (10.49)
6	8	45.7	0.01 (10.00)	0.03 (9.5)	0.10 (9.0)
3	11	62.8	3×10^{-4} (8.50)	10^{-3} (8.00)	3×10^{-3} (7.52)
0	14	79.9	$<10^{-5}$ (7.15)	$<10^{-5}$ (7.02)	10^{-5} (7.00)
−1	15	85.6	$<10^{-6}$ (7.02)	$<10^{-6}$ (7.00)	10^{-6} (7.00)
−2	16	91.3	$<10^{-7}$ (7.00)	$<10^{-7}$ (7.00)	10^{-7} (7.00)

reaction in which the base abstracts a proton from water, $\Delta_B G^{\ominus}_{298\,K}$ (eqn (5.14), standard state = 1 mol dm^{-3}), and the degree of protonation of the base, according to eqn (5.14), at different total base concentrations. The pH of each solution is also given.

$$B + H_2O \underset{T}{\overset{H_2O}{\rightleftharpoons}} BH^+ + OH^- \quad (5.14)$$

$$\Delta_B G^{\ominus}_T = -RT \ln K^{\ominus}_a.$$

Taking $\ln K^{\ominus}_a = 2.303 \log K^{\ominus}_a$ and $K^{\ominus}_a \sim K_B$,

$$\Delta_B G^{\ominus}_T \sim 2.303RT\, pK_B.$$

But $pK_B + pK_{BH} = 14$ at 25 °C, therefore

$$\Delta_B G^{\ominus}_T \sim 2.303RT(14 - pK_{BH}),$$

or

$$\underline{\Delta_B G^{\ominus} \sim 5.71(14 - pK_{BH}) \text{ at } 298.15\text{ K}.}$$

If $[B]_0$ = the concentration of B at extent of reaction (eqn (5.14)) = 0, then $[B]_0 = [B] + [BH^+]$ where $[B]$ and $[BH^+]$ are the concentrations of the base B and its conjugate acid BH^+ at equilibrium, and the percentage protonation (eqn (5.14)) $= \dfrac{[BH^+]}{[B]_0} \times 100$.

When $pK_{BH} \geqslant \sim 16$, the dissolved base is virtually completely protonated by water in dilute solution to give an equivalent concentration of OH^-. Consequently, such strong bases cannot exist in the unprotonated form to any appreciable extent in dilute aqueous solution, and the pH of their solutions depends only upon concentration being independent of the nature or pK_{BH} of the base. There are, however, very few neutral organic bases as strong as this. At the other extreme, water to which is added a base with $pK_{BH} \leqslant \sim -1$ will not appear alkaline at all (pH = 7) since the compound cannot deprotonate water to any noticeable extent and so does not generate any OH^-.

Example. Calculate the pH of an aqueous solution of glycine ($pK_{AH} = 9.78$, $pK_{BH} = 2.35$).

Glycine (XH) acting as a base:

$$\overset{+}{N}H_3CH_2CO_2H + H_2O \rightleftharpoons \overset{+}{N}H_3CH_2CO_2^- + H_3O^+$$
$$XH_2^+ \quad + H_2O \rightleftharpoons \quad XH \quad + H_3O^+$$

Glycine (XH) acting as an acid:

$$\overset{+}{N}H_3CH_2CO_2^- + H_2O \rightleftharpoons NH_2CH_2CO_2^- + H_3O^+$$
$$XH \quad + H_2O \rightleftharpoons \quad X^- \quad + H_3O^+$$

$$K_{BH} = \frac{[XH][H_3O^+]}{[XH_2^+]} = \text{antilog}(-2.35) = 4.47 \times 10^{-3}$$

$$K_{AH} = \frac{[X^-][H_3O^+]}{[XH]} = \text{antilog}(-9.78) = 1.66 \times 10^{-10}.$$

If the total glycine concentration is M mol dm^{-3}, then

$$[XH] + [XH_2^+] + [X^-] = M,$$

and, since the solution is overall electrically neutral,

$$[XH_2^+] + [H_3O^+] = [X^-] + [OH^-] \tag{5.20}$$

and

$$[H_3O^+][OH^-] = 10^{-14}.$$

These five simultaneous equations may be solved exactly to obtain $[H_3O^+]$ and hence the pH for a particular value of M but, mathematically, it is not simple. A *chemical* approximation makes the problem very much easier. We assume that $[HX] \sim M$ i.e. that when added to pure water, only a very small proportion of the glycine suffers protonation or deprotonation. This is reasonable in view of the magnitudes of K_{BH} (glycine is a very weak base) and K_{AH} (it is also a weak acid). Thus

$$K_{BH} = \frac{M \times [H_3O^+]}{[XH_2^+]}, \quad \text{so} \quad [XH_2^+] = \frac{M \times [H_3O^+]}{4.47 \times 10^{-3}}$$

and

$$K_{AH} = \frac{[X^-][H_3O^+]}{M}, \quad \text{so} \quad [X^-] = \frac{M \times 1.66 \times 10^{-10}}{[H_3O^+]}.$$

From eqn (5.20), therefore,

$$\frac{M \times [H_3O^+]}{4.47 \times 10^{-3}} + [H_3O^+] = \frac{M \times 1.66 \times 10^{-10}}{[H_3O^+]} + \frac{10^{-14}}{[H_3O^+]}$$

or

$$[H_3O^+]^2(1 + 224M) = 1.66 \times M \times 10^{-10} + 10^{-14}. \tag{5.21}$$

We see that, as long as M is not too low, this equation may be approximated to

$$224[H_3O^+]^2 = 1.66 \times 10^{-10}$$
$$[H_3O^+]^2 = 7.41 \times 10^{-13}$$
$$[H_3O^+] = 8.61 \times 10^{-7}$$
$$\underline{pH = 6.07.}$$

This demonstrates that as long as the above approximations remain good (low extents of protonation and deprotonation, and a total concentration which is not too low, say $M \geqslant \sim 0.05$ mol dm^{-3}), the pH of the glycine solution is independent of its concentration. In general, an amphoteric material in water gives a buffered solution the pH of which has the value of the arithmetic mean of the compound's pK_{AH} and pK_{BH}.

If we do not wish to make the last approximation regarding the concentration, we simply put the value for M into eqn (5.21) and solve it for $[H_3O^+]$.

We can also check the quality of the first approximation involved in this calculation. From the calculated pH of the solution, glycine is somewhat stronger as an acid than as a base, so $[\overset{+}{N}H_3CH_2CO_2H] < [NH_2CH_2CO_2^-]$. And since

$$K_{AH} = \frac{[X^-][H_3O^+]}{[HX]},$$

$$1.66 \times 10^{-10} = \frac{[X^-] \times \text{antilog}(-6.07)}{[HX]},$$

therefore

$$1.66 \times 10^{-10} = \frac{[X^-] \times 8.51 \times 10^{-7}}{M - [X^-]},$$

or

$$\frac{[X^-]}{(M - [X^-])} = 1.95 \times 10^{-4}.$$

At $M = 0.1$ mol dm^{-3}, this gives $[X^-] \sim 2 \times 10^{-5}$ mol dm^{-3} so glycine is only 2×10^{-2} per cent dissociated as an acid; and protonation of glycine to give XH_2^+ takes place to an even smaller extent. The first approximation, therefore, is sound.

5.3.4 Solvent effects upon Brönsted acid–base equilibria[17]

The range of acids and bases which may be used in a solvent is restricted by the solvent's own amphoteric properties. We have already seen that it is not possible to have an appreciable concentration of an acid in water which is a stronger proton donor than H_3O^+ or a base which is a stronger proton acceptor than OH^-. Acids with $pK_{AH} \leqslant \sim -2$ are completely dissociated in water; water, therefore, is said to have a *levelling effect* upon strong acids because it is too basic to differentiate them. Thus sulphuric and hydriodic acids with a difference in acidity of 10^6 ($pK_{AH} = -3$ and -9 respectively) appear equally strong in dilute aqueous solution. In contrast, monochloro- and dichloro-acetic acids have very much closer pK_{AH} values (2.86 and 1.26 respectively), yet at 0.1 M the former is 11 per cent dissociated (pH = 1.96) and the latter 52 per cent (pH = 1.29). These similar but more weakly acidic compounds are, therefore, easily differentiated by water.

In acetic acid, a solvent which is less basic than water, only perchloric acid of the common ones is completely dissociated. Nitric acid and hydrogen chloride are dissociated to only small extents (which, of course, increase with dilution) and are, in this weakly basic solvent, only weak acids.

Correspondingly, acids which are weak in water may dissociate fully in a solvent which is more basic (with a greater acid-levelling effect) such as liquid ammonia or a simple alkylamine. Indeed, in such liquids, all acids with aqueous $pK_{AH} \leqslant 5$, i.e. anything stronger than acetic acid will appear equally strong and fully dissociated. It is important to realize, therefore, that when an acid is termed strong or weak on the basis of its pK_{AH} related to dilute aqueous solution, we are told only about its dissociation in water. In a solvent of different base strength, the dissociation will be different, and for an accurate calculation of a degree of dissociation of an acid at a particular concentration in a non-aqueous solvent, the pK_{AH} of the acid in that solvent must be known. Unfortunately, such data are available for only a few solvents; representative pK_{AH}s are given in Table 5.8.

Solvents exert corresponding effects upon reactions of bases. Water has a levelling effect upon bases with $pK_{BH} \geqslant \sim 16$. So $NH_2^-K^+$, $C_2H_5O^-Li^+$, and $CH_3SOCH_2^-Na^+$ all react with water to give their conjugate acids, NH_3, C_2H_5OH, and CH_3SOCH_3, plus stoichiometric amounts of OH^-. But in a much less acidic solvent, e.g. $C_2H_5NH_2$ ($pK_{AH} \sim 33$), bases stronger than OH^- may be used without being protonated by the solvent.

Correspondingly, if a more acidic solvent than water is used, bases which are weak in water may be very extensively protonated. For example, in H_2SO_4,[17] bases as weak as aromatic amines are almost completely protonated:

$$H_2SO_4 + ArNH_2 \xrightleftharpoons{H_2SO_4} HSO_4^- + ArNH_3^+.$$

Indeed, compounds such as nitric acid and carboxylic acids act as bases in

TABLE 5.8
pK_{AH} values of some compounds in different solvents 25 °C

$$AH + S \overset{S}{\rightleftharpoons} SH^+ + A^-$$

$$K_{AH}(S) = \frac{[SH^+][A^-]}{[AH]}$$

$$pK_{AH}(S) = -\log K_{AH}(S)$$

Acid	$pK_{AH}(S)$		
	H_2O[a]	CH_3OH	DMSO[b]
HCl[c]	−7	1.2	2.0
Picric acid[c]	0.71	3.8	−0.3
$CHCl_2CO_2H$[c]	1.26	6.4	5.9
$3\text{-}NO_2\text{-}C_6H_4\overset{+}{N}H_3$[c,d]	2.47	2.89	1.30
Benzoic acid[c]	4.20	9.1	11.0
CH_3CO_2H[c]	4.76	9.6	12.6
$4\text{-}NO_2\text{-}C_6H_4OH$[e]	7.15	11.2	10.4
$\overset{+}{N}H_4$[e]	9.25	10.8	10.5
Phenol[c]	10.00	14.2	16.4
$(C_2H_5)_3\overset{+}{N}H$[e]	10.72	10.9	9.0
CH_3OH	15.5	18.2[f]	29.0[g]
H_2O	15.7	18.6[h]	31.4[g]
CH_3SOCH_3	33	—	35.1[g]

[a] $pK_{AH}(H_2O)$ results taken from Tables 5.3 and 5.5.
[b] DMSO = dimethyl sulphoxide, CH_3SOCH_3.
[c] Ref. 18, p. 249.
[d] Ref. 19. p. 268.
[e] C. D. Ritchie, *J. Am. chem. Soc.* **91,** 6749 (1969).
[f] C. H. Rochester, *J. chem. Soc., Dalton Trans.* 5 (1972).
[g] W. N. Olmstead, Z. Margolin, and F. G. Bordwell, *J. org. Chem.* **45,** 3295 (1980).
[h] Value estimated from results given in note *f*.

H_2SO_4:

$$H_2SO_4 + HNO_3 \xrightleftharpoons{H_2SO_4} HSO_4^- + ON(OH)_2^+$$

$$H_2SO_4 + RCO_2H \xrightleftharpoons{H_2SO_4} HSO_4^- + RC(OH)_2^+.$$

Other acidic solvents include anhydrous acetic and hydrofluoric acids.[17]

These effects of the acidic and basic properties of solvents upon the dissociation of acids and bases, and the relationships with the aqueous system are illustrated in Fig. 5.1.

Because of the amphoteric nature of many common solvents, one which is more acidic than water is often the same as one which is less

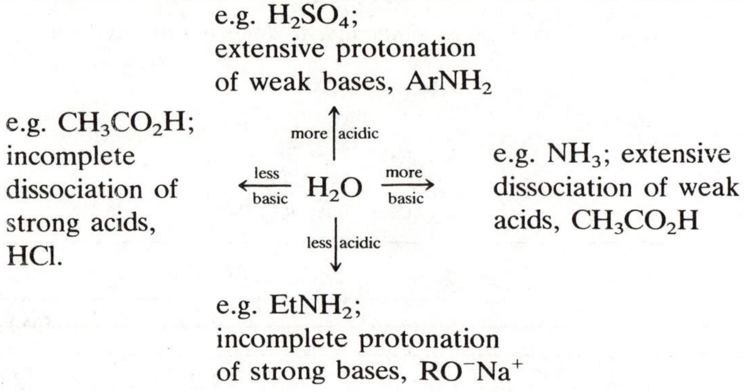

e.g. H_2SO_4;
extensive protonation
of weak bases, $ArNH_2$

more acidic

e.g. CH_3CO_2H;
incomplete
dissociation of
strong acids,
HCl.

less basic $\leftarrow$ H_2O $\rightarrow$ more basic

e.g. NH_3; extensive
dissociation of weak
acids, CH_3CO_2H

less acidic

e.g. $EtNH_2$;
incomplete protonation
of strong bases, RO^-Na^+

Fig. 5.1. Effect of solvents upon acid–base reactions.

basic, e.g. CH_3CO_2H, HF, H_2SO_4; and a solvent which is more basic than water is usually less acidic also, e.g. NH_3, $C_2H_5NH_2$.

As mentioned at the beginning of this section, it is the amphoteric nature of a solvent which restricts the range of acids and bases which may be used in it. If a solvent is not appreciably acidic, then it could be used for a very wide range of bases (provided they dissolve in it). Analogously, a solvent with no significant basic properties does not restrict the range of acids which may be used in it. However, liquids without any acidic or basic properties are unlikely to be very good solvents, and this practical aspect may severely limit their use. For example, strong acids in water such as methanesulphonic acid, CH_3SO_3H, or HNO_3, and strong bases such as $(CH_3)_3CO^-K^+$ or $NH{=}C(NH_2)_2$ do not readily dissolve in hydrocarbons, solvents which, typically, show no appreciable acidic or basic tendencies.

The so-called dipolar aprotic solvents such as dimethyl sulphoxide or dimethylformamide are very useful in this respect, since they are good solvents and yet much less acidic or basic than water or other protic solvents.[17,18,20]

5.4 Gas-phase proton-transfer reactions

Investigations in the gas phase by a range of mass spectrometric techniques of diverse organic chemical reactions previously known only in solution have been an important recent development. Results indicate that many trends and relative reactivities in solution, formerly interpreted in terms of direct effects of intrinsic electronic and steric properties of molecules upon the nature of the reaction, are actually due to differing solvation properties of reactants and activated complexes. Furthermore, these trends and orders of reactivities in solution often reside in entropy rather than enthalpy (energy) effects. We shall not deal in this book with

$$\mathrm{HX(g)} \xrightleftharpoons[]{\text{gas phase dissociation}} \mathrm{H^+(g) + X^-(g)} \tag{5.22}$$

solvation ⇅ (i) desolvation desolvation ⇅ (ii) solvation

$$\mathrm{HX\ (solv.)} \xrightleftharpoons[]{\text{solution dissociation}} \mathrm{H^+\ (solv.) + X^-\ (solv.)} \tag{5.23}$$

Fig. 5.2. Dissection of acid dissociation in solution into gas-phase dissociation and solvation steps.

the range of reactions which have been, or are currently being, researched in the gas phase. But the following account of proton transfer, one of the most extensively investigated of these reactions, should provide a sound introduction and background for further study of other reactions in this important field.

Acid strengths of compounds in a solvent S are compared via the acidity constants K_{AH} (or pK_{AH}) which relate to equilibria of the type

$$\mathrm{S + HX} \overset{\mathrm{S}}{\rightleftharpoons} \mathrm{SH^+ + X^-}$$

and it is implicit that all species are solvated. In particular, the proton from the acid HX does not become free, but bonds covalently to a solvent molecule, and this complex cation is itself solvated by further solvent molecules.

We can dissect the overall solution dissociation of an acid XH as shown in Fig. 5.2. This illustrates the relationship between the reaction in the gas phase, eqn (5.22), and the overall solution reaction, eqn (5.23). In the case of an acid which is extensively dissociated in solution (eqn (5.23) favourable from left to right), the decrease in standard free energy brought about by the solvation of H^+ and X^- (step (ii)) more than compensates for the increase in standard free energy required for the initial desolvation of HX (step (i)) and the heterolytic gas-phase dissociation (eqn (5.22) from left to right).

Instrumental developments in the general area of mass spectrometry during the 1960s and 1970s allowed gas-phase proton-transfer equilibria to be investigated.[21,22] In the gas phase at low pressure (but not so low that chemical equilibrium is not achieved), there is no solvation. Consequently, the relationship between a compound's *intrinsic acidity* and its molecular structure can be isolated. This has the effect of exposing the dependence of the compound's solution acidity upon solvation.

One measure of the gas-phase acidity of a compound is the *proton affinity* (PA) of its conjugate base.[2] Originally, it was defined as the negative of the standard heat change when the base combines with a proton,

$$\mathrm{X^-(g) + H^+(g)} \xrightarrow{298.15\ \mathrm{K}} \mathrm{XH(g)}; \qquad \Delta H^\circ = -\mathrm{PA(X^-)},$$

TABLE 5.9

Proton affinities of some bases in the gas phase, 25 °C[a]

$$X + H^+ \rightarrow HX^+; \quad \Delta H° = -PA(X)$$

or

$$X^- + H^+ \rightarrow HX; \quad \Delta H° = -PA(X^-)$$

Base	PA/kJ mol^{-1}	Base	PA/kJ mol^{-1}
CH_4	527	CH_3^-	1743[b]
NH_3	866	NH_2^-	1672[b]
H_2O	686	OH^-	1635[b]
CH_3OH	761	CH_3O^-	1587[c]
HF	548	F^-	1554[c]
HCl	586	Cl^-	1395[b]
HBr	590	Br^-	1354[b]
HI	607	I^-	1315[b]
H_2S	711	H_2CO	703
PH_3	774	$(CH_3)_2CO$	845
AsH_3	732	$CH_3CO_2C_2H_5$	858
C_2H_5OH	782	$(CH_3)_3COH$	828
CH_3CO_2H	787	$(C_2H_5)_2O$	858
CF_3CO_2H	699	Cyclopropane	749
$HC{\equiv}CH$	636	$CH_2{=}CH_2$	665
Benzene	745	Toluene	782

[a] All values taken from ref. 2, pp. 150 and 152 except where otherwise indicated.
[b] Ref. 23, p. 6050.
[c] Ref. 23, p. 6052.

and may be measured mass spectrometrically.[21,22] It is equal to the standard enthalpy change for the deprotonation of the acid in the gas phase:

$$XH(g) \xrightarrow{298.15\ K} H^+(g) + X^-(g); \quad \Delta H° = PA(X^-) \tag{5.24}$$

$$[\text{or } XH^+(g) \xrightarrow{298.15\ K} H^+(g) + X(g); \quad \Delta H° = PA(X)].$$

Some representative results are given in Table 5.9.

One immediate conclusion from the data in Table 5.9 is that combination of a proton with any anion or neutral molecule (even methane) is strongly exothermic in the gas phase. The corollary is that all neutral molecules (and even molecular cations) are only very weakly acidic in the absolute sense represented by eqn (5.24).

Because of the very low absolute acidities in the gas phase, we seldom deal with a simple deprotonation (eqn (5.22)). Usually we are concerned with the position of equilibrium in the reversible proton-transfer reaction, eqn (5.25).

$$XH + Y^- \rightleftharpoons X^- + HY; \quad \Delta_r G°; \quad K° \tag{5.25}$$

$$(\text{or } XH^+ + Y \rightleftharpoons X + HY^+).$$

This may be regarded as a competition between X^- and Y^- (or X and Y) for a proton. Equation (5.25) is, therefore, the resultant of two reactions set against each other:

$$XH \rightleftharpoons X^- + H^+; \quad \Delta_{AH}G^\circ(HX),$$

and

$$YH \rightleftharpoons Y^- + H^+; \quad \Delta_{AH}G^\circ(HY),$$

where $\Delta_{AH}G^\circ$ = the standard molar free-energy change of the gas-phase deprotonation; and $\Delta_r G^\circ$, the sign and magnitude of which describe the position of equilibrium in eqn (5.25), is given by

$$\Delta_r G^\circ = \Delta_{AH}G^\circ(HX) - \Delta_{AH}G^\circ(HY).$$

Equilibrium amounts of the different species in the gas-phase chemical equilibrium of eqn (5.25) may be measured by ion cyclotron resonance spectroscopy (a mass spectrometric technique) so an equilibrium constant, K°, is determined experimentally; this leads to $\Delta_r G^\circ$ for the reaction via $\Delta_r G^\circ = -RT \ln K^\circ$. By using an extensive range of compounds, a series of relative acidities has been built up from successive equilibria like eqn (5.25).

To put these on to an *absolute* scale of gas-phase acidities, $\Delta_{AH}G^\circ(HX)$ for at least one compound must be established. This has been done, in fact, for several compounds including hydrogen fluoride,[23] eqn (5.26), from $\Delta_{AH}H^\circ$ and $\Delta_{AH}S^\circ$ results obtained from thermochemical and statistical thermodynamic data.

$$HF \underset{298.15\,K}{\overset{gas}{\rightleftharpoons}} H^+ + F^-; \tag{5.26}$$

$$\Delta_{AH}H^\circ(HF) = 1554\ kJ\ mol^{-1}$$

$$\Delta_{AH}S^\circ(HF) = 81\ J\ K^{-1}\ mol^{-1}$$

$$\therefore\ \Delta_{AH}G^\circ_{298\,K}(HF) = 1530\ kJ\ mol^{-1}.$$

$\Delta_{AH}H^\circ$ for this reaction is, of course, the proton affinity of F^-, eqn (5.24) and Table 5.9. The standard free energy for the dissociation of hydrogen fluoride corresponds to pK_{AH}(gas phase) = 268, compared with $pK_{AH} = 3.18$ in water ($\Delta_{AH}G^\ominus_{298\,K} = 18.2\ kJ\ mol^{-1}$).

Absolute values of $\Delta_{AH}G^\circ$, $\Delta_{AH}H^\circ$, and $\Delta_{AH}S^\circ$ for the gas-phase deprotonation of a range of compounds are included in Table 5.10.

Table 5.10 shows that $\Delta_{AH}S^\circ$ values are relatively small and much the same for a wide range of compounds. Consequently, $T.\Delta_{AH}S^\circ$ contributions to $\Delta_{AH}G^\circ_T$ are quite small compared with $\Delta_{AH}H^\circ$, and differences in the gas-phase acidity between compounds reside almost entirely in the $\Delta_{AH}H^\circ$ values. This is why, as we anticipated earlier, the proton affinities of the conjugate bases shown in Table 5.9 (PA = $\Delta_{AH}H^\circ$) may be taken as good measures of gas-phase acidities.

TABLE 5.10
Thermodynamics of gas-phase acid dissociation[a]

$$XH(g) \xrightarrow{298.15\,K} H^+(g) + X^-(g); \quad \Delta_{AH}S^\circ, \Delta_{AH}H^\circ, \text{ and } \Delta_{AH}G^\circ_{298.15\,K}$$

Acid	$\Delta_{AH}S^\circ$/J K^{-1} mol^{-1}	$\Delta_{AH}H^\circ$/kJ mol^{-1}	$\Delta_{AH}G^\circ_{298.15\,K}$/kJ mol^{-1}
NH_3[b]	105	1672	1641
H_2O[c]	93	1635	1607
CH_3OH	92	1587	1559
$C_6H_5CH_3$	94	1586	1558
$CH_3C{\equiv}C{-}H$	109	1588	1556
C_2H_5OH	92	1574	1546
$(CH_3)_2CHOH$	92	1565	1538
$(CH_3)_3COH$	92	1562	1534
HF	81	1554	1530
CH_3SOCH_3	99	1559	1530
CH_3CN	110	1557	1525
$C_6H_5CH_2OH$	92	1546	1519
$C_6H_5C{\equiv}C{-}H$	109	1549	1517
CH_3COCH_3	100	1543	1513
$C_6H_5NH_2$	102	1536	1505
CH_3CHO	95	1533	1505
$(C_6H_5)_2CH_2$	75	1525	1502
CF_3CH_2OH	107	1525	1493
$C_6H_5COCH_3$	95	1520	1491
CH_3NO_2	94	1501	1473
Cyclopentadiene	87	1490	1464
H_2S	89	1479	1452
HCN	103	1477	1447
C_6H_5OH	96	1470	1441
HCl[c]	75	1395	1373
HNO_3[d]	98	1358	1328

[a] Results taken from ref. 23 except where otherwise indicated.
[b] $\Delta_{AH}H^\circ$ value from ref. 23 and $\Delta_{AH}S^\circ$ value from ref. 22(*a*), p. 103.
[c] $\Delta_{AH}H^\circ$ value from ref. 23 and $\Delta_{AH}S^\circ$ calculated from data in ref. 24.
[d] $\Delta_{AH}H^\circ$ and $\Delta_{AH}S^\circ$ values taken from ref. 22(*a*), p. 103.

The $\Delta_{AH}G^\circ$ values of Table 5.10 are so huge and positive that they correspond to minute absolute gas-phase acidity constants. From the data in Table 5.10, however, *relative* acidities are easily calculated.

Example. Calculate from the data in Table 5.10 the equilibrium constant for the reaction

$$CH_3OH + PhNH^- \underset{298\,K}{\overset{gas}{\rightleftharpoons}} CH_3O^- + PhNH_2; \quad \Delta_r G^\circ; \quad K^\circ.$$

This constitutes a direct comparision of the gas-phase acidities of CH_3OH and $PhNH_2$ (or base strengths of CH_3O^- and $PhNH^-$). From the data in Table 5.10 $\Delta_r G^\circ_{298\,K} = (1559 - 1505) = 54$ kJ mol^{-1} so equilibrium lies overwhelmingly on the left-hand side ($K^{\circ}_{298\,K} = 3.46 \times 10^{-10}$) for the reaction as written. Methanol, therefore, is *intrinsically* a very much weaker acid than aniline and the reason for this lies only in their respective molecular properties. For comparison, pK_{AH} values of aniline and methanol in water are 27 and 15.5 (Table 5.3) so, in aqueous solution, methanol is the stronger acid by $10^{11.5}$ (although both are very weak). The cause of this complete change around in relative acidity between gas phase and water must be due to the aqueous solvation properties of the two compounds and their respective conjugate bases.

5.5 Thermodynamics of Brönsted acid–base reactions in water

The common measure of the acid or base strength of a compound is the pK_{AH} (or pK_{BH}) which is directly related to a standard free-energy term:

$$XH + H_2O \underset{H_2O}{\overset{T}{\rightleftharpoons}} H_3O^+ + X^-; \tag{5.27}$$

$$\Delta_{AH}G^{\ominus}_T = -RT \ln K^{\ominus}_a,$$

or, to the extent that pK_{AH} approximates the thermodynamic parameter (see pp. 163 and 169),

$$\Delta_{AH}G^{\ominus}_T = 2.303\,RT\mathrm{p}K_{AH}$$
$$= 5.71\,\mathrm{p}K_{AH}\ \mathrm{kJ\,mol^{-1}}\ \text{at}\ 25\,°\mathrm{C},$$

and

$$\Delta_{AH}G^{\ominus}_T = \Delta_{AH}H^{\ominus} - T\,.\,\Delta_{AH}S^{\ominus},$$

where $\Delta_{AH}G^{\ominus}_T$, $\Delta_{AH}H^{\ominus}$, and $\Delta_{AH}S^{\ominus}$ = the standard molar free-energy, enthalpy, and entropy changes for the reaction described by eqn (5.27) at temperature T; the standard state of each solute = 1 mol dm^{-3}.

Although $\Delta_{AH}H^{\ominus}$ and $\Delta_{AH}S^{\ominus}$ are in no sense more fundamental than $\Delta_{AH}G^{\ominus}$, the resolution of the free-energy term into enthalpy and entropy contributions usually helps our understanding of the overall process. Some results are given in Table 5.11.

For the carboxylic acids, $\Delta_{AH}H^{\ominus}$ does not vary much and is close to zero. The enthalpy of solvation of the anion and proton in each case, therefore, almost exactly balances the enthalpy of desolvation of the acid and the gas-phase heterolytic dissociation (see Fig. 5.2, p. 187). It is also evident that the change in acid strength along the series from trimethylacetic acid to trifluoroacetic, a range of $10^{4.8}$ in acidity constants, is due almost entirely to the trend in the entropy terms. $\Delta_{AH}S^{\ominus}$ for solution dissociation will be dominated by the different extents to which the undissociated acid on the one hand, and the dissociated ions on the other, either disrupt the structure of water or impose further order upon it.

The position is different for the phenols; $\Delta_{AH}H^{\ominus}$ for the more weakly acidic members is substantially positive, so, for these compounds, ion

TABLE 5.11
Thermodynamic parameters for the dissociation of some organic acids in water, 25 °C

Acid	pK_{AH} [a]	$\Delta_{AH}H^{\ominus}/$ kJ mol^{-1}	$\Delta_{AH}S^{\ominus}/$ J K^{-1} mol^{-1}	$-T\Delta_{AH}S^{\ominus}/$ kJ mol^{-1}	$\Delta_{AH}G^{\ominus}_{298\,K}/$ kJ mol^{-1}
Carboxylic acids					
$(CH_3)_3CCO_2H$ [b]	5.03	−3.0	−106	31.7	28.7
CH_3CO_2H [b]	4.76	−0.08	−91.6	27.3	27.2
$C_6H_5CO_2H$ [c]	4.20	0.63	−78.2	23.4	24.0
HCO_2H [b]	3.75	−0.33	−72.8	21.7	21.4
$CNCH_2CO_2H$ [d]	2.47	−3.72	−59.8	17.8	14.1
$CHCl_2CO_2H$ [d]	1.26	−0.4	−25	7.6	7.2
CF_3CO_2H [d]	0.23	0	−4.2	1.3	1.3
Phenols [d]					
C_6H_5OH	10.00	22.9	−114	34.1	57.1
3-NO_2-C_6H_4OH	8.36	20.5	−91.2	27.2	47.7
4-NO_2-C_6H_4OH	7.15	19.5	−71.5	21.3	40.8
2,4-$(NO_2)_2C_6H_3OH$	4.09	11.0	−41	12.3	23.3
2,4,6-$(NO_2)_3C_6H_2OH$	0.71	−6.2	−25.2	7.5	1.3 [g]
Ammonium ions					
$C_2H_5NH_3^+$ [e]	10.68	57.2	−13	3.8	61.0
NH_4^+ [d]	9.25	52.2	−2	0.6	52.8
$C_6H_5NH_3^+$ [f]	4.60	30.9	15.6	−4.7	26.2
2-Cl-$C_6H_4NH_3^+$ [d]	2.64	25.1	34	−10	15.1

[a] pK_{AH} values from Table 5.3 (carboxylic acids and phenols) and Table 5.5 (ammonium ions).
$\Delta_{AH}H^{\ominus}$ and $\Delta_{AH}S^{\ominus}$ results taken from:
[b] J. J. Christensen, M. D. Slade, D. E. Smith, R. M. Izatt, and J. Tsang, *J. Am. chem. Soc.* **92,** 4164 (1970).
[c] J. J. Christensen, R. M. Izatt, and L. D. Hansen, *J. Am. chem. Soc.* **89,** 213 (1967).
[d] J. W. Larsen and L. G. Hepler, Ch. 1 'Heats and entropies of ionization' in ref. 20, Vol. 1.
[e] Ref. 15.
[f] P. D. Bolton and F. M. Hall, *Aust. J. Chem.* **20,** 1797 (1967).
[g] This result corresponds to $pK_{AH} = 0.2$ rather than 0.71.
Further results are given in ref. 19, pp. 371–373.

solvation does not compensate for the desolvation of undissociated acid and the gas-phase deprotonation (Fig. 5.2, p. 187). However, upon the introduction of electron-withdrawing substituents, $\Delta_{AH}H^{\ominus}$ becomes less unfavourable, and dissociation is actually exothermic for picric acid. The trend in the unfavourable $\Delta_{AH}S^{\ominus}$ values is similar to that for the carboxylic acids. However, because this now augments a strong trend in $\Delta_{AH}H^{\ominus}$ values, these phenols span a range in acidity of $10^{9.3}$ compared with $10^{4.8}$ for the carboxylic acids.

The low acidity of the ammonium ions is clearly seen to be due mainly to the very substantial positive $\Delta_{AH}H^{\ominus}$ terms, the entropy values being fairly small and varying over a relatively narrow range around zero. Table 5.11 illustrates, therefore, a difference between the neutral acids with their smaller $\Delta_{AH}H^{\ominus}$ values but more unfavourable $\Delta_{AH}S^{\ominus}$ terms, and the ammonium cations which have much more unfavourable $\Delta_{AH}H^{\ominus}$ terms but only small $\Delta_{AH}S^{\ominus}$ values. This difference probably resides in the compounds being of different charge types:

$$AH + H_2O \rightleftharpoons H_3O^+ + A^-, \qquad (5.6)$$

compared with

$$BH^+ + H_2O \rightleftharpoons H_3O^+ + B. \qquad (5.16)$$

Perhaps the clearest message of all from the results in Table 5.11, however, is that even substantial changes in solution acidity, either from one class of compounds to another, or along a series of quite closely related compounds, may be due to either differential entropy or enthalpy effects, or to both. And detailed molecular interpretations of relative pK_{AH} ($\Delta_{AH}G^{\ominus}$) values in solution without reliable $\Delta_{AH}H^{\ominus}$ and $\Delta_{AH}S^{\ominus}$ results are usually inadvisable.

5.6 Factors affecting Brönsted acid and base strengths

5.6.1 The medium

Table 5.12 contains a comparison of the acidities of a range of compounds in the gas phase, in dimethyl sulphoxide (a dipolar solvent with no hydrogens which will form hydrogen bonds), and in water. The results are given as the standard molar free-energy changes for the reactions:

$$XH(g) \xrightarrow{25\,°C} X^-(g) + H^+(g); \quad \Delta_{AH}G^{\circ}_{298\,K} \qquad (5.22)$$

(standard state = the pure gas at 1 standard atmosphere pressure), and

$$XH + S \underset{S}{\xrightarrow{25\,°C}} SH^+ + X^-; \quad \Delta_{AH}G^{\ominus}_{298\,K} \qquad (5.23)$$

(standard state of the solute = 1 mol dm^{-3}).

Strictly, the gas-phase reactions (standard state = the pure gas at 1 standard atmosphere pressure) should be converted to a standard state activity of 1 mol dm^{-3} to allow a proper comparison with the solution results.

$$\Delta G^{\ominus}(g) = \Delta H^{\ominus}(g) - T\Delta S^{\ominus}(g); \text{ standard state} = 1 \text{ mol dm}^{-3},$$

and $\Delta G^{\circ}(g) = \Delta H^{\circ}(g) - T\Delta S^{\circ}(g)$; standard state = 1 standard atmosphere pressure, therefore

$$\Delta G^{\ominus}(g) = \Delta G^{\circ}(g) + \{\Delta H^{\ominus}(g) - \Delta H^{\circ}(g)\} - T\{\Delta S^{\ominus}(g) - \Delta S^{\circ}(g)\}.$$

For a gas-phase reaction at 1 mol dm^{-3},

$$\Delta H^{\ominus} = \Delta U^{\circ} \quad \text{(see p. 147)},$$

TABLE 5.12

Relative acidities in the gas phase, dimethyl sulphoxide, and water

$$HX(g) \xrightarrow{\text{gas}} H^+(g) + X^-(g); \qquad \Delta_{AH}G^{\circ}_{298\,K} \qquad (5.22)$$

$$HX(solv.) \xrightarrow{\text{solvent}} H^+(solv.) + X^-(solv.); \qquad \Delta_{AH}G^{\ominus}_{298\,K} \qquad (5.23)$$

Acid	$\Delta_{AH}G^{\circ}_{298\,K}$/kJ mol^{-1}	$\Delta_{AH}G^{\ominus}_{298\,K}$/kJ mol^{-1}	
	gas[a]	DMSO[b]	H_2O[c]
NH_3	1641	—	~200
H_2O	1607	179[d]	90
CH_3OH	1559	166[d]	89
Toluene	1558	—	~234
C_2H_5OH	1546	170[e]	91
$(CH_3)_2CHOH$	1538	173[e]	97
$(CH_3)_3COH$	1534	185[e]	103
HF	1530	—	18
CH_3SOCH_3	1530	201	188
CH_3CN	1525	179	143
$C_6H_5C{\equiv}C{-}H$	1517	165	120
$C_6H_5NH_2$	1505	176	154
CF_3CH_2OH	1493	131	71
$C_6H_5COCH_3$	1492	141	108
CH_3NO_2	1473	102	58
Cyclopentadiene	1464	103	91
HCN	1447	—	53
C_6H_5OH	1441	94[d]	57
CH_3CO_2H	1429[f]	72[d]	27
$PhCO_2H$	1388[f]	63[d]	24
HCl	1373	11[d]	−40
$CHCl_2CO_2H$	1347[f]	34[d]	7

[a] Results from Table 5.10 except where otherwise indicated.
[b] Results from J. E. Bartmess, J. A. Scott, and R. T. McIver, *J. Am. chem. Soc.* **101,** 6056 (1979), except where otherwise indicated.
[c] Calculated from $\Delta_{AH}G^{\ominus}_{298\,K} = 5.71pK_{AH}$ (see p. 169) using $pK_{AH}(H_2O)$ results given in Table 5.3.
[d] Calculated from $\Delta_{AH}G^{\ominus}_{298\,K} = 5.71pK_{AH}$ using pK_{AH}(DMSO) results given in Table 5.8.
[e] Calculated from $\Delta_{AH}G^{\ominus}_{298\,K} = 5.71pK_{AH}$ using pK_{AH}(DMSO) results given by W. N. Olmstead, Z. Margolin, and F. G. Bordwell, *J. org. Chem.* **45,** 3295 (1980).
[f] Ref. 22(*a*), p. 101.

and

$$\Delta H^{\circ} = \Delta U^{\circ} + \Delta nRT \quad \text{(see p. 44)},$$

therefore

$$\Delta H^{\ominus} - \Delta H^{\circ} = -RT \text{ for a gas-phase dissociation (since } \Delta n = 1)$$
$$= -2.48 \text{ kJ mol}^{-1} \text{ at } 298.15 \text{ K}.$$

From eqn (3.13) of Chapter 3 (p. 95) $S^{\ominus}_{T} = S^{\circ}_{T} - R\ln(0.0821T)$ for an ideal gas, so, for a reaction in which the number of gas moles increases by 1 for each mole of reaction as in a gas-phase dissociation,

$$\Delta S^{\ominus} - \Delta S^{\circ} = -R\ln(0.0821T)$$
$$= -26.6\ \text{J K}^{-1}\ \text{mol}^{-1}\ \text{at}\ 298.15\ \text{K}.$$

Combining the enthalpy and entropy terms,

$$\Delta G^{\ominus}_{298\ \text{K}} - \Delta G^{\circ}_{298\ \text{K}} = (-2.48 + 298.15 \times 26.6 \times 10^{-3})\ \text{kJ mol}^{-1}$$
$$= 5.45\ \text{kJ mol}^{-1}$$

or

$$\Delta G^{\ominus}_{298\ \text{K}} = \Delta G^{\circ}_{298\ \text{K}} + 5.5\ \text{kJ mol}^{-1}.$$

This small correction is the same for all (ideal) gases and numerically insignificant compared with the absolute values of $\Delta G^{\circ}_{298\ \text{K}}$.

One gross feature is immediately evident. All compounds are, as we noted earlier, very much more acidic in solution (eqn (5.23)) than in the gas phase (eqn (5.22)). This is principally due to the very favourable enthalpies of solvation of ions. There is also some sort of correlation between the three phases. For example, the acid strengths decrease in the order

$$CHCl_2CO_2H > C_6H_5CO_2H > CH_3CO_2H > C_6H_5OH > CF_3CH_2OH > CH_3OH$$

in all three. But there are some spectacular exceptions. As acids in water,

$$HCl \gg CHCl_2CO_2H,$$

and

$$HF > \text{cyclopentadiene},$$

but in the gas phase both of these are the other way round by large margins; and we gave the example of methanol and aniline earlier (p. 190). Perhaps most striking of all is the finding that toluene is a stronger acid than water in the gas phase (although both are, of course, very weak).

A comparison of the effect of alkyl substitution upon the OH acidity of alcohols in the gas phase and in aqueous solution is illuminating.

$$ROH + H_2O \underset{298\ \text{K}}{\rightleftharpoons} RO^- + H_3O^+. \qquad (5.28)$$

The order of decreasing acid strength in water (Table 5.3, p. 166) is:

	$CH_3OH >$	$C_2H_5OH >$	$(CH_3)_2CHOH >$	$(CH_3)_3COH$
$pK_{AH}(H_2O)$:	15.5	15.9	17	18

This trend was thought at one time to be due to a molecular electronic effect. The larger and more branched alkyl groups along the series were believed to have increasing electron-donating inductive (polar) effects which increasingly destabilize the alkoxide conjugate bases.

The gas-phase acidities are in the opposite order, however, (Table 5.10, p. 190):

	CH_3OH <	C_2H_5OH <	$(CH_3)_2CHOH$ <	$(CH_3)_3COH$
$\Delta_{AH}G^\circ_{298\,K}$/ kJ mol^{-1}:	1559	1546	1538	1534

The gas-phase results, which reflect the inherent acidities, invalidate the earlier molecular electronic interpretation and establish that the inversion in water must be due to solvation. But in view of the results for carboxylic acids and phenols, Table 5.11 discussed on p. 191, it is not clear whether we are seeing an enthalpy or entropy solvation effect (or both) without separate $\Delta_{AH}H^\ominus$ and $\Delta_{AH}S^\ominus$ results.

Notice that water in the gas phase is less acidic than methanol ($\Delta_{AH}G^\circ_{298\,K} = 1607$ kJ mol^{-1}) so it extends the series shown above, and this is quite reasonable. In aqueous solution, its pK_{AH}(15.74) lies anomalously between the values for methanol and ethanol; again, this must be due to solvation effects.

5.6.2 Molecular structure

5.6.2.1 Acid strength

We shall consider first the effect upon acidity of changing the atom to which the proton is bonded. Two trends have long been recognized. First, the acidity of hydrides in both the gas phase and solution increases from left to right across the second period of the classification of elements.

	XH			
	CH_4	NH_3	H_2O	HF
$\Delta_{AH}H^\circ$/kJ mol^{-1} (Table 5.9)	1743	1672	1635	1554
$pK_{AH}(H_2O)$ (Table 5.3)	~48	~33	15.7	3.18

In a dissection of the overall gas-phase process:

$$HX \rightarrow H^+ + X^-; \qquad \Delta H^\circ = \Delta_{AH}H^\circ(HX),$$

we can write

$$HX \rightarrow H^\cdot + X^\cdot; \qquad \Delta H^\circ = DH^\circ(HX)$$
$$H^\cdot \rightarrow H^+ + e; \qquad \Delta H^\circ = IP(H)$$
$$X^\cdot + e \rightarrow X^-; \qquad \Delta H^\circ = -EA(X)$$

where D$H°$(HX) = the *bond dissociation enthalpy* of HX (see Chapter 2, p. 53)
IP(H) = the *ionization potential* of the hydrogen atom, and
EA(X) = the *electron affinity* of the free radical X$^{\cdot}$.

Consequently,

$$\Delta_{AH}H°(HX) = DH°(HX) + IP(H) - EA(X).$$

Clearly, IP(H) (1312 kJ mol^{-1})[24] is common to all the hydrides so differences between $\Delta_{AH}H°$(HX) values, which we shall take as measures of gas-phase (intrinsic) acidity, reside in either the bond dissociation enthalpies of HX or the electron affinities of the groups X$^{\cdot}$. From the following results, it is clear that the increase in acidity from CH_4 to HF is due principally to the increasing electron affinities along the isoelectronic series $CH_3^{\cdot}$, $NH_2^{\cdot}$, OH$^{\cdot}$, and F$^{\cdot}$. The effect may, therefore, be loosely ascribed to the increasing electronegativity of the elements C, N, O, and F.

	XH			
	CH_4	NH_3	H_2O	HF
D$H°$(HX)/kJ mol^{-1} (ref. 23)	438	432	499	569
EA(X)/kJ mol^{-1} (ref. 23)	7.7	72	176	328
$\Delta_{AH}H°$(HX)/kJ mol^{-1} (Table 5.9)	1743	1672	1635	1554
	←—— range = 189 kJ mol^{-1} ——→			

The effect of converting these enthalpies into free energies by taking the entropies of the gas-phase reactions into account is to contract the overall range somewhat (to about 180 kJ mol^{-1}). However, if the solvation of HX, H^+ and X^- is taken into consideration, the acidity range in water widens again:

	CH_4	NH_3	H_2O	HF
pK_{AH}(H_2O, 25 °C, Table 5.3)	~48	~33	15.7	3.18
$\Delta_{AH}G^{\ominus}_{298\,K}$/kJ mol^{-1} (water)	~274	~188	90	18
	←—— range ~256 kJ mol^{-1} ——→			

In the second well-known trend (the increasing acidity of hydrides *down* a group of the periodic table which follows *decreasing* electronegativity) the bond dissociation enthalpy dominates as the gas-phase results for the halogen hydrides illustrate:

	HX			
	HF	HCl	HBr	HI
D$H°$(HX)/kJ mol^{-1} (ref. 23)	569	431	367	298
EA(X)/kJ mol^{-1} (ref. 22(a))	328	349	325	295
$\Delta_{AH}H°$(HX)/kJ mol^{-1} (Table 5.9)	1554	1395	1354	1315
	←—— range 239 kJ mol^{-1} ——→			

The range in gas-phase standard free energies is slightly smaller (about 235 kJ mol^{-1}) and the effect of water as solvent this time is to compress the range enormously, but not to alter the order:

	HF	HCl	HBr	HI
pK_{AH}(H_2O, 25 °C, Table 5.3)	3.18	~−7	~−8	~−9
$\Delta_{AH}G^{\ominus}_{298\,K}$/kJ mol^{-1}($H_2O$)	18	~−40	~−46	~−51

⟵ range ~69 kJ mol^{-1} ⟶

The second aspect to consider in this section is the less drastic effect of substituents upon the gas-phase acidity of a compound, i.e. the effect of changes elsewhere in a molecule upon the acidity of a hydrogen bound to a particular element. As we noted above, increasing alkyl substitution in an alcohol has an acid-enhancing effect. According to current views, this is purely a polarizability effect; a larger group with many polarizable bonds is better able to accommodate an overall charge than a small molecular residue with only few such bonds.

Groups with electron-attracting polar (inductive) effects also enhance the acidity of a compound. This is illustrated for N—H acidity of the following arylamines:[23]

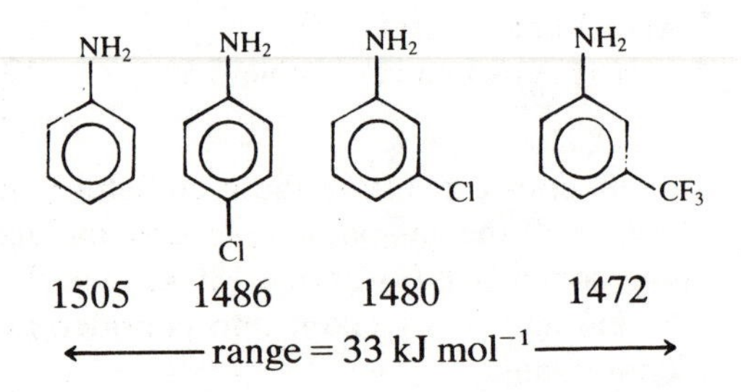

$\Delta_{AH}G^{\circ}_{298\,K}$/kJ mol^{-1} 1505 1486 1480 1472

⟵ range = 33 kJ mol^{-1} ⟶

But substituents which are able to delocalize electron density through a π-system (resonance effect) have a much larger gas-phase acid-enhancing capability. This is expected from simple theory and demonstrated by the C—H acidity of methyl and methylene hydrogens in the following series:[23]

$\underline{CH_3}$—X

Group X	C_6H_5	CN	$COCH_3$	CHO	COC_6H_5	NO_2
$\Delta_{AH}G^{\circ}_{298\,K}$/kJ mol^{-1}	1558	1525	1513	1505	1491	1473

⟵ range = 85 kJ mol^{-1} ⟶

CH_3—$\underline{CH_2}$—X

Group X	C_6H_5	CN	—	CHO	COC_6H_5	NO_2
$\Delta_{AH}G^{\circ}_{298\,K}$/kJ mol^{-1}	1556	1532	—	1504	1489	1472

⟵ range = 84 kJ mol^{-1} ⟶

5.6.2.2 *Base strength*

Base strengths of neutral molecules relative to ammonia are compared in the gas phase using equation (5.29).[22(b),25]

$$B + NH_4^+ \underset{298\,K}{\overset{gas}{\rightleftharpoons}} BH^+ + NH_3; \quad \Delta_B G^\circ_{298\,K}. \tag{5.29}$$

The smaller or the more negative the value of $\Delta_B G^\circ_{298\,K}$ for this reaction, the further over to the right-hand side it is at equilibrium, and the stronger B is as a base.

Some results and corresponding values for aqueous solution are given in Table 5.13.

We see from these results that increasing the alkyl size causes an increase in the base strength of primary alkylamines in the gas phase:

$$\text{as bases,} \quad (CH_3)_3CNH_2 > (CH_3)_2CHNH_2 > EtNH_2 > MeNH_2.$$

The base strength of amines and phosphines is also increased by alkyl substitution on the heteroatom:

$$\text{as bases,} \quad Me_3N > Me_2NH > MeNH_2 > NH_3,$$

$$PhNMe_2 > PhNHMe > PhNH_2,$$

and

$$Me_3P > Me_2PH > MePH_2 > PH_3.$$

Alkyl substitution, therefore, increases both the base strength and, as we saw earlier (p. 195), the acid strength of compounds in the gas phase. This is convincing evidence that in both trends the effect is due to increased stability of the conjugate ion caused by the higher polarizability of the more substituted alkyl residues.

As expected, electron-withdrawing substituents decrease the base strength of a compound (in both the gas phase and solution) regardless of whether the effect is transmitted via a π-system or otherwise,[26]

e.g. as bases

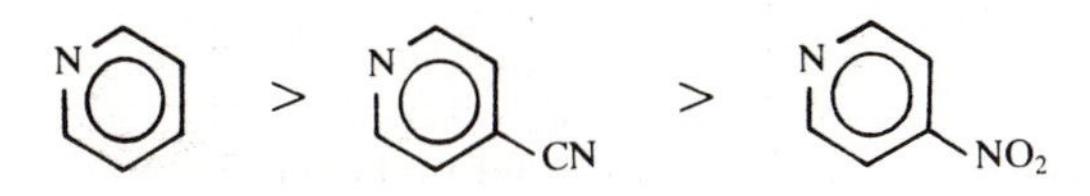

and

$$CH_3CH_2NH_2 > CH_2FCH_2NH_2 > CHF_2CH_2NH_2 > CF_3CH_2NH_2.$$

Early explanations in terms of electron-withdrawing effects rendering the nitrogen lone pair less available for protonation still seem valid.

An important effect is demonstrated by a comparison of the base strengths of ammonia, cyclohexylamine, and aniline.[26] The much greater gas-phase base strength of cyclohexylamine compared with ammonia is due to the polarizability of the large carbocyclic residue. This is the effect

TABLE 5.13
Comparison of bases with ammonia in the gas phase and in aqueous solution[a]

$$B + NH_4^+ \overset{gas}{\rightleftharpoons} BH^+ + NH_3; \quad \Delta_r G^\circ_{298\,K} \tag{5.29}$$

$$B + NH_4^+ \overset{H_2O}{\rightleftharpoons} BH^+ + NH_3; \quad \Delta_r G^\ominus_{298\,K} \tag{5.30}$$

Base	$\Delta_r G^\circ_{298\,K}$/kJ mol^{-1}	$\Delta_r G^\ominus_{298\,K}$/kJ mol^{-1}
	Gas phase	Water
CH_3OH	76.1	64[b]
PH_3	58.6	132
C_2H_5OH	57.3	82[b]
CH_3CN	56.5	110[b]
$(CH_3)_2CO$	27.2	94[b]
$(C_2H_5)_2O$	13.4	82[b]
$CF_3CH_2NH_2$[c]	5.9	20.8
CH_3PH_2	0.4	—
NH_3	0	0
$CHF_2CH_2NH_2$[c]	−16.7	12.3
4-Nitropyridine[c]	−21	44.8
4-Cyanopyridine[c]	−26	42.1
$C_6H_5NH_2$	−28.0	26.5[b]
$CFH_2CH_2NH_2$[c]	−33.5	2.6
CH_3NH_2	−38	−8.1[b]
$C_2H_5NH_2$	−49.4	−8.2[b]
$(CH_3)_2PH$	−49.8	30.4
$C_6H_5NHCH_3$	−54.0	24.9
$(CH_3)_2CHNH_2$	−59.0	−8.1
$(CH_3)_2NH$	−64.9	−8.7
Pyridine[c]	−66.9	23.0[b]
t-$BuNH_2$	−67.4	−7.8
Cyclohexylamine	−68.2	−7.6
$C_6H_5N(CH_3)_2$	−81.6	23.8
$(CH_3)_3N$	−83.7	−3.2[b]
$(CH_3)_3P$	−88.7	3.3
4-Methoxypyridine[c]	−95.0	15.2
N-Methylpyrrolidine[c]	−102	−6.9
Quinuclidine	−113	−10.9[b]

[a] Results taken from ref. 25 except where otherwise indicated.
[b] Calculated from pK_{BH} values given in Table 5.5.
[c] Ref. 26.

we noted above. Aniline is also much more basic than ammonia for the same reason, but it is very considerably less basic than cyclohexylamine which it resembles in size and shape. The nitrogen lone pair in aniline, however, unlike that in cyclohexylamine, is conjugated with the π-system of the benzene ring and, to some extent, is delocalized. Protonation of the nitrogen of aniline effectively localizes this electron pair and, therefore, causes some loss of electronic delocalization energy, hence aniline's reduced base strength compared with cyclohexylamine.

Table 5.13 indicates a modest correlation between base strengths in the gas phase and aqueous solution, and apparent anomalies must be due to differences in the solvation properties of the species involved. Equilibrium in the reaction of eqn (5.29) for a base B with many polarizable C–C and C–H bonds will be far over to the right-hand side for the reasons discussed above. In aqueous solution, however, the much smaller and very strongly polarizing ammonium cation NH_4^+ is very extensively solvated via hydrogen bonding. Solvation of BH^+ (eqn (5.30)) in water will be much poorer partly due to its lower polarizing ability (BH^+ is more diffusely charged) and partly for steric reasons. Consequently, there is an extra enthalpy contribution favouring the left-hand side of eqn (5.30) in aqueous solution which is absent for the gas-phase reaction of eqn (5.29). And the extent to which the base strength of B changes compared with ammonia upon switching from the gas phase to aqueous solution depends largely upon the extent to which this solvation enthalpy effect counterbalances the intrinsic polarizability effect. (Equations (5.29) and (5.30) are of the same charge type as the reactions of substituted ammonium ions, eqn (5.16), described in Table 5.11. As discussed on p. 192, enthalpy rather than entropy effects dominate these reactions.)

The standard free-energy change of eqn (5.29) for simple primary alkylamines (e.g. CH_3NH_2, $C_2H_5NH_2$) is reduced in magnitude substantially upon changing from the gas phase to water as seen from Table 5.13, but it remains negative. Such compounds are, therefore, still more basic than ammonia in water. The very strong gas-phase bases such as quinuclidine, trimethylamine, N-methylpyrrolidine, and trimethylphosphine suffer huge (relative) drops in base strength upon changing to water because their conjugate acids, BH^+, are much more poorly solvated than NH_4^+. The first three of these remain just stronger than ammonia as bases but trimethylphosphine becomes actually weaker.

The aqueous solvent effect upon the base strength of the arylamines is particularly interesting. These aromatic compounds, which are not as basic in the gas phase as corresponding alkylamines for the reason given above, still suffer a very large change in the standard free energy for the reaction of eqn (5.29) upon switching from the gas phase to aqueous solution. This is because their conjugate acids BH^+, with a very polarizable aromatic group, are even less well solvated than the corresponding cations from alkylamines. As seen from Table 5.13, $\Delta_{AH}G^{\ominus}_{298\,K}$ actually becomes substantially positive for the reactions of these compounds in eqn (5.30) in water; so these bases are much weaker than ammonia and simple alkylamines in aqueous solution.

5.7 The Hammett equation and chemical equilibria[2,27,28]

In previous sections of this chapter, we looked at how fairly gross changes either in experimental conditions or to the structure of a compound affect its acidic and basic properties. We shall now focus down and consider the effects of relatively modest structural changes upon acid–base and other equilibria of closely related aromatic compounds.

5.7.1 Acid–base equilibria

Equation (5.31) represents the dissociation of a family of *meta*- and *para*-substituted benzoic acids in water at 25 °C and, as expected, the magnitude of K_{AH} is affected by the nature and position of the substituent X.

$$X\text{-}C_6H_4\text{-}CO_2H + H_2O \underset{25°C}{\overset{H_2O}{\rightleftharpoons}} X\text{-}C_6H_4\text{-}CO_2^- + H_3O^+;\ K_{AH} \qquad (5.31)$$

We arbitrarily use the effect of substituent X upon the acidity of benzoic acid as a measure of some property of the substituent (though at this stage we may not be sure about what exactly this property is, or of how the effect is transmitted to the carboxyl group). The measure is given the Greek symbol, σ, and is defined by eqn (5.32)

$$\sigma_X = pK_{AH}(C_6H_5CO_2H) - pK_{AH}(XC_6H_4CO_2H) \qquad (5.32)$$

or

$$\sigma_X = \log\left\{\frac{K_{AH}(XC_6H_4CO_2H)}{K_{AH}(C_6H_5CO_2H)}\right\}. \qquad (5.33)$$

A positive σ-value indicates an acid-enhancing effect and this invariably corresponds to an electron-withdrawing capability of the substituent. Substituents with negative σ-values decrease the acidity of benzoic acid. Some experimental results are given in Table 5.14 and a selected list of substituent constants in Table 5.15.

TABLE 5.14
pK_{AH} *values for some substituted benzoic acids* (H_2O, 25 °C) *and* σ*-values of the substituents*[29,30]

Substituent X	$pK_{AH}(X\text{-}C_6H_4CO_2H)$	σ_X
H	4.20	0
m-OCH_3	4.09	0.11
m-F	3.86	0.34
m-NO_2	3.49	0.71
p-NO_2	3.42	0.78
p-CH_3	4.37	−0.17
p-OCH_3	4.48	−0.28

TABLE 5.15
Substituent constants defined by eqn (5.32), pK_{AH} *of benzoic acid* = 4.20 (H_2O, 25 °C)[29,30]

Substituent	σ_m	σ_p
NO_2	0.71	0.78
CN	0.61	0.70
CF_3	0.43	0.54
CH_3CO_2	0.39	0.31
Br	0.39	0.23
CH_3CO	0.38	0.48
$CO_2C_2H_5$	0.37	0.45
Cl	0.37	0.22
CHO	0.36	0.44
CO_2H	0.35	0.44
I	0.35	0.28
F	0.34	0.06
C≡CH	0.20	0.23
SCH_3	0.15	~0
OH	0.13	−0.38
OCH_3	0.11	−0.28
C_6H_5	0.05	~0
H	0	0
CH_3	−0.06	−0.17
C_2H_5	−0.07	−0.15
$CH(CH_3)_2$	−0.07	−0.15
$C(CH_3)_3$	−0.10	−0.20
$N(CH_3)_2$	−0.15	−0.63
NH_2	−0.16	−0.57

If we now look at the effects of some of these substituents upon the acid strengths of phenylacetic and 3-phenylpropionic acids, we find analogous trends.

Figure 5.3 comprises plots of $\log K_{AH}(XC_6H_4CH_2CO_2H)$ and $\log K_{AH}(XC_6H_4CH_2CH_2CO_2H)$ against $\log K_{AH}(XC_6H_4CO_2H)$. But since, by eqn (5.32), a plot against $\log K_{AH}(XC_6H_4CO_2H)$ is equivalent to one against the substituents' σ-values, the x-coordinate is doubly labelled.

Since both correlations are linear, they may be represented by

$$\log K_{AH}(\text{sub. acid}) = \rho \,.\, \sigma + \text{const.} \qquad (5.34)$$

where ρ = the gradient, 0.49 for $XC_6H_4CH_2CO_2H$ and 0.21 for $XC_6H_4CH_2CH_2CO_2H$. When we put the values for the unsubstituted acids into this equation ($X = H$, $\sigma_H = 0$), we evaluate the constant for each series:

$$\log K_{AH}(\text{unsub. acid}) = \text{constant},$$

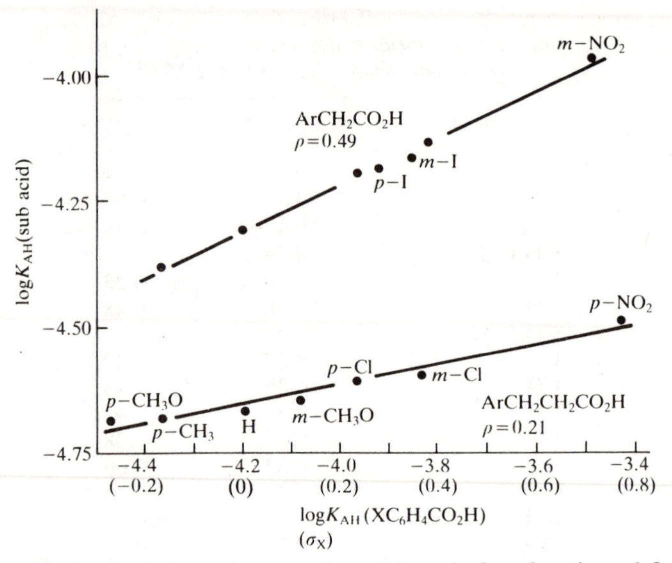

Fig. 5.3. Effects of substituents upon the acidity of phenylacetic and 3-phenylpropionic acids (acidity constants from ref. 12, σ-values from Table 5.15).

therefore eqn (5.34) may be re-written

$$\log K_{AH}(\text{sub. acid}) = \rho \, . \, \sigma + \log K_{AH}(\text{unsub. acid})$$

or

$$\log\left\{\frac{K_{AH}(\text{sub. acid})}{K_{AH}(\text{unsub. acid})}\right\} = \rho \, . \, \sigma. \qquad (5.35)$$

Instead of having used the dissociation of substituted phenylacetic or phenylpropionic acids to obtain eqn (5.35), we could equally well have used the dissociation of substituted cinnamic acids

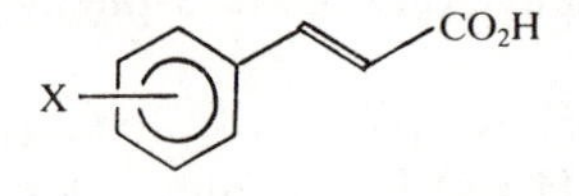

or arylphosphonic acids

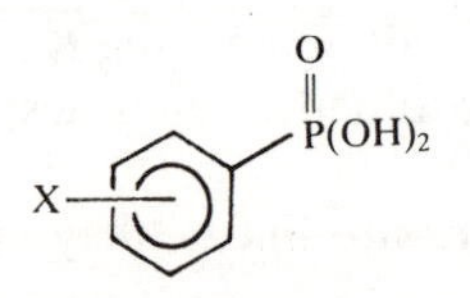

or one of several other reversible reactions of benzene derivatives whose equilibrium constants K_X are affected by the nature of the substituent X in the benzene ring.[31] We would have found in each case some sort of correlation between log K_X and σ_X. Consequently, we may generalize eqn (5.35) and write

$$\log\left\{\frac{K_X}{K_0}\right\} = \rho \, . \, \sigma_X, \qquad (5.36)$$

where K_0 = the equilibrium constant for the unsubstituted benzene derivative, and
K_X = the equilibrium constant for the *meta* or *para* X-substituted benzene derivative in its *corresponding reaction.*

Equation (5.36) is one expression of the *Hammett equation.* It includes two parameters. One is a *substituent parameter,* σ_X, which, via eqn (5.32), is a measure of the effect of X upon the acidity of benzoic acid. The other is the *reaction parameter,* ρ, which is a measure of the *relative* sensitivity of the reaction series under consideration to the introduction of substituents into the aromatic ring (relative to the effect of the same substituents upon the acidity of benzoic acid since, if we compare eqn (5.35) with eqn (5.33), we see that, by definition, $\rho = 1$ for the dissociation of benzoic acid at 25 °C in water).

For arylacetic acids, $\rho = 0.49$ so this reaction series is less sensitive to the introduction of substituents than the dissociation of benzoic acids; and for 3-arylpropionic acids $\rho = 0.21$ so the effect here is smaller still. These results are not surprising since the *reaction site* of the acetic and propionic acids (—CO_2H) becomes increasingly insulated from the effect of substituents in the aromatic ring, i.e. the electronic perturbation, by the successive intervening —CH_2— groups. But in both cases, ρ is positive, so both dissociations are affected in the same sense as the dissociation of benzoic acid. *The equilibrium constant in a reaction series such as an acid dissociation with a positive ρ-value, therefore, is increased by electron-withdrawing substituents, and the larger the positive ρ-value, the greater the sensitivity of the equilibrium constant to such substituents.*

Table 5.16 shows the ρ-values for the dissociation of a range of aromatic acids; as expected all the ρ-values are positive.

Notice that the ρ-values for the dissociation of benzoic acids *increase* along the solvent series water, ethanol, and dimethylformamide as the medium becomes less well able to solvate ions, but then remains constant for the other dipolar aprotic solvents. Plots of the gas-phase acidity of phenols, anilinium and pyridinium ions, as well as benzoic acids, against σ show even larger ρ-values.[23]

Example. Given that the pK_{AH} of benzoic acid is 11.0 in DMSO (Table 5.8, p. 185), calculate the pK_{AH} of *p*-nitrobenzoic acid in DMSO from the data in Tables 5.15 and 5.16.

$$\log \frac{K_X}{K_0} = \rho \,.\, \sigma_X,$$

or

$$\log K_X - \log K_0 = \rho \,.\, \sigma_X$$

$$\log K_0 = -11.0$$

$$\rho = 2.48 \text{ (Table 5.16)},$$

TABLE 5.16
Hammett ρ-values for the dissociation of aromatic acids, 25 °C

$$AH + S \underset{25\,°C}{\overset{S}{\rightleftharpoons}} SH^+ + A^-$$

or

$$BH^+ + S \underset{25\,°C}{\overset{S}{\rightleftharpoons}} SH^+ + B$$

Acid[a]	Solvent	ρ	Note or Ref.
$ArCO_2H$	H_2O	1.00	see text
$ArCO_2H$	C_2H_5OH	1.65	*b*
$ArCO_2H$	DMF	2.36	*c*
$ArCO_2H$	CH_3CN	2.41	*c*
$ArCO_2H$	DMSO	2.48	*c*
$ArCH_2CO_2H$	H_2O	0.49	Fig. 5.3
$ArCH_2CH_2CO_2H$	H_2O	0.21	Fig. 5.3
trans-$ArCH{=}CHCO_2H$	H_2O	0.45	ref. 31
$ArPO(OH)_2$	H_2O	0.76	ref. 31
ArOH	H_2O	2.11	ref. 31
$ArSO_2H$	H_2O	0.98	ref. 32
$ArNH_3^+$	H_2O	2.77	ref. 31
$ArNH_3^+$	C_2H_5OH	3.54	ref. 31
$ArCH_2NH_3^+$	H_2O	0.72	ref. 31

[a] Ar = X-C_6H_5- where X = the substituent.
[b] C. D. Ritchie and R. E. Uschold, *J. Am. chem. Soc.* **90,** 2821 (1968).
[c] I. M. Kolthoff and M. K. Chantooni, *J. Am. chem. Soc.* **93,** 3843 (1971); DMF = $HCON(CH_3)_2$, DMSO = CH_3SOCH_3.

and

$$\sigma(pNO_2) = 0.78 \text{ (Table 5.15)},$$

therefore

$$\log K_X + 11.0 = 2.48 \times 0.78$$
$$\log K_X = 1.93 - 11.0$$
$$= -9.1.$$

The pK_{AH} of *p*-nitrobenzoic acid in DMSO = 9.1

(in good agreement with the experimental value of 9.0[18]).

This example illustrates one use of the Hammett equation. If the ρ-value for a reaction type is established by measuring K_X for several (preferably, many) members of the series, the equilibrium constants for other members of the series can be easily estimated by interpolation using eqn (5.36) and σ-values for the substituents (of which well over 500 are now known[30]). The accuracy of the estimated pK_{AH} depends principally upon the quality of the correlation (which is expressed by the correlation coefficient) in the determination of ρ.[28]

5.7.2 The Hammett equation as a linear standard free-energy relationship

We have in eqn (5.36) a linear correlation between *logarithms* of equilibrium constants K_X for a reaction series and substituent parameters σ. But σ is really a derived parameter based upon the dissociation of a substituted benzoic acid, consequently the Hammett equation could equally well, though less conveniently, be written as

$$\log \frac{K_X}{K_0} = \rho \, . \log \frac{K'_X}{K'_0}$$

or

$$\log K_X = \rho \, . \log K'_X + \text{constant}, \qquad (5.37)$$

where K'_X refers to the dissociation of X-substituted benzoic acid.

Equation (5.37) demonstrates more clearly that the Hammett equation is a *linear relationship* between *logarithms of equilibrium constants* of different reaction series. This relationship was established from experimental results[27,33] and did not have a prior theoretical basis. An obvious question is why should the linear relationship be between *logarithms* of equilibrium constants?

Since $\Delta G^\ominus = -RT \ln K_a^\ominus$, eqn (5.37) may be transformed into

$$\Delta G^\ominus = \rho \, . \Delta G^{\ominus\prime} + \text{constant}.$$

So the Hammett equation at its most basic is a *linear relationship* between *standard free energies* of two reaction series. It may also be expressed in its differential form:

$$\delta\Delta G_X^\ominus = \rho \, . \, \delta\Delta G_X^{\ominus\prime},$$

which we can interpret as follows.

The standard free energy of dissociation of benzoic acid itself $\Delta G^{\ominus\prime}$ is changed by the introduction of a substituent X into the aromatic ring by an amount $\delta\Delta G_X^{\ominus\prime}$. The same substituent also causes a change $\delta\Delta G_X^\ominus$ in the standard free energy of reaction $\Delta G^\ominus$ of another aromatic compound, e.g. the dissociation of phenylacetic acid. The Hammett equation tells us that these changes in the standard free energies of different reactions caused by the same substituent are proportional to each other:

$$\delta\Delta G_X^\ominus \propto \delta\Delta G_X^{\ominus\prime}$$

and the constant of proportionality, ρ, is the measure of the sensitivity of the second reaction, *relative to the first,* to the introduction of the same substituent.

As implied on p. 205, simple Hammett correlations are obtained for *m*- and *p*-substituted aromatic compounds only. If σ-values for *ortho*-substituents

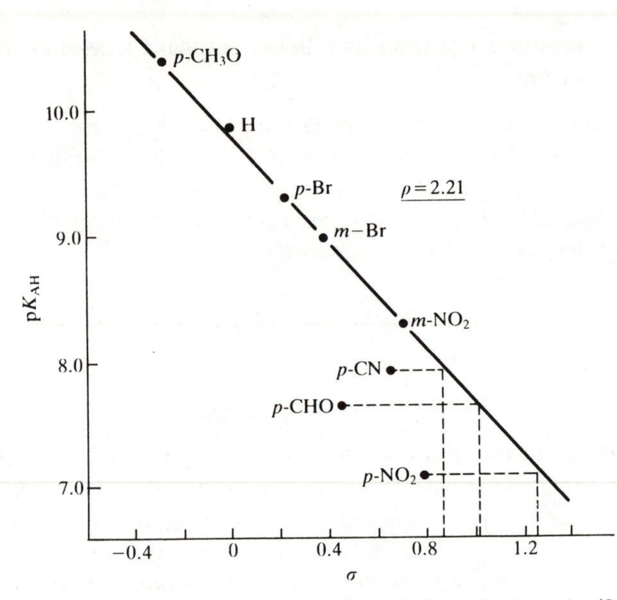

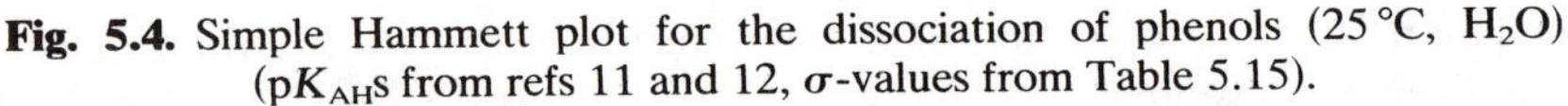

Fig. 5.4. Simple Hammett plot for the dissociation of phenols (25 °C, H_2O) (pK_{AH}s from refs 11 and 12, σ-values from Table 5.15).

are calculated from the pK_{AH} of *o*-substituted benzoic acids, they do not fit very well into the linear plots of σ against $\log K_X$ values for other equilibria.[34]

5.7.3 Modified substituent parameters

Figure 5.4 shows a simple Hammett plot for the dissociation of phenols. Notice that pK_{AH} values rather than $\log K_{AH}$ are plotted against σ, so the negative slope corresponds to a positive ρ.

Some points clearly deviate from the straight line by amounts which are outside experimental error, and the deviations are all in the same sense. Those phenols with substituents which can conjugatively withdraw electron density and which are *para* to the OH group have smaller pK_{AH} values i.e. they are more acidic than is anticipated from the correlation based upon the other compounds. In other words, substituents such as *p*-NO_2, *p*-CHO, and *p*-CN have an extra acid-enhancing effect upon phenol which is not predicted from the results of the other compounds and the effects of these same substituents upon the acidity of benzoic acid.

The anions of these phenols have their negative charge more extensively delocalized through the electron-withdrawing conjugative effects of these substituents. This is not possible when the same substituents are

TABLE 5.17
Substituent σ- and σ^--values[30]

Substituent	σ	σ^-
p-NO_2	0.78	1.27
p-CN	0.70	0.88
p-CH_3CO	0.48	0.84
p-$CO_2C_2H_5$	0.45	0.74
p-CHO	0.44	1.04
p-CO_2H	0.44	0.78
p-C≡CH	0.23	0.52
p-C_6H_5	~0	0.08

meta to the OH (the pK_{AH} for *m*-nitrophenol is on the line), or if the substituent does not have a conjugative electron-withdrawing capacity (the result for *p*-bromophenol also falls on the line). And this extra electronic effect of groups such as NO_2, CHO, and CN is only possible when the reaction site itself involves π-electron density conjugated with the substituent through the benzene ring. For example, it is not possible for the negative charge of the carboxylate of benzoate, arylacetate, or arylpropionate to be *conjugatively* delocalized to a *p*-nitro group in the benzene ring.

For those reactions in which a reaction site with π-electron density is conjugated with a substituent which has a conjugative electron-withdrawing capacity, another substituent parameter (rather than σ) is employed and given the symbol σ^-. Its numerical value is allocated such that the pK_{AH} of the *p*-substituted phenol falls back on the straight line as indicated in Fig. 5.4 for NO_2 ($\sigma^- = 1.27$), CHO ($\sigma^- = 1.04$), and CN ($\sigma^- = 0.88$). Further σ^--values are given in Table 5.17. These σ^--values may then be used in other analogous Hammett plots. For example, a better correlation of the pK_{AH} of anilinium cations is obtained with σ^- than with σ, Fig. 5.5.

5.7.4 Other equilibria

So far we have considered only the dissociation of acids,

$$AH + H_2O \rightleftharpoons A^- + H_3O^+,$$

and the substituents have been in an aromatic residue of A. Not surprisingly, electron-withdrawing substituents facilitate the generation of A^- from AH by dispersing the negative charge. Consequently, all these ionization–dissociation reactions have positive ρ-values.

If the reaction is of the type

$$Z{-}Y \rightleftharpoons Z^+ + Y^-,$$

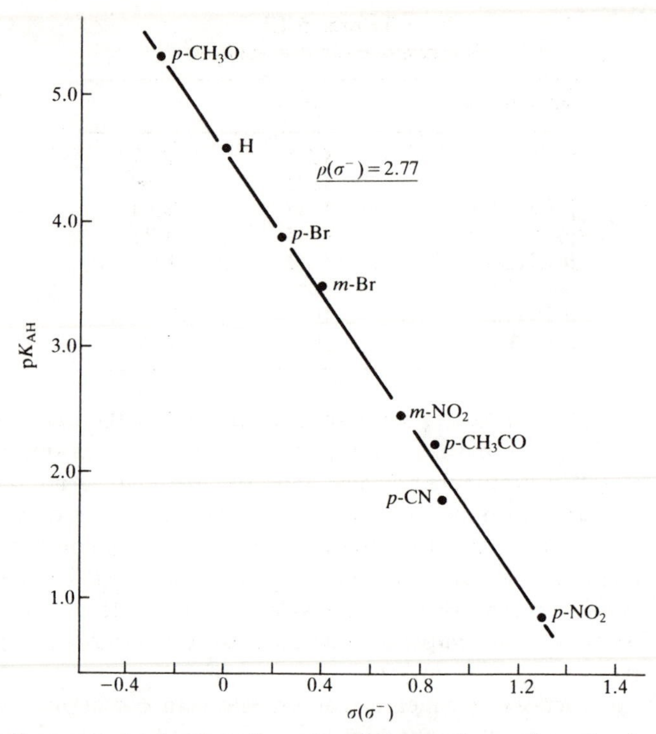

Fig. 5.5. Hammett plot for the pK_{AH} of anilinium cations using σ^--values (pK_{AH} data from ref. 10, σ^--values from Table 5.17).

and the substituted aromatic residue is in Y, then again we anticipate a positive ρ for the equilibrium as written. However, if the aromatic group is in Z, then the introduction of an electron-withdrawing substituent (one with a positive σ-value) *inhibits* the ionization–dissociation and causes the equilibrium constant to become smaller. The reaction as written, therefore, has a negative ρ-value for substitution in Z. But if the reaction is written the other way round,

$$Z^+ + Y^- \rightleftharpoons Z{-}Y$$

then electron-withdrawing substituents in Z now *facilitate* the reaction from left to right and the forward reaction has a positive ρ-value.

It is important, therefore, to appreciate fully just where substituents are being introduced in the molecule undergoing heterolysis, and that writing an equilibrium the other way round corresponds to inverting its equilibrium constant and reversing the sign of the ρ-value (but not altering its numerical value).

The reaction of aryldiazonium cations with arylsulphinate anions to give diazosulphones illustrates both aspects.[32]

$$\mathrm{ArN_2^+ + C_6H_5SO_2^- \underset{25\,°C}{\overset{\text{methanol}}{\rightleftharpoons}} ArN{=}NSO_2{-}C_6H_5}$$

$$\rho = 3.76$$

and

$$p\mathrm{ClC_6H_4N_2^+ + ArSO_2^- \underset{25\,°C}{\overset{\text{methanol}}{\rightleftharpoons}}} p\mathrm{ClC_6H_4N{=}NSO_2{-}Ar}$$

$$\rho = -2.07$$

or

$$p\mathrm{ClC_6H_4N{=}NSO_2{-}Ar \underset{25\,°C}{\overset{\text{methanol}}{\rightleftharpoons}}} p\mathrm{ClC_6H_4N_2^+ + ArSO_2^-}$$

$$\rho = +2.07.$$

Compared with a very extensive literature on the Hammett equation applied to rate processes (which we shall discuss in Chapter 10), only a few organic equilibria, besides the dissociation of acids, have been investigated by this approach.[31] Some results are shown below:

$$\mathrm{ArCHO + HCN \underset{20\,°C}{\overset{95\%\ \text{aqueous}\ C_2H_5OH}{\rightleftharpoons}} ArCH(OH)CN}$$

$$\rho = -1.49;^{31,35}$$

$$\mathrm{ArCOCH_3 + H_2 \underset{25\,°C}{\overset{H_2O}{\rightleftharpoons}} ArCH(OH)CH_3}$$

$$\rho = 1.63;^{31,36}$$

$$\mathrm{ArN_2^+ + CN^- \underset{23\,°C}{\overset{H_2O}{\rightleftharpoons}}}\ syn\text{-}\mathrm{ArN{=}NCN}$$

$$\rho = 3.53.^{37}$$

If $\rho = 0$ for a reaction series, then introducing electron-withdrawing substituents has no effect at all upon the equilibrium constant of the reaction. This could indicate that no electronic charge redistribution accompanies the conversion of reactant into product as is the case for many reactions controlled by orbital symmetry considerations.[38] Alternatively, it could be that the reaction site is exceedingly remote from the substituents in the aromatic ring, or otherwise well insulated from their effects.

Further uses of the Hammett equation and its extended versions are considered in Chapter 10 after applications to kinetics have been discussed.

5.8 Problems

1. Calculate the standard molar free energies and associated equilibrium constants for the following proton-transfer reactions (i) in the gas phase using data in Table 5.10 (p. 190), and (ii) in aqueous solution using pK_{AH} values in Table 5.3 (p. 166).

   (a) $F^- + (CH_3)_2SO \rightleftharpoons HF + CH_3SOCH_2^-$
   (b) $H_2S + F^- \rightleftharpoons HS^- + HF$
   (c) $CH_3CN + C_2H_5O^- \rightleftharpoons CH_2CN^- + C_2H_5OH$
   (d) $H_2O + (CH_3)_3CO^- \rightleftharpoons OH^- + (CH_3)_3COH$
   (e) $(CH_3)_2CO + OH^- \rightleftharpoons CH_3COCH_2^- + H_2O$.

2. Calculate the pH of aqueous solutions made up to 1 dm^3 from the following weak acids and their salts.

   (i) 0.05 mol CH_3CO_2H ($pK_{AH} = 4.76$) + 0.1 mol potassium acetate.
   (ii) 0.10 mol HCO_2H ($pK_{AH} = 3.75$) + 0.05 mol sodium formate.

3. Small concentrations of the following weak acids are added to buffered aqueous solutions as indicated. Calculate the percentage dissociation of the acid in each case.

Acid	pK_{AH}	pH *of buffer*
(i) $ClCH_2CO_2H$	2.86	4.10
(ii) 4-nitrophenol	7.15	6.5
(iii) C_6H_5OH	10.0	9.5.

4. Calculate (i) the pH and (ii) the percentage dissociation for 0.02 molar aqueous solutions of the following weak acids.

Acid	pK_{AH}
(a) $NH_3OH^+Cl^-$	6.0
(b) CH_3CO_2H	4.76
(c) HCO_2H	3.75.

5. Calculate (i) the percentage protonation of pyridine ($pK_{BH} = 5.23$) in an aqueous solution buffered at pH = 4.75, and (ii) the pH of the buffered aqueous solution which causes pyridine to be 90 per cent protonated.

6. Calculate (i) the pH and (ii) the percentage protonation for 0.01 molar aqueous solutions of the following bases.

Base	pK_{BH}
(a) $(CH_3)_3N$	9.80
(b) Na^+SH^-	7.0
(c) $C_6H_5NH_2$	4.60.

7. Calculate the pH of 0.1 molar solutions of the following compounds in water.

Compound	pK_{BH}	pK_{AH}
(i) pyridine-3-carboxylic acid	2.07	4.73
(ii) 4-(4′-imidazolyl)-butanoic acid	4.26	7.62
(iii) 4-aminophenol	5.50	10.3.

8. Calculate (i) the pK_{AH} and (ii) the percentage dissociation of each of the following acids from the information provided:

	pH at 0.05 mol dm^{-3}
(a) $BrCH_2CO_2H$	2.13
(b) $HOCH_2CO_2H$	2.58
(c) cyclopropanecarboxylic acid	3.07.

9. Calculate the proportions of neutral, anionic, and cationic forms of 5-aminopentanoic acid (pK_{BH} = 4.27, pK_{AH} = 10.8) in aqueous solutions buffered at (i) pH = 4.9, and (ii) pH = 10.1.

10. Calculate (i) the pK_{BH} and (ii) the percentage protonation of each of the following bases in water from the information provided:

	pH of 0.005 mol dm^{-3} solution
(a) ethanolamine	10.58
(b) tris(hydroxymethyl)methylamine	9.89
(c) N,N-dimethylaniline	8.28.

11. Acidity constants of substituted *anti* benzaldehyde oximes have been measured (H_2O, 25 °C; O. L. Brady and N. M. Chukshi, *J. chem. Soc.* 946 (1929); O. L. Brady and R. F. Goldstein, ibid. 1918 (1926)).

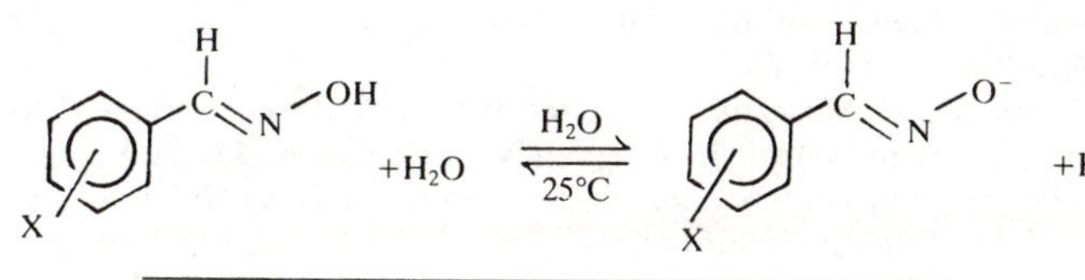

X	σ_X	pK_{AH}
p-$N(CH_3)_2$	−0.63	11.25
p-CH_3O	−0.28	10.92
H	0	10.68
m-CH_3O	0.11	10.59
m-NO_2	0.71	10.16
p-NO_2	0.78	9.96

Calculate ρ for this reaction series. Assuming that the dissociation of the *syn* isomers has the same ρ-value, calculate the pK_{AH} of *syn* *m*-nitrobenzaldehyde oxime given that the pK_{AH} of *syn* benzaldehyde oxime itself is 11.33. The experimental result is 10.74.

5.9 References

1. W. B. Jensen, *The Lewis acid–base concepts*, Wiley-Interscience, New York (1980).
2. J. Hine, *Structural effects on equilibria in organic chemistry*, Wiley-Interscience, New York (1975).

3. T-L. Ho, *Hard and soft acids and bases principle in organic chemistry*, Academic Press, New York (1977); I. Fleming, *Frontier orbitals and organic chemical reactions*, Wiley, Chichester (1976).
4. K calculated from $\Delta_f G^\circ_{298\,K}$ data taken from D. R. Stull, E. F. Westrum, and G. C. Sinke, *The chemical thermodynamics of organic compounds*, Wiley, New York (1969); see also R. G. Pearson and J. Songstad, *J. Am. chem. Soc.* **89,** 1827 (1967); J. Hine and R. D. Weimar, ibid. **87,** 3387 (1965).
5. Y. Wolman, Chapter 14 in *The chemistry of the thiol group* Part 2 (ed. S. Patai), Wiley, London (1974); M. J. Janssen, Chapt. 15 in *The chemistry of carboxylic acids and esters* (ed. S. Patai), Interscience, London (1969).
6. R. P. Bell, *The proton in chemistry*, 2nd edn, Chapman & Hall, London (1973).
7. E. J. King, *Acid-base equilibria*, Pergamon, Oxford (1965).
8. R. J. Gillespie, Ch. 1, in *Proton–transfer reactions* (eds. E. F. Caldin and V. Gold), Chapman & Hall, London (1975).
9. P. A. Giguère, *J. chem. Educ.* **56,** 571 (1979); P. A. Giguère and S. Turrell, *J. Am. chem. Soc.* **102,** 5473 (1980).
10. D. D. Perrin, *Dissociation constants of inorganic acids and bases in aqueous solution*, Butterworths, London (1969). Published as *Pure and Applied Chemistry* **20,** 133 (1969).
11. A. Albert and E. P. Serjeant, *Ionization constants of acids and bases*, Methuen, London (1962).
12. G. Kortüm, W. Vogel, and K. Andrussow, *Dissociation constants of organic acids in aqueous solution*, Butterworths, London (1961).
13. K. Bowden, *Chem. Rev.* **66,** 119 (1966).
14. R. F. Cookson, *Chem. Rev.* **74,** 5 (1974).
15. F. M. Jones and E. M. Arnett, 'Thermodynamics of ionization and solution of aliphatic amines in water', in *Progr. phys. org. Chem.* **11,** 263 (1974).
16. D. D. Perrin, *Dissociation constants of organic bases in aqueous solution*, Butterworths, London (1965).
17. *The chemistry of non-aqueous solvents* (ed. J. J. Lagowski), Academic Press, Vol. 1 (1966), *Principles and techniques*, Vol. 2 (1967), *Acidic and basic solvents*, Vol. 3 (1970), *Inert, aprotic, and acidic solvents.*
18. B. G. Cox, *Annual Reports*, Part A, The Chemical Society, London (1973).
19. J. E. Leffler and E. Grunwald, *Rates and equilibria of organic reactions*, Wiley, New York (1963).
20. *Solute–solvent interactions*, (eds. J. F. Coetzee and C. D. Ritchie), Marcel Dekker, Vol. 1 (1969) and Vol. 2 (1976).
21. J. L. Beauchamp, *A. Rev. phys. Chem.* **22,** 527 (1971); P. Kebarle, ibid. **28,** 445 (1977).
22. *Gas phase ion chemistry*, Vol. 2 (ed. M. T. Bowers), Academic Press, New York (1979): (*a*) Chapt. 11 'The gas phase acidity scale' by J. E. Bartmess and R. T. McIver; (*b*) Chapt. 9 'Stabilities of positive ions from equilibrium gas-phase basicity measurements' by D. H. Aue and M. T. Bowers.
23. J. E. Bartmess, J. A. Scott, and R. T. McIver, *J. Am. chem. Soc.* **101,** 6046 (1979).
24. D. R. Stull and H. Prophet, *JANEF Thermochemical tables*, 2nd edn, NSRDS-NBS37, National Bureau of Standards, Washington, DC (1971).
25. R. W. Taft, Ch. 2, 'Gas-phase proton-transfer equilibria' in *Proton-transfer reactions* (eds. E. F. Caldin and V. Gold), Chapman & Hall, London (1975).
26. E. M. Arnett, Ch. 3, 'Proton transfer and the solvation of ammonium ions' in *Proton-transfer reactions*, (eds. E. F. Caldin and V. Gold), Chapman & Hall (1975).

27. C. D. Johnson, *The Hammett equation*, Cambridge University Press (1973).
28. J. Shorter, *Correlation analysis in organic chemistry*, Oxford University Press (1973).
29. D. H. McDaniel and H. C. Brown, *J. org. Chem.* **23,** 420 (1958).
30. O. Exner, 'A critical compilation of substituent constants', Ch. 10 of *Correlation analysis in chemistry: Recent advances* (eds. N. B. Chapman and J. Shorter), Plenum Press, New York (1978).
31. H. H. Jaffe, *Chem. Rev.* **53,** 191 (1953).
32. C. D. Ritchie, J. D. Saltiel, and E. S. Lewis, *J. Am. chem. Soc.* **83,** 4601 (1961).
33. L. P. Hammett, *Physical organic chemistry*, 2nd edn, McGraw-Hill, New York (1970).
34. There have been attempts to incorporate *ortho* substituents into a more generalized version of the Hammett equation (M. Charton, *Progr. phys. org. Chem.* **8,** 235 (1971); T. Fujita and T. Nishioka, ibid. **12,** 49 (1976)).
35. J. W. Baker and H. B. Hopkins, *J. chem. Soc.* 1089 (1949); J. W. Baker, G. F. C. Barrett, and W. T. Tweed, ibid. 2831 (1952).
36. H. Adkins, R. M. Elofson, A. G. Rossow, and C. C. Robinson, *J. Am. chem. Soc.* **71,** 3622 (1949).
37. C. D. Ritchie and D. J. Wright, *J. Am. chem. Soc.* **93,** 6574 (1971).
38. R. B. Woodward and R. Hoffmann, *The conservation of orbital symmetry*, Verlag Chemie-Academic Press (1970). R. E. Lehr and A. P. Marchand, *Orbital symmetry*, Academic Press, New York (1972).

Supplementary references

R. S. Drago, 'A modern approach to acid–base chemistry', *J. chem. Educ.* **51,** 300 (1974).

T-L. Ho, 'Analysis of some synthetic reactions by the HSAB principle', *J. chem. Educ.* **55,** 355 (1978).

R. G. Pearson, *Hard and soft acids and bases*, Dowden, Hutchinson & Ross, Stroudsburg, Penn., USA (1973).

E. M. Arnett, 'Gas-phase proton transfer – A breakthrough for solution chemistry', *Acc. chem. Res.* **6,** 404 (1973).

E. M. Arnett and G. Scorrano, 'Protonation and solvation in strong aqueous acids', *Adv. phys. org. Chem.* **13,** 83 (1976).

J. R. Jones, *The ionization of carbon acids*, Academic Press, London (1973).

6

Rates of simple chemical reactions

6.1 Introduction

Study of the rates at which elements and compounds react is the field of chemical kinetics. Measurements of rates and investigation of the various factors which affect the rate of a given reaction provide information which is of practical and technological importance. This information also constitutes the bulk of the evidence which leads to a conception of how, at a molecular level, a given reaction proceeds.

6.2 Rate of reaction defined

An important preliminary to any discussion of reaction kinetics is to define precisely what we mean by the rate of a chemical reaction. In Chapter 4, the *extent of reaction*, ξ, was introduced (p. 117); for any general reaction

$$\nu_A . A + \nu_B . B + \ldots \rightleftharpoons \nu_X . X + \nu_Y . Y + \ldots,$$

ξ is defined as

$$\xi = \frac{n_i(\xi) - n_i(0)}{\nu_i},$$

where i represents any reactant or product,

ν_i = the number of moles of i in the balanced chemical equation (positive for products, negative for reactants),

$n_i(\xi)$ = number of moles of i at extent of reaction ξ, and

$n_i(0)$ = number of moles of i at extent of reaction = 0 (when the reaction is represented by the left-hand side of the equation as written).

We now define the *reaction rate* (or *velocity*) as the rate of change of ξ with time, t, and we shall deal only with isothermal processes.

$$\text{Rate of reaction at constant temperature} = \left(\frac{d\xi}{dt}\right)_T$$

so it follows that

$$\left(\frac{d\xi}{dt}\right)_T = \frac{1}{\nu_i} \cdot \left(\frac{dn_i}{dt}\right)_T.$$

If we wish to measure a reaction rate, we need to be able to monitor n_i with time and it does not matter whether i is a reactant or a product. Frequently, it is more convenient to measure (or refer to) the molar concentration, $[i]$, rather than the absolute number of moles, n_i, of a component i in a reaction. If the total volume of the reaction is V,

$$[i] = \frac{n_i}{V},$$

and if V remains constant throughout the reaction,

$$\frac{d[i]}{dt} = \frac{1}{V} \cdot \frac{dn_i}{dt},$$

therefore

$$\left(\frac{d\xi}{dt}\right)_{V,T} = \frac{V}{\nu_i} \cdot \left(\frac{d[i]}{dt}\right)_{V,T},$$

and the reaction rate can be investigated by monitoring the molar concentration of any component i with time. This applies to virtually all reactions in solution, and to gas-phase reactions at constant volume.

Occasionally, the partial pressure P_i of component i in a gas-phase reaction at constant volume V can be monitored. If the gaseous mixture behaves ideally,

$$P_i V = n_i \,.\, RT,$$

therefore

$$n_i = P_i \,.\, \frac{V}{RT},$$

and so at constant volume (and temperature),

$$\left(\frac{dn_i}{dt}\right)_{V,T} = \frac{V}{RT}\left(\frac{dP_i}{dt}\right)_{V,T},$$

and

$$\left(\frac{d\xi}{dt}\right)_{V,T} = \frac{V}{\nu_i RT}\left(\frac{dP_i}{dt}\right)_{V,T}.$$

But if a gas-phase reaction for which $\sum \nu_i \neq 0$ is investigated at constant total pressure, the volume changes as the reaction proceeds. Consequently, for such reactions P_i and $[i]$ are *not* proportional to n_i (or ξ) and dP_i/dt and $d[i]/dt$ are *not* proportional to the rate of the reaction $d\xi/dt$. Some property of the gaseous system which is directly proportional to the extent of the reaction at constant total pressure needs to be identified. In fact, the total volume of the gaseous system, V, is just such a property: at constant pressure (and temperature) $V \propto \xi$, and

$$\left(\frac{d\xi}{dt}\right)_{P,T} \propto \left(\frac{dV}{dt}\right)_{P,T} \quad \text{when} \quad \sum \nu_i \neq 0.$$

In what follows, however, we shall be dealing primarily with isothermal reactions in solution or gas-phase reactions as they are usually investigated in the laboratory – at constant volume. And in such reactions, the rate of change of the molar concentration (or partial pressure in the gas phase) of any component, reactant or product, is a proper measure of the reaction rate.

As the above definitions and equations show, the rate of a reaction is not strictly *equal* to the rate of change of molar concentration of a component, but *proportional* to it. However, it has been the habit of chemists to be rather imprecise in this respect and we shall occasionally follow imprecise usage by referring loosely to $-\frac{d[\text{reactant}]}{dt}$ or $+\frac{d[\text{product}]}{dt}$ as the rate of a reaction.

6.3 An experimental approach

At any instant, the rate of a simple reaction between, for example, two reactants will depend upon the experimental conditions and upon the momentary concentrations of the reactants. Similarly, a compound which reacts in a solvent, or in the gas phase, with the apparent involvement of no other reagents will have an instantaneous reaction rate which depends upon the concentration of the reactant (or its partial pressure in the gas phase), and upon the experimental conditions. In both cases, as the reaction proceeds, the concentrations of the reactants decrease, and consequently the rates of the chemical reactions slow down. The experimental measure of a reaction rate, which is independent of the extent of the reaction and of the scale on which the reaction is carried out is the *rate constant.*

6.3.1 First-order reactions[1]

If the instantaneous rate of reaction of compound A in eqn (6.1) under particular experimental conditions (including constant volume and temperature) is directly proportional to the instantaneous concentration of A as defined by eqn (6.2), the reaction is described as *first order* in A.

$$\mathrm{A} \rightarrow \text{Products} \tag{6.1}$$

$$\text{rate of reaction} = \frac{\mathrm{d}\xi}{\mathrm{d}t} = -\frac{\mathrm{d}n_{\mathrm{A}}}{\mathrm{d}t}$$

and, at constant volume,

$$-\frac{\mathrm{d}[\mathrm{A}]}{\mathrm{d}t} \propto [\mathrm{A}],$$

therefore

$$-\frac{\mathrm{d}[\mathrm{A}]}{\mathrm{d}t} = k_1[\mathrm{A}]; \tag{6.2}$$

so, upon rearrangement and integration,

$$\ln\left(\frac{[\mathrm{A}]_t}{[\mathrm{A}]_0}\right) = -k_1 t \quad \text{or} \quad \log\left(\frac{[\mathrm{A}]_t}{[\mathrm{A}]_0}\right) = -\frac{k_1 t}{2.303}, \tag{6.3}$$

where $[\mathrm{A}]_0$ = concentration of A when time = 0,
$[\mathrm{A}]_t$ = concentration of A at time = t, and
k_1 = *first-order rate constant.*

It follows from eqn (6.3) that the time required for a given proportion of a first-order reaction to take place is independent of the concentration of the reactant. For example, if $t_{1/2}$ = the *half-life* of the reaction (the time required for the amount or concentration of reactant to decrease by 50 per cent),

$$\ln\left(\frac{0.5[\mathrm{A}]_0}{[\mathrm{A}]_0}\right) = \ln(0.5) = -k_1 t_{1/2},$$

therefore

$$k_1 t_{1/2} = \ln 2,$$

or

$$\underline{k_1 t_{1/2} = 0.693.}$$

This shows that the time for the amount of reactant to decrease by a further one-half remains constant throughout a first-order reaction, and knowing $t_{1/2}$ is equivalent to knowing k_1.

From eqn (6.3) and a set of concentration versus time readings, the first-order rate constant can be calculated. Good approximate values may be obtained by a simple graphical method as illustrated below, but the availability of standard programmes for computers or even computing calculators makes the processing of experimental results less tedious and more reliable.[2]

Example. *Kinetics of the gas-phase isomerization of 1-ethylcyclobutene at 179.8* °C. (Results taken from H. M. Frey and R. F. Skinner, *Trans. Faraday Soc.* **61,** 1918 (1965).)

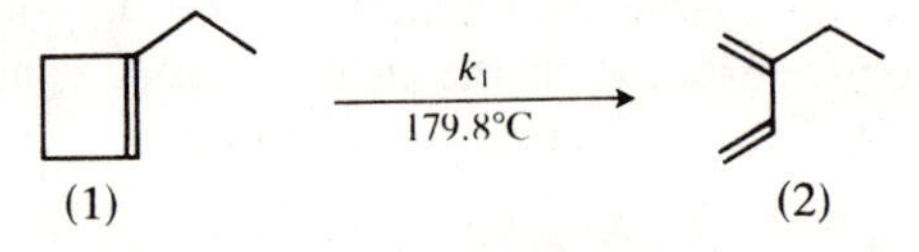

The percentage of 1-ethylcyclobutene (1), $\frac{[1]\times 100}{([1]+[2])}$, in the reaction mixture with 2-ethylbuta-1,3-diene (2) was measured at various times during the reaction at constant volume by gas–liquid chromatography. By the first-order rate law (eqn (6.3)), $\log\left(\frac{[1]_t}{[1]_0}\right)$ may be plotted against time, but $([1]_t+[2]_t)=[1]_0$ therefore log(% 1) can be plotted as shown; note the use of decadic logarithms.

Time/min	5	9	12	15	18.5	22	25	30
% (1) in the mixture	76.77	61.42	51.88	44.93	36.67	30.15	26.16	19.98
log(% (1))	1.885	1.788	1.715	1.653	1.564	1.479	1.418	1.301

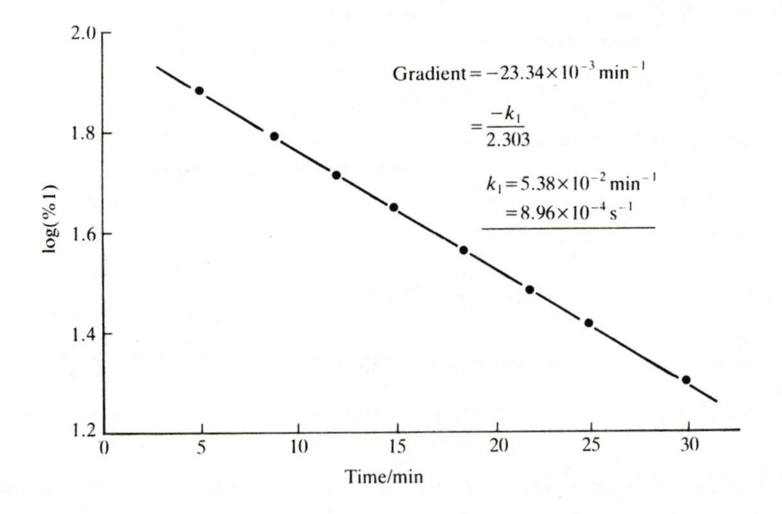

The half-life of this reaction at 179.8 °C is given by

$$8.96\times 10^{-4}\ \text{s}^{-1}\times t_{1/2}=0.693$$

$$t_{1/2}=773\ \text{s}$$

$$=12.9\ \text{min.}$$

So after 12.9 min, the reaction is 50 per cent complete, after a further 12.9 min (a total of 25.8 min), it is 75 per cent complete, after a total of 38.7 min it is 87.5 per cent complete, and so on. After 10 half-lives (a total of 129 min), the reaction is almost wholly complete (>99.9 per cent).

When the reactant A in eqn (6.1) does not lend itself to easy analysis, some other means of following the reaction must be devised. If there is evidence that the products form from A without the intervention of long-lived intermediates, and that they are stable to the experimental conditions, then the formation of a product whose concentration can easily be followed may be monitored.

A reaction may also be followed by monitoring any property of the whole system which is directly proportional to the extent of reaction. In gas-phase reactions at constant volume, the total pressure of the system is such a property if one mole of reactant yields either more or less than one mole of gaseous product.

Example. *Kinetics of the gas-phase thermal elimination of hydrogen chloride from cyclohexyl chloride at 350.2 °C.* (Results taken from E. S. Swinbourne, *Aust. J. Chem.* **11,** 314 (1958).)

The reaction was investigated by following the increase in total pressure at constant volume.

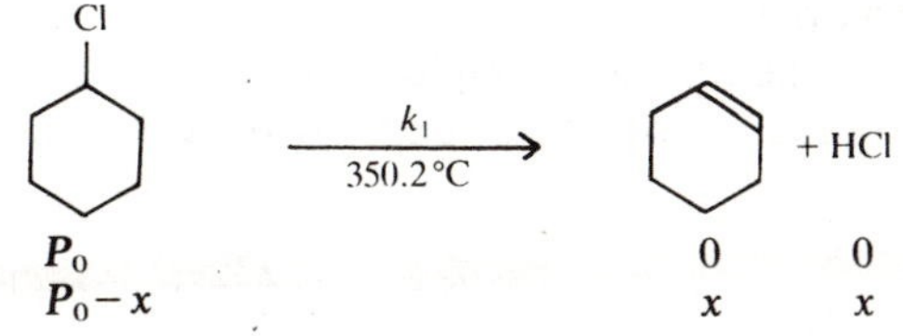

Initial partial pressures:	P_0	0	0
partial pressures at time t:	$P_0 - x$	x	x

So at time t, total pressure, $P_t = P_0 + x$ therefore $x = P_t - P_0$ and partial pressure of reactant at time t $(P_0 - x) = 2P_0 - P_t$. By eqn (6.3),

$$\ln\left(\frac{2P_0 - P_t}{P_0}\right)$$

is plotted against time, now using natural logarithms, and the rate constant is obtained directly from the gradient.

Time/min	0	2	10	30	50	70	90	110	130
Total pressure, P_t/Torr	4.83	4.94	5.29	6.14	6.78	7.32	7.75	8.12	8.39
$(2P_0 - P_t)$/Torr	4.83	4.72	4.37	3.52	2.88	2.34	1.91	1.54	1.27
$(2P_0 - P_t)/P_0$	1	0.977	0.905	0.729	0.596	0.484	0.395	0.319	0.263
$\ln\left(\frac{2P_0 - P_t}{P_0}\right)$	0	−0.0233	−0.0998	−0.316	−0.518	−0.726	−0.929	−1.14	−1.34

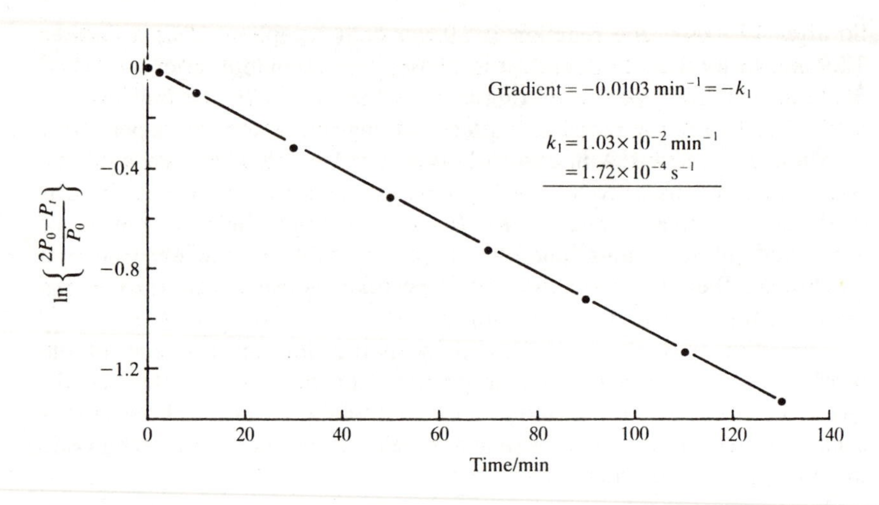

For reactions in solution, the property to be followed may be the electrical conductivity if ions are formed or replaced as the reaction proceeds. Other physical properties of a system which may be monitored in suitable cases include spectrophotometric absorbance, optical activity, and pH.

Some further examples of first-order reactions are shown below including the temperature ranges over which rates were measured.

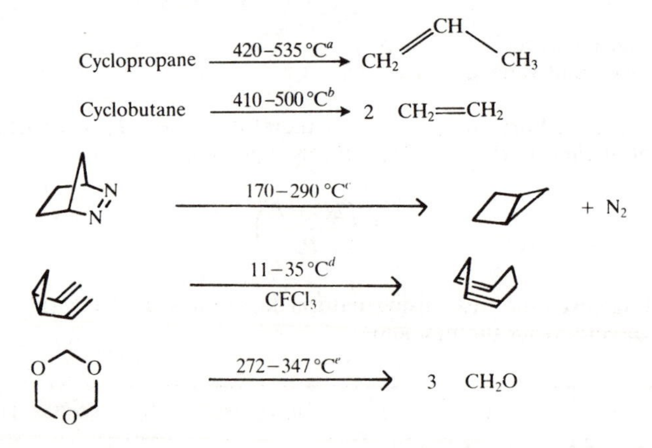

[a] W. E. Falconer, T. F. Hunter, and A. F. Trotman-Dickenson, *J. chem. Soc.* 609 (1961).

[b] R. W. Carr and W. D. Walters, *J. phys. Chem.* **67,** 1370 (1963).

[c] R. J. Crawford, R. J. Dummel, and A. Mishra, *J. Am. chem. Soc.* **87,** 3023 (1965).

[d] J. M. Brown, B. T. Golding, and J. J. Stofko, *J. chem. Soc., Perkin 2,* 436 (1978).

[e] W. Hogg, D. M. McKinnon, A. F. Trotman-Dickenson, and G. J. O. Verbeke, *J. chem. Soc.* 1403 (1961).

One important matter requires emphasis. Because of the form of the integrated rate law shown in eqn (6.3), the first-order rate constant k_1 will have the dimension only of $(\text{time})^{-1}$, usually s^{-1}. The numerical magnitude of the rate constant of a given first-order reaction at a particular temperature will be affected only by this choice of units of time, and will be independent of the property which was monitored to follow the progress of the reaction, and of the units of that property.

If it is found that the rate constant of a reaction measured by monitoring the disappearance of reactant appears higher than by monitoring the formation of product, or for example that the rate of loss of optical activity of a solution of a chiral reactant is faster than the rate of formation of the achiral product, then the reaction does not involve the exclusive conversion of reactant *directly* into products. A more complicated and perhaps more interesting mechanism is indicated.

6.3.2 Second-order reactions[1]

The reaction in eqn (6.1) is described as *second order* in A if the instantaneous rate is proportional to the square of the instantaneous concentration of A. This rate law is described by eqn (6.4)

$$\text{A} \rightarrow \text{products} \tag{6.1}$$

$$\text{Rate of reaction} = \frac{\mathrm{d}\xi}{\mathrm{d}t} = -\frac{\mathrm{d}n_\text{A}}{\mathrm{d}t}$$

and, at constant volume,

$$-\frac{\mathrm{d}[\text{A}]}{\mathrm{d}t} \propto [\text{A}]^2$$

therefore

$$-\frac{\mathrm{d}[\text{A}]}{\mathrm{d}t} = k_2[\text{A}]^2 \tag{6.4}$$

so, upon rearrangement and integration,

$$\frac{1}{[\text{A}]_t} - \frac{1}{[\text{A}]_0} = k_2 t, \tag{6.5}$$

where $[\text{A}]_0$ = concentration of A when time = 0,
$[\text{A}]_t$ = concentration of A at time = t, and
k_2 = *second-order rate constant.*

If $t_{1/2}$ is the half-life of this second-order reaction (the time required for the amount of A to be reduced by 50 per cent), then by eqn (6.5),

$$\frac{2}{[\text{A}]} - \frac{1}{[\text{A}]} = k_2 t_{1/2}$$

or

$$k_2 t_{1/2} = \frac{1}{[A]}.$$

This shows that $t_{1/2}$ of a second-order reaction at constant volume is inversely proportional to the amount or concentration of the reactant present. As the second-order reaction proceeds, the successive periods required for the concentration of reactant to decrease further by one half get progressively longer.

The integrated rate law, eqn (6.5), also shows that the second-order rate constant may be obtained graphically from a set of concentration versus time results.

Example. *Kinetics of the gas-phase dimerization of buta-1,3-diene at 326 °C.* (Results taken from W. E. Vaughan, *J. Am. chem. Soc.* **54,** 3863 (1932).)

The reaction was investigated by following the decrease in the total pressure at constant volume

$$\text{buta-1,3-diene} \xrightarrow[326\,°\text{C}]{k_2} 0.5\ \text{4-vinylcyclohexene}$$

Initial partial pressures:	P_0	0
partial pressures at time t:	$P_0 - x$	$x/2$

So at time t, total presure, $P_t = P_0 - x/2$ therefore

$$x = 2(P_0 - P_t)$$

and partial pressure of reactant at time t $(P_0 - x) = 2P_t - P_0$. By the second-order rate law (eqn (6.5)), $1/(2P_t - P_0)$ is plotted against t.

Time/min	0	3.25	10.08	17.3	24.5	36.38	49.50	60.87
pressure, P_t/Torr	632.0	618.5	591.6	567.3	546.8	521.2	498.1	482.8
$\frac{10^4}{(2P_t - P_0)}\Big/\text{Torr}^{-1}$	—	16.5	18.5	19.9	21.7	24.4	27.5	30.0

Time/min	90.05	135.72
pressure, P_t/Torr	453.3	422.8
$\frac{10^4}{(2P_t - P_0)}\Big/\text{Torr}^{-1}$	36.4	46.8

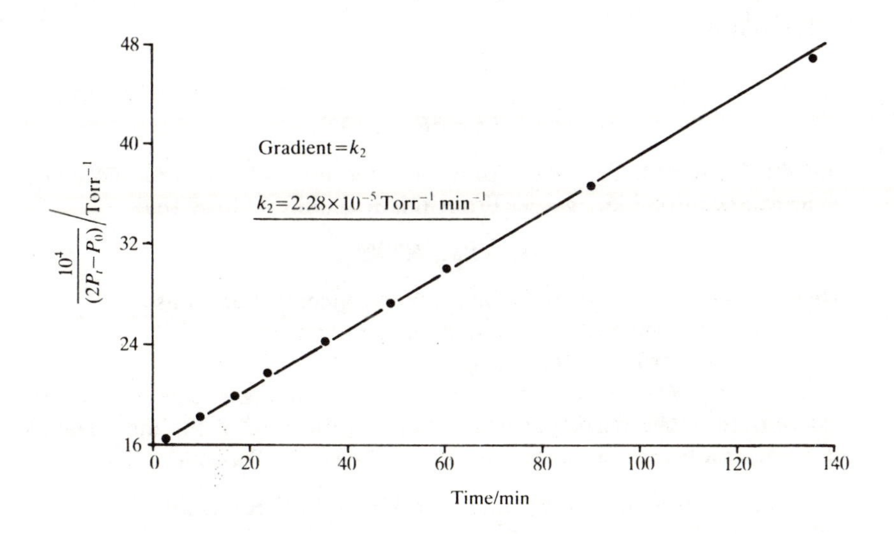

It is evident from the form of eqn (6.5) that the dimensions of a second-order rate constant of a reaction in solution or the gas phase are those of (molar concentration$\times$time)$^{-1}$ or, for a gas-phase reaction, (pressure$\times$time)$^{-1}$. The numerical value of the second-order rate constant of a particular reaction will, therefore, be affected both by the choice of time units (usually seconds), and by the units chosen to express molar concentration (or pressure). Regardless of how a rate is measured, it is now customary to express a second-order rate constant in units $dm^3\,mol^{-1}\,s^{-1}$ (sometimes abbreviated to $M^{-1}\,s^{-1}$). Consequently, we need to be able to convert the dimensions (pressure)$^{-1}$ to $dm^3\,mol^{-1}$ for rate constants of gas-phase reactions which have been measured by monitoring pressure at constant volume, for example the dimerization of buta-1,3-diene considered above. This is easily accomplished if we can assume ideal gas behaviour.

The partial pressure P_i of n_i moles of an ideal gas in a total volume V is given by

$$P_i = \frac{n_i}{V}\,.\,RT.$$

Therefore if there is a change δn in the total number of moles of gases present in a reaction at constant volume V at temperature T, there will be a total pressure change δP where

$$\delta P = \frac{\delta n}{V}\,.\,RT.$$

But $\delta n/V = \delta c$, the change in total gas concentration expressed in moles per unit volume.

Therefore

$$\delta P = \delta c \,.\, RT,$$

or

$$\delta c = \delta P \,.\, (RT)^{-1},$$

and this represents the relationship between units of pressure and molar concentration for ideal gases. From this it readily follows that

$$k_c = k_P \,.\, (RT)^{n-1}, \tag{6.6}$$

where k_c = rate constant including molar concentration units,
k_P = rate constant including pressure units, and
n = the order of the reaction.

It is important to ensure that the units of the gas constant R are appropriate to the particular application. In the case of the dimerization of butadiene given above, we must use $R = 62.36\ \text{Torr dm}^3\ \text{K}^{-1}\ \text{mol}^{-1}$:

$$k_P = 2.28 \times 10^{-5}\ \text{Torr}^{-1}\ \text{min}^{-1}\ \text{at } 326\ °\text{C } (599\ \text{K}),$$

$$n = 2,$$

therefore

$$k_c = 2.28 \times 10^{-5}\ \text{Torr}^{-1}\ \text{min}^{-1} \times (62.36\ \text{Torr dm}^3\ \text{K}^{-1}\ \text{mol}^{-1} \times 599\ \text{K})$$
$$= 0.852\ \text{dm}^3\ \text{mol}^{-1}\ \text{min}^{-1},$$

or

$$k_c = 1.42 \times 10^{-2}\ \text{dm}^3\ \text{mol}^{-1}\ \text{s}^{-1}.$$

Chemical reactions between two different compounds as in eqn (6.7) may be first order in either A or B, or both A and B.

$$\text{A} + \text{B} \rightarrow \text{Products} \tag{6.7}$$

If the reaction is first order in both, it is overall second order; and because A and B react stoichiometrically as described by eqn (6.7) the amounts of both must decrease at the same rate so the progress of the reaction at constant volume may be followed by monitoring the decrease in concentration of either A or B. The differential second-order rate law, therefore, is:

$$-\frac{d[\text{A}]}{dt} = -\frac{d[\text{B}]}{dt} = k_2[\text{A}][\text{B}] \tag{6.8}$$

and the integrated rate law is given by eqn (6.9):[1]

$$\frac{1}{([\text{A}]_0 - [\text{B}]_0)} \,.\, \ln\left(\frac{[\text{A}]_t[\text{B}]_0}{[\text{A}]_0[\text{B}]_t}\right) = k_2 t \tag{6.9}$$

or

$$\ln\left(\frac{[\text{A}]_t}{[\text{B}]_t}\right) = ([\text{A}]_0 - [\text{B}]_0)k_2 t + \ln\left(\frac{[\text{A}]_0}{[\text{B}]_0}\right)$$

where $[A]_0$ and $[B]_0$ = concentrations of A and B when time = 0,
$[A]_t$ and $[B]_t$ = concentrations of A and B when time = t, and
k_2 = second-order rate constant.

Example. *Kinetics of the elimination of* HCl *from the insecticide DDT* (4) *using sodium ethoxide in ethanol at* 25 °C. (Results taken from B. D. England and D. J. McLennan, *J. chem. Soc.* B 696 (1966).)

$$C_2H_5O^- + CCl_3\text{—}CHAr_2 \xrightarrow[Ar=pClC_6H_4]{k_2,\ C_2H_5OH} Cl^- + Cl_2C{=}CAr_2 + C_2H_5OH$$

	(3)	(4)		
Initial concentration/ mol dm^{-3}	0.0246	0.0129	0	0

The rate was measured by monitoring the concentration of chloride ion which increases as the reaction proceeds. The increase in $[Cl^-]$ may be related by the stoichiometry of the chemical reaction to the decrease in the concentrations of the reactants, and $\ln([3]_t/[4]_t)$ is plotted against time.

Time/s	0	75	135	225	400	665	900	1200	1800	∞
$100[Cl^-]$/ mol dm^{-3}	0	0.13	0.23	0.38	0.52	0.73	0.85	0.98	1.11	1.29
$100[3]_t$/ mol dm^{-3}	2.46	2.33	2.23	2.08	1.94	1.73	1.61	1.48	1.35	1.17
$100[4]_t$/ mol dm^{-3}	1.29	1.16	1.06	0.91	0.77	0.56	0.44	0.31	0.18	0
$\ln\left(\frac{[3]_t}{[4]_t}\right)$	0.645	0.697	0.744	0.827	0.924	1.13	1.30	1.56	2.01	∞

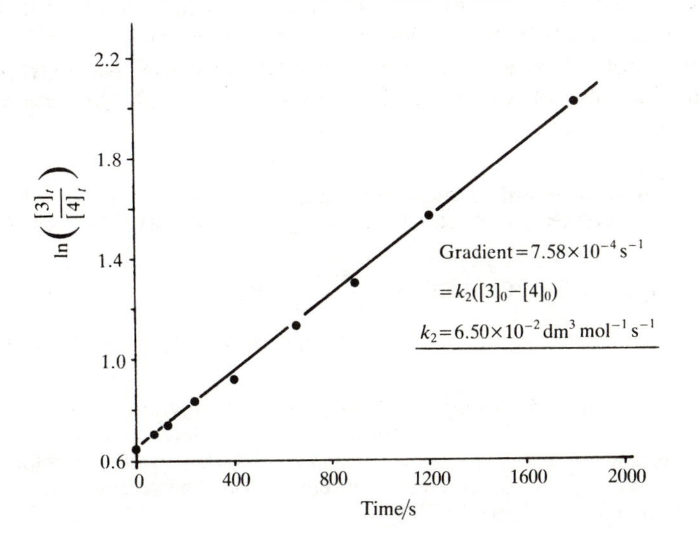

If $[A]_0 = [B]_0$, eqn (6.9) cannot be derived from eqn (6.8). In this event the reaction may be treated mathematically as the second-order reaction of a single compound and eqn (6.5) can be applied.[1]

If A is in vast excess over B, the complete reaction of B, in consuming an equimolar amount of A as required by eqn (6.7), will deplete the total amount of A by a negligible proportion. The concentration of A remains virtually constant, therefore, during the complete reaction. Expressing this mathematically, if $[A]_0 \gg [B]_0$, then $[A]_t \sim [A]_0$ and $[A]_0 - [B]_0 \sim [A]_0$, and eqn (6.9) becomes

$$\frac{1}{[A]_0} \cdot \ln\left(\frac{[B]_0}{[B]_t}\right) = k_2 t$$

or

$$\ln\left(\frac{[B]_t}{[B]_0}\right) = -k_2[A]_0 t$$

and

$$\ln\left(\frac{[B]_t}{[B]_0}\right) = -k_1' t \tag{6.10}$$

where

$$k_2[A]_0 \doteq k_1' \tag{6.11}$$

and k_1' = *pseudo* first-order rate constant.

Equation (6.10) contains the *pseudo first-order rate constant* $k_1' = k_2[A]_0$ and is identical in form with eqn (6.3). By this simple device of having one reactant in the overall second-order process present in large excess, the kinetics may be investigated more easily by using first-order rate law techniques. If the (initial) concentration of the reactant present in excess is known, the real second-order rate constant, k_2, may be calculated from the observed *pseudo* first-order rate constant k_1 by eqn (6.11). If a graphical method is used as in the following example, the extrapolated straight line plot of k_1' against $[A]_0$ must go through the origin since $k_1' = 0$ when $[A]_0 = 0$.

Example. *Base-promoted dehydrochlorination of N-chlorobenzylmethylamine.* (From R. A. Bartsch and B. R. Cho, *J. Am. chem. Soc.* **101,** 3587 (1979).)

$$PhCH_2NClCH_3 + (CH_3)_3CO^-K^+ \xrightarrow[(CH_3)_3COH]{39.0\,°C} PhCH{=}NCH_3 + K^+Cl^- + (CH_3)_3COH.$$

The rate was followed by monitoring the increase in u.v. absorption at 244 nm due to the imine. The reaction mixture contained a large excess of potassium t-butoxide and so the reaction is *pseudo* first order in the N-chloroamine. Calculate the second-order rate constant from the results given when $[PhCH_2NClCH_3] \sim 10^{-5}$ mol dm^{-3}.

$10^3[(CH_3)_3CO^-K^+]$/mol dm^{-3}	$10^3k'_1$/s^{-1}
1.39	1.28
6.50	6.08
13.9	14.3
14.5	14.9

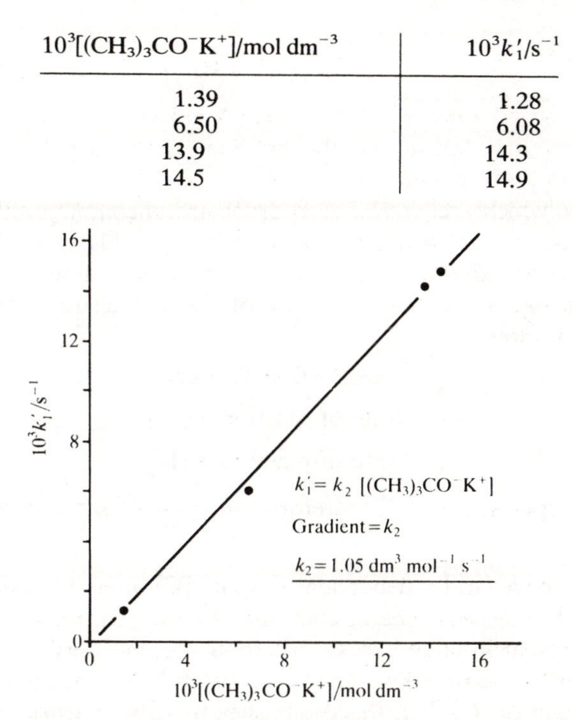

Below are more examples of some well known second-order reaction types (substitutions, cycloadditions, and eliminations).

$$C_2H_5{-}Br + Cl^- \xrightarrow[\text{acetone}^3]{25\text{–}65\ ^\circ C} C_2H_5{-}Cl + Br^-$$

$$PhCH_2CH_2Br + C_2H_5O^- \xrightarrow[C_2H_5OH^5]{30\text{–}60\ ^\circ C} PhCH{=}CH_2 + C_2H_5OH + Br^-.$$

6.3.3 Zero-order reactions

The rates of some reactions are independent of the amount or concentration of reactant present and, therefore, do not depend upon the extent of the reaction.

$$A \rightarrow \text{Products}$$

Rate of reaction $= d\xi/dt$ is independent of ξ, so $-(dn_A/dt)$ is independent of n_A. At constant volume, therefore, $-(d[A]/dt)$ is independent of [A], or $-(d[A]/dt) \propto [A]^0$, therefore, $-(d[A]/dt) = k_0$.

Upon integration,

$$[A]_0 - [A]_t = k_0 t$$

where k_0 = the *zero-order rate constant.* A graph of $[A]_t$ against time for such a reaction is linear and the gradient is the negative of the rate constant with units mol $dm^{-3} s^{-1}$.

Reactions which are *overall* zero order are uncommon although some heterogeneously-catalyzed gas reactions are of this type. In reactions between two or more substrates, the rate may have a zero-order dependence upon the concentration of one of them; such reactions are common. For example,

$$A + B + C \rightarrow \text{Product}$$

$$\text{Rate of reaction} \propto [A]$$

$$\text{Rate of reaction} \propto [B]^2$$

but rate independent of [C] therefore rate = $k_3[A][B]^2$ and zero order in [C].

The zero-order rate dependence upon [C] may be established by monitoring [C] against time at constant [A] and [B], for example having A and B present in large excess. Alternatively, the third-order rate constant k_3 can be measured at different constant concentrations of C. If k_3 is independent of [C] then this establishes that the reaction is zero-order in [C]. The hydrolysis of *t*-butyl chloride provides a simple illustration:[6]

$$(CH_3)_3C\text{—}Cl + OH^- \xrightarrow{H_2O} (CH_3)_3C\text{—}OH + Cl^-$$

$$-\frac{d[(CH_3)_3CCl]}{dt} = -\frac{d[OH^-]}{dt} = k_1[(CH_3)_3CCl] \text{ and zero order in } [OH^-].$$

The reaction is, therefore, first order in *t*-butyl chloride but zero order in hydroxide even though the overall reaction consumes OH^-, and the rate may be measured by monitoring the decrease in $[OH^-]$.

When a rate law is zero order in one of several reactants, we may conclude that the overall reaction proceeds in more than one step. And the reactant with the zero-order rate influence must be involved in a later step which does not restrict the overall rate. We shall consider such matters further in the next chapter.

We have not dealt with rate laws which are other than zero, first, or second order, or with the important experimental question of how one initially establishes the order of a chemical reaction. As these matters are fully described elsewhere,[1,2] we shall not cover them here except to emphasize that knowledge of the reaction order with respect to each reactant is in itself important evidence towards understanding the mechanism, as well as a prerequisite of measuring a rate constant.

As mentioned above, current computer programmes may be used with advantage to calculate a rate constant from knowledge of the reaction order and a set of experimental measurements.[2] Such programmes also usually provide the *standard deviation*, a statistical term which reflects the experimental uncertainty in the parameter which has been calculated from the particular results.[7] The probable error expressed in this way depends upon the experimental technique used, the temperature stability during the reaction, and the quality of the instrumentation, as well as the nature of the reaction and the skill of the investigator. In propitious cases it may be less than 1 per cent but more commonly is of the order of 1–5 per cent. It will not include *systematic errors* which may be more difficult to detect and cope with.[2,7]

Occasionally it is desirable to determine a rate constant several times and take a mean value. The standard deviations on individual rate constants and the spread of the rate constants around their mean value together indicate the precision and reliability of the result.

6.3.4 Effect of temperature upon reaction rates

Arrhenius deduced that the variation of a rate constant with temperature may be described by eqns (6.12) and (6.13). It has been shown over many years that these equations fit the experimental results for the rate constants of many reactions. Each such reaction has a *pre-exponential factor*, A, and an *activation energy*, E_a, which are approximately temperature-independent over a limited temperature range. Empirically, the activation energy is the parameter which expresses the temperature dependence of the rate constant;[8] a small E_a corresponds to a rate constant which does not increase rapidly with temperature whereas a reaction which has a very strong temperature dependence has a large E_a.

$$k = A\,.\,\mathrm{e}^{-E_a/RT} \qquad (6.12)$$

$$\ln k = \ln A - \frac{E_a}{RT} \quad \text{or} \quad \log k = \log A - \frac{E_a}{2.303RT}, \qquad (6.13)$$

where k = rate constant,
E_a = activation energy,
T = absolute temperature,
R = gas constant which includes the same energy units as E_a, and
A = the A-factor, or pre-exponential factor.

The pre-exponential factor, A, has the same units as the rate constant. These cancel in eqn (6.12) and it is understood that eqn (6.13) includes only the numerical values of k and A.[9] Mathematically, A is the value of the rate constant in the absence of the activation energy constraint. In other words, if E_a were zero or T infinite, then $k = A$. The activation energy and the pre-exponential factor may be readily obtained graphically by determining the rate constant at several temperatures and plotting ln k or log k against $1/T$ according to eqn (6.13). Because of the usually rapid increase in reaction rates with temperature, it is seldom practicable to obtain rate constants for a particular reaction over a temperature range greater than about 60 °C and occasionally one has to be satisfied with results over only about 30 °C.

Example. *Arrhenius plot for the first-order gas-phase isomerization of bicyclo*[4.2.0]*oct-7-ene.* (Results taken from G. R. Brandon, H. M. Frey, and R. F. Skinner, *Trans. Faraday Soc.* **62,** 1546 (1966).)

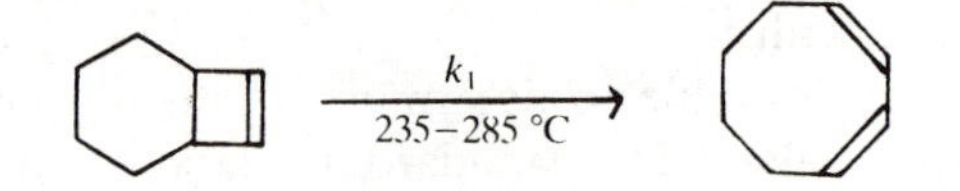

Temp./°C	235.6	244.0	248.4	252.9	256.0	259.3	262.4	267.2	278.7	285.0
Temp./K	508.7	517.1	521.5	526.0	529.1	532.4	535.5	540.3	551.8	558.1
$10^4 T^{-1}/K^{-1}$	19.66	19.34	19.18	19.01	18.90	18.78	18.67	18.51	18.12	17.92
$10^4\, k_1/s^{-1}$	0.376	0.763	1.08	1.53	1.93	2.44	3.32	4.61	10.5	16.6
$-\log(k_1/s^{-1})$	4.425	4.117	3.967	3.815	3.714	3.613	3.479	3.336	2.979	2.780

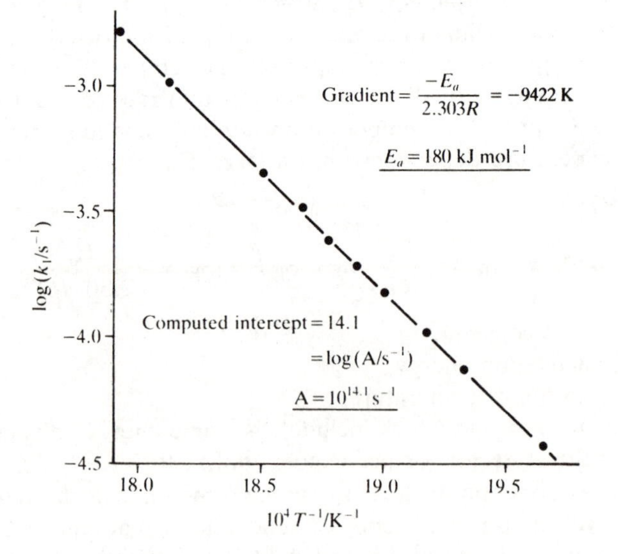

Example. From the above results, estimate the rate constant for the isomerization of bicyclo[4.2.0]oct-7-ene at 250 °C.

From eqn (6.13),

$$\log(k_{523.1\,\mathrm{K}}/\mathrm{s}^{-1}) = 14.1 - \frac{180\times 10^{3}\,\mathrm{J\,mol^{-1}}}{2.303\times 8.314\,\mathrm{J\,K^{-1}\,mol^{-1}}\times 523.1\,\mathrm{K}}$$

$$= 14.1 - 18.0$$

$$= -3.9$$

$$\underline{k_{523.1\,\mathrm{K}} = 1.3\times 10^{-4}\,\mathrm{s}^{-1}}.$$

The pre-exponential factor and the activation energy are experimental quantities which are important, not only because they allow calculation of a rate constant at one temperature from experimental results at other temperatures, but principally because they may be related to other terms which originate in a theory of reaction rates. This theory provides a self-consistent description of simple reactions in molecular terms which meshes with a wealth of knowledge of structure of organic compounds. It is this intimate relationship between structure of reactants and products, experimental rate measurements, and theory which is the basis of current mechanistic organic chemistry.

6.4 A theoretical approach

There are of course various theories involved in chemical kinetics including those such as quantum theory which pervade the whole of modern chemistry. One model, collision theory, originated in the kinetic theory of gases. But whilst this may be usefully applied to reactions of simple compounds in the gas phase, it has not significantly contributed to an understanding of how reactions of more complex molecules occur in solution. A more fruitful model, both quantitatively and qualitatively in the sense of allowing a description of a chemical reaction in terms of physical processes, is embodied in *Transition State Theory.*[1,2] The concept of relating the macroscopic conversion of reactant into product in a chemical reaction to a reorganization of the internal structure of individual molecules has been a major triumph of scientific imagination.

6.4.1 Reaction coordinate and activated complex

In the reversible isomerization reaction illustrated in eqn (6.14), the conversion of 1 mole of A into 1 mole of B involves a change in free energy, and if the initial and final states are standard states of A and B, this change is the standard molar free energy of the reaction, $\Delta_r G^\ominus$. When the standard molar free energy of the product is lower than that of the reactant, the reaction is thermodynamically favourable, $\Delta_r G^\ominus$ is

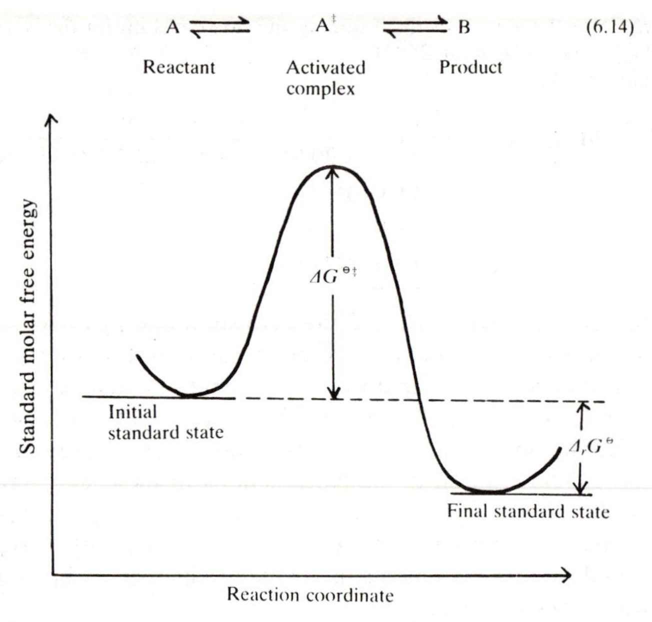

Fig. 6.1. Standard free-energy reaction profile of a simple reversible isomerization.

negative as shown in Fig. 6.1, and the thermodynamic equilibrium constant $K_a^{\ominus} > 1$. However, the process is not instantaneous and we can consider initial and final states to be separated in the *reaction coordinate* by a standard free-energy maximum or barrier.

The reaction coordinate in Fig. 6.1 must not be confused with the degree of reaction, α, encountered in Figs 4.1(a) and (b) (Chapter 4). The earlier diagrams on p. 119 represent the real change in total free energy of the system as reactant, molecule by molecule, is converted into product. The degree of reaction, therefore, measures the proportion of the reaction which has taken place; at a value of α between 0 and 1 in Fig. 4.1 (Chapter 4), some molecules will be unreacted, some will already have reacted, and a tiny number will be in the process of reacting. Figure 6.1 describes the *hypothetical* conversion of one whole mole of reactant into product, all molecules reacting synchronously. At some position along the reaction coordinate between initial and final states in Fig. 6.1, one mole of material is neither reactant, nor product but something between the two.

A basic assumption of the transition state theory of reaction rates is that the molecular species in the real reaction corresponding to the standard free-energy maximum, the *activated complexes*, may be treated statistical-thermodynamically as though they are in equilibrium with

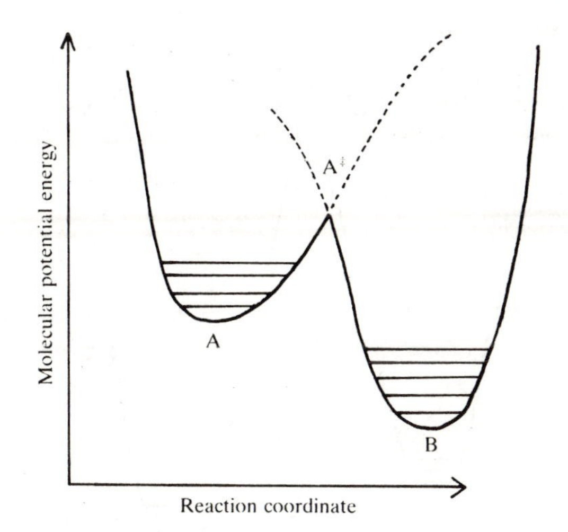

Fig. 6.2. Potential-energy reaction profile of a simple reversible isomerization.

reactant molecules. But it cannot be a real equilibrium because an activated complex is not an ordinary molecule. We can elucidate this by considering a molecular potential-energy reaction profile, Fig. 6.2. This diagram shows the change in potential energy of a *single molecule* of A as it is transformed into one of B. It may be considered as the superposition of two potential-energy curves, one for A and the other for B, and the activated complex $A^{\ddagger}$ which intervenes in the interconversion of the two molecules is represented by the intersection of the two curves. Plotted against potential energy in Fig. 6.2 is a coordinate which describes the extent of internal molecular reorganization which corresponds to the conversion of a molecule of A into one of B, i.e. the reaction coordinate. An activated complex $A^{\ddagger}$ will be structurally related to A and to B; but the reaction coordinate will almost never be a single structural parameter such as a bond angle or an interatomic distance. It will be some composite term which, because of dimensional restrictions, we represent as a single coordinate.

An example is the potential-energy reaction profile for the reversible thermal conversion of one molecule of *cis*-2-butene into *trans*-2-butene, eqn (6.15) and Fig. 6.3. We may represent the reaction coordinate for this isomerization approximately as the dihedral angle between the two methyl groups. This angle is 0 in the *cis* isomer, 180° in the *trans* isomer, and approximately 90° within the activated complex. Clearly, the size of the angle describes the progress of a single molecule undergoing reaction. But we say that the dihedral angle only approximates the reaction coordinate because other structural changes also occur. The length of the central C–C bond, for example, or the C–H bond lengths at C-2 and C-3

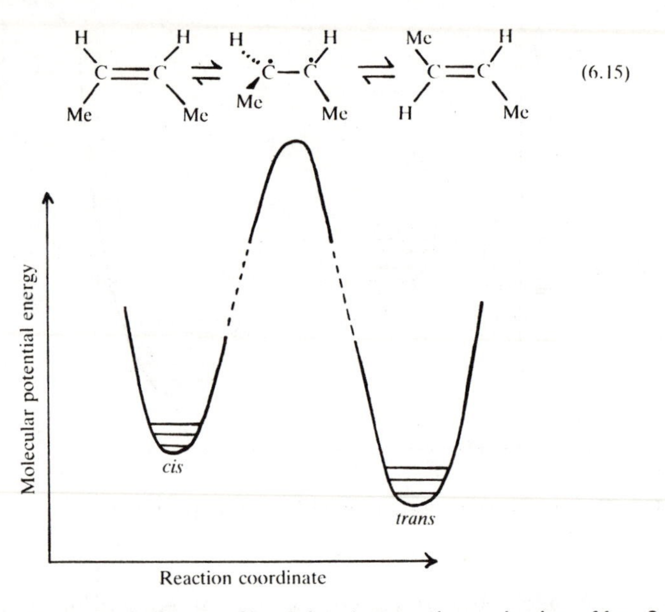

Fig. 6.3. Potential-energy profile of the *cis-trans*-isomerization of but-2-ene.

will not in principle remain unchanged between reactant and activated complex. Consequently, because we want the *total potential energy* of the reacting molecule to be represented and not just the torsional potential energy, the dihedral angle does not fully describe the change in the molecule as it is transformed from reactant into activated complex and from there to product. This necessitates an imprecision about the definition of reaction coordinate, but this should not be mistaken for a vagueness in the concept. Frequently, as in this *cis-trans* isomerization, the early stages of a reaction coordinate are assumed to be a vibrational normal coordinate (see p. 14).

Whereas reactant and product may undergo molecular vibrations *within* the reaction coordinate, and quantized energy levels for these are included in the potential-energy diagrams, an activated complex may not. This is the crucial difference between the activated complex, $A^{\ddagger}$, and the ordinary molecules of reactant and product, A and B in eqn (6.14) and Fig. 6.2. There is no restoring force for a vibration of the activated complex within the reaction coordinate because $A^{\ddagger}$ corresponds to a potential-energy maximum; it may only undergo a translation within the reaction coordinate and thereby start to become either reactant or product.

A third dimension representing one of the vibrational degrees of freedom which are not components of the reaction coordinate may be

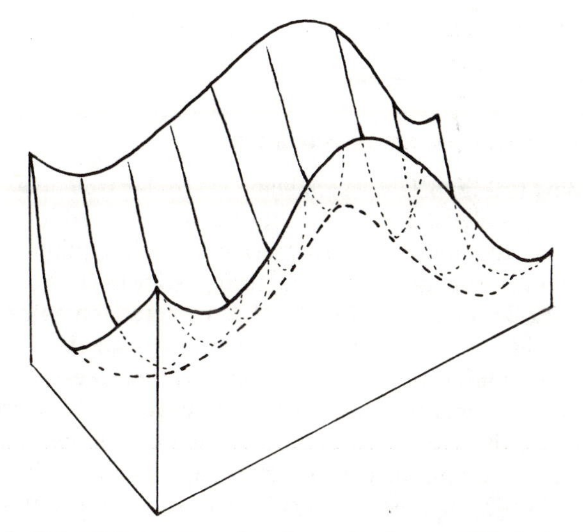

Fig. 6.4. Potential-energy surface for a simple reversible isomerization.

added to Fig. 6.2. We then have in Fig. 6.4 a potential-energy *surface* which includes a representation of how a molecular vibration of the reactant which is not the reaction coordinate is transformed into a vibration of the activated complex. A section along the trough of this energy surface in Fig. 6.4 would give the two-dimensional reaction profile of Fig. 6.2. A section through the surface perpendicular to the reaction coordinate would give a familiar potential-energy curve for the particular molecular vibration included in Fig. 6.4 of the molecule whose configuration is otherwise determined by the position along the reaction coordinate where the section is taken. Figure 6.4 shows that, in the region describing the activated complex, the energy surface is a saddle-point. With a hypothetical multi-dimensional potential-energy hypersurface, we could include all degrees of freedom of the molecule as well as the reaction coordinate. We can perceive that (just as in Figs 6.2–6.4) the reaction coordinate in the region corresponding to the activated complex shows an energy maximum; but in all other dimensions, the curvature of the energy hypersurface is concave.

This hypothetical potential-energy hypersurface in the region of the maximum in the reaction coordinate is the *transition state* of the reaction. The transition state of course has no physical existence; it is a multi-dimensional mathematical relationship between potential energy and atomic configuration. On the other hand, to the extent that transition state theory is valid, an activated complex is a real molecule and is the configuration of atoms corresponding to the potential energy maximum in the reaction coordinate. This is true regardless of how transient its existence may be, and regardless of how incompletely we understand its properties. As used here, the terms 'activated complex' and 'transition state' are not synonyms.

6.4.2 Standard free energy of activation

We return now to Fig. 6.1 in which standard molar free energy is plotted against the same molecular reaction coordinate as is used in the molecular potential-energy diagram. A quasi-thermodynamic equilibrium constant $K_c^{\ominus\ddagger}$ may in principle be calculated, using statistical thermodynamics, relating the molar concentration of the activated complex to that of reactant A during the real reaction. Treating A and $A^\ddagger$ as though they are in dynamic equilibrium requires that the lifetime of the activated complex is long compared with the time required for equilibrium to be achieved among the quantized energy levels of its molecular vibrations and rotations (excluding, of course, the absent vibration in the reaction coordinate), though it must be very short within the timescale of the chemical reaction.

The *standard molar free energy of activation*, $\Delta G^{\ominus\ddagger}$, corresponding to this quasi-thermodynamic equilibrium constant ($\Delta G^{\ominus\ddagger} = -RT \ln K_c^{\ominus\ddagger}$) is included in Fig. 6.1. Formally, $\Delta G^{\ominus\ddagger}$ is the change in the free energy of the system associated with the hypothetical formation of 1 mole of activated complex from reactant at constant temperature, both species being in the defined standard state (normally 1 mol dm^{-3}). This is a theoretical quantity directly related, as we shall see later, to the rate constant.

A reaction profile of another type of reversible reaction (eqn (6.16)) may also be constructed either in terms of molecular potential energy or standard molar free energy as shown in Fig. 6.5. The A–B and C–D intermolecular distances approximate the reaction coordinate and we shall consider a molecule of A and one of B as they react. In the initial (standard) state, A and B have their equilibrium structures, the intermolecular distance A–B is large, and no C or D molecules exist. As the molecule of A approaches the one of B, the internal structures of both begin to be distorted in order to minimize the total potential energy of the A–B bimolecular system in all dimensions other than the reaction coordinate. The total potential energy of the incipient complex continues to increase as the A–B distance decreases and is maximal for the activated complex $AB^\ddagger$. The two original molecules, A and B, now no longer exist independently, but $AB^\ddagger$ has the same atomic composition as A plus B. Progress further along the reaction coordinate cannot be represented in terms of A and B; it must be represented (approximately) as the intermolecular distance between the two nascent molecules, C and D. As C and D separate, the total potential energy of the C–D bimolecular system decreases and the internal structures of both product molecules approach their equilibrium configurations appropriate to the final (standard) state.

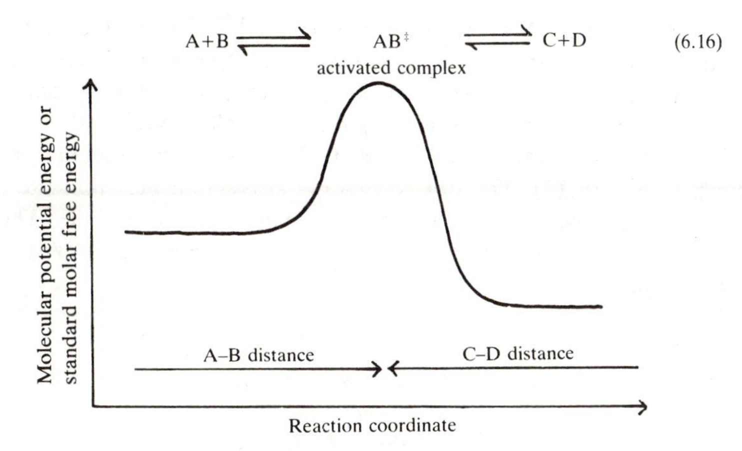

Fig. 6.5. Reaction profile of a reversible bimolecular reaction.

Because the reaction coordinate in an intermolecular reaction is essentially a translational degree of freedom, the initial and final states in the potential energy diagram are not properly represented as minima (cf. Figs 6.2 and 6.3) and quantized levels are not included.

The reaction coordinate encountered in both the standard molar free-energy and the molecular potential-energy diagrams introduces the notion of *mechanism*, or *how* a reaction occurs at a molecular level. In the example $A \rightarrow A^{\ddagger} \rightarrow B$ an activated complex is derived from a single molecule and such a mechanism is described as *unimolecular.* Simple reactions of the type $A \rightarrow A^{\ddagger} \rightarrow 2B$, or $A \rightarrow A^{\ddagger} \rightarrow B+C$ are also unimolecular. In reactions such as $A+B \rightarrow AB^{\ddagger} \rightarrow C$, or $2A \rightarrow A_2^{\ddagger} \rightarrow B$, or $A+B \rightarrow AB^{\ddagger} \rightarrow C+D$, an activated complex (which must be a single molecular species, admittedly of a special kind) is formed from two reactant molecules and the mechanism is *bimolecular.* Unimolecular and bimolecular are mechanistic terms which describe the composition, though not the structure, of an activated complex in an elementary reaction. As we shall see, they may be related to experimental quantities such as reaction order but they arise from a conception of how a reaction occurs and as such are based in theory.

6.5 Activation parameters from experimental results

According to the precepts of transition state theory,[1,2] the rate of an elementary chemical reaction is given by

$$\text{rate} = \frac{k_B T}{h} \, . \, [A^{\ddagger}],$$

where $k_B T/h$ is a universal factor comprising the Boltzmann and Planck constants, and the absolute temperature; $[A^{\ddagger}]$ is the molar concentration of the activated complex. (Usually a 'transmission coefficient' κ is included in the above expression. It is generally assumed to be unity and is seldom determinable, consequently we have left it out.)

We shall apply the above equation first to the bimolecular reaction shown in eqn (6.17). The dotted reverse arrow indicates that the formation of the activated complex from reactants is reversible. This is a postulate of transition state theory and is not amenable to experimental verification.

$$A + B \rightleftharpoons AB^{\ddagger} \rightarrow \text{Products.} \tag{6.17}$$

By eqn (4.16) (Chapter 4, p. 125) we can write

$$K_c^{\ominus\ddagger} = \frac{[AB^{\ddagger}]}{[A][B]} \,.\, (1\ \text{mol dm}^{-3}),$$

where $K_c^{\ominus\ddagger}$ = the quasi-thermodynamic equilibrium constant for the formation of the activated complex from reactants *related to standard state activities of 1 mol dm*$^{-3}$, and [A], [B], and $[AB^{\ddagger}]$ = the molar concentrations of A, B, and $AB^{\ddagger}$ which we are taking to be identical with activities on the molar scale – all activity coefficients in this hypothetical system being unity. From transition state theory,

$$-\frac{d[A]}{dt} = -\frac{d[B]}{dt} = \frac{k_B T}{h} \,.\, [AB^{\ddagger}]$$

$$= \frac{k_B T}{h} \,.\, K_c^{\ominus\ddagger}[A][B] \,.\, (1\ \text{mol dm}^{-3})^{-1}.$$

In practice, the rate of this simple bimolecular forward reaction will be given by the second-order rate law:

$$-\frac{d[A]}{dt} = -\frac{d[B]}{dt} = k_2[A][B]$$

where k_2 = second-order rate constant. Consequently, by equating the rate derived from transition state theory with the experimental rate, we can write

$$k_2 = \frac{k_B T}{h} \,.\, K_c^{\ominus\ddagger} \,.\, (1\ \text{mol dm}^{-3})^{-1}.$$

Analogously, for a unimolecular elementary reaction with a first-order rate constant k_1,

$$A \rightleftharpoons A^{\ddagger} \rightarrow \text{Products,}$$

$$K_c^{\ominus\ddagger} = \frac{[A^{\ddagger}]}{[A]},$$

the rate, according to transition state theory, is given by

$$-\frac{d[A]}{dt}=\frac{k_BT}{h}.[A^{\ddagger}]$$

$$=\frac{k_BT}{h}.[A].K_c^{\ominus\ddagger}.$$

The experimental rate law will be

$$-\frac{d[A]}{dt}=k_1[A],$$

therefore

$$k_1=\frac{k_BT}{h}.K_c^{\ominus\ddagger}.$$

In general, therefore,

$$k_c=\frac{k_BT}{h}.K_c^{\ominus\ddagger}.(1\ \text{mol dm}^{-3})^{1-n} \quad (6.18)$$

where k_c = rate constant for the elementary reaction including units with the dimensions $(\text{mol dm}^{-3})^{1-n}\times(\text{time})^{-1}$,

$K_c^{\ominus\ddagger}$ = quasi-thermodynamic equilibrium constant related to a hypothetical unit molar concentration standard state for the postulated equilibrium between reactant(s) and activated complex in the elementary reaction, and

n = molecularity and kinetic order of the elementary reaction (realistically, 1 or 2).[10]

There are two principal levels of approximation implicit in eqn (6.18) (and consequently in all that follows from it). At a basic level, there is the question of whether the assumption of 'equilibrium' between reactant(s) and activated complex is reasonable. Only when a reaction as a whole is at equilibrium could this obviously be so, and such a state is usually of little interest to anyone concerned with rates of reaction. But a theory should be judged by how well it describes a real system and by whether it leads to predictions which are subsequently confirmed by experiment. By such criteria, Transition State Theory appears well-founded.

At a less basic level, there is the question of rigour in the subsequent derivation of equations from the theory and its postulates. Our simple treatment has led to eqn (6.18) which includes $K_c^{\ominus\ddagger}$, formally a 'proper' thermodynamic equilibrium constant even though it relates to a process – the formation of activated complex from reactants – which may not strictly be a proper equilibrium. A more rigorous derivation leads to an equation which is similar in form, e.g. for a bimolecular reaction;[1]

$$k_c=\frac{k_BT}{h}.\frac{Q_{\ddagger}}{Q_A.Q_B}.e^{-\Delta E_0^{\ddagger}/RT}.\text{dm}^3\,\text{mol}^{-1}$$

or

$$k_c=\frac{k_BT}{h}.K_{\ddagger}.\text{dm}^3\,\text{mol}^{-1}$$

where $K_{\ddagger} = \dfrac{Q_{\ddagger}}{Q_A \cdot Q_B} \cdot e^{-\Delta E_0^{\ddagger}/RT}$,

Q_A, Q_B = the total molecular partition functions for reactants A and B,

$Q_{\ddagger}$ = the total molecular partition function for the activated complex *excluding the contribution for the degree of freedom which is the reaction coordinate,* and

$\Delta E_0^{\ddagger}$ = the difference (on a molar basis) in total zero-point energy between activated complex and reactants.

The difference between this and eqn (6.18) for a bimolecular reaction is evident if $K_c^{\ominus\ddagger}$ of eqn (6.18) is factorized according to standard statistical thermodynamics.

$$K_c^{\ominus\ddagger} = \frac{Q^{\ddagger}}{Q_A \cdot Q_B} \cdot e^{-\Delta E_0^{\ddagger}/RT}$$

or

$$k_c = \frac{k_B T}{h} \cdot \frac{Q^{\ddagger}}{Q_A \cdot Q_B} \cdot e^{-\Delta E_0^{\ddagger}/RT} \cdot \text{dm}^3\,\text{mol}^{-1},$$

where $Q^{\ddagger}$ = the total molecular partition function for the activated complex *including* the contribution for the degree of freedom which is the reaction coordinate.

The difference between eqn (6.18) and its more rigorously based alternative is largely inconsequential as far as results are concerned. Equation (6.18), however, has the advantage of leading more simply to further helpful equations. The main pitfall is to assume that these further quasi-thermodynamic relationships are more soundly based than they really are.

6.5.1 Standard free energy of activation

By analogy with the familiar thermodynamic relationship $\Delta G^{\ominus} = -RT \ln K_c^{\ominus}$ (p. 124) we may write $\Delta G^{\ominus\ddagger} = -RT \ln K_c^{\ominus\ddagger}$ and eqn (6.18) can be rewritten as eqn (6.19):

$$k_c = \frac{k_B T}{h} \cdot e^{-\Delta G^{\ominus\ddagger}/RT} \cdot (1\,\text{mol}\,\text{dm}^{-3})^{1-n}, \qquad (6.19)$$

where $\Delta G^{\ominus\ddagger}$ = standard molar free energy of activation for the elementary reaction and is one of the *activation parameters.*

The standard free energy of activation $\Delta G^{\ominus\ddagger}$ of a reaction at a particular temperature may be calculated from the experimental rate constant and the temperature using eqn (6.19). Knowledge of a rate constant is, therefore, equivalent to knowledge of the standard free energy of activation of the reaction.

We have already observed (see p. 223) that a first-order rate constant has the dimensions time^{-1} and consequently its magnitude is independent of the units of any property other than time which was monitored in the rate measurement. Correspondingly, although we have restricted ourselves to a standard state activity of 1 mol dm^{-3} in eqn (6.19) for the general case in accordance with modern usage, the standard molar free energy of activation of a unimolecular first-order reaction is, in fact, quite independent of the scale which is chosen for activity and of the choice of standard state. For unimolecular first-order gas reactions, therefore, we could write $\Delta G^{\circ\ddagger}$ rather than $\Delta G^{\ominus\ddagger}$ and relate to the thermochemical standard state (pure gas at 1 standard atmosphere pressure) if we wish; numerically, the value is the same. Second-order rate constants may be measured in various units, e.g. $dm^3\,mol^{-1}\,s^{-1}$ in gas or solution reactions or $Torr^{-1}\,s^{-1}$ for gas-phase reactions, and used empirically to compare rates of different reactions. But if eqn (6.19) is to be used to derive the standard molar free energy of activation of a second-order bimolecular reaction, the units of the rate constant must be $dm^3\,mol^{-1}\,s^{-1}$.[10] This is never a problem for reactions in solution. However, the rate constant of a second-order bimolecular gas reaction could have $(\text{pressure})^{-1}(\text{time})^{-1}$ dimensions. These must be converted to $dm^3\,mol^{-1}\,s^{-1}$ by eqn (6.6) (see p. 226) before eqn (6.19) may be used.

$$k_c = k_P\,.\,(RT)^{n-1} \tag{6.6}$$

To avoid possible ambiguity, it is always desirable to state the units of the second-order rate constant when a $\Delta G^{\ominus\ddagger}$ is quoted.

Example. Calculate $\Delta G^{\ominus\ddagger}$ at 326 °C for the gas-phase dimerization of buta-1,3-diene from the result given on p. 224.

k_P → 326 °C → 0.5

$$k_P = 2.28 \times 10^{-5}\,\text{Torr}^{-1}\,\text{min}^{-1}$$

As shown on p. 226, this may be converted into

$$k_c = 1.42 \times 10^{-2}\,\text{dm}^3\,\text{mol}^{-1}\,\text{s}^{-1} \text{ by eqn (6.6).}$$

By eqn (6.19),

$$1.42 \times 10^{-2}\,\text{dm}^3\,\text{mol}^{-1}\,\text{s}^{-1} = \frac{1.38 \times 10^{-23}\,\text{J K}^{-1} \times 599\,\text{K}}{6.626 \times 10^{-34}\,\text{J s}}\,.\,e^{-\Delta G^{\ominus\ddagger}/RT}\,.\,(1\,\text{mol dm}^{-3})^{-1}$$

$$\therefore\quad e^{-\Delta G^{\ominus\ddagger}/RT} = 1.138 \times 10^{-15}$$

$$\frac{\Delta G^{\ominus\ddagger}}{8.314\,\text{J K}^{-1}\,\text{mol}^{-1} \times 599\,\text{K}} = 34.4$$

$$\Delta G^{\ominus\ddagger}_{599\,\text{K}} = 171\,\text{kJ mol}^{-1}.$$

From the relationship $\Delta G^{\ominus\ddagger} = \Delta H^{\ominus\ddagger} - T\,\Delta S^{\ominus\ddagger}$, eqn (6.19) may be written as eqn (6.20) where $\Delta H^{\ominus\ddagger}$ and $\Delta S^{\ominus\ddagger}$ are the *standard molar enthalpy of activation* and the *standard molar entropy of activation* respectively.

$$k_c = \frac{k_B T}{h}\,.\,e^{\Delta S^{\ominus\ddagger}/R}\,.\,e^{-\Delta H^{\ominus\ddagger}/RT}\,.\,(1\,\text{mol dm}^{-3})^{1-n}. \tag{6.20}$$

Formally, these activation parameters correspond to the change in enthalpy and entropy of the system associated with the hypothetical formation of 1 mole of activated complex from constituent reactant(s) at constant temperature, the reaction being at $1\,mol\,dm^{-3}$.[10]

$\Delta H^{\ominus\ddagger}$ and $\Delta S^{\ominus\ddagger}$ of a unimolecular first-order reaction are (like $\Delta G^{\ominus\ddagger}$, see above) independent of the choice of standard state. So if the reaction is in the gas phase, we may refer to the thermochemical standard state and write $\Delta H^{\circ\ddagger}$ and $\Delta S^{\circ\ddagger}$ should this be more convenient. Commonly, the standard state superscript is omitted altogether from the activation parameter symbols for first-order reactions.

6.5.2 Standard enthalpy of activation

When a result for k_c, the rate constant of a reaction of order $\boldsymbol{n}$ in the proper molar concentration units, is put into eqn (6.20), all units cancel. Consequently, eqn (6.20) may be expressed logarithmically as

$$\ln k_c = \ln\left\{\frac{k_B}{h}\right\} + \ln T + \frac{\Delta S^{\ominus\ddagger}}{R} - \frac{\Delta H^{\ominus\ddagger}}{RT},$$

where k_c, k_B/h, and T in the logarithmic terms now refer only to their numerical values.

Differentiation with respect to temperature gives

$$\frac{d(\ln k_c)}{dT} = \frac{1}{T} + \frac{1}{R} \cdot \frac{d(\Delta S^{\ominus\ddagger})}{dT} - \frac{1}{RT} \cdot \frac{d(\Delta H^{\ominus\ddagger})}{dT} + \frac{\Delta H^{\ominus\ddagger}}{RT^2}$$

But

$$\frac{d(\Delta S^{\ominus\ddagger})}{dT} = \frac{1}{T} \cdot \frac{d(\Delta H^{\ominus\ddagger})}{dT}, \quad \text{(see p. 142)}$$

so

$$\frac{d(\ln k_c)}{dT} = \frac{1}{T} + \frac{\Delta H^{\ominus\ddagger}}{RT^2},$$

or

$$\frac{d(\ln k_c)}{dT} = \frac{\Delta H^{\ominus\ddagger} + RT}{RT^2}. \tag{6.21}$$

We may also re-write eqn (6.12) (p. 231) as

$$k_c = A(k_c) \,.\, e^{-E_a(k_c)/RT}, \tag{6.22}$$

to indicate that $E_a(k_c)$ and $A(k_c)$ are the activation energy and pre-exponential factor obtained from an Arrhenius plot with rate constants, k_c, in molar concentration units (if the reaction is other than first order).

If natural logarithms are taken (again, units on both sides of eqn (6.22) cancel) and the result is differentiated with respect to temperature (assuming that $A(k_c)$ and $E_a(k_c)$ are approximately constant over a limited temperature range), we obtain eqn (6.23)

$$\frac{d(\ln k_c)}{dt} = \frac{E_a(k_c)}{RT^2}. \tag{6.23}$$

Comparison of eqns (6.21) and (6.23), the one derived from application of transition state theory to a reaction and the other describing the experimentally determined temperature dependence of the rate constant, allows us to write

$$E_a(k_c) = \Delta H^{\ominus\ddagger} + RT. \tag{6.24}$$

Non-first-order gas-phase reactions investigated at constant volume may have rate constants k_P with dimensions $(\text{pressure})^{1-n} \times (\text{time})^{-1}$ where n = the reaction order. These must be converted into k_c rate constants with units $(\text{mol dm}^{-3})^{1-n}(\text{time})^{-1}$ using eqn (6.6) before the Arrhenius plot is constructed in order to obtain $E_a(k_c)$ for direct use in eqn (6.24).

We now see from eqn (6.24) that the Arrhenius activation energy $E_a(k_c)$ is closely related to the standard enthalpy of activation. In eqn (6.24) the mean temperature of the range covered in the Arrhenius plot should be used to calculate $\Delta H^{\ominus\ddagger}$ from the experimental $E_a(k_c)$. At room temperature and 100 °C, $RT = 2.5$ and 3.1 kJ mol^{-1} respectively and these are frequently comparable in magnitude with the uncertainty in a typical $E_a(k_c)$ result.

For the first-order unimolecular gas-phase isomerization of bicyclo[4.2.0]oct-7-ene considered on p. 232, $E_a = 180$ kJ mol^{-1} (509–558 K), $RT = 8.314 \times 533$ J therefore, by eqn (6.24), $\Delta H^{\ominus\ddagger} = \Delta H^{\circ\ddagger} =$ 176 kJ mol^{-1}.

6.5.3 Standard entropy of activation

Just as it has been shown that the activation energy can be related to the standard molar enthalpy of activation, the pre-exponential factor in the Arrhenius equation may be related to the standard molar entropy of activation of eqn (6.20). If eqn (6.24) is substituted back into eqn (6.20), we obtain

$$k_c = \frac{k_B T}{h} \,.\, e^{\Delta S^{\ominus\ddagger}/R} \,.\, e^{-(E_a(k_c) - RT)/RT} \,.\, (1\ \text{mol dm}^{-3})^{1-n},$$

therefore

$$k_c = \frac{k_B T e}{h} \,.\, e^{\Delta S^{\ominus\ddagger}/R} \,.\, e^{-E_a(k_c)/RT} \,.\, (1\ \text{mol dm}^{-3})^{1-n}. \tag{6.25}$$

We now use the Arrhenius equation (p. 231) written again as eqn (6.22)

$$k_c = A(k_c) \,.\, e^{-E_a(k_c)/RT} \tag{6.22}$$

to emphasize that all rate constants must include molar concentration units (if the reaction is other than first order). Comparison of eqns (6.22) and (6.25) shows that

$$A(k_c) = \frac{k_B Te}{h} \cdot e^{\Delta S^{\ominus\ddagger}/R}(1 \text{ mol dm}^{-3})^{1-n}, \tag{6.26}$$

from which we see that $\Delta S^{\ominus\ddagger}$ may be calculated from $A(k_c)$, the intercept in an Arrhenius plot. Again, T in eqn (6.26) is the mean absolute temperature of the range over which the rate constants were determined.

Example. To illustrate the use of eqn (6.26), we shall first return to the Arrhenius plot for the unimolecular isomerization of bicyclo[4.2.0]oct-7-ene shown on p. 232.

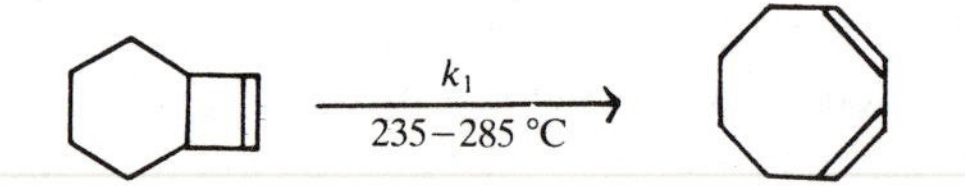

The mean temperature of the plot is 533 K and the computed intercept gives $A = 10^{14.1}\ \text{s}^{-1}$. From this and eqn (6.26) we can calculate $\Delta S^{\ominus\ddagger}$ which, since the reaction is first order, is equal to $\Delta S^{\circ\ddagger}$.

$$10^{14.1}\ \text{s}^{-1} = \frac{1.38 \times 10^{-23}\ \text{JK}^{-1} \times 533\ \text{K} \times 2.718}{6.626 \times 10^{-34}\ \text{J s}} \cdot e^{\Delta S^{\circ\ddagger}/R}$$

$$e^{\Delta S^{\circ\ddagger}/R} = 3.646$$

$$\underline{\Delta S^{\circ\ddagger} = 10.8\ \text{J K}^{-1}\ \text{mol}^{-1}.}$$

Example. *Arrhenius plot for the gas-phase second-order cycloaddition reaction of acrolein and buta-*1,3-*diene.* (Results taken from G. B. Kistiakowsky and J. R. Lacher, *J. Am. chem. Soc.* **58,** 123 (1936).)

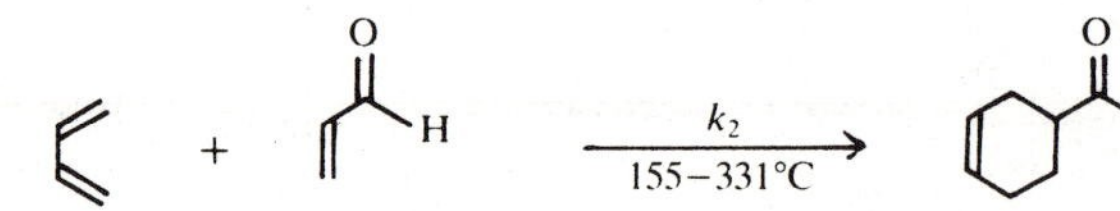

Temp./°C	155.3	208.3	246.5	295.8	330.8
Temp./K	428.5	481.5	519.7	569.0	604.0
$10^4 T^{-1}/\text{K}^{-1}$	23.34	20.77	19.24	17.57	16.56
$10^3 k_2/\text{dm}^3\ \text{mol}^{-1}\ \text{s}^{-1}$	0.138	1.63	7.2	36.8	81
$-\log(k_2/\text{dm}^3\ \text{mol}^{-1}\ \text{s}^{-1})$	3.86	2.79	2.14	1.43	1.09

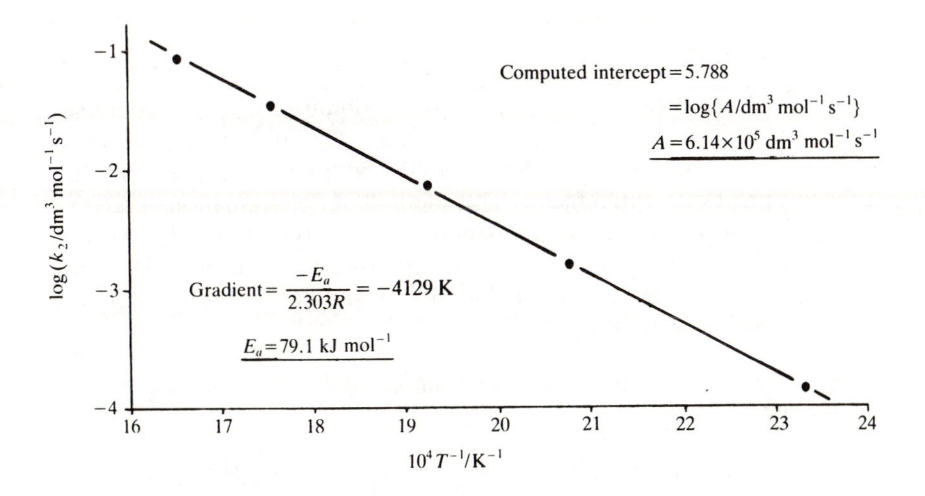

From the pre-exponential factor, we can now calculate $\Delta S^{\ominus\ddagger}$ using eqn (6.26).

$$6.14\times 10^{5}\,\text{dm}^{3}\,\text{mol}^{-1}\,\text{s}^{-1} = \frac{1.38\times 10^{-23}\,\text{JK}^{-1}\times 516\,\text{K}\times 2.718}{6.626\times 10^{-34}\,\text{J s}}\,e^{\Delta S^{\ominus\ddagger}/R}(\text{mol dm}^{-3})^{-1}$$

(mean temperature of the Arrhenius plot = 516 K)

$$e^{\Delta S^{\ominus\ddagger}/R} = 2.102\times 10^{-8}$$

$$\Delta S^{\ominus\ddagger} = -147\ \text{J K}^{-1}\,\text{mol}^{-1}.$$

From the gradient,

$$E_a = 79.1\ \text{kJ mol}^{-1};$$

by eqn (6.24)

$$\Delta H^{\ominus\ddagger} = (79.1 - 8.314\times 10^{-3}\times 516)\ \text{kJ mol}^{-1}$$

$$\Delta H^{\ominus\ddagger} = 74.8\ \text{kJ mol}^{-1}.$$

6.5.4 The Eyring equation

If eqn (6.20) is expressed logarithmically, we get eqn (6.27), a form of the Eyring equation.[11]

$$\ln\left(\frac{k_c\,.\,h}{k_B T(\text{mol dm}^{-3})^{1-n}}\right) = \frac{\Delta S^{\ominus\ddagger}}{R} - \frac{\Delta H^{\ominus\ddagger}}{RT} \qquad (6.27)$$

An Eyring plot of $\ln\left(\dfrac{k_c \,.\, h}{k_B T(\text{mol dm}^{-3})^{1-n}}\right)$ versus $1/T$ will be linear, and the standard enthalpy of activation may be obtained directly from the gradient without an intermediate step to obtain E_a. Further, the intercept in such a correlation gives $\Delta S^{\ominus\ddagger}$ directly without calculation of the Arrhenius pre-exponential factor. This method of determination of $\Delta H^{\ominus\ddagger}$ and $\Delta S^{\ominus\ddagger}$ is subject to the same restrictions as the use of eqns (6.24) and (6.26) for Arrhenius parameters. Rate constants must be in molar concentration units (if they are other than first order) before the Eyring plot is constructed.

Example. *Eyring plot for the second-order reaction of 2-phenylethyl bromide with sodium ethoxide in ethanol.*[5]

$$\text{Ph—CH}_2\text{—CH}_2\text{—Br} + \text{C}_2\text{H}_5\text{O}^- \xrightarrow[\text{C}_2\text{H}_5\text{OH}]{k_2} \text{Ph—CH=CH}_2 + \text{C}_2\text{H}_5\text{OH} + \text{Br}^-$$

Temp./°C	30.05	40.75	50.20	59.40
Temp./K	303.21	313.91	323.36	332.56
$10^4\,T^{-1}/\text{K}^{-1}$	32.98	31.86	30.93	30.07
$10^4 k_2/\text{dm}^3\ \text{mol}^{-1}\ \text{s}^{-1}$	4.17	13.3	34.2	81.4
$-\ln\left(\dfrac{k_2 \,.\, h}{k_B T\ \text{dm}^3\ \text{mol}^{-1}}\right)$	37.26	36.13	35.22	34.38

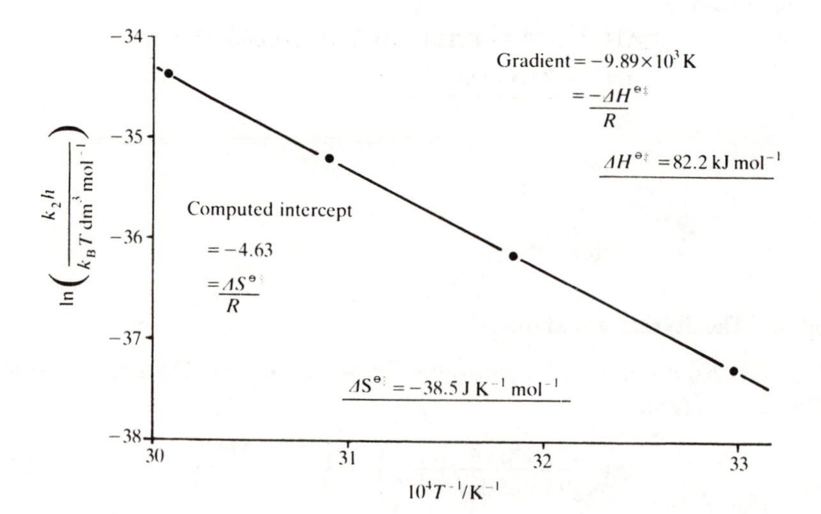

Although $\Delta H^{\ominus\ddagger}$ and $\Delta S^{\ominus\ddagger}$ may be obtained by simple graphical plots, either directly or from $E_a(k_c)$ and $A(k_c)$, more precise results and the standard deviations on the parameters are obtained by the use of standard computer programmes.[2]

Regardless of which method of calculating $\Delta H^{\ominus\ddagger}$ and $\Delta S^{\ominus\ddagger}$ is used, rate constants must be determined at several temperatures. It is necessary, therefore, to be able to measure the temperature of the reaction accurately as well as guarantee its constancy. The errors in the rate constants and in the temperature measurements are compounded in the determination of $\Delta H^{\ominus\ddagger}$ and $\Delta S^{\ominus\ddagger}$ and the magnitude of the experimental uncertainties must be taken into account when mechanistic deductions are being made from activation parameters. Experimental results which are reported without any indication of probable errors should be treated with scepticism.[2,7]

6.6 The physical significance of activation parameters of simple reactions

6.6.1 Standard enthalpy of activation

The standard molar enthalpy of activation related to a standard state of 1 mol dm^{-3} corresponds to the energy change between reactants and activated complex ascribable to the breaking and reforming of bonds in one mole of reaction per dm^3. Unlike rate constants, or $\Delta G^{\ominus\ddagger}$ values, $\Delta H^{\ominus\ddagger}$ results for quite different types of reactions, and reactions of different orders, may be directly compared meaningfully one with another.

In a reversible process, the standard enthalpy of reaction $\Delta H^{\ominus}$ is the difference between the standard enthalpies of activation of forward and reverse reactions. This is indicated in Fig. 6.6 (which includes standard molar enthalpy as the function plotted against reaction coordinate as opposed to standard molar free energy or molecular potential energy as in earlier reaction profiles) for a dissociation which is exothermic in the forward direction (eqn (6.28)).

$$\text{A—B} \underset{k_2}{\overset{k_1}{\rightleftharpoons}} \text{A}+\text{B}; \qquad \Delta H^{\ominus} = \Delta H_1^{\ominus\ddagger} - \Delta H_2^{\ominus\ddagger} \tag{6.28}$$

The standard enthalpy of a reaction is not in principle independent of temperature and its variation with temperature is given by the standard molar heat capacity of the reaction (see p. 45). We have already seen that

$$\Delta C_P^{\circ} = \frac{d(\Delta H^{\circ})}{dT},$$

for a gas-phase reaction at constant pressure where the superscript °

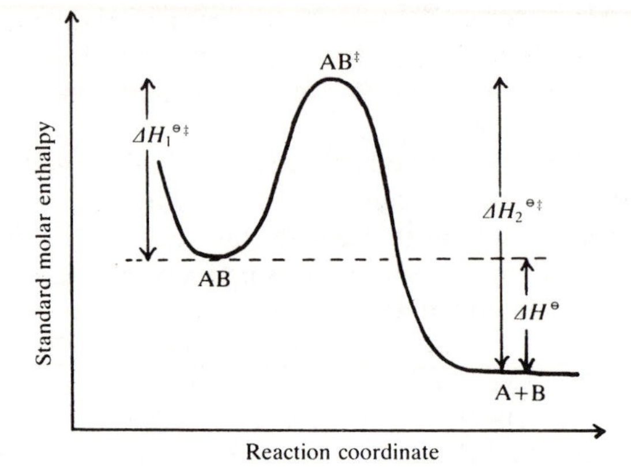

Fig. 6.6. Standard molar enthalpy reaction profile for a reversible exothermic dissociation.

indicates the thermochemical standard state. Analogously,

$$\Delta C^{\ominus} = \frac{d(\Delta H^{\ominus})}{dT},$$

for a gas standard state = 1 mol dm^{-3} where $\Delta H^{\ominus}$ = the heat change of the reaction carried out at constant volume (equal to $\Delta U°$, see p. 147)

$$\Delta C^{\ominus} = \sum C^{\ominus}(\text{products}) - \sum C^{\ominus}(\text{reactants})$$

and $C^{\ominus}$ = the heat capacity of a gas at 1 mol dm^{-3} (equal to C_V° from which the standard state superscript is normally omitted).

Reactions in solution are normally carried out at constant (atmospheric) pressure but also, effectively, at constant volume. So for reactions in solution we may also write

$$\Delta C^{\ominus} = \frac{d(\Delta H^{\ominus})}{dT},$$

where the superscript $^{\ominus}$ indicates a solution standard state of 1 mol dm^{-3} under 1 atmosphere pressure.

It is evident, therefore, that $\Delta H_1^{\ominus\ddagger}$ and $\Delta H_2^{\ominus\ddagger}$ in eqn (6.28) cannot, in principle, be temperature-independent terms and formally the variation with temperature of the standard enthalpy of activation is expressed by the standard heat capacity of activation $\Delta C^{\ominus\ddagger}$:

$$\Delta C^{\ominus\ddagger} = \frac{d(\Delta H^{\ominus\ddagger})}{dT}$$

where $\Delta C^{\ominus\ddagger} = C^{\ominus\ddagger} - \sum C^{\ominus}(\text{reactants})$,
$C^{\ominus\ddagger}$ = standard molar heat capacity of the activated complex,
$C^{\ominus}$ = standard molar heat capacity of a reactant,
and the standard state = 1 mol dm^{-3} under 1 standard atmosphere pressure. For many reactions in practice, the temperature variation of $\Delta H^{\ominus\ddagger}$ over a narrow temperature range is small ($\Delta C^{\ominus\ddagger}$ is small) and, except in the most

precise work[12] or when a large temperature difference is involved, it is ignored.

The reaction shown in eqn (6.29) is the simplest thermal *cis-trans* isomerization of an alkene and, because it is a unimolecular first-order gas-phase reaction, $\Delta H^{\ominus\ddagger} = \Delta H^{\circ\ddagger}$. A possible mechanism is indicated which involves an activated complex containing only a sigma bond between the carbon atoms because the unhybridized p-orbitals are orthogonal.

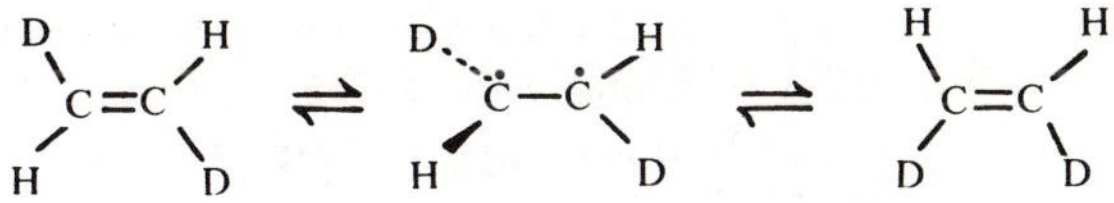

$$\text{DHC=CHD} \rightleftharpoons \text{DH}\dot{\text{C}}\text{—}\dot{\text{C}}\text{HD} \rightleftharpoons \text{DHC=CHD} \qquad (6.29)$$

The standard enthalpy of activation of this reaction, $\Delta H^{\circ\ddagger}_{773\,\mathrm{K}} = 266\ \mathrm{kJ\,mol^{-1}}$ from E_a in Table 6.1 using eqn (6.24), gives a measure of the π-bond dissociation enthalpy, πDH°, in ethene. Considering the temperature difference, this is in reasonable agreement with the estimated value of $\pi DH^{\circ}_{298\,\mathrm{K}} = 248\ \mathrm{kJ\,mol^{-1}}$ obtained by indirect methods.[13]

TABLE 6.1
Gas-phase unimolecular reactions

Reaction	$\log(A/\mathrm{s^{-1}})$	$E_a/\mathrm{kJ\,mol^{-1}}$	Temp. Range/K	Note
trans-CHD=CHD → *cis*-CHD=CHD	13.00	272	723–823	*a*
cis-Stilbene → *trans*-Stilbene	12.78	179	553–615	*b*
$C_2H_6 \rightarrow 2CH_3^{\cdot}$	16.7	368	823–893	c
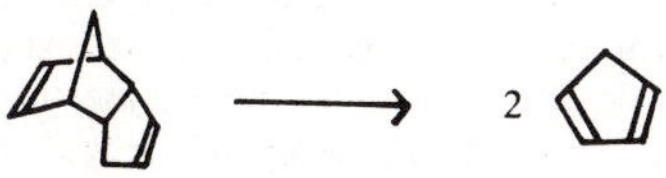 *endo*-Dicyclopentadiene → 2 Cyclopentadiene	13.01	142	426–484	*d*
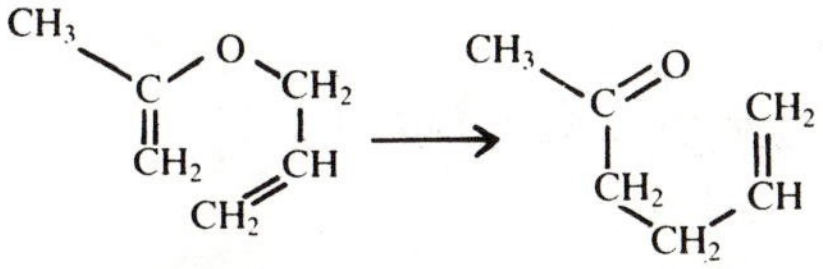 *iso*-Propenyl allyl ether ($CH_3C(=CH_2)OCH_2CH=CH_2$) → Allyl acetone ($CH_3COCH_2CH_2CH=CH_2$)	11.73	123	416–467	*e*

Notes

[a] B. S. Rabinovitch, J. E. Douglas, and F. S. Looney, *J. chem. Phys.* **20,** 1807 (1952); J. E. Douglas, B. S. Rabinovitch, and F. S. Looney, ibid. **23,** 315 (1955).
[b] G. B. Kistiakowsky and W. R. Smith, *J. Am. chem. Soc.* **56,** 638 (1934).
[c] M. C. Lin and M. H. Back, *Canad. J. Chem.* **44,** 505 (1966); C. P. Quinn, *Proc. R. Soc.* A**275,** 190 (1963). Values ranging from 332–384 kJ mol^{-1} may be found in the literature; see A. B. Trenwith, *J. chem. Soc. Faraday Trans. 1,* **75,** 614 (1979).
[d] W. C. Herndon, C. R. Grayson and J. M. Manion, *J. org. Chem.* **32,** 526 (1967).
[e] L. Stein and G. W. Murphy, *J. Am. chem. Soc.* **74,** 1041 (1952).

The result $\Delta H^{\circ\ddagger}_{584\,K} = 174\ \text{kJ mol}^{-1}$ for the isomerization of *cis*- to *trans*-stilbene obtained from E_a given in Table 6.1 is much lower. This is because the central bond of stilbene is not isolated but part of an extensively conjugated system, and the two unpaired electrons of the activated complex are not localized in atomic p-orbitals but delocalized within orthogonal benzylic residues. Therefore, although the reaction corresponds to a change in a particular attribute – the configuration about an easily identifiable group, the central bond of the molecule – it is still a reaction of the *whole molecule* and this must be taken into account when the enthalpy of activation is being considered.

We can calculate the standard enthalpy of activation for the dissociation of ethane into methyl radicals,

$$CH_3\text{—}CH_3 \xrightarrow[823\text{–}893\ K]{k_1} 2CH_3^{\cdot} \qquad \Delta H^{\ominus\ddagger}_{858\,K} = 361\ \text{kJ mol}^{-1}, \qquad (6.30)$$

from the activation energy given in Table 6.1 ($E_a = 368\ \text{kJ mol}^{-1}$). When the probable error is taken into account (see footnote *c* to Table 6.1), this result is very close to the calculated molar enthalpy change for the complete dissociation of ethane into methyl free radicals:[14]

$$CH_3CH_3 \rightleftharpoons 2CH_3^{\cdot}; \qquad \Delta H^{\circ}_{298\,K} = 369\ \text{kJ mol}^{-1}. \qquad (6.31)$$

A correction may be applied to the result in eqn (6.31) to convert it to the higher temperature of eqn (6.30) using heat capacity data for ethane[15] and methyl radical.[16] Secondly, it can be corrected for reaction at constant volume rather than constant pressure: in other words, to relate to a standard state of $1\ \text{mol dm}^{-3}$ rather than 1 standard atmosphere pressure. These have the effect of making the thermodynamic parameter and the activation parameter more properly comparable and even closer numerically:

$$CH_3\text{—}CH_3 \underset{k_2}{\overset{k_1}{\rightleftharpoons}} 2CH_3^{\cdot}; \qquad \Delta H^{\ominus}_{858\,K} = 364\ \text{kJ mol}^{-1}.$$

This result requires that the activation energy for the reverse reaction – the combination of methyl radicals to give ethane – be virtually zero, and its second-order rate constant, k_2, temperature-independent. This has been established experimentally for a number of alkyl radical combinations of this type.[17] We conclude that the central bond of the ethane molecule is almost completely broken at the transition state for the homolysis, and the activated complex resembles two very loosely-bonded methyl groups.

The enthalpies of activation for the following representative bimolecular reactions were given earlier:

$$\text{butadiene} + \text{acrolein (CH}_2\text{=CHCHO)} \xrightarrow[155-331^\circ\text{C}]{k_2} \text{cyclohex-3-enecarbaldehyde} \qquad \text{(p. 247)}$$

$$\Delta H^{\ominus\ddagger} = 74.8 \text{ kJ mol}^{-1}$$

$$PhCH_2CH_2Br + EtO^- \xrightarrow[30-60\,^\circ\text{C}]{k_2} PhCH{=}CH_2 + EtOH + Br^- \quad \text{(p. 248)}$$

$$\Delta H^{\ominus\ddagger} = 82.2 \text{ kJ mol}^{-1}.$$

It is immediately evident that these $\Delta H^{\ominus\ddagger}$ results are appreciably lower than those of the unimolecular reactions considered above. They are also lower than dissociation enthalpies of many bonds in organic compounds. In bimolecular reactions, the breaking of old bonds (which requires energy) and the forming of new bonds (which releases energy) in the generation of the activated complex are highly concerted and usually synchronous. Consequently, the overall energy requirements in the formation of the activated complexes are fairly modest so $\Delta H^{\ominus\ddagger}$ values for such reactions are relatively low. This does not mean that the reactions will always be very fast because, as we shall see, the molecules have to come together first in order that rebonding may take place and this may be slow. Furthermore, they may have to come together in a very particular way and this would make the process even slower. These aspects are discussed more fully in the next section. It does follow, however, that the rates of second-order bimolecular reactions with low $\Delta H^{\ominus\ddagger}$ and E_a values are less temperature dependent than typical first-order unimolecular reactions.

6.6.2 Standard entropy of activation

It is more difficult to express the significance of the standard entropy of activation in general terms, but it has been described vaguely as a probability or geometrical term. This inability to define precisely its meaning lies not so much in the inadequacy of chemical theory or terminology, but in the nature of entropy. There are quite disparate causes of $\Delta S^{\ominus\ddagger}$ being positive or negative and of some particular magnitude, and the mechanistic interpretation of a result can usually be made only within the context of the particular chemical reaction. Again, in this section, some simple examples will be given and these should clarify what is difficult to envisage in the abstract.

We see from eqn (6.25) (p. 245) that if a reaction has $\Delta S^{\ominus\ddagger} = 0$ (when the standard molar entropy of the activated complex is equal to the total standard entropy of reactants, the molar quantities being defined by the stoichiometry of the process) the rate constant is determined by an exponential energy term and the universal factor $k_B Te/h$. The entropy of activation term $e^{\Delta S^{\ominus\ddagger}/R}$, therefore, may be regarded as a pre-exponential component which modifies this universal factor.

At 100 °C, $k_B Te/h = 2.1 \times 10^{13}\ s^{-1}$ so, for reactions in the region of this temperature, $\log(A(k_c)/(\text{mol dm}^{-3})^{1-n}\ s^{-1})$ will be approximately 13.3 when $\Delta S^{\ominus\ddagger} = 0$ using eqn (6.26) (p. 246). If it is significantly less than this, then $\Delta S^{\ominus\ddagger}$ is negative and there is a loss of entropy in the activation process. Alternatively, if $\log(A(k_c)/(\text{mol dm}^{-3})^{1-n}\ s^{-1})$ is greater than approximately 13.3, the formation of the activated complex is associated with an increase in entropy and $\Delta S^{\ominus\ddagger}$ is positive

From the result shown in Table 6.1 for the dissociation of *endo*-dicyclopentadiene, we can calculate (using eqn (6.26)) $\Delta S^{\ominus\ddagger}_{455\,K} = -8\ J\ K^{-1}\ mol^{-1}$ which, since the reaction is unimolecular, is equal to $\Delta S^{o\ddagger}_{455\,K}$. When the probable uncertainty is taken into account ($\sim \pm 10\ J\ K^{-1}\ mol^{-1}$), this is numerically very small and indicates that no appreciable change in entropy is associated with the formation of the activated complex. The isomerizations of the two alkenes also have small negative $\Delta S^{o\ddagger}$ values.

The A-factor for the dissociation of ethane into methyl radicals shown in Table 6.1 is distinctly high and corresponds to $\Delta S^{o\ddagger}_{858\,K} = 58\ J\ K^{-1}\ mol^{-1}$. There can be no gain in translational entropy as the activated complex is still a single molecular species. However, we saw from the consideration of the enthalpy of activation that the two methyl fragments are only weakly bonded together at the transition state. Consequently, some molecular vibrations of the activated complex have low force constants corresponding to shallow potential energy curves with closely spaced quantized energy levels. This is illustrated for a general case $X_2 \rightarrow 2X$ in Fig. 6.7. Superimposed upon the line representing the total potential energy of the reacting molecule in the reaction coordinate is a potential energy curve A for some (unspecified) molecular vibration of the X_2 molecule which transforms, as the reaction proceeds, into a vibration of the activated complex represented by the shallower potential energy curve B. Because the quantum states of curve B are closer together than those of A, there will be a greater population of the higher states of vibration B than of the higher states of vibration A at a given temperature. Vibration B has, therefore, a higher entropy than vibration A and consequently the activated complex has a higher standard molar entropy than the reactant (see Chapter 3).

In general, the magnitude of this effect will depend both upon the extent to which a given molecular vibration is 'loosened' (the extent to

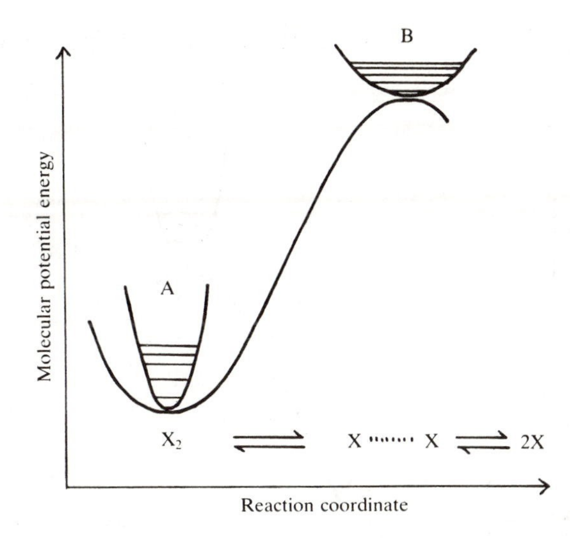

Fig. 6.7. Potential-energy profile of a dissociation with a positive standard entropy of activation.

which the force constant is reduced or weakened) in the activated complex, and upon the number of vibrations which are so affected. This latter aspect is dramatically illustrated by the exceptionally large positive entropies of activation associated with the denaturation of some proteins.[11,18] In such reactions innumerable specific intramolecular hydrogen bonds and many salt bridges are broken and the highly organized complex native protein molecule becomes a polypeptide with more random and mobile conformational freedom. There is, therefore, in the formation of the activated complex from the native protein, a loosening of a colossal number of vibrations and consequently a large increase in entropy.

We see from Table 6.1 that the pre-exponential factor for the gas-phase isomerization of *iso*-propenyl allyl ether is low; it corresponds to $\Delta S^{\circ\ddagger}_{442\mathrm{K}} = -32\ \mathrm{J\ K^{-1}\ mol^{-1}}$. A mechanism which is compatible with this involves a cyclic activated complex as shown in eqn (6.32).

$$CH_3C(=CH_2)\text{–}O\text{–}CH_2\text{–}CH{=}CH_2 \rightleftharpoons \left[CH_3C(\cdots CH_2)\cdots O\cdots CH_2\cdots CH\cdots CH_2\right]^{\ddagger} \rightleftharpoons CH_3C(=O)\text{–}CH_2\text{–}CH_2\text{–}CH{=}CH_2 \qquad (6.32)$$

The relatively low enthalpy of activation for this reaction indicated in Table 6.1 supports this mechanistic interpretation. The concerted rebonding which takes place in the activated complex involves some bond breaking but also some bond making. So overall, the formation of the activated complex is not energetically very costly.

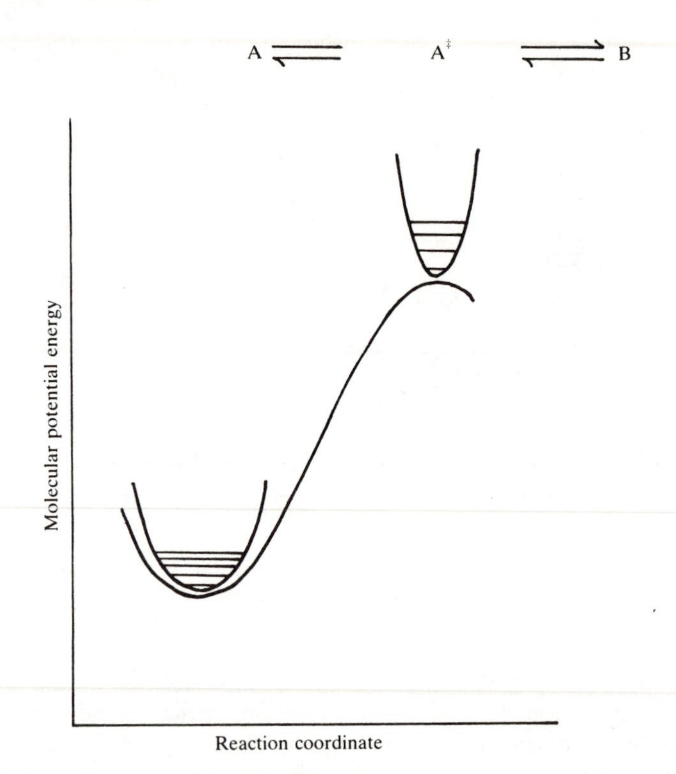

Fig. 6.8. Potential-energy profile for an isomerization with a negative standard entropy of activation.

An electrocyclic mechanism such as this clearly has a stringent stereo-electronic requirement as the number of conformations available to the activated complex in which the electronic redistribution (rebonding) may occur is limited. The reactant, however, being acyclic, has enormous conformational freedom. In other words, many internal rotations and floppy vibrations in the reactant with closely spaced energy levels and low force constants are replaced by stiffer vibrational degrees of freedom with higher force constants and more widely spaced energy levels in the activated complex.

Such a reaction is illustrated in Fig. 6.8 and the potential-energy curve superimposed upon the vibration of the reactant in the reaction coordinate is for a single molecular vibration with a low force constant which transforms into the stiffer vibration of the activated complex.

The activated complex in a reaction of this sort is frequently called 'tight', as it is formed with an increase in vibrational force constants (other than that of the reaction coordinate) and a corresponding decrease in vibrational entropy. It is in contrast to a 'loose' activated complex such as those implicated in the dissociation of ethane into methyl radicals, and the denaturation of proteins.

So far in this section we have considered first-order unimolecular reactions in which the entropy of activation is due to intramolecular atomic reorganization as molecules of reactant are transformed into molecules of activated complex. In a bimolecular reaction an activated complex is formed from two reactant molecules, so three translational and up to three rotational degrees of freedom (depending upon the symmetry of the molecules) are replaced by vibrations.

We have seen already in Chapter 3 that, quantitatively, $S(\text{trans}) > S(\text{rot}) > S(\text{vib})$, consequently we can expect that the norm for a bimolecular reaction will be a substantial negative entropy of activation.

This is found for simple S_N2 reactions[3] and Diels–Alder cycloadditions.[4]

$$C_2H_5Br + Cl^- \xrightarrow[k_2,\ \text{acetone}]{25\text{–}65\ °C} C_2H_5Cl + Br^-$$

$$E_a = 73.6\ \text{kJ mol}^{-1}$$

$$\log(A/\text{dm}^3\ \text{mol}^{-1}\ \text{s}^{-1}) = 8.9,$$

therefore

$$A = 7.9 \times 10^8\ \text{dm}^3\ \text{mol}^{-1}\ \text{s}^{-1},$$

and, by eqn (6.26),

$$\Delta S^{\ominus\ddagger}_{318\ \text{K}} = -83\ \text{J K}^{-1}\ \text{mol}^{-1}.$$

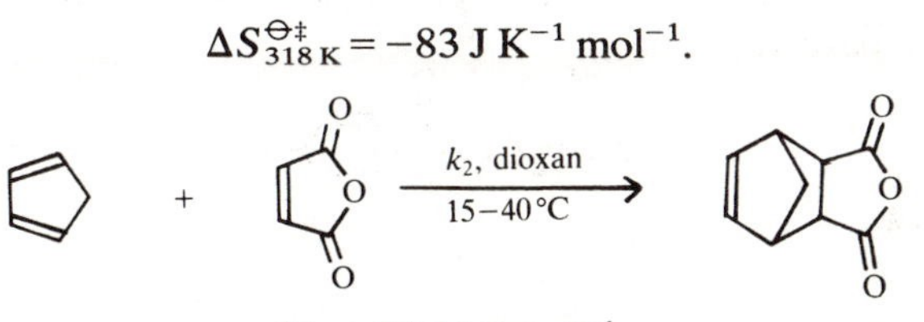

$$E_a = 37.2\ \text{kJ mol}^{-1}$$

$$\log(A/\text{dm}^3\ \text{mol}^{-1}\ \text{s}^{-1}) = 5.4,$$

therefore

$$A = 2.5 \times 10^5\ \text{dm}^3\ \text{mol}^{-1}\ \text{s}^{-1},$$

and, by eqn (6.26),

$$\Delta S^{\ominus\ddagger}_{300\ \text{K}} = -150\ \text{J K}^{-1}\ \text{mol}^{-1}.$$

It is important to realize that the much lower A-values for these reactions are due to the bimolecular mechanisms and not to some arithmetic curiosity arising from the fact that they are second order rather than first. The hydrolysis of methyl and simple primary alkyl halides and sulphonates are *pseudo* first order but the consistently large negative standard entropies of activation (Table 6.2) constitute part of the evidence that these are solvent-induced bimolecular substitutions.

TABLE 6.2
Activation parameters for the hydrolysis of methyl and simple primary alkyl halides and sulphonates[a]

Compound	$\Delta H^{\ominus\ddagger}$/kJ mol^{-1}	$\Delta S^{\ominus\ddagger}$/J K^{-1} mol^{-1}
CH_3Cl	100.4	−51.5
CH_3Br	96.2	−42.3
CH_3I	102.7	−33.9
C_2H_5Br	97.9	−37.2
n-C_3H_7Br	92.5	−56.9
CH_3OMs[b]	88.7	−49.0
C_2H_5OMs[b]	88.4	−51.5
n-C_3H_7OMs[b]	86.4	−60.2
n-C_4H_9OMs[b]	80.3	−72.0
CH_3OTs[c]	86.9	−50.6
C_2H_5OTs[c]	89.0	−44.8

[a] Data taken from R. E. Robertson, *Progr. phys. org. Chem.* **4,** 213 (1967).
[b] OMs = methanesulphonate —OSO_2CH_3
[c] OTs = toluene-*p*-sulphonate $-OSO_2-C_6H_4-CH_3$

For comparison, the first-order unimolecular solvolyses of t-butyl chloride and 2-adamantyl toluene-*p*-sulphonate (5) shown in Table 6.3 have $\Delta S^{\ominus\ddagger}$ values which are much less negative, indeed two of the results are actually positive.

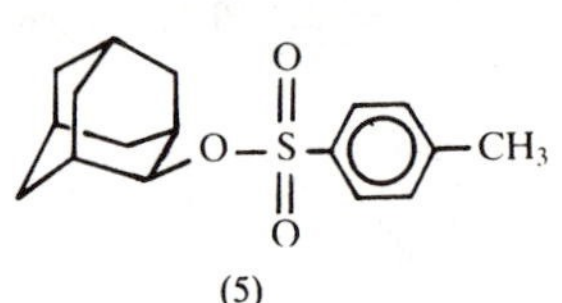

(5)

TABLE 6.3
Activation parameters for solvolyses of t-butyl chloride and 2-adamantyl toluene-p-sulphonate[a]

Compound	Solvent	$\Delta H^{\ominus\ddagger}$/kJ mol^{-1}	$\Delta S^{\ominus\ddagger}$/J K^{-1} mol^{-1}
t-Butyl chloride	H_2O	97.15	51.0
	CH_3OH	104.1	−13
	C_2H_5OH	109.3	−13
	HCO_2H	87.9	−7.1
	CH_3CO_2H	107.9	−10
2-Adamantyl tosylate[b] (5)	C_2H_5OH	124	−9.2
	HCO_2H	104	15
	CH_3CO_2H	118	−8.9

[a] Data taken from T. W. Bentley and P. von R. Schleyer, *Adv. phys. org. Chem.* **14,** 1 (1977).
[b] tosylate = toluene-*p*-sulphonate.

The second-order rate law for the reaction of chlorine atoms with molecular hydrogen, eqn (6.33), is given by eqn (6.34) over a wide range of temperature (273–1071 K).[19]

$$\mathrm{Cl^{\cdot} + H_2 \xrightarrow{k_2} HCl + H^{\cdot}} \tag{6.33}$$

$$k_2/\mathrm{dm^3\,mol^{-1}\,s^{-1}} = 8.3\times 10^{10}\mathrm{e}^{-2754/T} \tag{6.34}$$

$$(E_a = 22.93\ \mathrm{kJ\,mol^{-1}};\qquad A/\mathrm{dm^3\,mol^{-1}\,s^{-1}} = 8.3\times 10^{10}).$$

Using eqn (6.26),

$$\Delta S^{\ominus\ddagger}_{672\,\mathrm{K}} = -51\ \mathrm{J\,K^{-1}\,mol^{-1}}.$$

We can calculate the translational contribution towards $\Delta S^{\ominus\ddagger}_{672\,\mathrm{K}}$ for this gas-phase reaction using the Sackur–Tetrode equation, i.e. assuming ideal gas behaviour, see p. 99.

$$S(\text{trans}) = R\,.\ln\left\{\frac{\mathrm{e}^{5/2}\,.\,V}{N_\mathrm{A}}\left(\frac{2\pi m k_\mathrm{B}T}{h^2}\right)^{3/2}\right\}$$

$$\text{mass of } \mathrm{H_2} = 2.0160\ \text{amu}$$

$$= 3.3478\times 10^{-27}\ \text{kg}.$$

Volume of 1 mole $= 1\ \mathrm{dm^3}$ ($1\ \mathrm{mol\,dm^{-3}}$ standard state rather than 1 atmosphere as in the examples of p. 100) $= 10^{-3}\ \mathrm{m^3}$.

$\left(\dfrac{2\pi m k_\mathrm{B}T}{h^2}\right)^{3/2}$ for $\mathrm{H_2}$ at 672 K

$$= \left(\frac{2\pi\times 3.3478\times 10^{-27}\ \text{kg}\times 1.3807\times 10^{-23}\ \mathrm{J\,K^{-1}}\times 672\ \mathrm{K}}{6.6262^2\times 10^{-68}\ \mathrm{J^2\,s^2}}\right)^{3/2}$$

$$= (4.4451\times 10^{20}\ \mathrm{m^{-2}})^{3/2}$$

$$= 9.372\times 10^{30}\ \mathrm{m^{-3}}.$$

$$S^{\ominus}_{672\,\mathrm{K}}(\text{trans}) = 8.314\ \mathrm{J\,K^{-1}\,mol^{-1}}\,.\ln\left\{\frac{\mathrm{e}^{5/2}\times 10^{-3}\ \mathrm{m^3}\times 9.372\times 10^{30}\ \mathrm{m^{-3}}}{6.022\times 10^{23}}\right\}$$

therefore

$$S^{\ominus}_{672\,\mathrm{K}}(\text{trans})\mathrm{H_2(g)} = 101\ \mathrm{J\,K^{-1}\,mol^{-1}}.$$

Exactly analogously,

$$S^{\ominus}_{672\,\mathrm{K}}(\text{trans})\mathrm{Cl^{\cdot}(g)} = 137\ \mathrm{J\,K^{-1}\,mol^{-1}}$$

and

$$S^{\ominus}_{672\,\mathrm{K}}(\text{trans})\mathrm{H_2Cl^{\cdot\ddagger}(g)} = 138\ \mathrm{J\,K^{-1}\,mol^{-1}}$$

therefore

$$\Delta S^{\ominus\ddagger}_{672\,\mathrm{K}}(\text{trans}) = 138-(137+101)\ \mathrm{J\,K^{-1}\,mol^{-1}}$$

$$= -100\ \mathrm{J\,K^{-1}\,mol^{-1}}.$$

There must therefore be a substantial *increase* in rotational and vibrational entropy in the formation of a very loosely bonded activated complex since the *overall* experimental standard entropy of activation is only $-51\,\mathrm{J\,K^{-1}\,mol^{-1}}$.

In contrast, the gas-phase dimerization of cyclopentadiene, eqn (6.35), has the rate law between 393 and 467 K shown in eqn (6.36)[20] from which we obtain $\Delta S^{\ominus\ddagger}_{430\,\mathrm{K}} = -162\,\mathrm{J\,K^{-1}\,mol^{-1}}$ by eqn (6.26).

$$2\ \text{(cyclopentadiene)} \underset{(k_1)}{\overset{k_2}{\rightleftharpoons}} \text{(dicyclopentadiene)} \tag{6.35}$$

$$k_2/\mathrm{dm^3\,mol^{-1}\,s^{-1}} = 8.5\times10^4 e^{-7505/T} \tag{6.36}$$

The translational contribution towards this, calculated exactly as illustrated above using the Sackur–Tetrode equation and assuming ideal gas behaviour, is $\Delta S^{\ominus\ddagger}_{430\,\mathrm{K}}(\mathrm{trans}) = -130\,\mathrm{J\,K^{-1}\,mol^{-1}}$. In this reaction, therefore, the total loss of entropy upon formation of the activated complex is greater than just the translational contribution. Interestingly, when the probable experimental uncertainty is taken into account, the standard entropy of activation is virtually identical with the standard entropy of dimerization of cyclopentadiene to give *endo*-dicyclopentadiene ($\Delta S^{\ominus}_{430\,\mathrm{K}} = -168\,\mathrm{J\,K^{-1}\,mol^{-1}}$, standard state $= 1\,\mathrm{mol\,dm^{-3}}$, calculated from $S°$ and C°_{P} data,[14] and eqn (3.13), Chapter 3, p. 95). From this we conclude that, geometrically, the activated complex is very similar to the dimer. In agreement, the standard entropy of activation for the reverse unimolecular process with first-order rate constant k_1, the dissociation of the dimer, is very small as was shown on p. 254 from the result for the pre-exponential factor given in Table 6.1.

6.6.3 Standard free energy of activation

In one sense, the standard free energy of activation has no special physical significance. It is a single parameter which conveys just the same information as the rate constant to which it is directly related by transition state theory in eqn (6.19)

$$k_c = \frac{k_B T}{h} . e^{-\Delta G^{\ominus\ddagger}/RT} . (1\,\mathrm{mol\,dm^{-3}})^{1-n}. \tag{6.19}$$

The lower the $\Delta G^{\ominus\ddagger}$ of a reaction, the larger the rate constant and, expressed somewhat loosely, the faster the reaction.

But $\Delta G^{\ominus\ddagger}$ has an importance which transcends simplistic physical interpretation. Equation (6.19) is the first stage in applying the methods of equilibrium thermodynamics to the dynamic transformation of reactants into products. This approach is developed by resolving $\Delta G^{\ominus\ddagger}$ into enthalpy and entropy contributions using

$$\Delta G^{\ominus\ddagger} = \Delta H^{\ominus\ddagger} - T\,\Delta S^{\ominus\ddagger}$$

and leading to eqn (6.20) (p. 243):

$$k_c = \frac{k_B T}{h} \,.\, e^{\Delta S^{\ominus\ddagger}/R} \,.\, e^{-\Delta H^{\ominus\ddagger}/RT} (1\ \text{mol dm}^{-3})^{1-n}. \qquad (6.20)$$

Knowledge of $\Delta H^{\ominus\ddagger}$ and $\Delta S^{\ominus\ddagger}$ from rate measurements at different temperatures allows an insight into the nature of a reaction which the single parameter $\Delta G^{\ominus\ddagger}$ (or a rate constant) does not.

In linking familiar equilibrium properties of elements and compounds (such as molecular motion, vibrations, and rotations) to the kinetic process, we develop an understanding of how chemical reactions take place. We see a chemical reaction in terms of stretches and bends of bonds and of collisions between molecules. The presence of certain structural features in a molecule indicates not only a particular vibration or bond dissociation enthalpy but also a propensity of the compound towards a particular type of reaction.

So far our approach has been largely analytical. We have resolved $\Delta G^{\ominus\ddagger}$ into $\Delta H^{\ominus\ddagger}$ and $\Delta S^{\ominus\ddagger}$ and then, for particular reactions, interpreted these two terms at a molecular level. $\Delta H^{\ominus\ddagger}$ is closely related to the ease of the bond breaking and reforming in the generation of the activated complex. The lower $\Delta H^{\ominus\ddagger}$, the faster the reaction. In contrast, but fully in accord with our understanding of thermodynamics, the more negative the $\Delta S^{\ominus\ddagger}$, the slower the reaction. And the composite $\Delta G^{\ominus\ddagger}$ term reflects the *overall* ease of forming the activated complex.

As we have seen, unimolecular reactions do not generally have a serious entropy constraint. Indeed some have a strong entropic driving force and such reactions proceed at appreciable rates even if $\Delta H^{\ominus\ddagger}$ is very high. On the other hand, bimolecular reactions always have negative (unfavourable) entropies of activation and are immeasurably slow unless they have much lower enthalpies of activation than are typically found in unimolecular reactions.

6.7 The Hammond Postulate

When molecules A and B of Fig. 6.2 (p. 235) are structurally modified in some minor way to become A′ and B′, the reaction profile for the new interconversion will not be identical with the original because A′ and B′ have potential-energy surfaces slightly different from those of A and B. If the minor structural modification lowers the potential energy of B more than it lowers that of A as shown in Fig. 6.9, the effect is that the new reaction from left to right will involve a greater decrease in standard

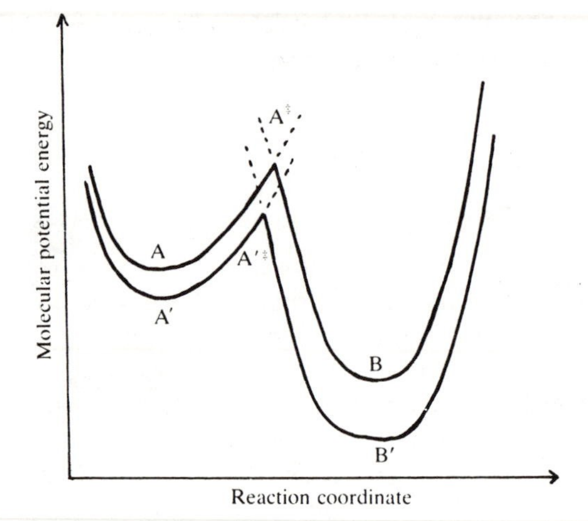

Fig. 6.9. Reaction profiles showing the effect of a substituent which increases the exothermicity of the reaction.

molar enthalpy than the original. The reaction has become more exothermic.

We see in Fig. 6.9 that the transition state for the conversion of A′ into B′ occurs not as far along the reaction coordinate as that for A into B. Also, the difference in molecular potential energy (or standard molar enthalpy) between A′ and $A'^{\ddagger}$ is less than that between A and $A^{\ddagger}$. In other words, the transition state in the more exothermic reaction occurs *earlier* in the reaction coordinate and at a (relatively) *lower* molecular potential-energy (or standard molar enthalpy) maximum. This seemingly trivial deduction based upon a particular way of representing an elementary reaction is the basis of a very useful qualitative guide, the *Hammond Postulate.*[21] It may be used when the conversions $A \rightarrow B$ and $A' \rightarrow B'$ occur by the same general mechanism and when the structural modification (for example, the introduction of a substituent) which perturbs the potential-energy hypersurface of the reaction (and hence alters the standard molar enthalpy of activation) does not also cause a substantial change in the standard molar entropy of activation.

This is unlikely to be a problem in gas-phase reactions but it could be for reactions in solution where differential solvation between reactant and activated complex could be drastically altered by particular substituents. In this event the free-energy change could, for example, be reduced whilst the enthalpy change is increased and the Hammond Postulate would lead to a wrong conclusion.

According to this model, a structural change which causes a simple reaction to become more exothermic should also cause it to become faster, and the activated complex should become more reactant-like. We may conclude, therefore, that a reaction which is fast and strongly

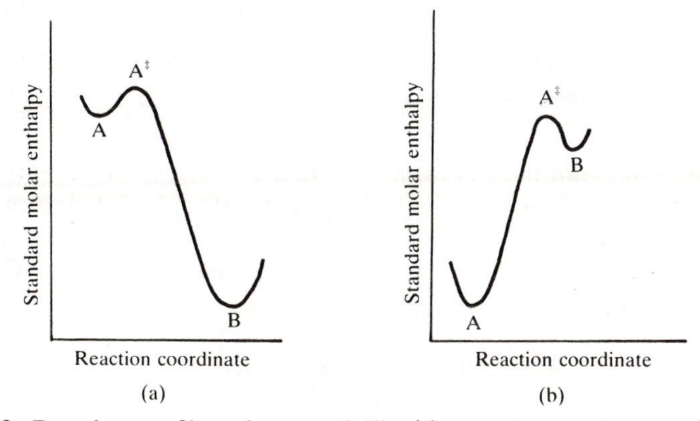

Fig. 6.10. Reaction profiles of energetically (a) very favourable, and (b) very unfavourable processes.

exothermic, as shown in Fig. 6.10(a), has a transition state which is not far along the reaction coordinate and the activated complex is *structurally* more similar to the reactant than it is to the product. In this case, most of the *structural* change which occurs between A and B takes place between $A^{\ddagger}$ and B. On the other hand a reaction as shown in Fig. 6.10(b) which is slow and endothermic (to give a product which may react further) involves a transition state far advanced along the reaction coordinate and, therefore, the activated complex is *structurally* very product-like. Here, most of the structural change required in the conversion of A into B takes place prior to the formation of $A^{\ddagger}$, the activated complex.

We shall see in subsequent chapters that kinetic analysis of a reaction provides information about the *composition* of the activated complex. The value of the Hammond Postulate is that it provides us with a means of deducing something qualitative about the *structure* of an activated complex. It must, however, be used with caution and may not be employed to compare reactions of different types. If a particular substituent is introduced into A to give a new reactant A″ and this causes a gross change in mechanism such that the product B″ is formed by a fundamentally different route which cannot be related to the original reaction profile in Fig. 6.9 (or perhaps reaction gives an entirely different product type, C), we cannot invoke the Hammond Postulate to compare the transition state between A and B with that between A″ and B″ (or A″ and C).

6.8 Problems

1. The rate of hydrolysis of $(CH_3)_3CCl$ in aqueous ethanol ($C_2H_5OH:H_2O = 70:30$) at 25 °C was monitored by periodically withdrawing 5.00 cm^3 aliquots and titrating them against 0.025 molar hydroxide. Calculate the first-order rate constant and $\Delta G^{\ominus\ddagger}_{298\,K}$ from the following results obtained from a reaction mixture initially 0.0465 molar in

$(CH_3)_3CCl$. (Results taken from E. D. Hughes, *J. chem. Soc.* 255 (1935).)

time/h	0.5	1.0	1.5	2.0	3.0	4.0	5.0	6.5	8.0	10.0
titre/cm^3	0.65	1.28	1.85	2.38	3.32	4.18	4.88	5.70	6.40	7.22

2. Calculate from the following results the first-order rate constant and $\Delta G^{\ominus\ddagger}_{703\,\mathrm{K}}$ for the gas-phase thermal decomposition of 1,1-dichloroethane which was monitored by following the increase in the total gas pressure P_t with time. (Data taken from H. Hartmann, H. Heydtmann, and G. Rinck, *Z. phys. Chem.* **28,** 71 (1961).)

$$CH_3CHCl_2 \xrightarrow{430\,°C} CH_2{=}CHCl + HCl$$

time/s	0	180	480	780	1080	1380	1680	1980	2280
P_t/Torr	158.5	173.1	194.3	213.2	228.9	242.2	253.6	262.8	270.8

3. Calculate the second-order rate constant for the reaction

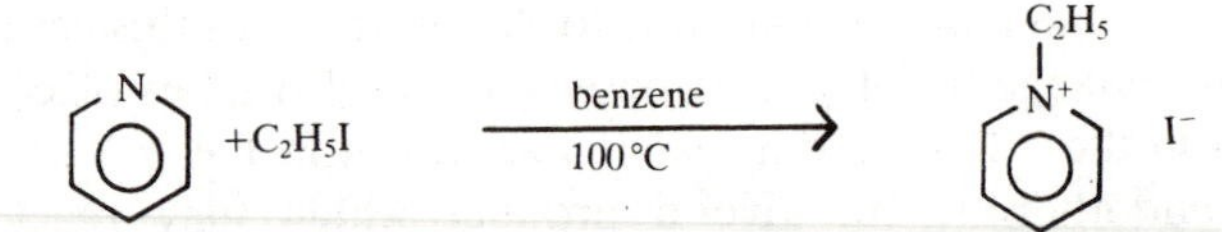

from the following results (reported by C. A. Winkler and C. N. Hinshelwood, *J. chem. Soc.* 1147 (1935)). Both reactants are initially at 0.10 mol dm^{-3}.

time/min	0	179	420	737	1054	1530	2497	4020
% reaction	0	12.0	24.0	35.0	44.5	53.0	65.5	75.0

4. The rate of dimerization of cyclopentadiene in the gas phase was investigated by monitoring the decrease in total pressure P_t at constant volume. Calculate the second-order rate constant (units, Torr^{-1} min^{-1}) from the following results obtained at 112 °C (by G. A. Benford and A. Wassermann, *J. chem. Soc.* 362 (1939)).

time/min	0	30	70	139	196	270	350
P_t/Torr	694.9	687.9	679.0	664.0	651.9	637.0	627.9

From your result, calculate the rate constant with units dm^3 mol^{-1} s^{-1} and $\Delta G^{\ominus\ddagger}_{385\,\mathrm{K}}$.

5. Calculate the second-order rate constant for the reaction

$$(CH_3)_3N + p\text{-}BrC_6H_4CH_2Cl \xrightarrow[100\,°C]{\text{benzene}} p\text{-}BrC_6H_4CH_2\overset{+}{N}(CH_3)_3Cl^-$$

from the following results (reported by C. G. Swain and W. P. Langsdorf, *J. Am. chem. Soc.* **73,** 2813 (1951)).

time/min	0	12	28	47	77	104	126	150	183	335
% reaction	0	10.5	21.9	33.1	46.5	54.4	60.8	65.6	71.8	86.4

Initial concentrations of amine and halide are 0.083 and 0.148 mol dm^{-3} respectively.

6. Calculate the Arrhenius parameters (E_a and A), $\Delta H^{\ominus\ddagger}$, and $\Delta S^{\ominus\ddagger}$ for the hydrolysis of t-butyl chloride from the following results (reported by E. A. Moelwyn-Hughes, R. E. Robertson, and S. Sugamori, *J. chem. Soc.* 1965 (1965)).

T/°C	1.01	5.02	10.02	15.02	20.01
$10^3\ k/s^{-1}$	0.842	1.62	3.57	7.49	15.19

7. Calculate the Arrhenius parameters (E_a and A), $\Delta H^{\ominus\ddagger}$, and $\Delta S^{\ominus\ddagger}$ (all related to an ideal gas standard state of 1 mol dm^{-3}) for the reaction:

$$C_2H_4 + H^{\cdot} \xrightarrow[T]{k} C_2H_5^{\cdot}\ (\rightarrow \text{further reaction})$$

from the following results (reported by J. H. Lee, J. V. Michael, W. A. Payne, and L. J. Stief, *J. chem. Phys.*, **68,** 1817 (1978)).

T/K	198	216	234	258	283	298	320
$10^{13}\ k/cm^3$ molecule^{-1} s^{-1}	2.00	2.98	4.07	6.31	9.43	11.34	14.52

8. Calculate $\Delta H^{\ominus\ddagger}$ and $\Delta S^{\ominus\ddagger}$ (standard state = ideal gas at 1 mol dm^{-3}) for the gas-phase dimerization of buta-1,3-diene from the results given below (reported by W. E. Vaughan, *J. Am. chem. Soc.* **54,** 3863 (1932)).

T/°C	326	342	370	388
$10^5 k_P$/Torr^{-1} min^{-1}	2.50	4.15	10.0	17.5

0.5

9. Calculate the Arrhenius parameters (E_a and A), $\Delta H^{\ominus\ddagger}$, and $\Delta S^{\ominus\ddagger}$ for the gas-phase thermal decomposition of *cis*-4-chloropent-2-ene from the following results (reported by P. J. Robinson, G. G. Skelhorne, and M. J. Waller, *J. chem. Soc., Perkin Trans. 2,* 349 (1978)).

$$CH_3CH{=}CH{-}CHCl{-}CH_3 \xrightarrow[T]{k} CH_3CH{=}CH{-}CH{=}CH_2 + HCl$$

T/K	542.5	550.6	562.6	574.9	589.8
$10^3\, k/\mathrm{s}^{-1}$	1.70	2.64	4.98	8.98	16.7

10. Calculate $\Delta H^{\ominus\ddagger}$ and $\Delta S^{\ominus\ddagger}$ from an Eyring plot of the results given below for the reaction[3]

$$\text{n-C}_3\text{H}_7\text{Br} + \text{Cl}^- \xrightarrow[k_2,\, T]{\text{acetone}} \text{n-C}_3\text{H}_7\text{Cl} + \text{Br}^-.$$

T/°C	25.00	34.62	44.50	55.20	64.83
$10^5\, k_2/\mathrm{dm^3\, mol^{-1}\, s^{-1}}$	6.45	16.4	41.0	106	215

6.9 References

1. A. A. Frost and R. G. Pearson, *Kinetics and mechanism* 2nd edn, Wiley (1961). This excellent book and its more recent edition by J. W. Moore and R. G. Pearson (1981) give an account of how reaction order may be determined. They also cover fully the derivation of the rate law which a given reaction mechanism will lead to, techniques involved in calculating rate constants from experimental results, and collision and transition state theories of reaction rates. See also S. W. Benson, *The foundation of chemical kinetics*, McGraw-Hill (1960), a comprehensive account of virtually all aspects of chemical kinetics.
2. A. Weissberger, *Technique of organic chemistry*, Vol. VIII, Pt. 1, Interscience (1961), and its updated version *Techniques of chemistry*, Vol. VI, Pt. 1, Wiley–Interscience (1974), contain authoritative chapters on theory, experimental techniques, and the evaluation and interpretation of results. In the more recent volume, Chapt. 13 'The use of computers' by K. B. Wiberg, and Chapt. 4 'Kinetics in solution' and Chapt. 8 'From kinetic data to reaction mechanism', both by J. F. Bunnett, are particularly useful.
3. E. D. Hughes, C. K. Ingold, and J. D. H. Mackie, *J. chem. Soc.* 3173 (1955).
4. J. Sauer, H. Wiest, and A. Mielert, *Ber.* **97,** 3183 (1964).
5. W. H. Saunders and R. A. Williams, *J. Am. chem. Soc.* **79,** 3712 (1957).
6. J. A. Landgrebe, *J. chem. Educ.* **41,** 567 (1964).
7. J. Topping, *Errors of observation and their treatment*, 4th edn, Chapman & Hall (1972); B. J. Brinkworth, *An introduction to experimentation*, 2nd edn, The English Universities Press (1973); Chapt. 5 'The treatment of experimental data', by D. Margerison in Vol. 1 of *Comprehensive chemical kinetics* (eds. C. H. Bamford and C. F. H. Tipper), Elsevier (1969).
8. M. Menzinger and R. Wolfgang, *Angew. Chem. int. edn* **8,** 438 (1969), give a general account of the meaning and significance of the Arrhenius Equation.
9. The general problem of dealing logarithmically with equations which involve dimensions is discussed by J. E. Boggs, *J. chem. Educ.* **35,** 30 (1958), and G. N. Copley, ibid., p. 366.
10. It is possible to derive more general versions of eqns (6.18)–(6.20) which admit any choice of standard state; see P. J. Robinson, *J. chem. Educ.* **55,** 509 (1978). We have restricted ourselves to 1 mol dm^{-3} because this is the most usual in the context of chemical kinetics. Less rigorous versions of eqns

(6.18)–(6.20) which omit the $(1\,\text{mol}\,\text{dm}^{-3})^{1-n}$ factor are frequently encountered. There are difficulties with units and dimensions in the use of such equations which we are able to avoid. It is important to realize that the $(1\,\text{mol}\,\text{dm}^{-3})^{1-n}$ is a factor in the equation which arises from the general thermodynamic derivation of the expression for the (dimensionless) thermodynamic equilibrium constant, see p. 125. It is not merely an appendage arbitrarily added.

11. S. Glastone, K. J. Laidler, and H. Eyring, *The theory of rate processes*, McGraw-Hill (1941).
12. G. Kohnstam, 'Heat capacities of activation and their uses in mechanistic studies', in *Adv. phys. org. Chem.* **5,** 121 (1967).
13. S. W. Benson, *Thermochemical kinetics*, 2nd edn, Wiley (1976). See also S. W. Benson, *J. chem. Educ.* **42,** 502 (1965), K. W. Egger and A. T. Cocks, *Helv. chim. Acta* **56,** 1516 (1973), and S. I. Miller, *J. chem. Educ.* **55,** 778 (1978).
14. S. W. Benson and H. E. O'Neal, *Kinetic data on gas phase unimolecular reactions*, US Dept. of Commerce, NSRDS-NBS21 (1971), a very useful compilation of experimental results fully referenced and accompanied by brief but helpful and expert comments.
15. US Dept. of the Interior, Bureau of Mines, Bulletin 666, 1974.
16. H. E. O'Neal and S. W. Benson, *Free Radicals* **2,** 275 (1973).
17. E. L. Metcalfe and A. F. Trotman-Dickenson, *J. chem. Soc.* 4620 (1962); p. 229 in Chapter 5 by M. Quack and J. Troe, Specialist Periodic Report (*Gas kinetics and energy transfer*), Vol. 2 (eds. P. G. Ashmore and R. J. Donovan) (1977).
18. C. Tanford, *Physical chemistry of macromolecules*, Wiley (1961); H. Morawetz *Adv. Protein Chem.* **26,** 243 (1972).
19. G. C. Fettis and J. H. Knox, in *Progr. Reaction Kinetics* **2,** 1 (1964).
20. J. B. Harkness, G. B. Kistiakowsky, and W. H. Mears, *J. chem. Phys.* **5,** 682 (1937).
21. G. S. Hammond, *J. Am. chem. Soc.* **77,** 334 (1955); D. Farçasiu, *J. chem. Educ.* **52,** 76 (1975).

Supplementary references

K. B. Wiberg, *Physical organic chemistry*, Wiley (1964).

L. P. Hammett, *Physical organic chemistry*, 2nd edn, McGraw-Hill (1970).

7

Rates of complex chemical reactions

7.1 Introduction

The theoretical basis for the interpretation of rates of chemical reactions which was presented in the previous chapter really applies only to what are described as *simple* reactions. These are reactions in which a single compound is converted directly to a product (or several products) by a unimolecular or bimolecular mechanism, or ones in which two different compounds react together with a bimolecular mechanism to give product (or products) directly. Such simple reactions always have either first- or second-order rate laws.

We may not assume, however, that every reaction which shows first- or second-order kinetics occurs by a simple mechanism; in practice it is necessary to establish whether or not a reaction of simple kinetic order is mechanistically really a simple reaction. The stoichiometry may be a guide. Oxidation of some aldehydes by potassium permanganate in buffered non-acidic aqueous solution has been shown[1] to be kinetically second order. The stoichiometry shown in eqn (7.1), however,

$$3RCHO+2MnO_4^- \xrightarrow[25\,°C]{H_2O} 2RCO_2^- + RCO_2H + 2MnO_2 + H_2O \quad (7.1)$$

$$-\frac{d[RCHO]}{dt} = k_2[RCHO][MnO_4^-],$$

immediately establishes that such a reaction involves more than a single bimolecular step, and that the simple rate law belies a more complicated mechanism. But this does not mean that simple reaction stoichiometry alone proves that the mechanism is simple. For example, the pyrolysis of acetaldehyde is not mechanistically simple.

$$CH_3CHO \xrightarrow{500\,°C} CH_4 + CO$$

We also saw in the previous chapter how first- and second-order rate constants can be obtained from experimental rate data. A problem we did not dwell on is how rate constants can be obtained when the reactions are not simple. For example, the dimerization of cyclopentadiene (p. 260) and the isomerization of stilbene (p. 252) are reversible yet we are able to obtain rate constants, under appropriate experimental conditions, for both forward and reverse reactions.

If the conversion of reactant (or reactants) into products is through several steps which may be parallel or consecutive (and any of these *elementary* steps could in principle be reversible), the *overall reaction* is described as *complex.* Complex reactions may occasionally have uncomplicated first- or second-order rate laws as, for example, the oxidation of aldehydes by permanganate illustrates. In such cases, the experimental rate constant is a composite term from which the rate constants of elementary steps may sometimes be derived by techniques which we shall consider in this chapter. Frequently, however, complex reactions have complicated rate expressions and sometimes may not even yield overall rate constants at all.

In considering complex reactions, it is encouraging to remember a mechanistic consequence of the quantized nature of matter: individual molecules can undergo only simple reactions, so even the most complicated of chemical transformations must be the outcome of competing and interacting simple elementary reactions. The kinetic investigation of a complex chemical reaction is, therefore, one probe into what is conceptually an analytical problem, and the objective is to determine the number and the nature of the elementary steps which together constitute the overall reaction.

The usual general procedure in handling the kinetics of complex reactions is to set up a tentative reaction scheme on the basis of initial, perhaps preliminary, evidence and then test whether this describes the experimental results. If it does not, the initial scheme has to be modified or replaced; this process is repeated until some reaction scheme has been developed which does account for the experimental results (both rates and products).

A prediction of the effect of some modification to the reaction, based upon the provisional mechanism, is next tested experimentally. The modification may be a change in the solvent, or it may be a change in the initial concentration of a reactant, or the addition of a non-reactant. If the outcome is as predicted by the tentative reaction scheme, we may be more confident in the hypothesis. Bear in mind, however, that a single piece of sound experimental evidence is sufficient to discredit a hypothetical reaction scheme however attractive it may be, and a mechanism, which is always a hypothesis, can never strictly be proven.

We shall now consider a few of the more common types of complex reactions, and then introduce an important general method of approximation which facilitates the elucidation of some complex reactions.

7.2 Complex reactions illustrated[2,3]

7.2.1 Parallel reactions of the same order

If compound A in Fig. 7.1 has several concurrent independent modes of reaction all of the same order $\boldsymbol{n}$, the experimental rate constant, k_{obs} in eqn (7.2), is the *sum* of the elementary (or component) rate constants k_x, k_y, and k_z. When $\boldsymbol{n} = 1$, the overall reaction, which yields several products, is first order in A; when $\boldsymbol{n} = 2$, it is second order in A. In either case, eqn (7.2) may be integrated exactly and the overall rate constant, k_{obs}, determined as described in the previous chapter.

$$\mathrm{A} \xrightarrow{k_x} \mathrm{X}$$

$$\mathrm{A} \xrightarrow{k_y} \mathrm{Y}$$

$$\mathrm{A} \xrightarrow{k_z} \mathrm{Z}$$

$$\text{Rate of decrease in [A]} = -\frac{\mathrm{d[A]}}{\mathrm{d}t} = k_x[\mathrm{A}]^{\boldsymbol{n}} + k_y[\mathrm{A}]^{\boldsymbol{n}} + k_z[\mathrm{A}]^{\boldsymbol{n}}$$

$$= (k_x + k_y + k_z)[\mathrm{A}]^{\boldsymbol{n}}$$

$$= k_{obs}[\mathrm{A}]^{\boldsymbol{n}} \qquad (7.2)$$

where the observed rate constant for the decrease in $[\mathrm{A}] = k_{obs} = k_x + k_y + k_z$ and $\boldsymbol{n}$ = reaction order.

Fig. 7.1. Parallel reactions of the same order from a single compound.

Similarly, if compound A in Fig. 7.2 undergoes several parallel second-order reactions with another reactant, B, the overall rate constant for the decrease in concentration of A (or B) is again easily related to the individual rate constants of the elementary steps.

$$\mathrm{A+B} \xrightarrow{k_u} \mathrm{U}$$

$$\mathrm{A+B} \xrightarrow{k_v} \mathrm{V}$$

$$\mathrm{A+B} \xrightarrow{k_w} \mathrm{W}$$

Rate of decrease in concentration of reactants

$$= -\frac{\mathrm{d[A]}}{\mathrm{d}t} = -\frac{\mathrm{d[B]}}{\mathrm{d}t}$$

$$= k_u[\mathrm{A}][\mathrm{B}] + k_v[\mathrm{A}][\mathrm{B}] + k_w[\mathrm{A}][\mathrm{B}]$$

$$= k_{\mathrm{obs}}[\mathrm{A}][\mathrm{B}],$$

where $k_{\mathrm{obs}} = k_u + k_v + k_w$ = the observed second-order rate constant for the decrease in [A] and [B]. If [B]≫[A] and, therefore, effectively constant, the reaction becomes *pseudo* first order with

$$-\frac{\mathrm{d[A]}}{\mathrm{d}t} = k'_{\mathrm{obs}}[\mathrm{A}]$$

and

$$k'_{\mathrm{obs}} = k_{\mathrm{obs}}[\mathrm{B}]$$

= the observed *pseudo* first-order rate constant.

Fig. 7.2. Parallel second-order reactions of two compounds.

Although single overall reaction rate constants are obtained, the several products in each of these hypothetical schemes do not arise from a common mechanism. They are products of independent parallel competing routes, and each has its own mechanism and individual rate constant. The reactions in Figs 7.1 and 7.2 are therefore quite different from a *simple* reaction which yields more than one product by a single mechanism, for example the *retro*-Diels–Alder reaction in eqn (7.3) at high temperatures.[4]

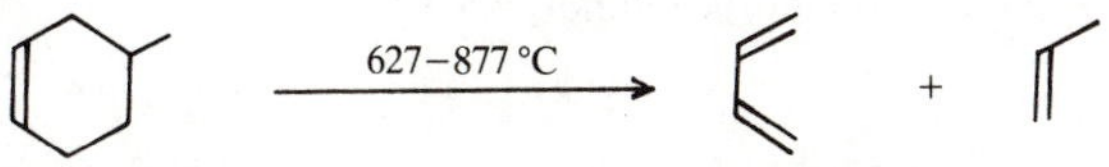

627–877 °C (7.3)

In the complex reactions of Figs 7.1 and 7.2, the component rate constants k_x, k_u, etc. will always be less than the rate constant for the disappearance of reactant. But the rate constants for the formation of butadiene and the formation of propene will be *identical* with the rate constant for the disappearance of 4-methylcyclohexene in eqn (7.3).

If the products of the complex reaction in Fig. 7.1 are stable to the reaction conditions, $[X]:[Y]:[Z] = k_x : k_y : k_z$ (and, in Fig. 7.2, $[U]:[V]:[W] = k_u : k_v : k_w$), so the ratio of products is independent of the extent of the reaction.[2] It is possible, therefore, to obtain the individual component rate constants from the overall reaction rate constant and an accurate product analysis. It is important, if this technique is used, to establish that indeed the products *are* stable to the reaction conditions, and that they do not interconvert or otherwise react. The best procedure is to subject each product independently to the reaction conditions for a time at least as long as is required for the reaction to reach more than 99 per cent completion, and show that it remains completely unchanged and especially that it is not converted into any of the other products. Such a chemical reaction is said to be *kinetically controlled*, the *ratios* of the products being determined strictly by their *relative rates of formation* and *not* according to their relative thermodynamic stabilities.

In the present context, a reaction at constant volume is complete when the concentrations of reactants and products have become constant with respect to time. In a kinetically-controlled process, this is when the concentration of any one reactant has decreased to zero. A reaction in which the products are able to interconvert will give, at completion, a mixture of products determined by their relative stabilities under the particular experimental conditions. Such a reaction is commonly described as *thermodynamically controlled*. It is of no consequence whether the product interconversions are via routes subsequent to the initial product formation or whether they all form reversibly from the original reactant. The important point is that a product analysis of such a reaction provides no information about the relative rates of formation of the various products.

Example. *Gas-phase thermal decomposition of t-butyl prop-2-yl ether.* The experimental rate constant, k_{obs}, was measured by monitoring the total pressure of the system at constant volume with time. The products were analyzed by gas–liquid chromatography and mass spectrometry and the analysis was shown to be independent of the extent of the reaction. (Results taken from N. J. Daly and F. J. Ziolkowsky, *Aust. J. Chem.* **23,** 541 (1970).)

$$(CH_3)_3COH + CH_2{=}CHCH_3 \xleftarrow{k_2} (CH_3)_3COCH(CH_3)_2 \xrightarrow{k_1} (CH_3)_2C{=}CH_2 + (CH_3)_2CHOH.$$

At 441.5 °C, the experimental first-order rate constant $k_{obs} = 3.02 \times 10^{-4}\ s^{-1}$, and product analysis gave [propanol]/[butanol] = 5.10.

$$k_{obs} = k_1 + k_2$$

and

$$k_1/k_2 = [propanol]/[butanol],$$

therefore we can calculate

$$\underline{k_1 = 2.52 \times 10^{-4}\ s^{-1}}$$

and

$$\underline{k_2 = 0.50 \times 10^{-4}\ s^{-1}}.$$

Many substitution reactions of aromatic compounds involve kinetically-controlled parallel second-order reactions which give isomeric products.

Example. *Chlorination of t-butylbenzene in acetic acid at 25 °C.* The overall second-order rate constant was determined by monitoring the concentration of unreacted chlorine by iodometric titration of aliquots after known time intervals starting from known concentrations of hydrocarbon and chlorine. The product analysis was by infra-red spectroscopy. (Results taken from L. M. Stock and H. C. Brown, *J. Am. chem. Soc.* **81,** 5615 (1959).)

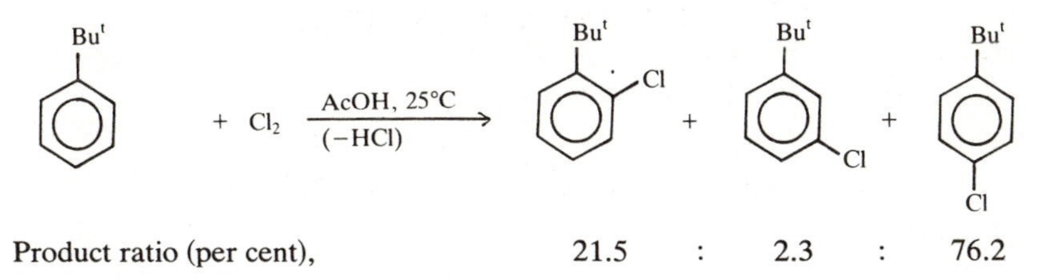

Product ratio (per cent), 21.5 : 2.3 : 76.2

The overall second-order rate constant, k_{obs},

$$= 1.35 \times 10^{-4}\ \mathrm{dm^3\ mol^{-1}\ s^{-1}}$$
$$= k_o + k_m + k_p$$

therefore $k_o = 0.215 \times k_{obs}$, $k_m = 0.023 \times k_{obs}$, and $k_p = 0.762 \times k_{obs}$, so

$$\underline{k_o = 0.29 \times 10^{-4}\ \mathrm{dm^3\ mol^{-1}\ s^{-1}},}$$

$$\underline{k_m = 0.03 \times 10^{-4}\ \mathrm{dm^3\ mol^{-1}\ s^{-1}},}$$

and

$$\underline{k_p = 1.03 \times 10^{-4}\ \mathrm{dm^3\ mol^{-1}\ s^{-1}}.}$$

In a kinetically-controlled complex reaction, individual elementary rate constants – k_x, k_y, and k_z (or k_u, k_v, and k_w) in the above – can only be obtained from the overall rate constant by a product analysis. If the rate of formation of one of several products formed in parallel competing reactions is monitored, this still gives the overall reaction rate constant – *not* the elementary rate constant for its own formation. This must obviously be so since the concentration of any product is a property of the whole system which changes as the overall reaction proceeds (at constant volume). We can show this as follows for those who prefer a more explicit proof related to a particular reaction.

In parallel first-order reactions of A,

$$A \xrightarrow{k_x} X$$
$$A \xrightarrow{k_y} Y$$
$$A \xrightarrow{k_z} Z$$
$$k_{obs} = k_x + k_y + k_z$$

$$\text{Rate of formation of X} = \frac{d[X]}{dt} = k_x[A].$$

But this differential equation cannot be integrated directly as it includes two concentration variables, [X] and [A]. However, they are not independently variable.

At time 0, $[A]=[A]_0$ and $[X]=[Y]=[Z]=0$,

at time ∞, $[A]=0$, $[X]=[X]_\infty$, $[Y]=[Y]_\infty$, and $[Z]=[Z]_\infty$, and

at some intermediate time, $[A]+[X]+[Y]+[Z]=[A]_0=[X]_\infty+[Y]_\infty+[Z]_\infty$. Consequently, substituting for [A],

$$\frac{d[X]}{dt}=k_x([X]_\infty+[Y]_\infty+[Z]_\infty-[X]-[Y]-[Z]).$$

And, as the reaction is kinetically controlled,

$$\frac{[Y]}{[X]}=\frac{[Y]_\infty}{[X]_\infty}=\frac{k_y}{k_x} \quad \text{and} \quad \frac{[Z]}{[X]}=\frac{[Z]_\infty}{[X]_\infty}=\frac{k_z}{k_x},$$

therefore

$$\frac{d[X]}{dt}=k_x\left([X]_\infty+[X]_\infty\,.\,\frac{k_y}{k_x}+[X]_\infty\,.\,\frac{k_z}{k_x}-[X]-[X]\,.\,\frac{k_y}{k_x}-[X]\,.\,\frac{k_z}{k_x}\right)$$

$$=(k_x+k_y+k_z)\,.\,[X]_\infty-(k_x+k_y+k_z)\,.\,[X],$$

or

$$\frac{d[X]}{dt}=k_{obs}([X]_\infty-[X]).$$

This differential equation can now be integrated to give the familiar rate law for a first-order reaction. It shows that we still obtain the overall reaction rate constant, k_{obs}, even when we monitor the rate of formation of the product of only one of several competing routes which together comprise the overall complex reaction.

7.2.2 Parallel first- and second-order reactions

7.2.2.1 Reactions of a single compound

Compound A in eqns (7.4) and (7.5) undergoes parallel reactions of first and second order to give B and C respectively.

$$A \xrightarrow{k_1} B \tag{7.4}$$

$$2A \xrightarrow{k_2} C. \tag{7.5}$$

k_1 and k_2 are the rate constants of the competing elementary processes. As the reactions proceed, the change in [A] at constant volume is given by

$$-\frac{d[A]}{dt}=k_1[A]+2k_2[A]^2.$$

This rate expression includes both k_1 and k_2 but only one concentration term, [A], so in principle we should be able to determine k_1 and k_2 by monitoring the decrease in [A] with time.

Rearrangement of the above rate expression gives

$$\frac{-\mathrm{d}[\mathrm{A}]}{(k_1[\mathrm{A}]+2k_2[\mathrm{A}]^2)}=\mathrm{d}t,$$

and the left-hand side of this equation has the form of a standard integral:[5]

$$\int\frac{\mathrm{d}x}{(ax+bx^2)}=\frac{1}{a}\,.\,\ln\left(\frac{x}{a+bx}\right).$$

Integration between time $= t$ when $[\mathrm{A}]=[\mathrm{A}]_t$ and time $= 0$ when $[\mathrm{A}]=[\mathrm{A}]_0$, gives

$$-\frac{1}{k_1}\,.\,\ln\left(\frac{[\mathrm{A}]_t}{k_1+2k_2[\mathrm{A}]_t}\right)+\frac{1}{k_1}\,.\,\ln\left(\frac{[\mathrm{A}]_0}{k_1+2k_2[\mathrm{A}]_0}\right)=t,$$

or

$$\ln\left(\frac{[\mathrm{A}]_0(k_1+2k_2[\mathrm{A}]_t)}{[\mathrm{A}]_t(k_1+2k_2[\mathrm{A}]_0)}\right)=k_1t,$$

which relates $[\mathrm{A}]_t$ with time in terms of the reaction parameters k_1 and k_2 and the initial concentration of reactant, $[\mathrm{A}]_0$.

The modern way to analyse the kinetics of a reaction of this type from a set of $[\mathrm{A}]_t$ versus time data is to use a standard computer programme and systematically find by a reiterative method the values of k_1 and k_2 which best fit the experimental results to the above analytical expression.

The concurrent isomerization and dimerization of *cis,trans*-cyclo-octa-1,5-diene in the gas phase are an example of this type of a complex reaction.[6]

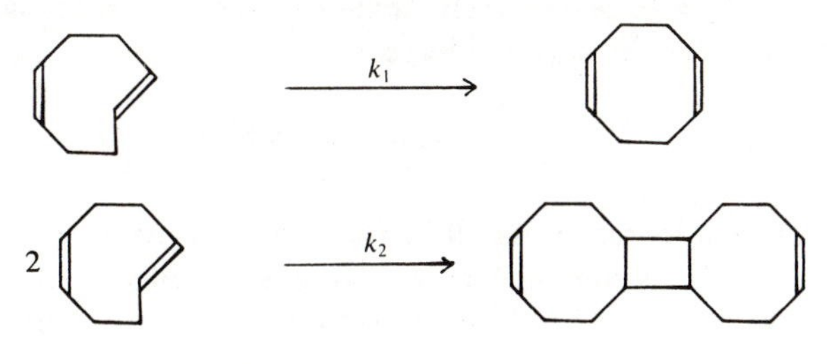

At 123 °C,

$$k_1=9.98(\mp 0.82)\times 10^{-6}\,\mathrm{s}^{-1}$$

and

$$k_2=5.69(\mp 0.65)\times 10^{-4}\,\mathrm{dm}^3\,\mathrm{mol}^{-1}\,\mathrm{s}^{-1}.$$

The partitioning of A between eqns (7.4) and (7.5) above gives the ratio of the first-order reaction to the second-order one:

$$\frac{\mathrm{d}[\mathrm{B}]}{\mathrm{d}t}\bigg/\frac{\mathrm{d}[\mathrm{C}]}{\mathrm{d}t}=\frac{k_1[\mathrm{A}]}{k_2[\mathrm{A}]^2}=\frac{k_1}{k_2[\mathrm{A}]}.$$

So, although numerically $k_2 \gg k_1$ for *cis,trans*-cyclo-octa-1,5-diene, the actual rate of isomerization under the particular experimental conditions was faster than the rate of dimerization because [*cis,trans*-diene] was so low (initially $\sim(2.12)\times 10^{-4}$ mol dm^{-3}). Furthermore, as the concentration of the reactant decreases during the overall reaction, the proportion of isomerization increases further. Consequently, [*cis,cis*-diene] > [dimer] and the ratio [*cis,cis*-diene]/[dimer] increases as the reaction proceeds. Such a variation in a product ratio is characteristic of a complex reaction comprising elementary reactions of different order in some reactant even though the overall reaction is kinetically controlled. This is in contrast to competing kinetically-controlled reactions of the same order which we dealt with earlier in this chapter. In those reactions, the product ratios (relative proportions of the parallel elementary reactions) do not vary as the overall reaction proceeds.

7.2.2.2 *Reactions involving two compounds*

We shall now consider a complex reaction in which one compound undergoes a first-order process and a competing second-order one with another compound.

$$\mathrm{A} \xrightarrow{k_1} \mathrm{B}$$

$$\mathrm{A}+\mathrm{Y} \xrightarrow{k_2} \mathrm{D},$$

where k_1 and k_2 are, respectively, first- and second-order rate constants. The expression for the rate of decrease in [A] is

$$-\frac{\mathrm{d}[\mathrm{A}]}{\mathrm{d}t}=k_1[\mathrm{A}]+k_2[\mathrm{A}][\mathrm{Y}],$$

and the best way, by far, of handling this type of a reaction is to transform it into an overall *pseudo* first-order process. In other words, we chose experimental conditions which enormously simplify the subsequent mathematics.

If $[\mathrm{Y}] \gg [\mathrm{A}]$, then [Y] remains effectively constant as the reaction proceeds, consequently

$$-\frac{\mathrm{d}[\mathrm{A}]}{\mathrm{d}t}=k_{\mathrm{obs}}[\mathrm{A}],$$

where $k_{obs} = k_1 + k_2[Y]$ = the observed overall *pseudo* first-order rate constant.

By the usual method of measuring k_{obs} at several known values of [Y], a linear plot of k_{obs} against [Y] allows the determination of k_1 (from the intercept) and k_2 (from the gradient).

Organic reactions with this type of rate law are very common in solution and include many substitution and elimination reactions of alkyl halides and arenesulphonates in hydroxylic solvents containing a dissolved base/nucleophile:

e.g. $$RX \xrightarrow[C_2H_5O^-Na^+]{C_2H_5OH} \text{substitution, elimination;}$$

$$-\frac{d[RX]}{dt} = k_1[RX] + k_2[RX][C_2H_5O^-Na^+].$$

Notice that in these reactions, the second-order mode can be suppressed completely simply by making [Y] = 0. This is not possible in the complex reaction of A alone by competing first- and second-order processes as in eqns (7.4) and (7.5) above.

Example. *Reaction of benzenesulphonyl chloride with sodium fluoride in aqueous solution,* 15 °C. (Results taken from O. Regne, *J. chem. Soc.* B 1056 (1970).) The *pseudo* first-order rate constants (k_{obs}) for the reaction of benzenesulphonyl chloride (initial concentration approximately 2×10^{-4} mol dm^{-3}) with sodium fluoride in aqueous solution (pH = 7.4, ionic strength made up to 0.05 M with sodium chloride) were measured at different fluoride concentrations.

$$C_6H_5SO_2Cl + NaF \xrightarrow[k_2]{H_2O,\ 15\,°C} C_6H_5SO_2F + NaCl$$

concurrent with

$$C_6H_5SO_2Cl + H_2O \xrightarrow{k_0} C_6H_5SO_3H + HCl$$

10^2[NaF]/mol dm^{-3}	$10^3\ k_{obs}$/s^{-1}
0	1.20
0.5	2.70
1.0	3.97
2.0	6.95
3.0	10.1
4.0	12.6
5.0	16.0

Calculate k_2 the second-order rate constant for this reaction.

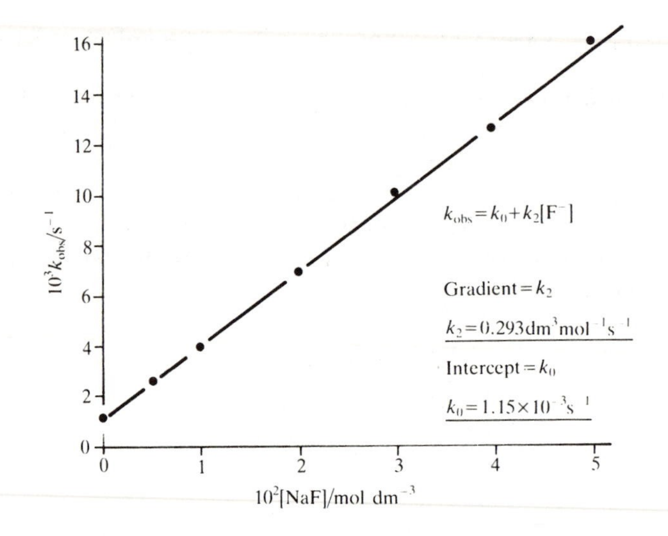

The solvolytic process alone was investigated by measuring the rate of reaction in the absence of F^-. The rate constant in this direct measurement, $k_0 = 1.20 \times 10^{-3}\,s^{-1}$, agrees with the value obtained by extrapolation of the best straight line through the points in the graph to the intercept at $[F^-] = 0$, $k_0 = 1.15 \times 10^{-3}\,s^{-1}$.

7.2.3 Reversible first-order reactions

The general case of a reversible first-order reaction at constant volume is shown in eqn (7.6) and the equilibrium of such a system could be treated thermodynamically as described in Chapter 4. We shall now consider the kinetics of the process.

$$A \underset{k_{-1}}{\overset{k_1}{\rightleftharpoons}} B; \qquad K = \frac{[B]_e}{[A]_e} \tag{7.6}$$

k_1 and k_{-1} = the first-order rate constants for forward and reverse reactions,
$[A]_e$ and $[B]_e$ = the molar concentrations of A and B at equilibrium, and
K = the equilibrium constant.

Before the system achieves equilibrium, the rate at which the concentration of A decreases is given by

$$-\frac{d[A]}{dt} = k_1[A] - k_{-1}[B]. \tag{7.7}$$

We again have a differential equation which cannot be integrated directly because it includes two concentration variables, [A] and [B]. They are not, however, independent so we can transform eqn (7.7) into one containing only a single concentration variable, x.

We define x by

$$x = [\mathrm{A}] - [\mathrm{A}]_e \qquad (7.8)$$
$$= [\mathrm{B}]_e - [\mathrm{B}],$$

then

$$[\mathrm{A}] = x + [\mathrm{A}]_e,$$

and

$$[\mathrm{B}] = [\mathrm{B}]_e - x.$$

Substitution into eqn (7.7) gives

$$-\frac{\mathrm{d}x}{\mathrm{d}t} = k_1(x + [\mathrm{A}]_e) - k_{-1}([\mathrm{B}]_e - x),$$

or

$$-\frac{\mathrm{d}x}{\mathrm{d}t} = (k_1 + k_{-1})x + k_1[\mathrm{A}]_e - k_{-1}[\mathrm{B}]_e. \qquad (7.9)$$

At equilibrium, $$\frac{\mathrm{d}[\mathrm{A}]}{\mathrm{d}t} = \frac{\mathrm{d}[\mathrm{B}]}{\mathrm{d}t} = \frac{\mathrm{d}x}{\mathrm{d}t} = 0, \quad \text{and} \quad x = 0,$$

therefore from eqn (7.9)

$$k_1[\mathrm{A}]_e - k_{-1}[\mathrm{B}]_e = 0$$

so

$$[\mathrm{B}]_e/[\mathrm{A}]_e = k_1/k_{-1} = K$$

and

$$-\frac{\mathrm{d}x}{\mathrm{d}t} = (k_1 + k_{-1})x.$$

By resubstituting from eqn (7.8), we obtain

$$-\frac{\mathrm{d}([\mathrm{A}] - [\mathrm{A}]_e)}{\mathrm{d}t} = (k_1 + k_{-1})([\mathrm{A}] - [\mathrm{A}]_e)$$

therefore upon rearrangement and integration

$$\ln\left\{\frac{[\mathrm{A}]_t - [\mathrm{A}]_e}{[\mathrm{A}]_0 - [\mathrm{A}]_e}\right\} = -(k_1 + k_{-1})t \qquad (7.10)$$

where $[\mathrm{A}] = [\mathrm{A}]_0$ when time $= 0$, and $[\mathrm{A}] = [\mathrm{A}]_t$ when time $= t$.

When $k_1 \gg k_{-1}$, i.e. when K is large and the equilibrium mixture is almost solely B, the reaction is essentially unidirectional, $[\mathrm{A}]_e \sim 0$, and eqn (7.10) may be simplified to give an equation we are already familiar with:

$$\ln\left(\frac{[\mathrm{A}]_t}{[\mathrm{A}]_0}\right) = -k_1 t,$$

the rate law of a simple unidirectional first-order reaction.

Equation (7.10), which may be re-written as

$$\ln\left\{\frac{[A]_t-[A]_e}{[A_0]-[A]_e}\right\}=-k_{obs}t$$

where $k_{obs}=k_1+k_{-1}$, also shows that the progress of A, or any non-equilibrium mixture of A plus B, towards equilibrium is an overall first-order process and the observed experimental rate constant, k_{obs}, is the sum of the individual first-order rate constants for forward and reverse reactions. If the equilibrium mixture can also be analysed, the equilibrium constant, K, can be measured and this is equal to the *ratio* of forward and reverse rate constants. From k_{obs}, (k_1+k_{-1}), and K, (k_1/k_{-1}), the individual rate constants, k_1 and k_{-1}, can be obtained.

Example. *Gas-phase isomerization of cis- and trans*-1-*ethyl*-2-*methylcyclopropane at* 425.6 °C. Gas-liquid chromatography was used to monitor progress towards equilibrium and also to measure the ratio of compounds at equilibrium. (Results taken from C. S. Elliott and H. M. Frey, *J. chem. Soc.* 900 (1964).)

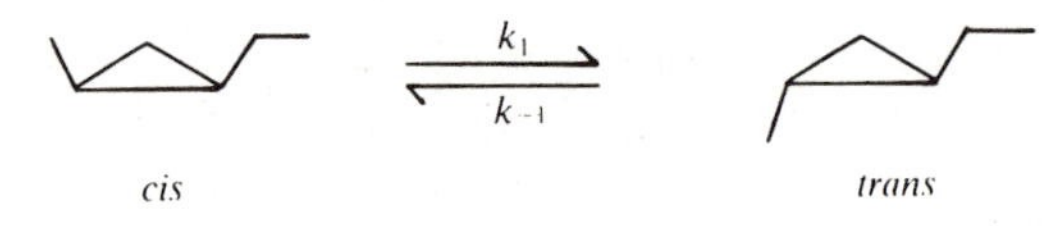

At 425.6 °C, a result of $k_{obs}=(k_1+k_{-1})=6.29\times10^{-4}\ s^{-1}$ was obtained using eqn (7.10). At equilibrium $[trans]_e/[cis]_e=2.79$ therefore

$$K=k_1/k_{-1}=2.79.$$

From these results we calculate

$$\underline{k_1=4.63\times10^{-4}\ s^{-1},}$$

and

$$\underline{k_{-1}=1.66\times10^{-4}\ s^{-1}.}$$

7.2.3.1 *Racemization*

The uncatalysed racemization of an optically-active compound involves a reversible first-order reaction in which the rate constants for the forward and reverse processes are numerically identical. Each of the opposed individual reactions in eqn (7.11) is an *enantiomerization* which we cannot observe in isolation. The overall reaction, starting from either enantiomer or an unequal mixture of both, to give the optically inactive 1:1 mixture at equilibrium is the *racemization*.

$$(+)\mathrm{A} \underset{k_1}{\overset{k_1}{\rightleftharpoons}} (-)\mathrm{A} \qquad (7.11)$$

$$(-)\mathrm{A}\ \text{or}\ (+)\mathrm{A} \xrightarrow{k_{rac.}} (+,-)\mathrm{A}. \qquad (7.12)$$

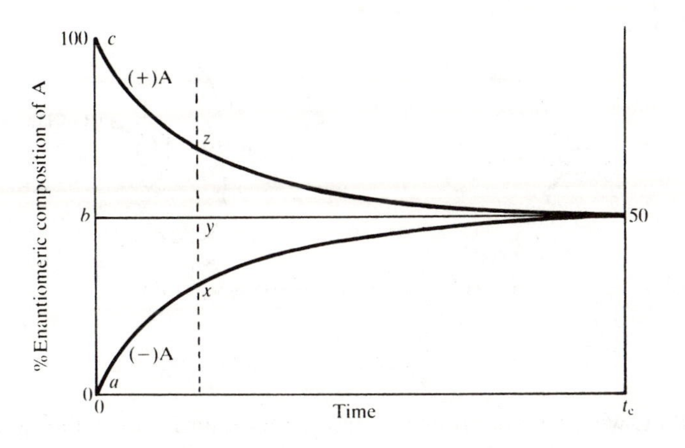

Fig. 7.3. Racemization of (+)A to (±)A.

Although racemization may be described by eqn (7.11), it is conventional to use eqn (7.12). Racemization can be monitored polarimetrically, the optical activity, α, of a mixture of enantiomers being proportional to the excess of one over the other:

$$+\alpha \propto ([(+)\text{A}]-[(-)\text{A}])$$

or

$$-\alpha \propto ([(-)\text{A}]-[(+)\text{A}]).$$

If we represent a racemization by Fig. 7.3, it is easily seen that

$$\left(\frac{[(+)\text{A}]_0-[(+)\text{A}]_e}{[(+)\text{A}]_t-[(+)\text{A}]_e}\right)=\frac{bc}{yz}=\frac{ac}{xz}=\left(\frac{[(+)\text{A}]_0}{[(+)\text{A}]_t-[(-)\text{A}]_t}\right)=\frac{\alpha_o}{\alpha_t},$$

where α_o = optical activity when time = 0,
α_t = optical activity when time = t,
and the subscript e refers to the equilibrium condition.
When eqn (7.10) is applied to racemization, it may, therefore, be rewritten as

$$\ln\left(\frac{\alpha_t}{\alpha_o}\right)=-2k_1t,$$

or

$$\ln\left(\frac{\alpha_t}{\alpha_o}\right)=-k_{\text{rac.}}t,$$

with $k_{\text{rac.}}$ = the rate constant for racemization
= $2k_1$ where k_1 = the rate constant for enantiomerization.

Examples of this type of reaction include the racemizations of 2,2′-diiodobiphenyl,[7] (1) and the methyl ester of a paracyclophanecarboxylic

acid,[8] (2)

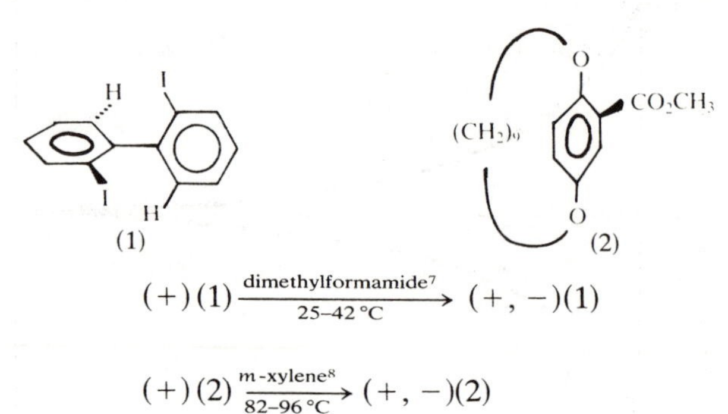

Note that, whereas enantiomerization is necessarily reversible, racemization is strictly uni-directional.

7.2.4 Consecutive first-order reactions

The general reaction illustrated in eqn (7.13) is the easiest complex reaction comprising consecutive steps to deal with. It can be seen that A yields C not directly, but through an *intermediate* B. Such reactions can be treated rigorously in terms of simultaneous differential equations.

$$\mathrm{A} \xrightarrow{k_1} \mathrm{B} \xrightarrow{k_2} \mathrm{C} \tag{7.13}$$

$$-\frac{\mathrm{d}[\mathrm{A}]}{\mathrm{d}t} = k_1[\mathrm{A}] \tag{7.14}$$

$$-\frac{\mathrm{d}[\mathrm{B}]}{\mathrm{d}t} = k_2[\mathrm{B}] - k_1[\mathrm{A}]. \tag{7.15}$$

The rate at which product is formed is given by

$$\frac{\mathrm{d}[\mathrm{C}]}{\mathrm{d}t} = k_2[\mathrm{B}].$$

But neither this nor eqn (7.15) can be integrated directly as both contain two concentration variables.

Equation (7.14) is the first-order rate law for the exponential decrease in the concentration of A as the reaction proceeds. By monitoring the concentration of A with time, k_1 may be determined using the logarithmic form of the integrated version of eqn (7.14) as described in the previous chapter.

If the integrated version of eqn (7.14) is expressed in its exponential form we have eqn (7.16)

$$[\mathrm{A}]_t = [\mathrm{A}]_0 \mathrm{e}^{-k_1 t}, \tag{7.16}$$

where $[A]=[A]_0$ when time = 0, and $[A]=[A]_t$ when time = t. By substituting eqn (7.16) into eqn (7.15) we obtain a linear differential equation (eqn (7.17))

$$-\frac{d[B]}{dt}=k_2[B]-k_1[A]_0e^{-k_1t}, \tag{7.17}$$

which can be integrated[3] to give eqn (7.18). Equation (7.18) describes the variation in the single concentration variable, [B], with time.

$$[B]_t=\frac{k_1}{(k_2-k_1)}\cdot[A]_0(e^{-k_1t}-e^{-k_2t}). \tag{7.18}$$

We now proceed to obtain an analogous expression for $[C]_t$. At any moment during the reaction,

$$[A]+[B]+[C]=[A]_0,$$

therefore at time t,

$$[C]_t=[A]_0-[A]_t-[B]_t.$$

Substituting for $[A]_t$ and $[B]_t$ using eqn (7.16) and (7.18),

$$[C]_t=[A]_0-[A]_0e^{-k_1t}-\frac{k_1}{(k_2-k_1)}[A]_0(e^{-k_1t}-e^{-k_2t}).$$

At completion ($t=\infty$), $[A]_\infty=[B]_\infty=0$ and $[C]_\infty=[A]_0$, therefore

$$[C]_t=[C]_\infty\left\{1-e^{-k_1t}-\frac{k_1}{(k_2-k_1)}(e^{-k_1t}-e^{-k_2t})\right\}. \tag{7.19}$$

Note that eqns (7.18) and (7.19) are complex functions of both rate constants, k_1 and k_2, whereas eqn (7.16) describes the variation in the concentration of A with time in terms of only the first rate constant, k_1.

It is clear from eqn (7.13) that, as the reaction proceeds, [A] decreases from $[A]_0$ to zero at completion as [C] increases to $[C]_\infty$ from its initial value of zero, whereas [B] is initially and finally zero, yet is finite during the reaction. The concentration of B therefore increases, reaches a maximum, and then decreases again.

The concentration–time relationships for A, B, and C are illustrated in Fig. 7.4 for a representative reaction in which $k_1 \sim k_2$. The position of the maximum in [B] with respect to the cross-over between the curves for [A] and [C] depends upon the relative magnitudes of k_1 and k_2. Obviously, if $k_2 \gg k_1$, B is a *reactive intermediate* and no large amount of it can ever accumulate. If $k_1 \gg k_2$, B is a relatively long-lived intermediate and an appreciable concentration of it may develop (the actual concentration depending upon the absolute values of k_1, k_2, and $[A]_0$).

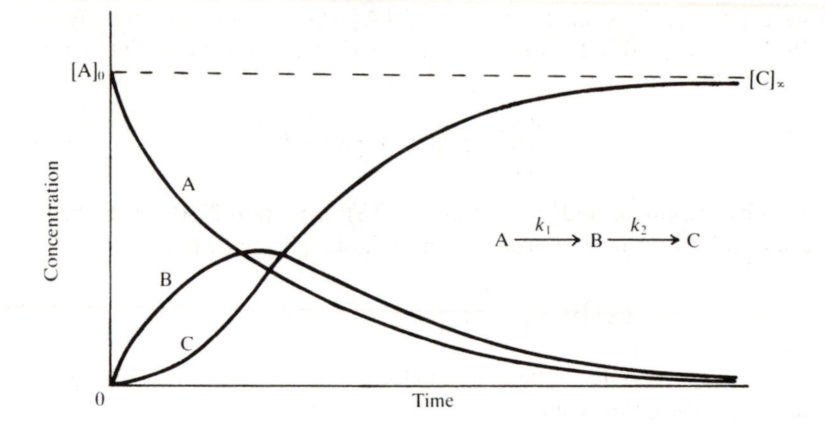

Fig. 7.4. Concentration–time relationships for consecutive first-order reactions.

The shape of the curve for [C] is particularly interesting and is characteristic of this type of complex reaction. During an initial *induction period* the rate of formation of C is low; it then increases as the curve turns upwards, and over the next short period the rate of formation of C is approximately linear, but then it slows down again as the curve levels off, and tends towards zero as [C] approaches its asymptotic value, $[C]_\infty$.

We saw from eqn (7.19) that if a reaction of this sort is monitored by following the rate of formation of product, we are unable to obtain a linear correlation by plotting some simple function of the concentration of the product against time. Strictly, this type of reaction does not have an apparently simple *overall* rate law and consequently we cannot determine a single overall experimental rate constant. Here is another example where a standard computer programme should be used, this time to find systematically the values of k_1 and k_2 which best fit the experimental results of $[C]_t$ and time to eqn (7.19).

As mentioned above, k_1 can be determined independently by treating the rate of disappearance of A as a simple first-order reaction.

7.3 The steady-state approximation for complex reactions

In the previous section, the *exact* kinetic analysis of a straightforward example of a complex chemical reaction – two consecutive first-order steps – required the integration of a linear differential equation and computer-assisted fitting of data to an equation to obtain the two rate constants. As complex reactions become more complicated, their mathematical analysis becomes increasingly difficult and can frequently be achieved only by making simplifying assumptions. One generally useful technique is the application of the *steady-state approximation* without which many kinetic problems would be quite intractable. The difficulties are in recognizing the chemical situation when the simplifying approximation has to be used, and in appreciating the limitations of the method and the conditions under which it may not be used.

In order to illustrate the method, we shall apply the approximation to the reaction in eqn (7.13) – the same complex mechanism (two consecutive first-order steps) the exact kinetic analysis of which we have already achieved. We shall then see the conditions under which the approximate and the exact solutions are in good agreement.

7.3.1 Consecutive first-order reactions

$$\mathrm{A} \xrightarrow{k_1} \mathrm{B} \xrightarrow{k_2} \mathrm{C} \tag{7.13}$$

$$-\frac{\mathrm{d[A]}}{\mathrm{d}t} = k_1[\mathrm{A}] \tag{7.14}$$

$$-\frac{\mathrm{d[B]}}{\mathrm{d}t} = k_2[\mathrm{B}] - k_1[\mathrm{A}] \tag{7.15}$$

$$+\frac{\mathrm{d[C]}}{\mathrm{d}t} = k_2[\mathrm{B}]. \tag{7.20}$$

The objective is to deduce an expression for the *overall* rate of the reaction – the rate at which C is formed – in terms of the initial concentration of A (or final concentration of C), the reaction parameters k_1 and k_2, and the single independent variable, t. We start from eqn (7.20) and, in order to obtain a simple expression for [B], we make an assumption which will obviate the need to integrate the linear differential equation, eqn (7.15). We *assume* that, after an initial brief period, the concentration of B achieves a steady state with its rate of formation and rate of disappearance just balanced, i.e.

$$-\frac{\mathrm{d[B]}}{\mathrm{d}t} = k_2[\mathrm{B}] - k_1[\mathrm{A}] = 0,$$

therefore

$$[\mathrm{B}] = \frac{k_1}{k_2}[\mathrm{A}],$$

or, since eqn (7.14) may be solved exactly (to give eqn (7.16), p. 282) we may write

$$[\mathrm{B}] = \frac{k_1}{k_2}[\mathrm{A}]_0 \mathrm{e}^{-k_1 t}. \tag{7.21}$$

Substituting for [B] in eqn (7.20) we now obtain

$$\frac{d[C]}{dt} = k_1[A], \tag{7.22}$$

or

$$\frac{d[C]}{dt} = k_1[A]_0 e^{-k_1 t},$$

therefore $d[C]/dt = k_1[C]_\infty e^{-k_1 t}$ since at completion

$$[A] = [B] = 0, \quad \text{and} \quad [C] = [C]_\infty = [A]_0.$$

As $[C] = 0$ when time $= 0$, integration gives

$$[C]_t = -[C]_\infty e^{-k_1 t} + [C]_\infty$$

or

$$[C]_t = [C]_\infty (1 - e^{-k_1 t}) \tag{7.23}$$

which may be rearranged and expressed logarithmically to give

$$\ln\left\{\frac{[C]_\infty - [C]_t}{[C]_\infty}\right\} = -k_1 t.$$

Equation (7.23) expresses the variation in the concentration of C with time, i.e. it describes the rate of the *overall* reaction.

We can now compare eqns (7.21) and (7.23), which have been obtained very easily with the help of the steady-state approximation, with eqns (7.18) and (7.19) respectively, which are exact, though less easily come by.

We find that when $k_2 \gg k_1$ eqn (7.18) gives eqn (7.21), and when $k_2 \gg k_1$ *and* $k_2 t \gg 1$, eqn (7.19) gives eqn (7.23). These, then, are the conditions under which the steady-state approximation introduces no serious errors.[9] The first of these conditions could have been predicted intuitively by consideration of eqn (7.13). The absolute variation in [B], which we can see from eqn (7.21) and Fig. 7.4 is never actually zero but decreases slowly after an initial build-up, would be lowest (in comparison with the concurrent changes in [A] and [C]) if [B] is exceedingly small. This is achieved when B is a very reactive, short-lived intermediate with $k_2 \gg k_1$.

The second condition ($k_2 t \gg 1$) merely indicates that the approximation is not valid in the very early part of the reaction (the induction period) whilst the concentration of B is being built up from zero to its steady-state concentration.

If the concentration of B remains virtually constant throughout most of the reaction of eqn (7.13), then the rate of disappearance of A is the same as the rate of formation of C as comparison of eqns (7.14) and (7.22)

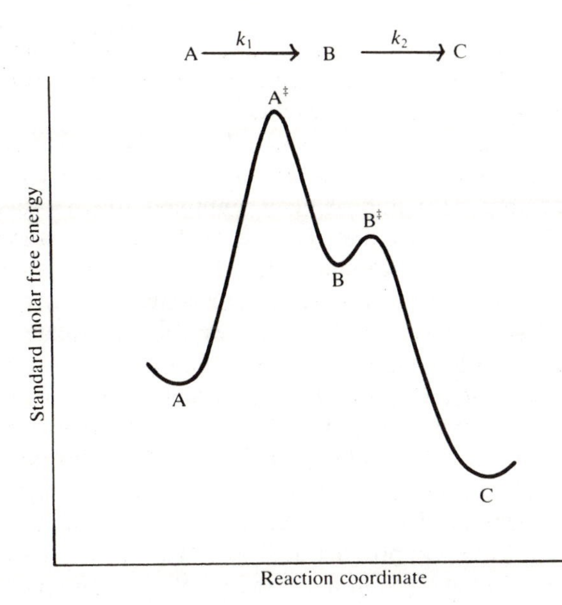

Fig. 7.5. Reaction profile of a complex reaction comprising two sequential first-order steps with $k_2 \gg k_1$.

shows. When, in a particular example of this type of complex mechanism, therefore, the same k_{obs} is obtained experimentally from monitoring the rate of formation of product (after the induction period) as from the rate of disappearance of reactant, we have *a posteriori* evidence in support of the use of the steady-state approximation. Furthermore, in such a reaction ($k_2 \gg k_1$) the rate of the overall reaction is effectively determined by the rate constant of the first step ($k_{obs} = k_1$). Consequently, this first step is frequently called the *rate-determining step* of the complex reaction (see also p. 308), and a standard free-energy profile is shown in Fig. 7.5.

It is commonly stated in descriptions of such a reaction that the first step is slow and the second step fast. This is not strictly correct because in a sequential process, the rate of a later step can never exceed the rate of an early step. What is really meant is that the second step has a much higher elementary rate constant than the first step as indicated in Fig. 7.5 by the low standard free-energy barrier which separates B and C compared with the high one between A and B. But because the concentration of B is so low, the *actual rate* of the step $B \rightarrow C$ (rate $= k_2[B]$) is low. It is important, therefore, to distinguish between rates and rate constants of the elementary steps in a series of consecutive processes.

In a complex reaction comprising coupled first-order reactions only, there is usually no need to use the steady-state approximation because such schemes can be solved exactly and the approximation involves the disadvantage that not all rate constants can be determined (for example

k_2 in the above). But if there is reason to believe that the intermediates are very short-lived in a reaction which is not amenable to exact analysis, the steady-state approximation is invaluable.

However, the assumptions upon which the steady-state approximation is based may not be valid in a particular case and although a hypothetical reaction scheme on paper will have been simplified, it may provide a poor description of the actual reaction. Failure of the rate equations derived from the simplified reaction scheme to describe the experimental results is adequate evidence that the steady-state approximation is not justified or the hypothetical scheme is in some other respect inadequate.

Conversely, a good kinetic description of the experimental results by a mechanistic scheme based upon the steady-state approximation is frequently taken as evidence of its validity. But whilst this is obviously a necessary criterion, it is not always sufficient. It is desirable to have independent evidence of the intervention of reactive intermediates. Sometimes they may be detected spectroscopically or trapped by added reagents and implicated by product analysis. The study of reactive intermediates is a substantial topic in itself and involves techniques additional to kinetics some of which we shall consider in later chapters.[10]

We shall now apply the steady-state approximation to a more complicated complex reaction.

7.3.2 Reversible first-order reaction followed by an irreversible second-order step

The complex reaction comprising eqns (7.24) and (7.25) should be regarded merely as a representative type: one which gives rise to an overall rate law, expressed as a differential equation, with too many concentration variables to allow an exact solution easily.

$$\mathrm{A} \underset{k_{-1}}{\overset{k_1}{\rightleftharpoons}} \mathrm{B} \tag{7.24}$$

$$\mathrm{B} + \mathrm{C} \xrightarrow{k_2} \mathrm{P} \tag{7.25}$$

The reactants are A and C, but A reacts via a reactive intermediate B. The rates of disappearance of the reactants are given by

$$-\frac{\mathrm{d[A]}}{\mathrm{d}t} = k_1[\mathrm{A}] - k_{-1}[\mathrm{B}],$$

and

$$-\frac{\mathrm{d[C]}}{\mathrm{d}t} = k_2[\mathrm{B}][\mathrm{C}].$$

The rate of formation of product P, the overall reaction rate, is given by

$$+\frac{d[P]}{dt}=k_2[B][C].$$

We now apply the steady-state approximation to [B] to obtain an approximate value for [B] to put into the above equations.

$$-\frac{d[B]}{dt}=k_{-1}[B]-k_1[A]+k_2[B][C].$$

If [B] is constant,

$$\frac{d[B]}{dt}=0,$$

therefore

$$k_{-1}[B]-k_1[A]+k_2[B][C]=0,$$

and

$$[B]=\frac{k_1[A]}{(k_{-1}+k_2[C])}.$$

We can now replace [B] in the above equations.

The rate of disappearance of reactant A becomes

$$-\frac{d[A]}{dt}=k_1[A]-\frac{k_{-1}k_1[A]}{(k_{-1}+k_2[C])},$$

or

$$-\frac{d[A]}{dt}=\frac{k_1k_2[A][C]}{(k_{-1}+k_2[C])}. \tag{7.26}$$

The rate of disappearance of C, which is equal to the rate of formation of product – the overall reaction rate, is given by

$$-\frac{d[C]}{dt}=\frac{d[P]}{dt}=k_2[B][C],$$

and this becomes

$$-\frac{d[C]}{dt}=\frac{d[P]}{dt}=\frac{k_1k_2[A][C]}{(k_{-1}+k_2[C])},$$

so, in the complex reaction of eqns (7.24) and (7.25)

$$-\frac{d[A]}{dt}=-\frac{d[C]}{dt}=+\frac{d[P]}{dt}=\frac{k_1k_2[A][C]}{(k_{-1}+k_2[C])}$$

when B is a reactive intermediate.

Because [A], [C], and [P] are not all independent concentration variables, the expression for d[P]/dt can be simplified. By the stoichiometry of the reaction, $[A]_0 = [A] + [B] + [P] = [P]_\infty$, and assuming that [B] is negligible compared with [A]+[P], we may write $[A] = [P]_\infty - [P]$ so the rate law for the formation of product becomes

$$\frac{d[P]}{dt} = \frac{k_1 k_2 [C]}{(k_{-1} + k_2[C])}([P]_\infty - [P]). \tag{7.27}$$

In contrast to rate laws obtained previously, eqns (7.26) and (7.27) are not simple because the concentration of one of the reactants, [C], appears in both numerator and denominator of the overall rate expressions. This reaction type provides us with a further example of a general technique in kinetics. It will be recalled that a second-order reaction may be converted into a *pseudo* first-order reaction by making one of the two reactants present in large excess and, therefore, effectively constant in concentration throughout the reaction (see p. 228). If [C] in the present example is kept constant simply by having it present in large excess, we see from eqns (7.26) and (7.27) that the disappearance of A and the formation of P are rendered *pseudo* first-order processes. Therefore, by monitoring the decrease in concentration of reactant A or the increase in concentration of product P with time, a *pseudo* first-order rate constant, k_{obs}, is obtained where

$$k_{obs} = \frac{k_1 k_2 [C]}{k_{-1} + k_2[C]}. \tag{7.28}$$

We can now go further because, upon rearrangement of eqn (7.28) we obtain

$$\frac{1}{k_{obs}} = \frac{k_{-1}}{k_1 k_2} \cdot \frac{1}{[C]} + \frac{1}{k_1}.$$

If the reaction is studied at several constant known concentrations of C and the *pseudo* first-order rate constant, k_{obs}, is obtained in each case, a plot of $1/k_{obs}$ versus $1/[C]$ will be linear. k_{-1}/k_1k_2 is obtained from the gradient, and $1/k_1$ from the intercept. We are able to calculate, therefore, the forward rate constant k_1 for the first step and, consequently, the ratio of the two rate constants involved in the partitioning of the reactive intermediate, k_{-1}/k_2. But we are still unable to determine k_{-1} and k_2 individually.

We can also consider two limiting extremes of the reaction described by eqns (7.24) and (7.25) which allow further simplifications of eqns (7.26) and (7.27).

7.3.2.1 *Consecutive irreversible first- and second-order reactions.*

When $k_2[C] \gg k_{-1}$ in eqns (7.24) and (7.25), the intermediate reacts further to give product much faster than it reverts to reactant, and the complex reaction approximates consecutive irreversible first- and second-order steps.

Equation (7.27) may be simplified to give

$$\frac{d[P]}{dt} = k_1([P]_\infty - [P]),$$

and we see that, *without* having to maintain [C] constant, the reaction is first order and the observed rate constant, k_{obs}, is equal to k_1, the rate constant of the initial and rate-determining step. Equation (7.26) may be simplified correspondingly.

7.3.2.2 *Pre-equilibrium followed by an irreversible second-order step*

When $k_{-1} \gg k_2[C]$ in eqns (7.24) and (7.25), the reactive intermediate returns to reactant much faster than it proceeds to give product. In this event, eqn (7.26) may be approximated to give

$$-\frac{d[A]}{dt} = \frac{k_1}{k_{-1}} \,.\, k_2[A][C].$$

We see that reactant and reactive intermediate constitute a *pre-equilibrium* and the reaction is overall second order: first order in [A] and first order in [C]. The experimentally-obtained second-order rate constant (see p. 226) is given by

$$k_{obs} = \frac{k_1}{k_{-1}} \,.\, k_2 = K \,.\, k_2$$

where K is the equilibrium constant ([B]/[A]) for the pre-equilibrium. Alternatively, by again making [C] large and therefore effectively constant, the reaction becomes *pseudo* first order in [A] and the observed rate constant is given by $k'_{obs} = k_2 \,.\, K[C]$. Either way, the second step may be regarded as rate determining even though the composite rate constant k_{obs} comprises k_1, k_{-1}, and k_2. The same approximation may be applied to eqn (7.27).

The reaction of acetone with chlorine in aqueous acidic solution provides a good example of the mechanistic possibilities under (7.3.2.1) and (7.3.2.2) above.

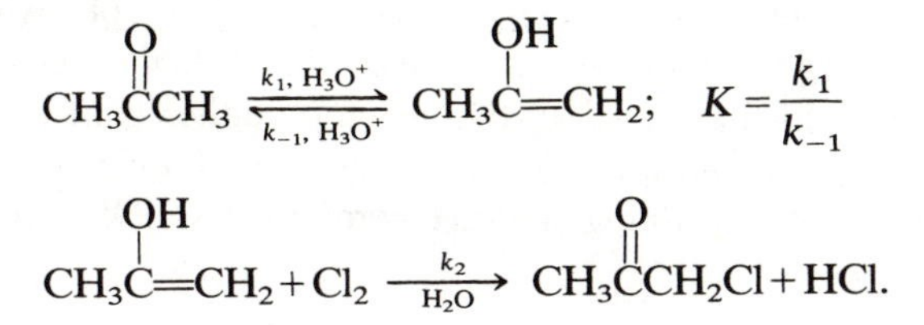

By application of the steady-state approximation to [enol], we obtain

$$-\frac{d[Cl_2]}{dt} = -\frac{d[CH_3COCH_3]}{dt} = \frac{k_1 k_2}{(k_{-1} + k_2[Cl_2])} \cdot [Cl_2][CH_3COCH_3].$$

The reaction was followed at 25 °C by monitoring $[Cl_2]$ in the presence of an excess of acetone in dilute perchloric acid.[11] (The enolization is acid catalysed; this is considered in detail in the next chapter.) At initial concentrations of Cl_2 in the range 0.001–0.005 mol dm^{-3} the rate law was found to be

$$-\frac{d[Cl_2]}{dt} = k_{obs}[CH_3COCH_3].$$

Under these conditions, therefore, the reaction is zero order in $[Cl_2]$ and the overall rate is determined by the rate of enolization of acetone with $k_{obs} = k_1$. As expected according to this mechanism, the rate of chlorination was identical with the previously measured rates of bromination and iodination.

But at much lower initial concentrations of chlorine ($[Cl_2] < 10^{-5}$ mol dm^{-3}) a different rate law was observed:

$$-\frac{d[Cl_2]}{dt} = k_{obs}[Cl_2][CH_3COCH_3].$$

Under these conditions, the first step constitutes a pre-equilibrium and the second-order second step is rate determining, so the reaction is overall second order with

$$k_{obs} = k_2 \cdot \frac{k_1}{k_{-1}}.$$

Since the equilibrium constant $K = k_1/k_{-1}$ for the keto-enol equilibrium was already known, the elementary second-order rate constant k_2 for the reaction between chlorine and the enol form of acetone was calculable from k_{obs}.

The notion that the overall rate of any complex reaction is determined simply by the smallest elementary rate constant in the forward direction is widespread but inaccurate as the examples above demonstrate. It is true only when the first of the sequential steps is rate determining. The reaction types in this section to which the steady-state approximation can be applied may be generalized as follows: individual rate constants of steps which *succeed* the rate-determining step neither contribute towards k_{obs} nor affect the overall rate, but rate constants of all steps prior to and including the rate-determining step are components of k_{obs} and do affect the overall rate.

7.4 The Curtin–Hammett Principle

Most organic compounds consist of molecules which have considerable conformational flexibility, and consequently the great majority of chemical reactions are preceded by much faster conformational pre-equilibria. We shall consider in detail the general case of a compound in a reaction which has strict stereochemical requirements. Figure 7.6 is the standard free-energy diagram of the complex reaction shown in eqn (7.29). Two conformers A′ and A″ of a compound A interconvert rapidly and undergo slower stereospecific reactions to give X and Y respectively.

A reaction is *stereospecific*[12] when stereoisomers, under the same experimental conditions, give different products, or the same product at different rates. In the present context, we can regard the different conformers of A as stereoisomers.

Our objective is to investigate the factors which affect the relative extents of these two reaction paths from A. Our assumptions are simply that the overall reactions are kinetically controlled and that, for A′, $k_1 \gg k_x$, and, for A″, $k_{-1} \gg k_y$.

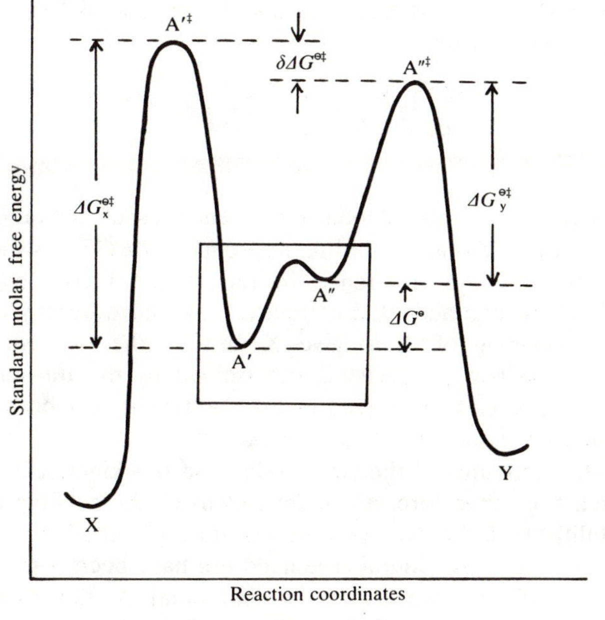

$$\mathrm{X} \xleftarrow{k_x} \mathrm{A'} \underset{k_{-1}}{\overset{k_1}{\rightleftharpoons}} \mathrm{A''} \xrightarrow{k_y} \mathrm{Y} \qquad (7.29)$$

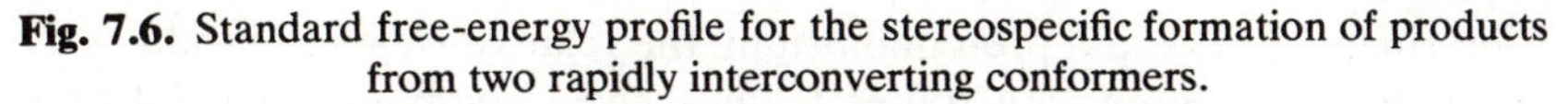

$$k_1 \gg k_x \quad \text{and} \quad k_{-1} \gg k_y$$

Fig. 7.6. Standard free-energy profile for the stereospecific formation of products from two rapidly interconverting conformers.

Consider first the part of the diagram within the rectangle; this relates to the conformational pre-equilibrium between A′ and A″ which is maintained during the complex reaction:

$$K = \frac{[\mathrm{A}'']}{[\mathrm{A}']} = \frac{k_1}{k_{-1}} = \mathrm{e}^{-\Delta G^{\ominus}/RT}$$

where $\Delta G^{\ominus}$ = the standard molar free-energy difference between the two conformers at temperature T.
At any time during the reaction,

$$\frac{\mathrm{d}[\mathrm{X}]}{\mathrm{d}t} = k_x[\mathrm{A}']$$

and

$$\frac{\mathrm{d}[\mathrm{Y}]}{\mathrm{d}t} = k_y[\mathrm{A}'']$$

therefore

$$\frac{\mathrm{d}[\mathrm{Y}]}{\mathrm{d}t} \Big/ \frac{\mathrm{d}[\mathrm{X}]}{\mathrm{d}t} = \frac{k_y}{k_x} \cdot \frac{[\mathrm{A}'']}{[\mathrm{A}']} = \frac{k_y}{k_x} \cdot K.$$

Replacing equilibrium and rate constants by exponential standard free-energy terms, we can write

$$\frac{\mathrm{d}[\mathrm{Y}]}{\mathrm{d}t} \Big/ \frac{\mathrm{d}[\mathrm{X}]}{\mathrm{d}t} = \frac{\mathrm{e}^{-\Delta G_y^{\ominus\ddagger}/RT}}{\mathrm{e}^{-\Delta G_x^{\ominus\ddagger}/RT}} \cdot \mathrm{e}^{-\Delta G^{\ominus}/RT}$$

$$= \mathrm{e}^{\delta\Delta G^{\ominus\ddagger}/RT}.$$

$\Delta G_x^{\ominus\ddagger}$ and $\Delta G_y^{\ominus\ddagger}$ are the standard free energies of activation for the formation of X and Y *from* A′ *and* A″ *respectively*; $\delta\Delta G^{\ominus\ddagger}$, as shown in the figure, is the absolute difference in the standard free energies (of formation) of the two activated complexes. We have hereby established that the partitioning of A to give X and Y (the rate *ratio* for the formation of the two products) is determined by the *difference* in the standard free energies of activation for the two routes both measured from the *ground state conformation of* A.

The relative amounts of the two products in this kinetically-controlled complex reaction, therefore, are independent of the relative concentrations (stabilities) of the two conformers through which the respective reactions proceed. This general case need not have been restricted to the reactions of only two conformers of compound A, nor to a complex reaction the irreversible steps of which were unimolecular. Furthermore, because the deduction did not rely upon a steady-state approximation, the non-ground state conformations need not be only very sparsely populated ones. As long as the products are kinetically controlled and the conformers are in rapid pre-equilibrium, the relative yields of products by any number of competing routes from a given compound depend only upon the standard free energies of the respective activated complexes all related to a common origin. This generalization is an expression of the so-called *Curtin–Hammett Principle.*[13]

7.4.1 Unimolecular thermolytic elimination

Thermolysis of *N,N*-dimethyl-3-phenylbut-2-ylamine oxides provides an illustration of a reaction whose interpretation is facilitated by an appreciation of the Curtin–Hammett Principle, and contributes to our understanding of the mechanism and stereochemical course of pyrolytic eliminations. The experimental results are given in Table 7.1 and mechanisms, including principal conformations of reactants and the activated complexes, are shown in Figs 7.7 and 7.8 for *erythro* and *threo* diastereoisomers, (3) and (4), respectively.

In Fig. 7.7, Newman projections of the main conformations of the *erythro* compound are (3a), (3b), and (3c) viewed along the C2–C3 bond, and (3d) is a view along the C2–C1 bond. The most credible mechanism which accounts for the stereospecific formation of Z-2-phenylbut-2-ene (7) involves a concerted *syn* elimination through the unstable conformer (3e) via the activated complex (5). A comparable mechanism for the concurrent formation of 3-phenylbut-1-ene (8) via (6) is also shown. The fact that conformers (3a–c) are much more stable than the fully eclipsed (3e), and that (3d) is more stable than (3f), is immaterial as (3a–d) do not lead directly to activated complexes of accessible standard molar free energy whereas (3e) and (3f) do. The but-2-ene (7) is formed in higher yield (faster) than the but-1-ene (8), therefore (5) must be of lower standard molar free energy than (6). (A minor statistical factor should also be considered as there are three equivalent conformations (3f) which may be interconverted by 120° rotation about the C2–C1 bond, but this does not qualitatively affect the argument.) This is most reasonably ascribable to the conjugated and more heavily substituted nature of the

TABLE 7.1

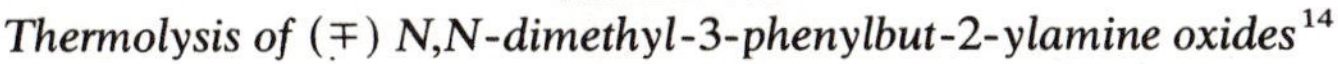

Thermolysis of (∓) N,N-dimethyl-3-phenylbut-2-ylamine oxides[14]

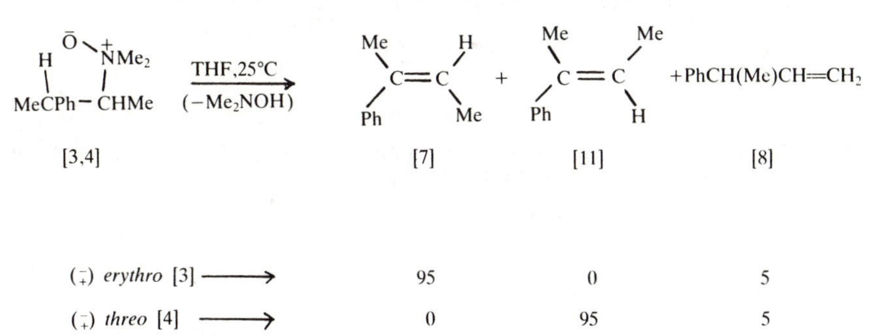

	[7]	[11]	[8]
(∓) *erythro* [3] ⟶	95	0	5
(∓) *threo* [4] ⟶	0	95	5

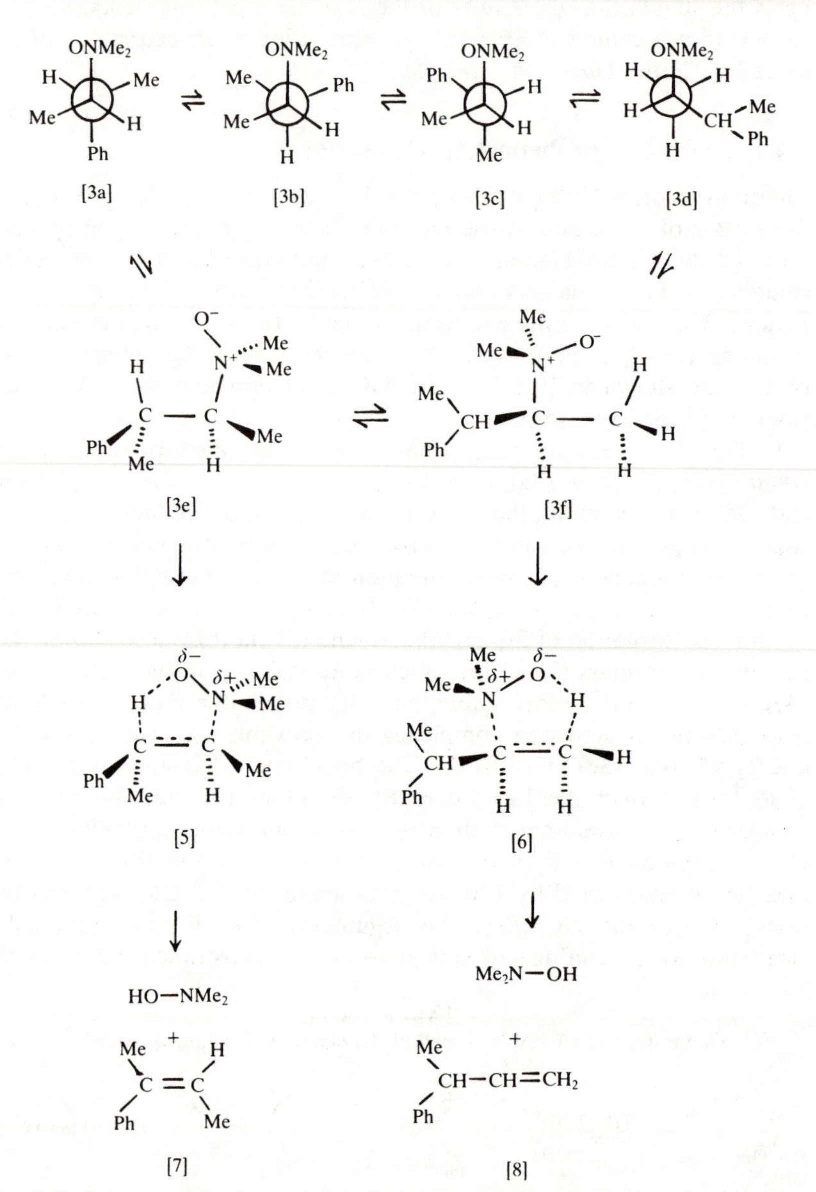

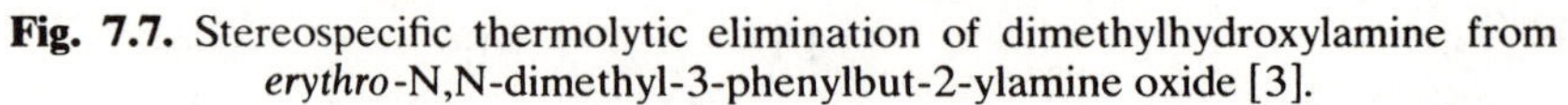

Fig. 7.7. Stereospecific thermolytic elimination of dimethylhydroxylamine from *erythro*-N,N-dimethyl-3-phenylbut-2-ylamine oxide [3].

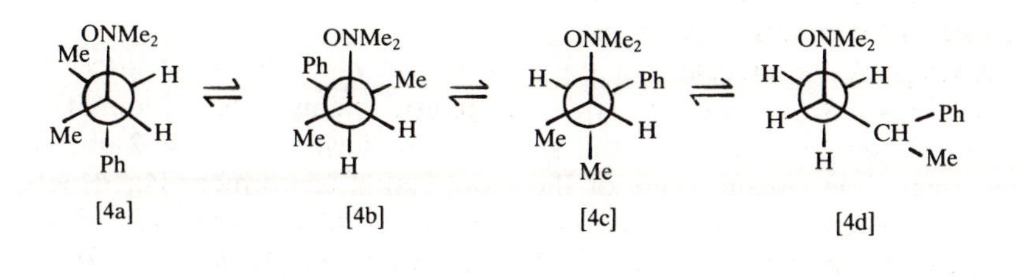

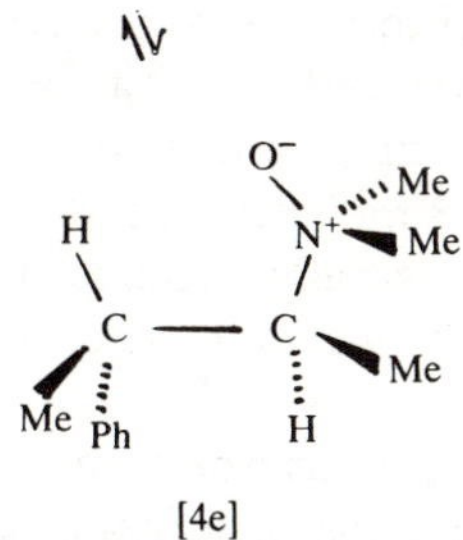

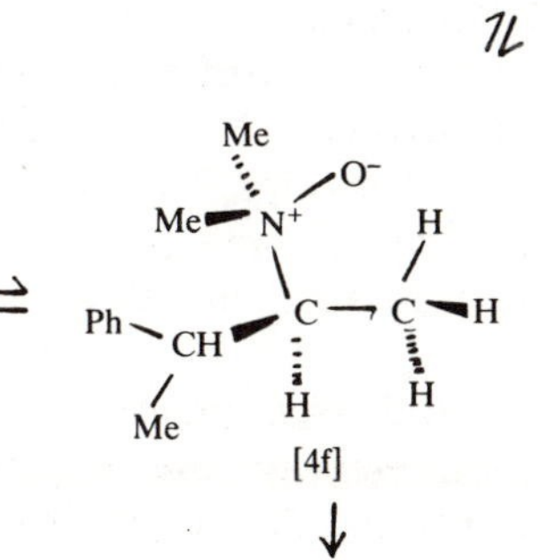

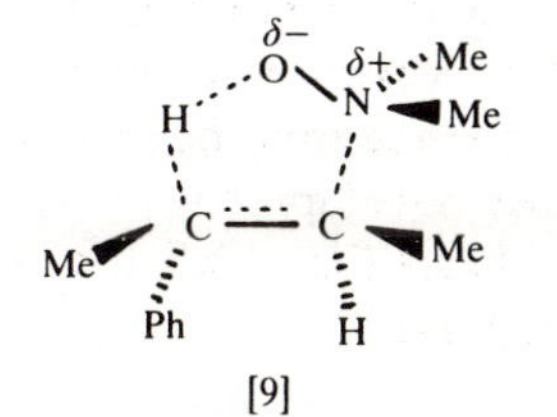

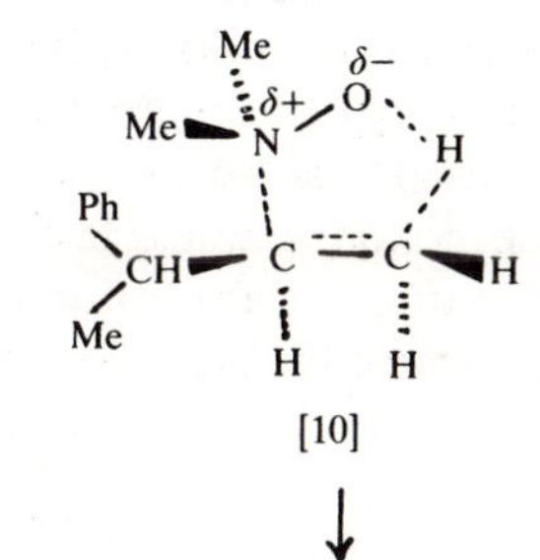

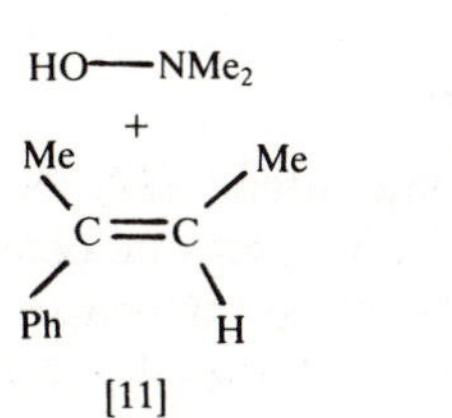

Me2N—OH

+

Ph

CH—CH=CH2

Me

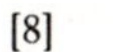

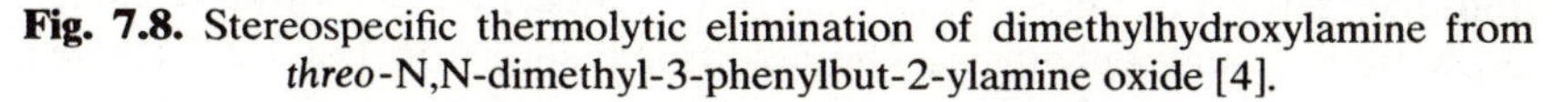

Fig. 7.8. Stereospecific thermolytic elimination of dimethylhydroxylamine from *threo*-N,N-dimethyl-3-phenylbut-2-ylamine oxide [4].

developing double bond outweighing the more adverse non-bonded interactions in (5) compared with (6).

Analogous consideration of Fig. 7.8 for the *threo* diastereoisomer (4) shows how the concerted *syn* thermolytic elimination of dimethylhydroxylamine accounts for the stereospecfic formation of E-2-phenylbut-2-ene (11). Again, none of the more stable conformers (4a–d) lead directly to activated complexes of accessible standard molar free energy whereas the unstable eclipsed conformers (4e) and (4f) lead to (9) and (10) respectively. The yield of (11) is higher than the yield of (8) therefore (9) must be of lower standard free energy than (10). (Again, a minor statistical factor is involved.) Note that although single enantiomers of (3) and (4) are included in Figs 7.7 and 7.8 for simplicity, the diastereoisomeric amine oxides (3) and (4) were both actually racemic. Consequently, the same racemic modification of (8) was obtained in both reactions.

Our confidence in these mechanisms is strengthened when we consider the activation parameters for the elimination reactions. For the reasons discussed in the previous chapter, the concerted *syn* elimination mechanisms shown in Figs 7.7 and 7.8 should have substantial negative standard entropies of activation because in each case a flexible ground state conformer is converted into an activated complex containing a five-membered ring with very little conformational freedom. Standard enthalpies of activation should, according to these mechanisms, be low since weakening of some bonds is concerted with the development of others. Results[14] obtained using tetrahydrofuran as solvent (in which the reactions are not complicated by hydrogen-bonding) are:

	$\Delta H^{\ominus\ddagger}$/kJ mol^{-1}	$\Delta S^{\ominus\ddagger}$/J K^{-1} mol^{-1}
Thermolysis of (3)	62 (∓6)	−116 (∓21)
Thermolysis of (4)	80 (∓3)	−44 (∓10)

7.4.2 Base-induced bimolecular elimination

The elimination of HBr from racemic 2-bromobutane (12) using potassium ethoxide in ethanol is another reaction[15] which may be considered in the context of the Curtin–Hammett Principle. Here, of course, the slow irreversible step is bimolecular. It was established that this E2 elimination, like those of almost ali acyclic alkyl halides, occurs virtually exclusively by an *anti* mechanism. The results are shown in Fig. 7.9. We see that the reaction to form *trans*-but-2-ene (13) through the conformer (12a) with the methyl groups *trans* to each other is approximately three times faster than the reaction to form *cis*-but-2-ene (14) through (12b) with the gauche arrangement of methyl groups. Conformer (12c) cannot give a but-2-ene by an *anti* mechanism. But-1-ene (15) can be formed by

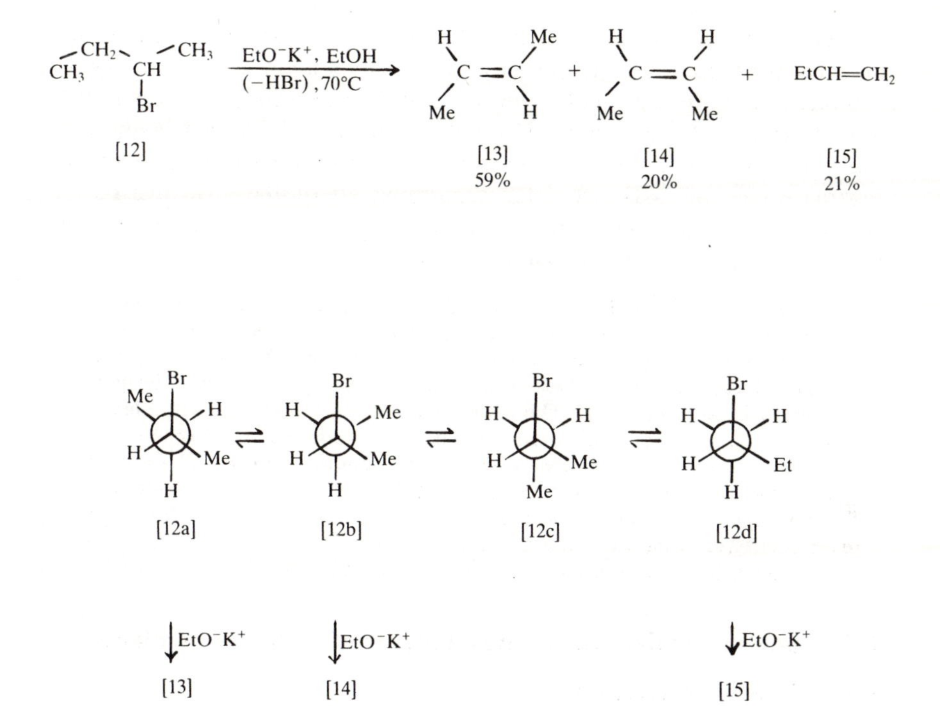

Fig. 7.9. Bimolecular elimination reaction of (±)2-bromobutane.[15]

an *anti* mechanism from (12a), (12b), and (12c) as the Newman projection (12d) (viewed along the C2–C1 bond) shows. But it is clear from the product analysis that the combined routes which produce (15) do so only as fast as (14) is obtained via the single conformer (12b).

We have again excluded the enantiomers of conformers (12a–d) for the sake of clarity and in accordance with convention but they would yield identical products in the same ratios. This is because the routes from, for example, (12a) and its enantiomer (12a′) are identical in all respects except chirality and consequently occur to exactly the same extent.

This reaction is quite different in an interesting respect from the thermolysis of the amine oxides considered previously. In the present case, the fastest product-forming route is the formation of the most stable product (13) via the ground state conformation (12a). However, the stereoselectivity in this elimination is *not* due to the relative free energies of the conformers of the reactant through which the reactions proceed *nor* is it because the *trans*-stereoisomer is the more stable product. It is caused by the lower standard free energy of the activated complex leading to the *trans*-isomer compared with that of the activated complex leading to the *cis*-isomer. It is true, however, that the structural feature which causes (14) to be less stable than (13), the adverse non-bonded interaction between the *cis* methyl groups, is developing in the activated complex which precedes (14), and this may be the principal factor in the destabilization of this activated complex compared with that leading to the *trans*-product. This reasoning is of course based upon the Hammond Postulate (see p. 261); we infer something about the structure of an activated complex from structural knowledge of a stable entity not far removed in the reaction coordinate.

A *stereoselective*[12] reaction is one in which a non-stereoisomeric compound or, as in the present case, a racemic modification reacts to give stereoisomeric products in unequal amounts. It may be qualified as strongly or weakly stereoselective according to whether the ratio of stereoisomers produced is large or small. The term should not be confused with *stereospecific* (see p. 293). An overall reaction is stereoselective because the parallel routes which yield the stereoisomeric products in unequal amounts are individually stereospecific and occur at different rates. Stereoselectivity is, therefore, always due to the stereospecificity of elementary reactions. An overall reaction which is stereospecific is a reaction of an individual stereoisomer and, by definition therefore, is not stereoselective.

7.5 Standard molar free-energy profiles of complex reactions

7.5.1 First-order reactions

The standard molar free energy of activation $\Delta G^{\ominus\ddagger}$ of a reaction, as we have already seen, corresponds to the rate constant in the way that a standard molar free energy of a reaction $\Delta G^{\ominus}$ corresponds to an equilibrium constant. The standard free-energy barrier in a reaction profile of a first-order reaction, therefore, is simply a graphical representation of the size of the rate constant and, indirectly, the speed of the reaction. Rates of different first-order reactions may be compared in juxtaposed standard free-energy profiles; indeed, profiles of competing first-order reactions of the same compound can be combined since they have a common point of reference – the standard molar free energy of formation of the reactant.

Figure 7.10 illustrates the combined reaction profile for the competing parallel first-order reactions shown in eqns (7.30) and (7.31) (where both X and Y may be one or more compounds).

$$\mathrm{A} \xrightarrow[(\leftarrow)]{k_x} \mathrm{X}; \qquad \Delta G^{\ominus} = \Delta_f G^{\ominus}(\mathrm{X}) - \Delta_f G^{\ominus}(\mathrm{A}) \tag{7.30}$$

$$\frac{\mathrm{d}[\mathrm{X}]}{\mathrm{d}t} = k_x[\mathrm{A}]$$

$$\mathrm{A} \xrightarrow[(\leftarrow)]{k_y} \mathrm{Y}; \qquad \Delta G^{\ominus} = \Delta_f G^{\ominus}(\mathrm{Y}) - \Delta_f G^{\ominus}(\mathrm{A}) \tag{7.31}$$

$$\frac{\mathrm{d}[\mathrm{Y}]}{\mathrm{d}t} = k_y[\mathrm{A}].$$

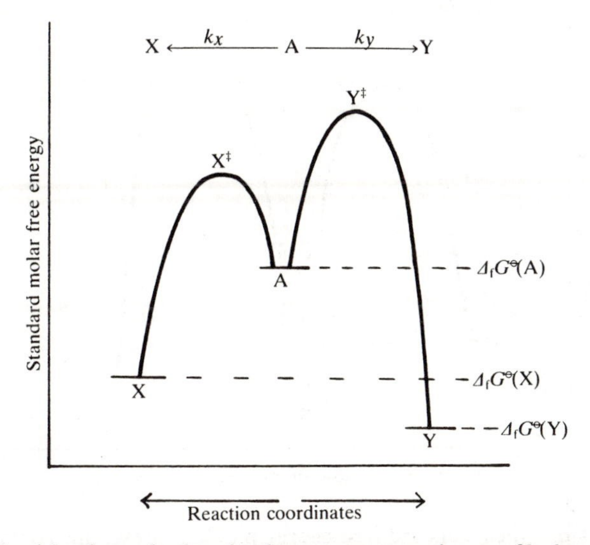

Fig. 7.10. Combined standard molar free-energy reaction profile for two parallel first-order reactions of compound A.

The formation of product(s) X is faster than the formation of product(s) Y but the latter is thermodynamically more favourable because $\Delta_fG^\ominus(Y) < \Delta_fG^\ominus(X)$. Under conditions of kinetic or thermodynamic control (p. 272), this diagram illustrates clearly how A partitions between the two routes.

Competing kinetically-controlled first-order unimolecular reactions such as the thermolysis of t-butyl prop-2-yl ether given earlier (p. 272), or the formation of isomeric cycloalkenes (17) and (18) from bicyclo[3.1.0]hexane[16] (16), are fairly common. In the gas phase at 418–491 °C, (17) is formed about three times faster than (18) ($k_1/k_2 \sim 3$) but (18) is more stable than (17) by about 7 kJ mol^{-1} under these conditions.[17]

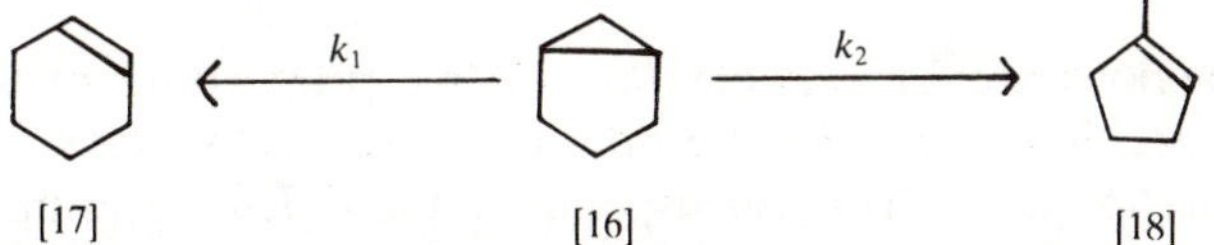

In contrast, parallel unimolecular reactions under thermodynamic control from a single compound are rarely encountered in organic chemistry.

The overall sequence of a complex reaction comprising consecutive unimolecular steps (which have first-order elementary rate constants) can also be represented in a single standard free-energy reaction profile. An example is shown in Fig. 7.5, p. 287.

Complex reactions, therefore, involving only first-order or *pseudo* first-order steps (either parallel or sequential) are satisfactorily represented by standard free-energy reaction profiles and their interpretation presents no difficulties.

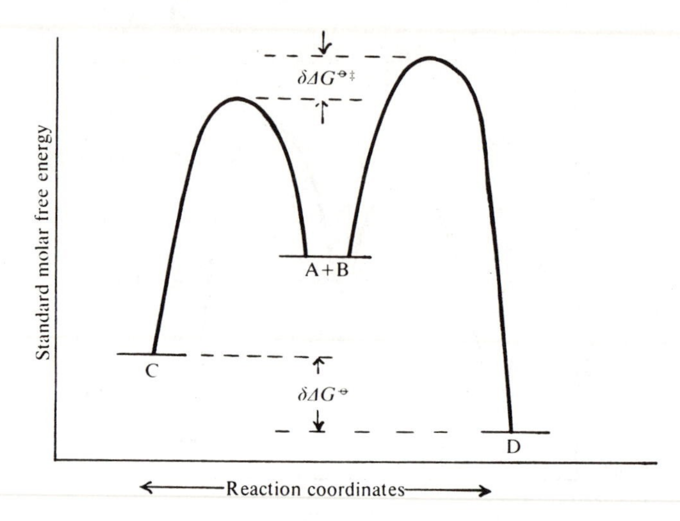

Fig. 7.11. Combined standard free-energy profile representing competing second-order reactions from two reactants.

7.5.2 Second-order reactions

Competing second-order reactions may also be described concisely by combined standard molar free-energy profiles whose interpretations are straightforward.

$$A+B \xrightarrow[(\leftarrow)]{k_c} \text{Products, C}$$

$$A+B \xrightarrow[(\leftarrow)]{k_d} \text{Products, D}$$

If the reactions are irreversible, the relative proportions of C and D in Fig. 7.11 are determined by the *ratio* of the respective rate constants for their formation k_c/k_d. This corresponds to the *difference* in the respective standard molar free energies of activation of the two processes, $\delta\,\Delta G^{\ominus\ddagger}$, and [C]/[D] does not change during the course of the reaction. The reactions of 2-propyl bromide with ethanolic sodium ethoxide[18] are examples of this type of complex reaction:

$$(CH_3)_2CHBr + C_2H_5O^- \xrightarrow[S_N2,\ \text{ethanol},\ 55\ ^\circ C]{k_S = 0.50\times10^{-4}\ dm^3\ mol^{-1}\ s^{-1}} (CH_3)_2CHOC_2H_5 + Br^-$$

$$(CH_3)_2CHBr + C_2H_5O^- \xrightarrow[E2,\ \text{ethanol},\ 55\ ^\circ C]{k_E = 1.5\times10^{-4}\ dm^3\ mol^{-1}\ s^{-1}} CH_3CH{=}CH_2 + C_2H_5OH + Br^-.$$

However, if the competing bimolecular reactions are reversible, then, in the early stages when back reactions are negligible, the product which is formed fastest predominates. As the reaction proceeds, the ratio of products changes and as completion is approached, the proportions of C and D are determined by the thermodynamic stabilities under the particular conditions: by $\delta\,\Delta G^{\ominus}$ in Fig. 7.11.

The Diels–Alder reaction of furan and maleic anhydride[19] is typical of many cycloadditions which come into this category.

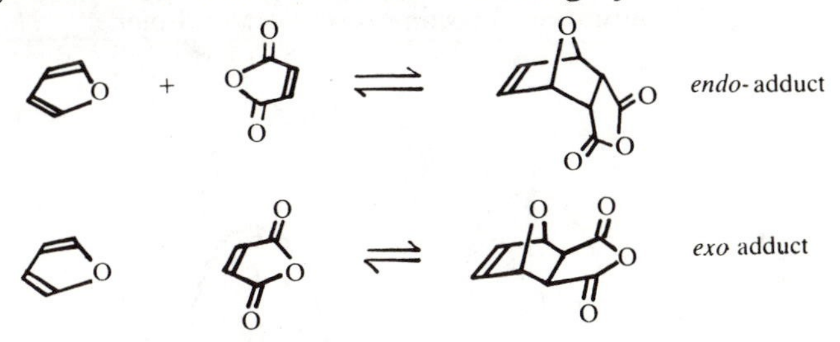

In the early stages of the reaction,
[*endo*-adduct] > [*exo*-adduct]; but towards completion,
[*exo*-adduct] ≫ [*endo*-adduct].

In principle, and occasionally in practice, one can select a temperature for such reactions at which progress to equilibrium is slow even though formation of the fastest-formed product is appreciable. Under these conditions it is possible to produce and isolate a good yield of the less stable 'kinetic' product preferentially. Of course, this is not possible if the product which is formed faster is also the more stable!

The reversible dimerization of cyclopentadiene[20] to give *endo* and *exo* adducts is easily represented by a diagram similar to Fig. 7.11 but with $A = B$.

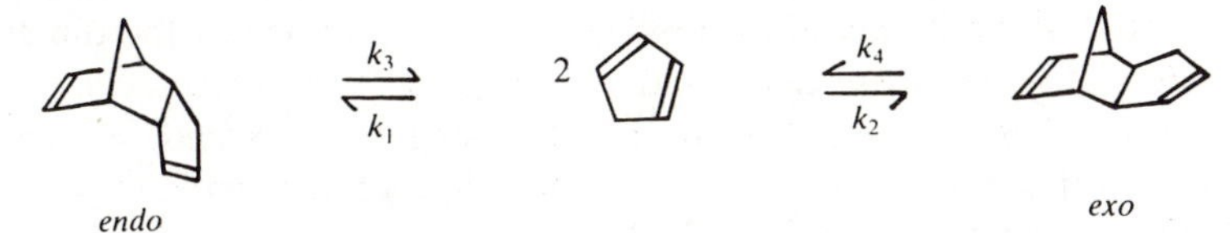

For this particular system at about 200 °C,

$$\Delta G^{\ominus}(exo) \sim \Delta G^{\ominus}(endo), \quad \text{but} \quad \Delta G^{\ominus\ddagger}(exo) > \Delta G^{\ominus\ddagger}(endo),$$

i.e. $k_1 > k_2$, and $k_3 > k_4$.

7.5.3 Parallel first- and second-order reactions

7.5.3.1 Reactions of a single compound

We have already seen (p. 275) that *cis,trans*-cyclo-octa-1,5-diene undergoes competing first-order isomerization and second-order dimerization.[6]

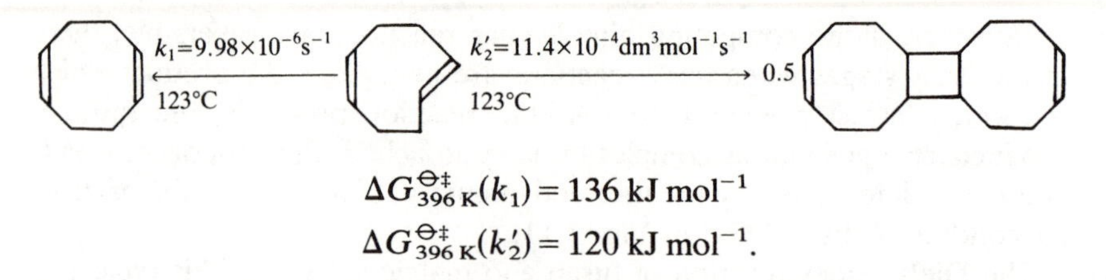

$$\Delta G^{\ominus\ddagger}_{396\,\mathrm{K}}(k_1) = 136 \text{ kJ mol}^{-1}$$
$$\Delta G^{\ominus\ddagger}_{396\,\mathrm{K}}(k_2') = 120 \text{ kJ mol}^{-1}.$$

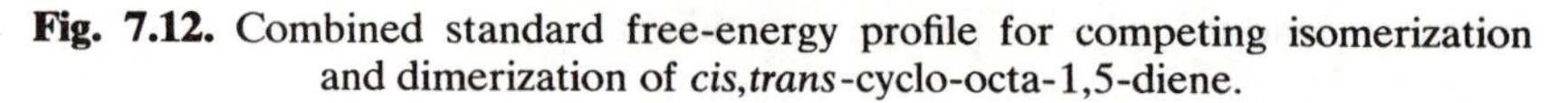

Fig. 7.12. Combined standard free-energy profile for competing isomerization and dimerization of *cis,trans*-cyclo-octa-1,5-diene.

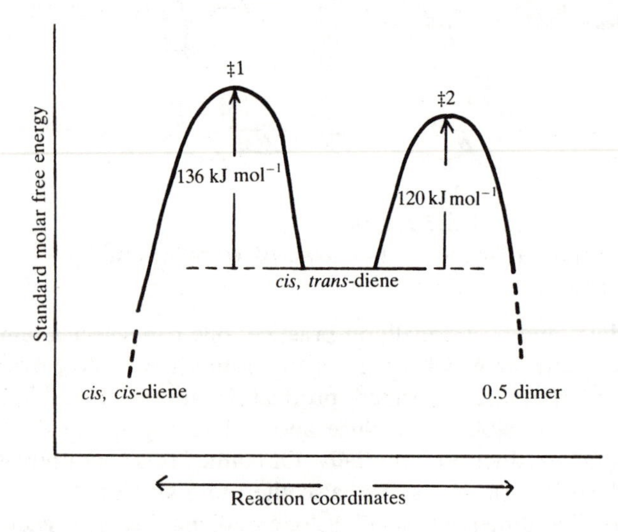

Rate constants at 123 °C are shown for the equations as written here (which is not exactly as they were written earlier; k_2' here = $2k_2$ on p. 275 due to the different way of expressing the stoichiometry of the dimerization). $\Delta G^{\ominus\ddagger}_{396\,\mathrm{K}}$ values are calculated using eqn (6.19) of Chapter 6, p. 242. Since these competing reactions are of a single compound, we have a common reference point for the construction of a combined standard molar free-energy profile, Fig. 7.12.

Formation of 1 mole of activated complex ‡1 in a hypothetical experiment starting from 1 mole of the *cis,trans*-diene in 1 dm^3 would involve an increase in free energy of the system of 136 kJ at 123 °C. In an alternative hypothetical experiment, formation of 0.5 mole of activated complex ‡2 from 1 mole of *cis,trans*-diene would correspond to an increase in free energy of only 120 kJ. None the less, as we saw on p. 276, the actual rate of isomerization of the *cis,trans*-diene under normal experimental conditions is faster than the rate of dimerization. There is no inconsistency here. The standard molar free-energy profile relates to (ideal) unit molar concentrations, so the free-energy barriers reflect

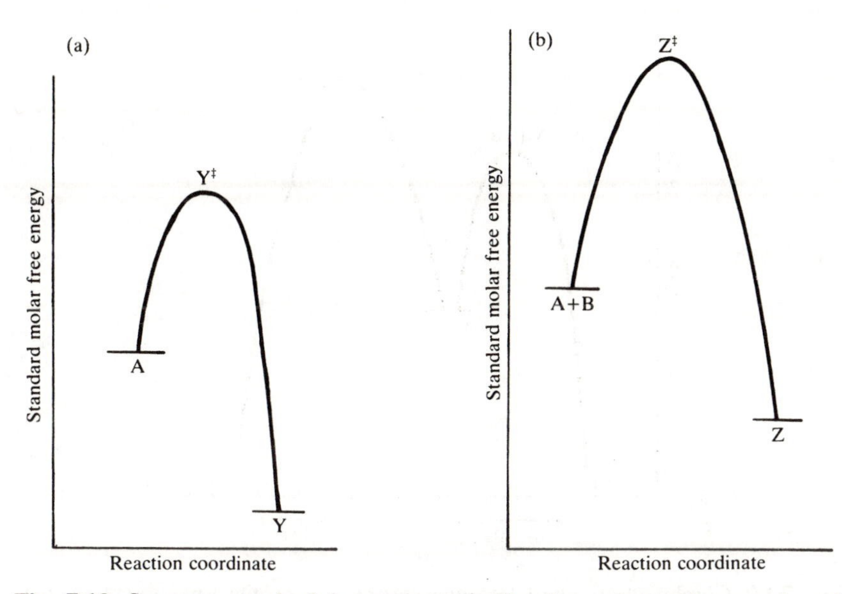

Fig. 7.13. Separate standard free-energy profiles for a first-order reaction of compound A and its second-order reaction with compound B.

relative rates of the competing routes only from reactant at 1 mol dm^{-3}. In the actual experiments, the initial concentration of the *cis,trans*-diene was only about 10^{-3} mol dm^{-3}. This aspect was considered in terms of the elementary rate constants on p. 276.

7.5.3.2 *Reactions involving two compounds*

Figures 7.13(a) and (b) illustrate first- and second-order reactions of compound A, the latter with compound B: eqns (7.31) and (7.32).

$$\mathrm{A} \xrightarrow{k_y} \mathrm{Y}; \qquad \Delta G^{\ominus} = \Delta_f G^{\ominus}(\mathrm{Y}) - \Delta_f G^{\ominus}(\mathrm{A}) \tag{7.31}$$

$$\frac{\mathrm{d}[\mathrm{Y}]}{\mathrm{d}t} = k_y[\mathrm{A}]$$

$$\mathrm{A+B} \xrightarrow{k_z} \mathrm{Z}; \qquad \Delta G^{\ominus} = \Delta_f G^{\ominus}(\mathrm{Z}) - (\Delta_f G^{\ominus}(\mathrm{A}) + \Delta_f G^{\ominus}(\mathrm{B}))$$

$$\frac{\mathrm{d}[\mathrm{Z}]}{\mathrm{d}t} = k_z[\mathrm{A}][\mathrm{B}]. \tag{7.32}$$

But now the two profiles do not share a common initial state; eqn (7.32) and Fig. 7.13(b) include B whereas eqn (7.31) and Fig. 7.13(a) do not. If we wish to represent these two reactions of A in the presence of B in competition using a single reaction profile, we must artificially add one

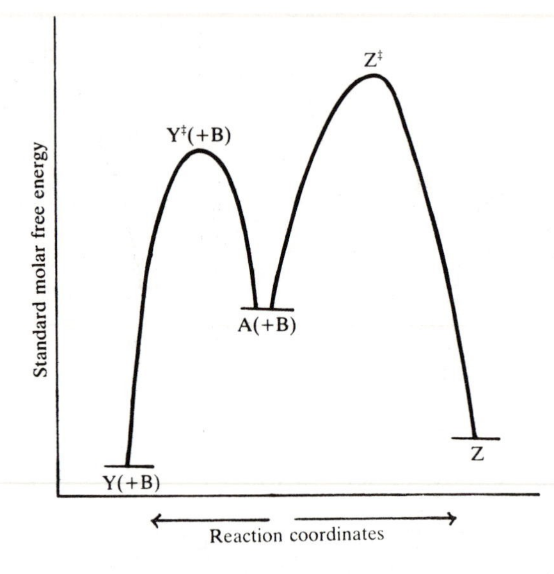

Fig. 7.14. Combined standard free-energy reaction profile for competing first- and second-order reactions of compound A in the presence of compound B.

mole of B to both sides of eqn (7.31) and correspondingly displace the whole of Fig. 7.13(a) vertically in the free-energy coordinate so that the initial state of Fig. 7.13(a) becomes identical with the initial state of Fig. 7.13(b). We can now combine the two and construct Fig. 7.14 showing B in parentheses.

Usually component B is simply omitted from such a profile and its involvement is understood. Unlike competing reactions of a single compound of the same order which occur at relative rates determined only by the ratio of the respective elementary rate constants, the relative rates of the two reactions of A in eqns (7.31) and (7.32) are determined not only by the respective rate constants, k_y and k_z, but also by [B]. And because Fig. 7.14 includes *standard* molar free energies, the relative heights of $Y^‡$ and $Z^‡$ above A (+B) accurately represent the relative rates of formation of Y and Z from A only when B is present at unit molar concentration. So although $Y^‡$ is lower than $Z^‡$ in the profile as drawn, the rate of formation of Y could be equal to or faster or slower than the rate of formation of Z according to [B]. We can generalize this. The relative heights of the barriers in a combined standard molar free-energy profile of reactions of different orders must not obscure the fact that the relative rates of the reactions depend upon concentrations of reactants as well as rate constants (barrier heights).

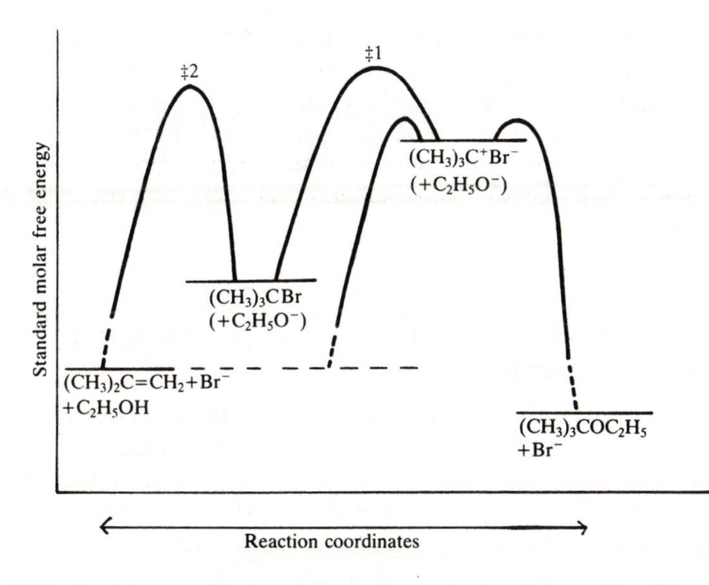

Fig. 7.15. Reaction profile for the reactions of t-butyl bromide with ethanolic sodium ethoxide at 55 °C.

The reactions of t-butyl bromide with ethanolic sodium ethoxide[18] are an example of this type of complex process.

$$(CH_3)_3CBr + C_2H_5O^- \xrightarrow[k_2 \sim 5\times10^{-4}\,\text{dm}^3\,\text{mol}^{-1}\,\text{s}^{-1}]{\text{ethanol, 55 °C}} \text{Elimination (E2)}$$

$$\Delta G^{\ominus\ddagger}_{328\,\text{K}}(k_2) \sim 101\ \text{kJ mol}^{-1}.$$

$$(CH_3)_3CBr \xrightarrow[k_1 \sim 2\times10^{-4}\,\text{s}^{-1}]{\text{ethanol, 55 °C}} \text{Substitution } (S_N1) \text{ and Elimination (E1)}$$

$$\Delta G^{\ominus\ddagger}_{328\,\text{K}}(k_1) \sim 104\ \text{kJ mol}^{-1}.$$

Figure 7.15 and the equations above indicate that the standard molar free energy of activation for bimolecular elimination via ‡2 is very similar to that for the rate-determining step of the unimolecular reaction via ‡1. None the less, the ratio of E2 to S_N1 plus E1 will be less than, equal to, or more than unity according to the concentration of the sodium ethoxide. Moreover, unless the concentration of sodium ethoxide remains approximately constant (*pseudo* first-order conditions) the partitioning of $(CH_3)_3CBr$ between unimolecular and bimolecular routes will change during the course of the reaction. In contrast, the partitioning of the intermediate ion-pair $(CH_3)_3C^+Br^-$ between E1 and S_N1 involves competing elementary routes of the same order and is unaffected by the concentration of ethoxide. These reactions of the reactive intermediate are *product-determining* steps subsequent to the heterolysis of the t-butyl bromide which is the rate-determining step of the unimolecular route.

Figures 7.12–7.15 demonstrate that profiles of complex reactions can be helpful but that they must be constructed and interpreted with caution when they include reactive intermediates and competing routes of different kinetic order.

7.6 The rate-determining step and composition of the activated complex

Each elementary step in a complex reaction has its own activated complex so any complex reaction comprising sequential steps proceeds via several, perhaps many, activated complexes. We have also seen that the rate expression for a complex reaction may not allow a kinetic order to be ascribed to the overall process (see, for example, eqn (7.19), p. 283). But when approximations are included in the construction of a hypothetical complex mechanistic scheme, an apparently simple kinetic order will generally ensue.

If the predicted kinetic order is confirmed experimentally, the conditions upon which the approximations were based are taken to be valid. Inevitably, several hypothetical complex mechanisms could lead to the same simplified overall kinetic order under particular experimental conditions, consequently we are seldom led unambiguously to a single mechanism by an apparently simple rate law.

If we were to reconsider previous sections and examine the chemical implications of the mathematical approximations involved in obtaining a simple rate law from a complex mechanism, we would realise that invariably the assumption is that just one of the elementary sequential steps involves an activated complex which is significantly less stable in an absolute sense than the activated complexes of all other elementary steps. This need not be in the step with the smallest elementary rate constant. This highest free-energy barrier in the overall reaction intervenes in the so-called rate-determining step and the multi-dimensional potential-energy hypersurface which describes this species is the transition state for the overall reaction.

Conversely, if the steady-state approximation cannot be applied to a complex reaction, no single activated complex is significantly less stable than all others, no single step can be regarded as rate determining, and we may not be able to ascribe a kinetic order to the reaction as a whole.

An important generalization now follows: if a rate constant and kinetic order can be ascribed to a complex reaction, the composition of the most unstable activated complex can be deduced from the overall rate law.

For example, in the complex reaction

$$A + 2B + C \xrightarrow{k_{obs}} P, \quad \text{Products,}$$

if

$$\frac{d[P]}{dt} = k_{obs}[A][B], \tag{7.33}$$

then one mole of the most unstable activated complex which intervenes between reactants and products has the same composition as one mole of A plus one mole of B. The third reactant, C, and the second mole of B must, therefore, be involved in steps which *succeed* the formation of the activated complex.

If, on the other hand,

$$\frac{d[P]}{dt} = k_{obs}[A], \tag{7.34}$$

and the reaction is zero order in [B] and [C], then neither B nor C are involved in the formation of the overall activated complex. Again, B and C, which are reactants as shown by the stoichiometric equation, must be involved in steps subsequent to the formation of the activated complex of the overall reaction.

An important corollary of this generalization concerning those complex mechanisms which give rise to an overall experimental rate constant is that we learn little simply from the overall rate law besides the composition of the most unstable activated complex which intervenes between reactants and products. We do not learn how many intermediates or pre-equilibria are involved prior to the formation of the activated complex, neither may we deduce the number of steps, or the reagents involved in the steps, subsequent to its decomposition simply from the rate law.

There are also at least two further caveats. It is not possible, by methods so far described, to determine how many moles of solvent (if any) are involved in the formation of one mole of activated complex in a solution reaction because we cannot determine reaction order with respect to solvent.

Secondly, it is only the *composition* of the activated complex which we deduce from the overall rate law, not the structure. Consequently, in the second case above (eqn (7.34)) the activated complex could be some strained conformer of A or it could be a structural isomer or a stereoisomer of A. Similarly, in the first case above (eqn (7.33)) the most that we can rigorously conclude from the rate law is that the activated complex of the overall reaction contains the same atoms as A plus B. (See, however, the discussion of the Hammond Postulate on p. 261).

These may seem very severe limitations to kinetics as a method of studying the mechanisms of complex reactions and indeed they are too frequently overlooked. Although we may, by a propitious choice of experimental conditions, reduce a very complicated chemical transformation to a system which will yield an apparently simple overall rate law, we can not establish unambiguously a single mechanism this easily. But by this method it is frequently possible to exclude wrong alternative schemes, and progress towards fitting a realistic mechanism to a complex reaction involves eliminating the plausible alternatives. Usually, the more techniques, kinetic and non-kinetic, which can be employed, the closer we get to a single credible but still hypothetical mechanism.

7.7 Problems

1. The mutarotation of glucose is the conversion of the α form into the equilibrium mixture of α and β forms

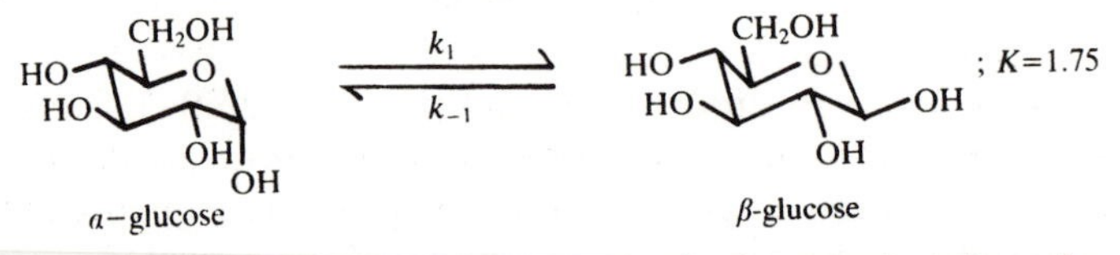

Calculate the individual rate constants for the interconversion of α and β anomers of glucose from the equilibrium constant given and the following polarimetric results obtained at 278 K (reported by J. C. Kendrew and E. A. Moelwyn-Hughes *Proc. R. Soc.* A **176,** 352 (1940)).

t/min	30	60	105	154	190	240	280	330	390	450
$(\alpha_t - \alpha_\infty)/°$	11.86	10.84	9.43	8.10	7.27	6.22	5.49	4.72	3.90	3.23

α_t and α_∞ = the polarimeter readings in degrees at time t and at equilibrium.

2. The rate of conversion of racemic 1,2,3,4-tetramethylcyclo-octatetraene into the equilibrium mixture of racemic plus *meso* valence isomers was monitored by n.m.r. at 120 °C. (L. A. Paquette, J. M. Gardlik, L. K. Johnson, and K. J. McCullough, *J. Am. chem. Soc.* **102,** 5026 (1980).)

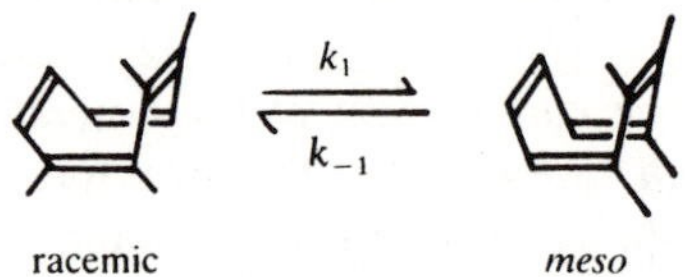

Calculate k_1 and k_{-1} for this interconversion from the following results:

time/10^3 s	0	7.2	29.6	50.4	64.8	86.4	108	∞
% *meso*	0	5.8	16	23	27	31	36	56

3. For a complex reaction of the type

$$\mathrm{A} \underset{k_{-1}}{\overset{k_1}{\rightleftharpoons}} \mathrm{B} \xrightarrow{k_2} \mathrm{C},$$

where B is a reactive intermediate, show that

$$-\frac{\mathrm{d[A]}}{\mathrm{d}t} = k_{\mathrm{obs}}[\mathrm{A}],$$

and

$$\frac{\mathrm{d[C]}}{\mathrm{d}t} = k_{\mathrm{obs}}([\mathrm{C}]_\infty - [\mathrm{C}]),$$

where

$$k_{\mathrm{obs}} = \frac{k_1 k_2}{(k_{-1} + k_2)}.$$

Identify the conditions under which (i) the initial step becomes rate determining, and (ii) the initial step constitutes a pre-equilibrium. Sketch standard free-energy reaction profiles and approximate the above expression for k_{obs} in these two limiting cases.

4. The temperature dependence of the rates of the reversible gas-phase

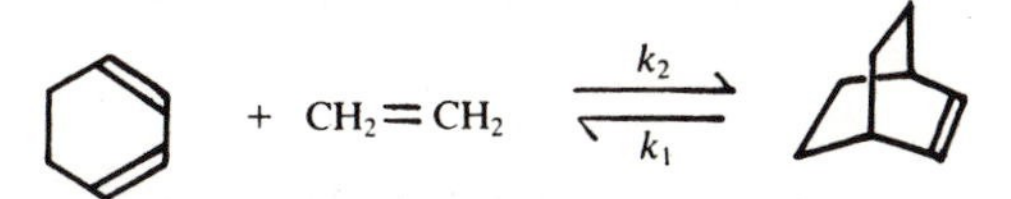

Diels–Alder reaction between cyclohexadiene and ethylene between 548–591 K is given by

$$\log(k_2/\mathrm{dm^3\,mol^{-1}\,s^{-1}}) = 6.66 - \frac{5675}{T},$$

and

$$\log(k_1/\mathrm{s^{-1}}) = 15.12 - \frac{12\,522}{T},$$

where T = the absolute temperature (reported by G. Huybrechts, D. Rigaux, J. Vankeerberghen, and B. Van Mele, *Int. J. chem. Kin.* **12,** 253 (1980)). Calculate (i) the practical equilibrium constant at 570 K, and (ii) mean values for $\Delta_r H^\circ$ and $\Delta_r S^\circ$ (standard state = ideal gas at 1 standard atmosphere pressure) over the above temperature range.

5. The second-order rate constant for the dimerization of *cis,trans*-cyclo-octa-1,5-diene is reported on p. 275.[6]

$$2\ \textit{cis,trans}\text{-diene} \xrightarrow[5.69\times10^{-4}\,\mathrm{dm^3\,mol^{-1}\,s^{-1}}]{k_2,\ 123\,^\circ\mathrm{C}} \text{dimers: X + Y + Z}$$

where X, Y, and Z are stereoisomers. The following product analysis was obtained for the reaction at 123 °C (and shown to be independent of the extent of the reaction)

$$X:Y:Z = 62:19:19.$$

Calculate the elementary second-order rate constants for the individual bimolecular reactions which yield X, Y, and Z.

The same relative yields of X:Y:Z were obtained at other temperatures. What do you conclude regarding the activation energies of the individual elementary reactions?

6. The following results were obtained for the racemization of 2,2′-diiodobiphenyl ((1), p. 282).[7]

Temp/°C	25.4	35.2	42.2
$10^4\ k_{rac.}/s^{-1}$	1.34	3.95	8.14

Calculate the standard molar enthalpy barrier to rotation about the central bond in this biphenyl.

7. The following results were obtained for the gas-phase thermal decomposition of t-butyl prop-2-yl ether (see p. 272).

$$(CH_3)_3COH + CH_2{=}CHCH_3 \xleftarrow{k_2} (CH_3)_3COCH(CH_3)_2 \xrightarrow{k_1} (CH_3)_2C{=}CH_2 + (CH_3)_2CHOH$$

Temp./°C	k_{obs}/s^{-1}	[Propanol]/[Butanol]
419.4	0.84	4.96
441.5	3.02	5.10
464.8	9.9	4.86
475.2	17.8	4.71

Calculate the Arrhenius parameters for k_1 and k_2.

8. The following polarimetric results (*m*-xylene solvent, 95.5 °C) were obtained for the racemization of the dextro-rotatory paracyclophane (2) on p. 282.[8]

time/min	0	15	240	302	422	542	662	1142	∞
$[\alpha]_t/°dm^{-1}\ g^{-1}\ cm^3$	109.0	106.0	74.3	67.7	56.8	49.4	41.4	21.1	0

$[\alpha]_t$ = the specific rotation of the sample at time *t*.

Calculate the rate constant for racemization and $\Delta G^{\ominus\ddagger}_{369\ K}$ for the enantiomerization.

9. Sketch a standard molar free-energy reaction profile for the interconversion of *exo*- and *endo*-dicyclopentadiene from the information given on p. 303.

10. The reaction of hydroxylamine and pyruvate at pH 6.9 and 25 °C

may be formulated as follows:[21]

$$\underset{\text{(pyruvate)}}{CH_3\overset{\overset{\displaystyle O}{\|}}{C}CO_2^-} + NH_2OH \underset{k_{-1}}{\overset{k_1}{\rightleftharpoons}} \underset{\displaystyle \overset{|}{HNOH}}{CH_3\overset{\overset{\displaystyle OH}{|}}{C}CO_2^-} \xrightarrow{k_2} \underset{\text{(oxime)}}{CH_3\overset{\overset{\displaystyle NOH}{\|}}{C}CO_2^-} + H_2O$$

When the first step constitutes a pre-equilibrium ($k_{-1} \gg k_2$; $k_1/k_{-1} = K = 57\ dm^3\ mol^{-1}$), show that the *pseudo* first-order rate law ($[NH_2OH] \gg [pyruvate]$) is given by

$$\frac{d[oxime]}{dt} = k_{obs}([oxime]_\infty - [oxime]),$$

where

$$k_{obs} = \frac{k_2K[NH_2OH]}{(1 + K[NH_2OH])},$$

and deduce the limiting forms of this expression for k_{obs} when (i) $[NH_2OH]$ is high, and (ii) $[NH_2OH]$ is very low (but still in excess over [pyruvate]).

7.8 References

1. K. B. Wiberg and R. Stewart, *J. Am. chem. Soc.* **77,** 1786 (1955).
2. A. A. Frost and R. G. Pearson, *Kinetics and mechanism*, 2nd edn, Wiley (1961). This book and its more recent edition by J. W. Moore and R. G. Pearson (1981) deal fully with many hypothetical complex reaction schemes and show how the rate laws may be derived.
3. S. W. Benson, *The foundations of chemical kinetics*, McGraw-Hill (1960). Chapter 3 of this book is particularly good at the mathematical aspects of complex chemical reactions.
4. W. Tsang, *J. chem. Phys.* **42,** 1802 (1965).
5. *Handbook of chemistry and physics*, 48th edn, The Chemical Rubber Co. (1968).
6. J. Leitich, *Int. J. chem. Kin.* **11,** 1249 (1979). The dimerization of *cis,trans*-cyclo-octa-1,5-diene gives three stereoisomeric products which shows that the second-order rate constant, k_2 on p. 275, is itself a composite term. This does not really detract from the usefulness of the example. The author in fact analysed the overall kinetics by a method which did not require computer-assisted fitting of a single set of reactant concentration versus time data to an analytical expression. Instead, he used a graphical method which required concentration versus time data for both the reactant and the product of the first-order reaction.
7. C. C. K. Ling and M. M. Harris, *J. chem. Soc.* 1825 (1964).
8. A. Lüttringhaus and G. Eyring, *Annalen* **604,** 111 (1957).
9. L. Volk, W. Richardson, K. H. Lau, M. Hall, and S. H. Lin, *J. chem. Educ.* **54,** 95 (1977). These authors give a critical account of the steady-state and equilibrium approximations in the simplification of the kinetics of complex reactions. In particular, they consider the error in the approximate result as a function of the ratio k_2/k_1 in the example of eqn (7.13), p. 282.

10. R. Huisgen, *Angew. Chem. int. edn* **9,** 751 (1970); R. P. Wayne, Chapter 4, 'The detection and estimation of intermediates' in *Comprehensive chemical kinetics,* (eds. C. H. Bamford and C. F. H. Tipper), Elsevier Publishing Company, Amsterdam (1969). These are general reviews of kinetics methods used for the detection of reactive intermediates. N. S. Isaacs, *Reactive intermediates in organic Chemistry,* Wiley (1974) is a comprehensive account of the organic chemistry of a range of reactive intermediates. Other books on particular types of reactive intermediates include D. Bethell and V. Gold, *Carbonium ions; an introduction,* Academic Press (1967); W. A. Pryor, *Free Radicals,* McGraw-Hill (1966); T. L. Gilchrist and C. W. Rees, *Carbenes, nitrenes, and arynes,* Nelson (1969).
11. R. P. Bell and K. Yates, *J. chem. Soc.* 1927 (1962).
12. A. Ault, 'Selectivity and specificity in organic reactions', *J. chem. Educ.* **54,** 614 (1977).
13. D. Y. Curtin, *Rec. chem. Progr.* **15,** 111 (1954); L. P. Hammett, *Physical organic chemistry* 2nd edn, McGraw-Hill (1970); N. S. Zefirov, *Tetrahedron* **33,** 2719 (1977), 'General equation of the relationship between products ratio and conformational equilibrium'. This paper shows the Curtin–Hammett Principle to be a limiting extreme of a general relationship.
14. M. R. V. Sahyun and D. J. Cram, *J. Am. chem. Soc.* **85,** 1263 (1963).
15. R. A. Bartsch, *J. Am. chem. Soc.* **93,** 3683 (1971).
16. H. M. Frey and R. C. Smith, *Trans. Faraday Soc.* **58,** 697 (1962).
17. D. R. Stull, E. F. Westrum, and G. C. Sinke, *The chemical thermodynamics of organic compounds,* Wiley, New York (1969).
18. E. D. Hughes, C. K. Ingold, S. Masterman, and B. J. McNulty, *J. chem. Soc.* 899 (1940).
19. M. W. Lee and W. C. Herndon, *J. org. Chem.* **43,** 518 (1978).
20. W. C. Herndon, C. R. Grayson, and J. M. Manion, *J. org. Chem.* **32,** 526 (1967).
21. W. P. Jencks, *Catalysis in chemistry and enzymology,* McGraw-Hill, New York (1969); *J. Am. chem. Soc.* **81,** 475 (1959).

Supplementary references

J. O. Edwards, E. F. Greene, and J. Ross, *J. chem. Educ.* **45,** 381 (1968): 'From stoichiometry and rate law to mechanism'.

K. B. Wiberg, *Physical organic chemistry,* Wiley (1964).

Techniques of chemistry, Vol. VI, *Investigation of rates and mechanisms of reactions,* Pt. 1 (ed. E. S. Lewis), Wiley–Interscience (1974).

F. Wilkinson, *Chemical kinetics and reaction mechanisms,* Van Nostrand-Reinhold, Wokingham, England (1980).

8

Catalysis

8.1 Introduction

There is no reason in principle why a chemical compound should react by only a single mechanism under given experimental conditions. We saw in the previous chapter that, if a compound undergoes parallel kinetically-controlled reactions, products will be formed in yields determined by the relative magnitudes of the respective rate constants and the concentrations of reactants if the competing routes are of different order. For example, compound A may react in the presence of compound B by a first-order route and two second-order routes to give products X, Y, and Z.

$$\mathrm{A+B} \xrightarrow{k_1[\mathrm{A}]} \mathrm{X},$$

$$\mathrm{A+B} \xrightarrow{k_2[\mathrm{A}]^2} \mathrm{Y},$$

$$\mathrm{A+B} \xrightarrow{k_3[\mathrm{A}][\mathrm{B}]} \mathrm{Z}.$$

The products under kinetic control will be formed in relative amounts determined by the ratios $k_1 : k_2[A] : k_3[B]$. Analogously, if the same product is formed from a reactant by more than one mechanism, the different routes will be followed according to the respective elementary rate constants and the experimental conditions.

Many reactions are known whose rates can be increased by the addition of a reagent which is not included in the stoichiometry of the reaction. Such a reagent is called a *catalyst* and it makes available an extra route to the same product(s). The uncatalysed reaction may be either slower or faster than the catalysed reaction. For example, if the conversion of A into B is catalysed by C,

$$A \xrightarrow{k_1[A]} B$$

$$A \xrightarrow{k_c[C][A]} B$$

$$-\frac{d[A]}{dt} = k_1[A] + k_c[C][A]$$

$$= k_{obs}[A] \text{ at constant } [C],$$

where k_{obs} = the observed experimental *pseudo* first-order rate constant $= k_1 + k_c[C]$,
k_1 = the first-order rate constant of the uncatalysed reaction, and
k_c = the second-order *catalytic constant* for the catalysed reaction.

We see, therefore, that in this simple case the catalysed and uncatalysed reactions take place concurrently in proportions given by

$$\text{catalysed reaction: uncatalysed reaction} = k_c[C] : k_1 .$$

For a given reaction of this type at a particular temperature, k_1 and k_c are constants, therefore only by altering [C] can the relative contributions of the two routes be changed.

In some reactions, a reactant may also act as a catalyst. The rate law of such a reaction will include the catalytic reactant to a higher power than would be expected on the basis of the stoichiometric equation alone. If a product of a reaction acts as a catalyst, the process is known as *autocatalytic.* Special measures may be required to simplify the kinetic investigation of a reaction when either a reactant or a product is catalytic to maintain the concentration of the catalyst constant.

Changes in standard molar free energy, enthalpy, and entropy associated with a conversion will be the same for catalysed and uncatalysed routes because these are functions determined only by the initial and final states, and are unaffected by the means of the conversion. This is illustrated in Fig. 8.1 for the reaction $A \rightarrow B$.

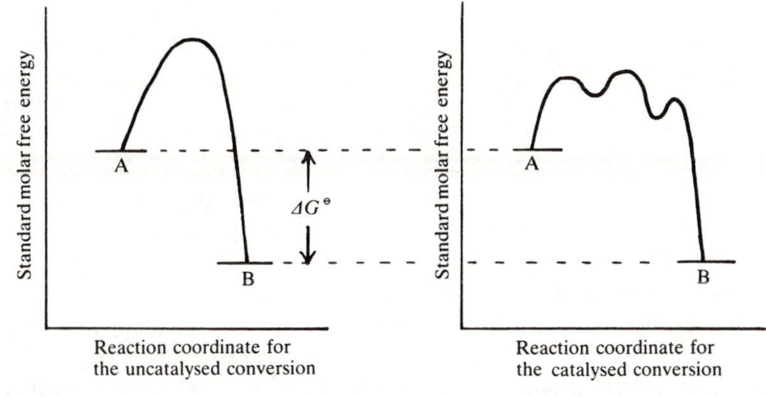

Fig. 8.1. Reaction profiles for uncatalysed and catalysed conversions A → B.

The uncatalysed reaction could be via one step, as illustrated, or several; but a catalysed reaction, which continually uses and regenerates the catalyst, must always involve more than one step. According to the definition presented earlier (p. 269) therefore, catalysed reactions are necessarily complex reactions. Furthermore, although initial and final states illustrated in Fig. 8.1 are identical for catalysed and uncatalysed routes, different reaction profiles are drawn because the two routes involve different reaction coordinates. Occasionally, they are represented on the same diagram for the sake of brevity, but this is not proper unless the axis which represents the reaction coordinate is doubly labelled.

It is also evident from Fig. 8.1 that a catalyst for the forward reaction of a reversible process must also catalyse the reverse reaction.

Catalysis is not restricted to only certain types of chemical reactions, indeed it appears to be a very common phenomenon. Catalysts are usually classified as heterogeneous (for example, metals in the hydrogenation of alkenes and other unsaturated compounds) or homogeneous (for example, a dissolved acid in the hydrolysis of an ester). The former are very important in many laboratory and industrial processes, but we shall consider here only homogeneous catalysis in solution.

8.2 Electrophile catalysis

Methyl esters of sterically-hindered carboxylic acids are not easily hydrolysed but they may be cleaved by anhydrous lithium iodide in pyridine[1] (eqn (8.1)).

$$I^{-}\ CH_3{-}O{-}C({=}O)R\ \ Li^{+} \xrightarrow{\text{pyridine}} I{-}CH_3 + RCO_2^{-}Li^{+}. \qquad (8.1)$$

The electrophilic lithium cation becomes complexed by the oxygen of the ester and this facilitates the nucleophilic displacement from methyl of an otherwise poor leaving group by iodide. This ability of Li^+ is ascribed to its small size and high charge density. It is an example of a *hard* Lewis acid acting as an *electrophile catalyst* in the displacement of a leaving group which is a hard Lewis base.[2] If an oxygen-containing protic co-solvent such as water or an alcohol were used, it would solvate the lithium and the iodide more effectively than the pyridine does, and the cation and anion would then be far less active as catalyst and nucleophile respectively. Potassium and sodium iodides are less effective in this reaction because K^+ and Na^+ are larger cations and consequently *softer* Lewis acids so they have a poorer interaction with the hard oxygen of the carboxylate leaving group. If the leaving group is a soft base, then a soft electrophile catalyst is required to facilitate displacement. The favourable soft–soft interaction between Ag^+ and halide anion is exploited in Ag^+-promoted solvolysis of alkyl halides; a hard electrophile such as Li^+ or H^+ would be much less effective in this reaction.

Besides nucleophilic substitutions, nucleophilic addition reactions may also be subject to electrophile catalysis. The formation of a thioketal shown in eqn (8.2) is accelerated by boron trifluoride. The boron trifluoride etherate transfers the hard Lewis acid BF_3 to the ketone to give a complex (eqn (8.3)) which is much more susceptible to nucleophilic attack by the ethanedithiol than is the free ketone.

$$\mathrm{R{-}\overset{\displaystyle O}{\overset{\|}{C}}{-}R + HSCH_2CH_2SH \xrightarrow{Et_2\overset{+}{O}{-}\overset{-}{B}F_3} \underset{R\ \ R}{\overset{S\frown S}{C}} + H_2O} \tag{8.2}$$

$$\mathrm{Et_2\overset{+}{O}{-}\overset{-}{B}F_3 + R_2C{=}O \rightleftharpoons Et_2O + R_2C{=}\overset{+}{O}{-}\overset{-}{B}F_3 \leftrightarrow R_2\overset{+}{C}{-}O{-}\overset{-}{B}F_3} \tag{8.3}$$

The hydration of alkynes to give, ultimately, carbonyl compounds (eqn (8.4)) is a well-known nucleophilic addition reaction which is catalysed by various heavy metal compounds including mercury(II) salts.[3]

$$\mathrm{R{-}C{\equiv}C{-}H \xrightarrow[H_2O]{Hg^{2+},\ H^+} \left[\begin{matrix} HO & & H \\ & C{=}C & \\ R & & H \end{matrix}\right] \longrightarrow R{-}\overset{\displaystyle O}{\overset{\|}{C}}{-}CH_3} \tag{8.4}$$

In the absence of metal ions, alkynes are not sufficiently electrophilic to undergo additions except with the most potent of nucleophiles. However, a labile organometallic species is generated from the alkyne and the soft Lewis acid Hg^{2+} which undergoes nucleophilic addition of water quite readily. The reaction is completed by the regeneration of the Hg^{2+}

catalyst and the tautomerization of the enol to the more stable carbonyl compound.

The above interpretations of the roles of Li^+, BF_3, and Hg^{2+} in the catalysis of ester cleavage, thioketal formation, and hydration of alkynes are entirely credible on the basis of current theories of organic chemistry. None the less they are speculative to the extent that they are not firmly based upon quantitative kinetics results.

In each case an electrophilic compound A—B (Fig. 8.2 for substitution) is rendered more electrophilic by a preliminary step involving the catalyst E^+. This preliminary step is in principle reversible, so it could constitute a pre-equilibrium. Whether or not it is genuinely a pre-equilibrium in any particular instance will depend upon the relative rates of the competing routes from the reactive intermediate A–$\overset{+}{B}$–E: the unimolecular reverse step with first-order rate constant k_{-1}, and the bimolecular forward reaction with the nucleophile with second-order rate constant k_2.

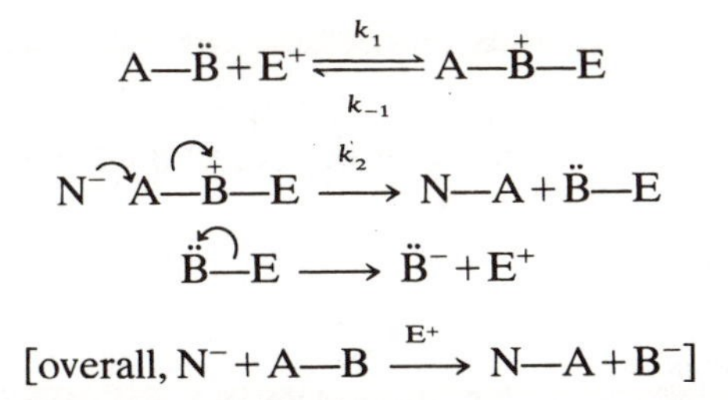

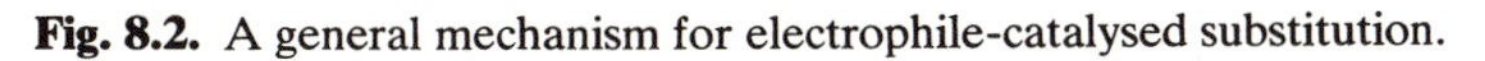

Fig. 8.2. A general mechanism for electrophile-catalysed substitution.

Reactions which have been extensively investigated because of their relationship to some processes of biological importance are the decarboxylations of oxaloacetic acid derivatives catalysed by metal ions.[4,5] A mechanism for the uncatalysed reaction of dimethyloxaloacetate is shown in Fig. 8.3.

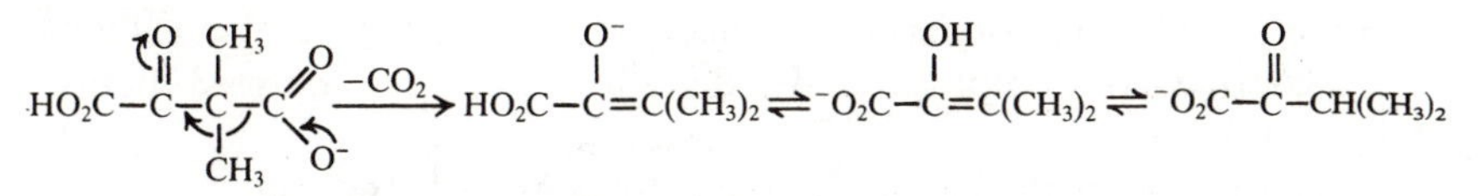

Fig. 8.3. A mechanism for the uncatalysed decarboxylation of dimethyloxaloacetate.

This mechanism is based upon the pH dependence of the *pseudo* first-order rate constant for dimethyloxaloacetic acid decarboxylation and the detection by chemical and spectroscopic methods of the enol intermediate. The decarboxylation is strongly accelerated by metal ions such as Cu^{2+}, Fe^{3+}, and Al^{3+} and the mechanism in Fig. 8.4 has been proposed for the Cu^{2+}-catalysed reaction.

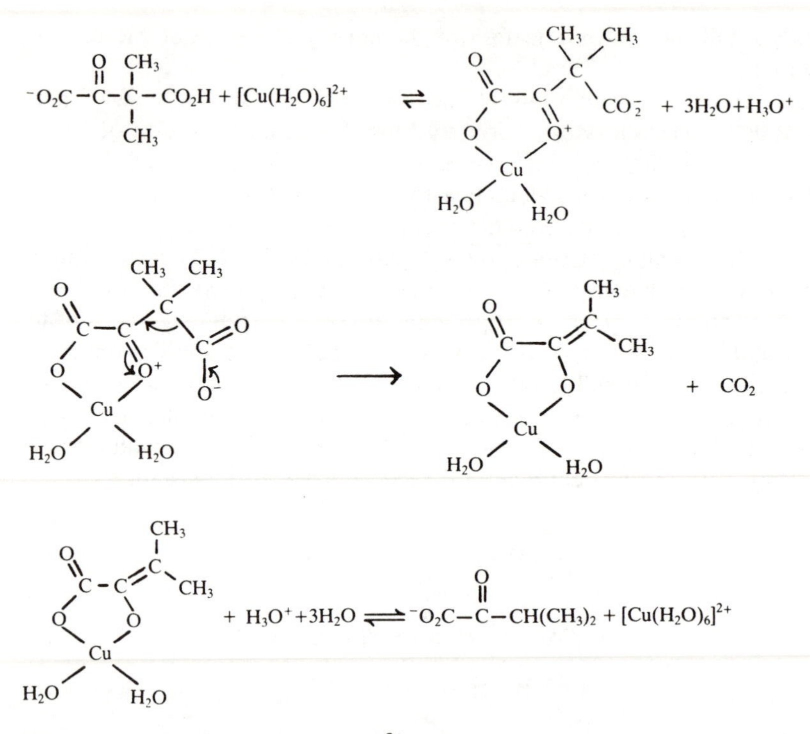

Fig. 8.4. A mechanism for the Cu^{2+}-catalysed decarboxylation of dimethyloxaloacetate.

The coordination of the Cu(II) in the complexes may not be four as shown. Each complex could include two more water molecules and be a Jahn–Teller distorted octahedron.

Dimethyloxaloacetate is a good ligand and readily forms the metal complex. The metal polarizes the carbonyl bond and thereby facilitates the decarboxylation and consequent electronic redistribution. The product of decarboxylation is the Cu(II) complex of the enolate form of dimethylpyruvate. Protonation and dissociation regenerate the catalytic aquo-copper(II) complex and dimethylpyruvate (2-oxo-3-methylbutanoate).

In accordance with this mechanism, it has been found that

(a) negatively-charged chelating ligands such as citrate, which form complexes with Cu^{2+} more strongly than oxaloacetate does, inhibit the catalysis,

(b) Fe^{3+}, having a higher charge and being better able to act as an 'electron sink' in the decarboxylation step, is a more effective catalyst than Fe^{2+},

(c) non-labile complex cations, even when highly charged such as $[Co(III)(NH_3)_6]^{3+}$, are not effective catalysts, and

(d) monocarboxylic acids and β-ketoacids (which do not have the second carboxylate group in the appropriate position) do not undergo metal-ion-catalysed decarboxylations.

The metal-ion-catalysed decarboxylation of oxaloacetate itself is believed to be by a related mechanism. Significantly, any of the ions which themselves catalyse this reaction (e.g. Mn^{2+}, Mg^{2+}, Zn^{2+}, Fe^{2+}) will activate the enzyme *oxaloacetate decarboxylase* which catalyses the reaction in nature.

8.3 Specific acid catalysis[5–8]

The reversible hydrolysis of simple esters, exemplified by eqn (8.5), is very slow in the absence of a catalyst.

$$CH_3CO_2C_2H_5 + H_2O \underset{k_r \atop H_2O}{\overset{k_f}{\rightleftharpoons}} CH_3CO_2H + C_2H_5OH. \qquad (8.5)$$

Any such reaction which does occur we shall for the present call the uncatalysed reaction (this aspect is amplified later). Various electrophiles such as metal ions catalyse these hydrolyses and a very common one, the proton, is particularly effective.

Many authors refer to H^+, proton, or hydrogen ion in solution. Almost invariably they mean the hydronium ion, H_3O^+;[9] see Chapter 5. However, since a proton is transferred to a substrate (from the hydronium ion) in many catalysed processes in aqueous solution, we may refer to such reactions as either hydronium ion catalysed, or hydrogen ion or proton catalysed.

In aqueous acidic solution, the hydrolysis of simple esters is first order in ester concentration [E] and experimental *pseudo* first-order rate constants, k_{obs}, are found to be linearly dependent upon the hydronium ion concentration $[H_3O^+]$.[8,10]

The rate laws for such reactions have the form of eqns (8.6) and (8.7).

$$-\frac{d[E]}{dt} = k_0[E] + k_H[H_3O^+][E] \qquad (8.6)$$

or, at constant pH,

$$-d[E]/dt = k_{obs}[E],$$

where $k_{obs} = k_0 + k_H[H_3O^+]$ (8.7)

$\quad = $ the experimental *pseudo* first-order rate constant,

$k_0 = $ the first-order rate constant for the uncatalysed reaction, and

$k_H = $ the second-order catalytic constant for the reaction catalysed by H_3O^+.

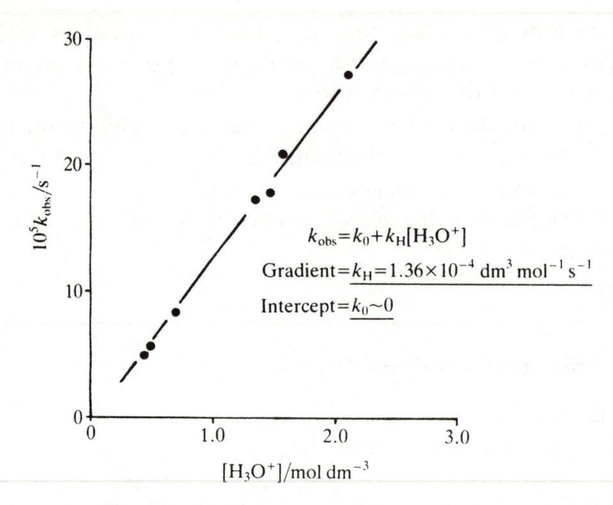

Fig. 8.5. The rate of hydrolysis of ethyl acetate in aqueous acidic solution, 25 °C. (Results taken from R. P. Bell, A. L. Dowding, and J. A. Noble, *J. chem. Soc.* 3106 (1955).)

Although the reaction is reversible, the rate of the second-order back reaction of the low concentrations of products is usually negligible compared with the *pseudo* first-order forward reaction of the reactant with the solvent, water. In other words, equilibrium in dilute aqueous solution (eqn (8.5)) is far over to the right-hand side, and the observed reaction is essentially unidirectional from left to right.

Results for the hydrolysis of ethyl acetate are shown graphically in Fig. 8.5; the experimental rate constants are plotted against $[H_3O^+]$ and, by eqn (8.7), the gradient of the straight line gives k_H. Extrapolation confirms that, for this particular reaction, when $[H_3O^+]=0$, $k_{obs}=k_0\sim0$; so the 'background' reaction uncatalysed by hydrogen ions is negligible.

The source of the protons which contribute to the total hydronium-ion concentration for incorporation into eqn (8.7) is inconsequential. Besides strong acids, all weak Brönsted acids which are present will contribute to extents determined by the magnitudes of their acidity constants. The catalytic effect of an *un-ionized* weak acid A–H must be assessed separately. We saw in Chapter 5 (p. 164) that the acidity constant K_{AH} of acid AH

$$AH+H_2O \overset{H_2O}{\rightleftharpoons} H_3O^+ + A^-$$

is given by

$$K_{AH}=\frac{[H_3O^+][A^-]}{[AH]},$$

where $[i]$ = numerical value of the molar concentration of i.

It follows from this that

$$[H_3O^+] = K_{AH}\frac{[AH]}{[A^-]},$$

therefore

$$pH = -\log[H_3O^+] = pK_{AH} - \log\frac{[AH]}{[A^-]},$$

or

$$pH = pK_{AH} - \log r,$$

where

$$r = \frac{[AH]}{[A^-]} = \text{the } \textit{buffer ratio.}$$

It is possible, therefore, to vary the absolute value of [AH] at constant pH as long as the buffer *ratio*, r, is kept constant. (Strictly, the ionic strength must also be kept constant.)

The rates of hydrolysis of many simple esters have been measured in buffered solutions of weak acids at constant buffer ratio, and the observed *pseudo* first-order rate constants, k_{obs}, are almost invariably independent of [AH] at constant pH (Fig. 8.6). The hydrolysis of a simple ester is, therefore, catalysed by H_3O^+ but not by undissociated weak Brönsted acids. This is *specific acid catalysis.*

We may conclude from the foregoing kinetic evidence and knowledge of the stoichiometry that an activated complex in the (very slow) uncatalysed hydrolysis involves a single ester molecule and water, whereas an activated complex in the catalysed reaction comprises ester plus proton and water.

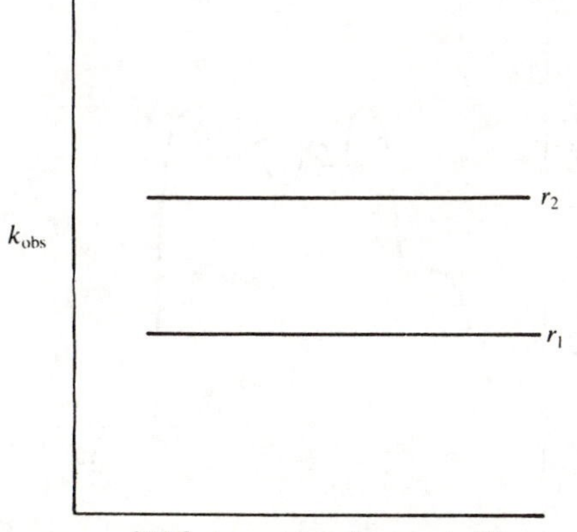

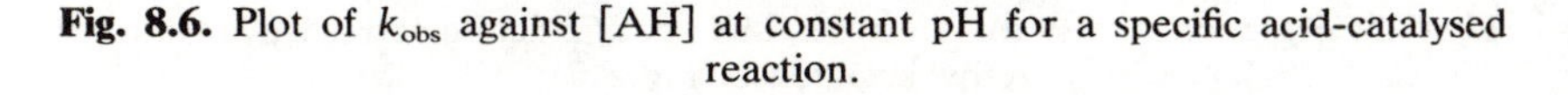
Fig. 8.6. Plot of k_{obs} against [AH] at constant pH for a specific acid-catalysed reaction.

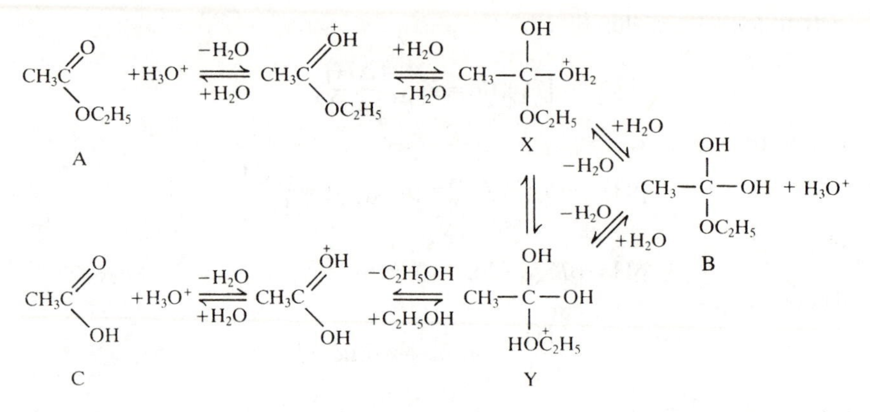

Fig. 8.7. Mechanism for the reversible specific acid-catalysed hydrolysis of ethyl acetate.

By methods so far considered, we have no means of determining the number of molecules of solvent in an activated complex. It is known that simple esters of highly branched carboxylic acids are hydrolysed only slowly and some, such as R_3CCO_2R', are very unreactive. Such decreasing reactivity with increasing molecular congestion near the reaction site is most reasonably ascribed to *steric hindrance* in a bimolecular rate-determining step. We deduce that the activated complex, therefore, in the (H^+-catalysed) hydrolysis of a simple ester is derived from a molecule of the (protonated) ester and at least one water molecule.

The so-called A2 mechanism for specific acid-catalysed ester hydrolysis shown in Figs 8.7 and 8.8 involves rapid protonation of the ester in a pre-equilibrium followed by slow nucleophilic attack by water. The protonated tetrahedral intermediate X may then undergo deprotonation

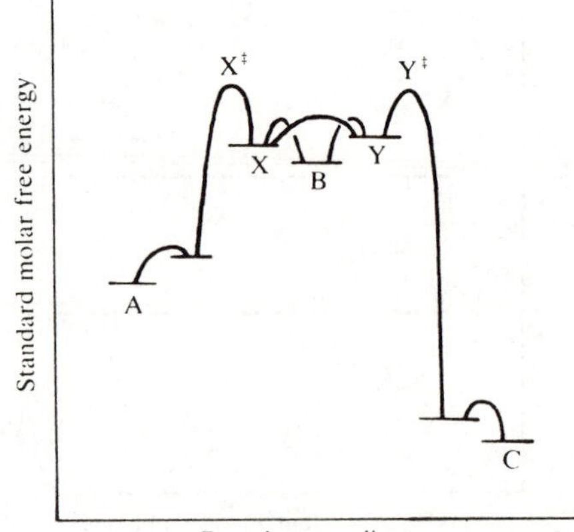

Fig. 8.8. Reaction profile for the reversible specific acid-catalysed hydrolysis of ethyl acetate (see Fig. 8.7).

to give a neutral tetrahedral intermediate, B. Reprotonation on the ether oxygen gives another charged tetrahedral intermediate, Y. Y could also be obtained directly from X by intramolecular proton transfer. So the reaction profile in Fig. 8.8 requires a third dimension in the region of X, B, and Y. Regardless of how it is formed, Y can expel ethanol to give the protonated carboxylic acid. Proton transfer to the solvent in the final equilibrium gives the carboxylic acid and regenerates the catalytic hydronium ion.

In this specific acid-catalysed reaction, which activated complex, $X^{\ddagger}$ or $Y^{\ddagger}$, has the higher standard free energy of formation? In other words, which is the faster process of the tetrahedral intermediates? – expulsion of ethanol to give protonated carboxylic acid (the forward reaction) or expulsion of water to regenerate the protonated ester (the reverse reaction).

This question has been investigated by using esters specifically labelled with ^{18}O in the carbonyl group and measuring both the rate of hydrolysis and the rate of exchange of ^{18}O between unhydrolysed ester and solvent under the hydrolysis conditions.[11] Return of the tetrahedral intermediate B in Figs 8.7 and 8.8 to ester causes scrambling of the OH groups since either the labelled or the unlabelled one may be protonated and lost as water. In the case of ethyl benzoate in aqueous acidic solution, hydrolysis is faster than ^{18}O scrambling, but not by a large factor ($k_{hyd}/k_{exch} = 5.2$, 99 °C). So expulsion of ethanol to give the protonated acid is somewhat faster than expulsion of water to give protonated ester. This factor will vary with the structure of the ester and the experimental conditions.

Note that this interpretation of the ^{18}O exchange results is based upon the *postulate* that ester hydrolysis and isotopic exchange from the ester occur through closely related mechanisms involving common tetrahedral intermediates.

Another well-investigated example of specific acid catalysis is the hydrolysis of simple acetals and ketals in dilute aqueous acidic solution; for example, eqn (8.8).[12,13(a)]

$$CH_3CH(OCH_3)_2 + H_2O \xrightarrow[k_{obs}]{H_3O^+,\ H_2O} CH_3CHO + 2CH_3OH \qquad (8.8)$$

This reaction is first order in acetal concentration [A] and, again, experimental *pseudo* first-order rate constants, k_{obs}, are linearly dependent upon the hydronium-ion concentration, $[H_3O^+]$. At constant pH,

$$-\frac{d[A]}{dt} = k_{obs}[A]$$

where $k_{obs} = k_O + k_H[H_3O^+]$,
k_O = the first-order rate constant for the uncatalysed reaction, and
k_H = the second-order catalytic constant.

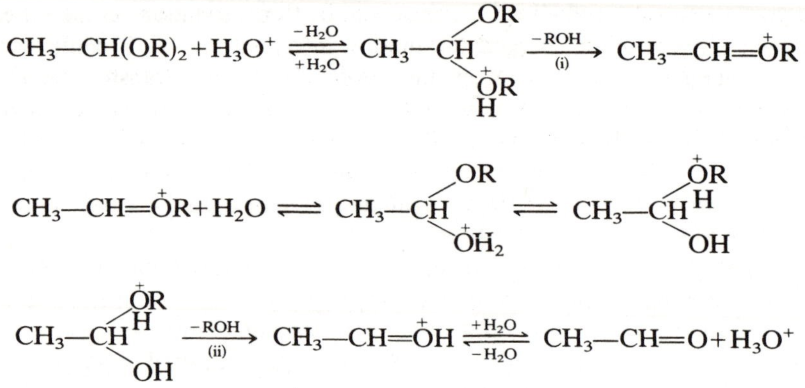

$$CH_3{-}CH(OR)_2 + H_3O^+ \underset{+H_2O}{\overset{-H_2O}{\rightleftharpoons}} CH_3{-}CH\begin{matrix}OR\\ \overset{+}{O}R\\ H\end{matrix} \xrightarrow[(i)]{-ROH} CH_3{-}CH{=}\overset{+}{O}R$$

$$CH_3{-}CH{=}\overset{+}{O}R + H_2O \rightleftharpoons CH_3{-}CH\begin{matrix}OR\\ \overset{+}{O}H_2\end{matrix} \rightleftharpoons CH_3{-}CH\begin{matrix}\overset{+}{O}R\\ H\\ OH\end{matrix}$$

$$CH_3{-}CH\begin{matrix}\overset{+}{O}R\\ H\\ OH\end{matrix} \xrightarrow[(ii)]{-ROH} CH_3{-}CH{=}\overset{+}{O}H \underset{-H_2O}{\overset{+H_2O}{\rightleftharpoons}} CH_3{-}CH{=}O + H_3O^+$$

Steps (i) and (ii) in this scheme which involve the expulsion of ROH appear, in principle, reversible. However if the reactant is in dilute aqueous solution, the forward reactions (*pseudo* first-order capture of the electrophilic intermediates by water) will be much faster than their reverse reactions (second-order capture by the very low concentration of alcohol). This assumes that H_2O and ROH have comparable nucleophilicities. Consequently, the overall reaction is effectively irreversible and virtually no acetal can be detected at equilibrium.

Fig. 8.9. A mechanism for the specific acid-catalysed hydrolysis of an acetal.

By plotting k_{obs} against $[H_3O^+]$ as in Fig. 8.5 for reactions in solutions of a strong acid, k_H may be obtained from the gradient and k_0 (which is again virtually zero) from the intercept when $[H_3O^+]=0$. When k_{obs} is plotted against [AH] from rates measured using buffered solutions of a weak acid A–H at constant buffer ratio $[AH]/[A^-]$ (constant pH), a graph analogous to that in Fig. 8.6 is obtained. This establishes that undissociated weak acids do not catalyse the reaction – it is specific acid catalysed.

Dramatically larger k_H values are found for compounds obtained by alkyl substitution of the methylene group in $CH_2(OR)_2$ (*ca.* $10^{3.5}$ per alkyl).[13(a)] Taken with the specific acid catalysis, this suggests the mechanism shown in Fig. 8.9 which incorporates two virtually irreversible unimolecular fragmentation steps in the forward direction, (i) and (ii), either of which could in principle be rate determining.

Fragmentation of the protonated substrate involves a tetrahedrally coordinated carbon atom becoming trigonal; consequently, unfavourable non-bonded interactions between bulky groups are relieved. The more sterically-congested the reactant, the greater the release of molecular strain in the heterolysis, and consequently the faster the reaction. This may be described as *steric acceleration* due to the relief of strain in the reactant, and is characteristic of a reaction the rate-determining step of which is unimolecular. It is in contrast to *steric hindrance* which characterizes reactions with bimolecular rate-determining steps; see p. 324.

This mechanism for acetal hydrolysis, like the one in Fig. 8.7 for simple ester hydrolysis, involves rapid protonation as a pre-equilibrium followed at some stage by the rate-determining step. In the so-called A1 mechanism of simple acetal hydrolysis, this step is a unimolecular fragmentation, either (i) or (ii) in Fig. 8.9. Hydrolysis of simple esters involves rate-determining nucleophilic attack by water upon the protonated substrate, the A2 mechanism. In general, any mechanism which involves rapid protonation of substrate E as a pre-equilibrium followed by slower steps *not* involving proton transfer (but either unimolecular or bimolecular) always leads to a specific acid-catalysis rate law. This is true even if undissociated weak acids contribute directly in the pre-equilibria:

$$E + AH \overset{\text{fast}}{\rightleftharpoons} EH^+ + A^-; \qquad K = \frac{[EH^+][A^-]}{[E][AH]}$$

$$EH^+ \xrightarrow[\text{r.d.s.}]{k} P, \quad \text{products.}$$

By this route,

$$\frac{d[P]}{dt} = k[EH^+]$$

$$= kK[E]\frac{[AH]}{[A^-]}.$$

But

$$AH + H_2O \overset{\text{fast}}{\rightleftharpoons} H_3O^+ + A^-$$

$$K_{AH} = \frac{[H_3O^+][A^-]}{[AH]}.$$

By this route, therefore,

$$\frac{d[P]}{dt} = kK[E]\frac{[H_3O^+]}{K_{AH}}$$

$$= k_H[E][H_3O^+].$$

8.4 General acid catalysis[5–7]

If a solvolytic reaction of compound S is first order in [S], and catalysed not only by H_3O^+ but also by an un-ionized weak acid A–H, the experimental rate law will have the form of eqns (8.9) and (8.10).

$$S \xrightarrow[k_{obs}]{H_3O^+,\ AH,\ H_2O} \text{Products.}$$

At constant $[H_3O^+]$ and [AH],

$$-\frac{d[S]}{dt} = k_{obs}[S], \tag{8.9}$$

where

$$k_{obs} = k_0 + k_H[H_3O^+] + k_{AH}[AH] \tag{8.10}$$

= the experimental *pseudo* first-order rate constant,

k_0 = the first-order rate constant of the uncatalysed reaction,

k_H = the second-order catalytic constant for the proton-catalysed reaction, and

k_{AH} = the second-order catalytic constant for the reaction catalysed by the *general acid*, A–H.

In the presence of another weak acid B–H, an additional term $k_{BH}[BH][S]$ would be included, and so on for all undissociated acids present. Generally, therefore, we may write

$$k_{obs} = k_0 + k_H[H_3O^+] + \sum k_{AH}[AH]. \tag{8.11}$$

Reactions whose rates are described by expressions of the form of eqns (8.10) and (8.11) are said to show *general acid catalysis*. Normally, they are easily distinguished experimentally from specific acid-catalysed reactions (reactions in which $\sum k_{AH} = 0$ therefore $\sum k_{AH}[AH] = 0$). The variation in the *pseudo* first-order rate constant, k_{obs}, with [AH] is investigated at constant buffer ratio (constant pH) as described on p. 323. According to eqn (8.10), a general acid-catalysed reaction in the presence of a single general acid AH yields a linear plot with gradient = k_{AH} as illustrated in Fig. 8.10 (cf. Fig. 8.6 for a specific acid-catalysed reaction). In principle, the intercept when [AH] = 0 gives $k_0 + k_H[H_3O^+]$, but the two parameters k_0 and k_H are most reliably determined by plotting k_{obs} against $[H_3O^+]$ using results obtained from solutions containing only a strong acid. (When $\sum[AH] = 0$, $\sum k_{AH}[AH] = 0$ therefore eqn (8.10) is reduced to $k_{obs} = k_0 + k_H[H_3O^+]$; see Problem 1, p. 362).

Needless to say, all Brönsted acids which are present will contribute towards $[H_3O^+]$ according to their limited dissociations, so catalysis by an aqueous solution of general acids is usually encountered against a background of catalysis by H_3O^+, and a $k_H[H_3O^+]$ term is present in the general acid-catalysis rate expression. However, if the solution is buffered to be near neutral or even alkaline, $[H_3O^+]$ will be negligibly small and the proton-catalysed reaction can then usually be ignored.

Several types of mechanism lead to general acid-catalysis rate laws[6,7] and we shall now consider three of them.

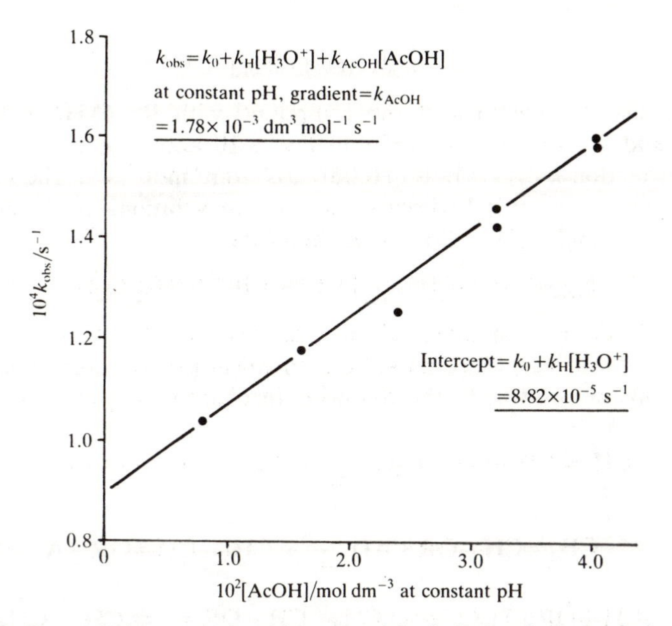

Fig. 8.10. Rate of the general acid-catalysed hydrolysis of vinyl ethyl ether in aqueous solution, 26.7 °C.[14]

8.4.1 Rate-determining proton transfer from an acid to the substrate

An example of this, the most obvious mechanistic type which leads to general acid catalysis, is the hydrolysis of vinyl ethers (enol ethers) in acidic aqueous solution, eqn (8.12).[14,15]

$$CH_2{=}CH{-}OR + H_2O \xrightarrow[k_{obs}]{AH,\ H_3O^+,\ H_2O} CH_3CHO + ROH. \tag{8.12}$$

In solutions of a strong acid, the reaction rates are first order both in the concentration of vinyl ether and in the concentration of hydronium ions. By plotting k_{obs} against $[H_3O^+]$, k_H may be obtained from the gradient and k_0 by extrapolation to $[H_3O^+]=0$ as in Fig. 8.5. When $R=C_2H_5$ in eqn (8.12), $k_H=2.08\ dm^3\ mol^{-1}\ s^{-1}$ and $k_0 \sim 0$.[14] There is, therefore, virtually no reaction in non-acidic solutions, and vinyl ethers are stable to hydrolysis in basic or neutral aqueous media. By investigating rates of reaction in buffered solutions of weak acids, the following rate law was deduced:

$$-\frac{d[\text{vinyl ether}]}{dt} = k_{obs}[\text{vinyl ether}]$$

where

$$k_{obs} = k_H[H_3O^+] + k_{AH}[AH]. \tag{8.13}$$

The results for reactions of the compound with $R = C_2H_5$ in buffered acetic acid are shown graphically in Fig. 8.10 and from the gradient of this correlation $k_{AcOH} = 1.78\ (\mp 0.06) \times 10^{-3}\ dm^3\ mol^{-1}\ s^{-1}$. The complete rate equation for this hydrolysis reaction in solutions containing both H_3O^+ and CH_3CO_2H at 26.7 °C is therefore:

$$k_{obs}/s^{-1} = 2.08[H_3O^+] + 1.78 \times 10^{-3}[CH_3CO_2H].$$

Clearly, H_3O^+ is a far more effective catalyst than CH_3CO_2H.

On the basis that there must be a chemical process corresponding to both terms in eqn (8.13), the following mechanism was proposed:

$$CH_2{=}CH{-}OR + H_3O^+ \xrightarrow[H_2O]{slow} CH_3{-}CH{=}\overset{+}{O}R + H_2O$$

$$CH_2{=}CH{-}OR + AH \xrightarrow[H_2O]{slow} CH_3{-}CH{=}\overset{+}{O}R + A^-$$

$$CH_3{-}CH{=}\overset{+}{O}R + H_2O \rightleftharpoons CH_3{-}\underset{^+OH_2}{\underset{|}{C}H}{-}OR \rightleftharpoons CH_3{-}\underset{OH}{\underset{|}{C}H}{-}\underset{H}{\overset{+}{O}R}$$

$$CH_3{-}\underset{OH}{\underset{|}{C}H}{-}\underset{H}{\overset{+}{O}R} \xrightarrow{-ROH} CH_3{-}CH{=}\overset{+}{O}H \underset{-H_2O}{\overset{+H_2O}{\rightleftharpoons}} CH_3{-}CHO + H_3O^+.$$

Equilibrium concentrations of AH, A^-, and H_3O^+ are maintained by:

$$AH + H_2O \overset{H_2O}{\rightleftharpoons} H_3O^+ + A^-.$$

Two parallel bimolecular slow steps with second-order elementary rate constants generate a reactive electrophilic intermediate $CH_3CH{=}\overset{+}{O}R$. This undergoes nucleophilic capture by water as fast as it is formed to give the protonated hemiacetal which is itself unstable in the acidic aqueous medium and fragments to give, finally, acetaldehyde (see also Fig. 8.9).

The details of the later steps in this mechanism are intuitively based upon analogy rather than upon direct evidence.

It is not easy to distinguish kinetically between the mechanism given, involving sequential proton transfer and nucleophilic attack by water, from one in which the protonation and nucleophilic attack are synchronous, since they differ only in the presence or otherwise of a solvent molecule in the activated complex. Furthermore, the mechanism may seem, in principle, reversible. However, at equilibrium in acidic aqueous solution no vinyl ether can be detected, so the reverse reaction is very much slower than the forward one and the reaction is essentially unidirectional. This is principally because [ROH] is so very low compared with the high concentration of water in an aqueous solution. See p. 326 for an analogous case.

8.4.2 Rapid protonation of the substrate in a pre-equilibrium followed by slow proton abstraction by any base

We shall demonstrate first that this mechanistic type leads to general acid catalysis using the enolization of a simple ketone, eqn (8.14), as an example.

$$CH_3\text{—}\overset{\overset{\displaystyle O}{\|}}{C}\text{—}CH_3 \underset{k_r}{\overset{H_3O^+,\, A\text{—}H,\, H_2O \atop k_f}{\rightleftharpoons}} CH_3\text{—}\overset{\overset{\displaystyle OH}{|}}{C}\text{=}CH_2 \qquad (8.14)$$

Mechanism:
rapidly maintained equilibria:

$$CH_3\text{—}\overset{\overset{\displaystyle O}{\|}}{C}\text{—}CH_3 + H_3O^+ \underset{+H_2O}{\overset{-H_2O}{\rightleftharpoons}} CH_3\text{—}\overset{\overset{\displaystyle {}^+OH}{\|}}{C}\text{—}CH_3; \quad K = \frac{[CH_3\text{—}\overset{\overset{\displaystyle {}^+OH}{\|}}{C}\text{—}CH_3]}{[CH_3COCH_3][H_3O^+]}$$

$$AH + H_2O \rightleftharpoons A^- + H_3O^+$$

$$2H_2O \rightleftharpoons OH^- + H_3O^+,$$

slow proton-abstraction from carbon steps:

$$CH_3\text{—}\overset{\overset{\displaystyle {}^+OH}{\|}}{C}\text{—}CH_3 + OH^- \xrightarrow{k_{OH}} CH_3\text{—}\overset{\overset{\displaystyle OH}{|}}{C}\text{=}CH_2 + H_2O$$

$$CH_3\text{—}\overset{\overset{\displaystyle {}^+OH}{\|}}{C}\text{—}CH_3 + H_2O \xrightarrow{k_{H_2O}} CH_3\text{—}\overset{\overset{\displaystyle OH}{|}}{C}\text{=}CH_2 + H_3O^+$$

$$CH_3\text{—}\overset{\overset{\displaystyle {}^+OH}{\|}}{C}\text{—}CH_3 + A^- \xrightarrow{k_A} CH_3\text{—}\overset{\overset{\displaystyle OH}{|}}{C}\text{=}CH_2 + A\text{—}H.$$

The rate of the forward reaction of eqn (8.14) will be the *sum* of these three parallel bimolecular elementary reactions which have second-order rate constants:

$$\frac{d[\text{enol}]}{dt} = [CH_3\text{—}\overset{\overset{\displaystyle {}^+OH}{\|}}{C}\text{—}CH_3]\{k_{OH}[OH^-] + k_{H_2O}[H_2O] + k_A[A^-]\}$$

$$= [CH_3COCH_3]K\{k_{OH}[OH^-][H_3O^+] + k_{H_2O}[H_2O][H_3O^+] + k_A[A^-][H_3O^+]\}.$$

But $[H_3O^+][OH^-] = K_{AP}$, the autoprotolysis constant of water, and $K_{AH} = [A^-][H_3O^+]/[AH]$, the acidity constant of the catalytic weak acid.

The rate of the forward reaction, therefore, is given by

$$\frac{d[\text{enol}]}{dt} = [CH_3COCH_3]\{k_{OH}KK_{AP} + k_{H_2O}K[H_2O][H_3O^+] + k_AKK_{AH}[AH]\},$$

and, by amalgamating constant factors, we may re-write this kinetic expression as

$$\frac{d[\text{enol}]}{dt} = [CH_3COCH_3](k_0 + k_H[H_3O^+] + k_{AH}[AH]) \quad (8.15)$$

$$= k_f[CH_3COCH_3],$$

where

$$k_f = k_0 + k_H[H_3O^+] + k_{AH}[AH].$$

It has been established that eqn (8.15) describes the rate laws for the enolization of many simple ketones in aqueous solution. Plots of the type shown in Figs 8.5 and 8.10 (in the absence and presence, respectively, of weak acid AH) may be obtained, and from these k_0, k_H, and k_{AH} can be calculated.

Direct measurement of enolization is seldom practicable largely because there is so little enol present in equilibrium with the ketone. However if the enol can be removed much more rapidly than it reverts to ketone (eqn (8.16)), the enolization becomes essentially unidirectional.

$$R{-}\overset{\overset{\displaystyle O}{\|}}{C}{-}CH_3 \xrightarrow[\not\leftarrow]{AH,\ H_3O^+,\ H_2O} R{-}\overset{\overset{\displaystyle OH}{|}}{C}{=}CH_2 \xrightarrow{\text{v. fast}} \begin{array}{c}\text{further}\\ \text{reaction}\end{array} \quad (8.16)$$

The actual rate of the 'further reaction' is obviously constrained to be equal to the rate of enolization, so if measurement of the overall reaction is possible, it gives the rate of the initial enolization.

Halogenation is a 'scavenging' reaction which has been used extensively to remove enols as rapidly as they are formed,[6] and it is easily monitored. The same rates are obtained regardless of which halogen is used, and the overall reaction is zero order in halogen concentration. These latter pieces of information are evidence that it is indeed the enolization and not the halogenation which is rate determining. The rate of racemization of an optically-active ketone whose enol tautomer is achiral gives the rate of enolization directly, eqn (8.17). The reverse reaction need not be suppressed in this case since it gives racemic ketone which does not complicate the polarimetric measurement of the forward reaction of chiral material.

$$R{-}\overset{\overset{\displaystyle O}{\|}}{C}{-}CHR^1R^2 \underset{k_r}{\overset{AH,\ H_3O^+,\ H_2O}{\underset{k_f = k_{rac}}{\rightleftharpoons}}} R{-}\overset{\overset{\displaystyle OH}{|}}{C}{=}CR^1R^2 \quad (8.17)$$

When rates of racemization and halogenation have been measured, for example using phenyl 2-butyl ketone, identical results within experimen-

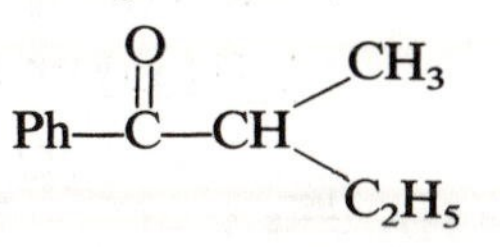

tal error were observed.[16] Again, this supports the contention that both racemization and halogenation occur via enolization and that it is the rate of formation of the enol from the ketone which, in both cases, is in effect being measured.

8.4.3 Rate-determining acid-assisted nucleophilic attack upon the substrate

We have already seen (Fig. 8.8, p. 324) that rapid reversible protonation of a substrate followed by slow nucleophilic attack by water leads to a specific acid catalysis rate law. If, *in addition*, there is a pre-equilibrium whereby substrate and undissociated weak general acid form a hydrogen-bonded complex (which will be more electrophilic than the substrate itself) we have a catalysis mechanism which requires another term in the rate equation. This is illustrated for the (reversible) hydration of simple aldehydes in aqueous acidic solution – reactions which experimentally are known to be general acid catalysed.[6,17]

$$R\text{—}CHO + H_2O \underset{k_r}{\overset{H_3O^+,\ AH,\ H_2O}{\underset{}{\overset{k_f}{\rightleftharpoons}}}} R\text{—}CH(OH)_2$$

k_r is a first-order rate constant and k_f is *pseudo* first order. The overall reaction is reversible therefore $k_{obs} = k_f + k_r$ and the proportion of an aldehyde which is in equilibrium with its hydrate depends upon the structure of the aldehyde and the nature and proportion of any organic co-solvent. If progress towards equilibrium were essentially the forward reaction, then $k_{obs} \approx k_f$. But usually equilibrium consists of a significant proportion of aldehyde, so allowance must be made for this. In the following mechanisms, however, we deal only with the forward reaction, i.e. the hydration.

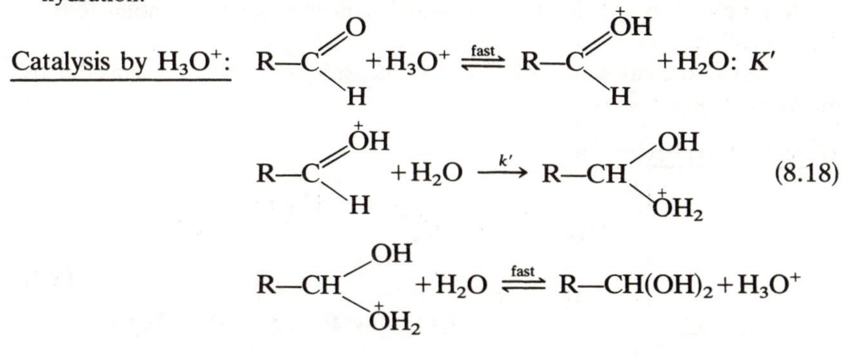

$$\text{partial forward rate} = k'[H_2O][RCHOH^+] = k'K'[H_3O^+][RCHO] = k_H[H_3\overset{+}{O}][RCHO].$$

Catalysis by HA:

$$R\text{—}C(=O)H + HA \overset{\text{fast}}{\rightleftharpoons} R\text{—}C(=O\cdots H\text{—}A)H \quad ; K''$$

$$R\text{—}C(=O\cdots H\text{—}A)H + H_2O \xrightarrow{k''} R\text{—}CH(OH)_2 + H\text{—}A \qquad (8.19)$$

$$\text{partial forward rate} = k''[H_2O][RCHO\,.\,HA] = k''K''[H_2O][RCHO][AH] = k_{AH}[AH][RCHO].$$

It may seem that the second step of the reaction catalysed by HA (eqns (8.19)) ought to be dissected to show formation of the contact ion-pair $R\text{—}CH(OH)(\overset{+}{O}H_2)\ A^-$ and then a subsequent proton-transfer step to give the products shown. Such a proton transfer between cation and anion within the same solvation shell would be so exceedingly fast in this thermodynamically-favourable direction that the nucleophilic attack and the proton transfer would be coupled, and the ion-pair would not correspond to a real intermediate.

Equations (8.18) give rise to the proton-catalysis term in the overall rate law, and are exactly analogous to the mechanism in Fig. 8.7 for the specific acid-catalysed hydrolysis of a simple ester. We shall call the particular mechanism outlined in eqns (8.19), whereby the undissociated general acid enhances the susceptibility of the substrate to nucleophilic attack by the solvent (water), *mechanistic general acid catalysis* to distinguish it from *kinetic* general acid catalysis. The latter is used of any *overall* reaction to which the rate law in eqn (8.10) applies; it is defined by an experimental rate law and implies nothing as far as mechanism is concerned. We shall use *mechanistic* general acid catalysis for only the single mechanism specified above which corresponds to a single term in the overall rate law.

We can also treat water itself as a weak Brönsted acid and consider the mechanism in eqns (8.20):

Water-induced reaction:

$$R\text{—}C(=O)H + H_2O \overset{\text{fast}}{\rightleftharpoons} R\text{—}C(=O\cdots H\text{—}OH)H \quad ; K'''$$

$$R\text{—}C(=O\cdots H\text{—}OH)H + H_2O \xrightarrow{k'''} R\text{—}CH(OH)_2 + H_2O \qquad (8.20)$$

$$\text{partial forward rate} = k'''[H_2O][RCHO\,.\,H_2O]$$
$$= k'''K'''[H_2O]^2[RCHO]$$
$$= k_0[RCHO].$$

We now see that, for a reaction in aqueous acidic solution, the mode which we previously have described as uncatalysed corresponds to a mechanism in which water itself acts as a general acid. For convenience, we shall retain use of the *pseudo* first-order rate constant k_0 for what we shall henceforth describe as a *solvent-induced* reaction when the solvent is also in the stoichiometric equation.

According to these mechanisms, the overall forward rate of the reversible hydration of an aldehyde in aqueous solution at constant $[H_3O^+]$ and [AH] will be given by

$$-\frac{d[RCHO]}{dt} = k_{obs}[RCHO]$$

where

$$k_{obs} = k_0 + k_H[H_3O^+] + k_{AH}[AH] \tag{8.10}$$

which is in agreement with experiment.

We shall see later (p. 348) that the solvent may also be regarded as a general base. Consequently, in *solvent-induced reaction*, we imply no mechanistic specification. Other common names are *spontaneous reaction* and *solvent-catalysed reaction*; we shall reserve the latter for those reactions in which the solvent acts as a catalyst but is not included in the stoichiometric equation.

8.5 Nucleophile catalysis[5,17]

A reaction is nucleophile catalysed if the catalyst reacts with some electrophilic centre of the substrate as a nucleophile but not as a base. So nucleophile catalysis does not include catalytic action via proton abstraction. Nucleophile catalysis of a variety of reaction types (addition, substitution, etc.) has been investigated. Some reactions are catalysed by only very few particular nucleophiles; in other reactions, a wide range of catalysts are effective. Regardless of detail, however, all nucleophile-catalysed reactions proceed via intermediates which occasionally may be detected spectroscopically or intercepted chemically.

The Benzoin condensation, eqn (8.21), familiar to many students as a laboratory preparation, is subject to fairly specific catalysis by only a few reagents including cyanide, CN^-. The reaction is third order overall and there is no reaction in the absence of a catalyst.[13(b),18]

$$2\,PhCHO \xrightarrow[k_{obs}]{CN^-,\,H_2O} PhC(=O){-}CH(OH)Ph \tag{8.21}$$

$$\frac{d[\text{benzoin}]}{dt} = k_{obs}[CN^-][PhCHO]^2.$$

The only reasonable mechanism which is compatible with this third-order rate law is shown in Fig. 8.11 with attack of the carbanion upon benzaldehyde as the rate-determining step (r.d.s.).

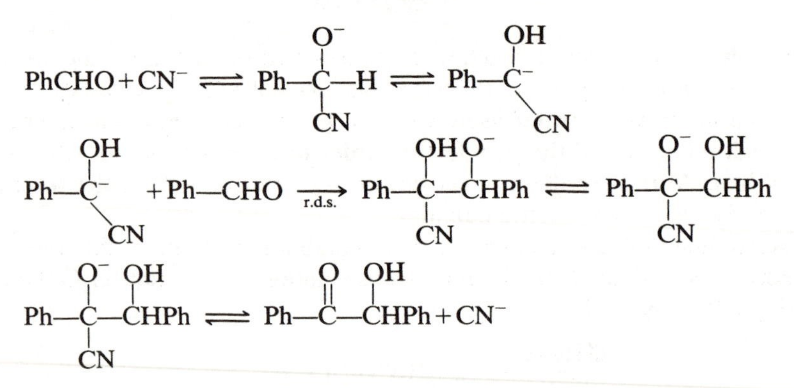

Fig. 8.11. A mechanism for the CN^--catalysed self-condensation of benzaldehyde (Benzoin condensation).

If either of the first two steps were unidirectional and rate determining, the rate law would be: rate $= k[PhCHO][CN^-]$, i.e. only first order in [PhCHO]. A rate-limiting final step is not ruled out by the kinetics but seems improbable on the basis of what is otherwise known of the reactions of cyanide with carbonyl compounds.

Clearly, the efficacy of the CN^- in this reaction depends upon its capacity to undergo reversible addition to carbonyl groups, and upon the ability of the cyano group as a substituent to stabilize the intermediate carbanion by delocalizing the negative charge.

The hydrolysis of simple primary alkyl chlorides is a displacement reaction which is catalysed by iodide, Fig. 8.12.

Solvent-induced reaction:

$$R{-}CH_2{-}Cl + 2H_2O \xrightarrow{H_2O} R{-}CH_2{-}OH + H_3O^+ + Cl^-$$

Nucleophile-catalysed reaction:

$$\text{(i)}\ R{-}CH_2{-}Cl + I^- \xrightarrow{H_2O} R{-}CH_2{-}I + Cl^-$$

$$\text{(ii)}\ R{-}CH_2{-}I + 2H_2O \xrightarrow{H_2O} R{-}CH_2{-}OH + H_3O^+ + I^-$$

Fig. 8.12. The uncatalysed and iodide-catalysed hydrolysis of primary alkyl chlorides.

In the solvent-induced process, water displaces chloride in a bimolecular reaction but only very slowly. Iodide, being a more powerful nucleophile than water, displaces chloride more effectively to give the alkyl

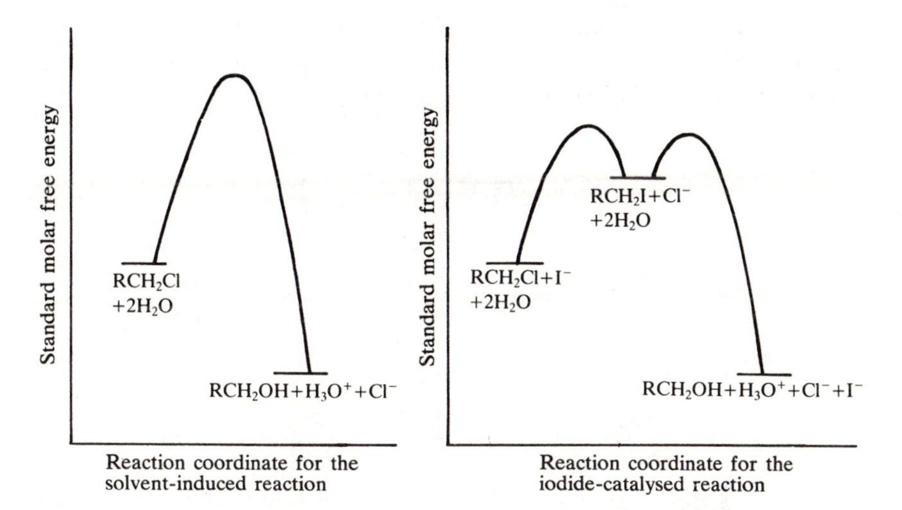

Fig. 8.13. Reaction profiles for the uncatalysed and iodide-catalysed hydrolysis of primary alkyl chlorides (see Fig. 8.12).

iodide in the first step of the catalysed reaction. The alkyl iodide in turn is readily hydrolysed by water as I^- is an easily-displaced leaving group. These mechanisms are represented by the reaction profiles in Fig. 8.13.

Three features of this simple example are important and relate to all cases in which the nucleophile-catalysed reaction is faster than its uncatalysed counterpart.[17]

1. The catalyst must be more nucleophilic than the reagent in the uncatalysed reaction or the solvent in the solvent-induced reaction (I^- is a better nucleophile than H_2O).
2. The catalyst must become a better leaving group in the intermediate compound than the original leaving group is in the initial substrate (I^- is a better leaving group than Cl^-).
3. The intermediate formed from the catalyst and the substrate must, *under the conditions of the reaction*, be thermodynamically less stable than the product (RCH_2I in water has a higher chemical potential than RCH_2OH in dilute aqueous acid). If this were not the case, the reaction would stop after the initial reaction of the substrate with the added nucleophile.

The hydrolysis of acetic anhydride is an example of an acyl transfer reaction which is nucleophile catalysed by pyridine[19] (eqn (8.22)).

$$CH_3\overset{\overset{O}{\|}}{C}{-}O{-}\overset{\overset{O}{\|}}{C}CH_3+H_2O \xrightarrow[\text{pyridine}]{H_2O,\ k_{obs}} 2CH_3CO_2H. \quad (8.22)$$

A mechanism which serves as a basis for kinetic predictions is shown. (We have not included tetrahedral intermediates so k_1, k_{-1}, and k_2 are really composite rate constants, and Ac = CH_3CO.)

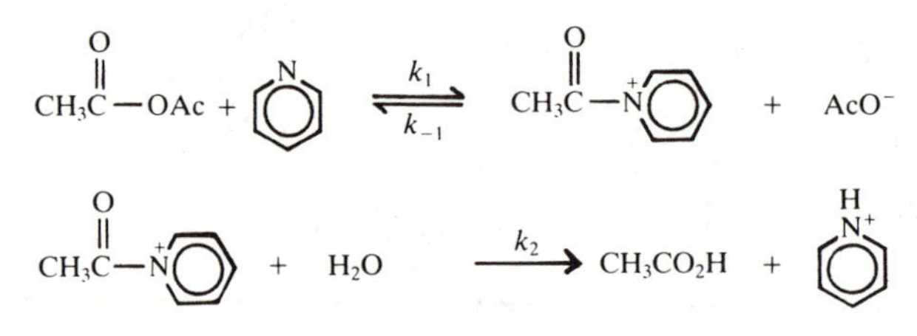

Rapidly-maintained equilibrium:

$$\text{PyrH}^+ + \text{AcO}^- \underset{}{\overset{H_2O}{\rightleftharpoons}} \text{Pyr} + \text{AcOH}$$

We can deduce, on the basis of this mechanism, the net rate of formation of the reactive intermediate (N-acetylpyridinium, $AcPyr^+$) in this catalysed reaction and apply the steady-state approximation.

$$\frac{d[AcPyr^+]}{dt} = k_1[Ac_2O][Pyr] - k_{-1}[AcPyr^+][AcO^-] - k_2[AcPyr^+][H_2O]$$

$$= 0 \text{ (steady-state approximation)},$$

therefore

$$[AcPyr^+] = \frac{k_1[Pyr][Ac_2O]}{k_{-1}[AcO^-] + k_2[H_2O]}.$$

Rate of formation of product via this mechanism

$$= k_2[AcPyr^+][H_2O]$$

$$= \frac{k_1 k_2[Pyr][Ac_2O][H_2O]}{k_{-1}[AcO^-] + k_2[H_2O]}. \tag{8.23}$$

If $k_{-1}[AcO^-] \gg k_2[H_2O]$, i.e. if the reactive intermediate returns to starting material much faster than it undergoes further reaction to give product, we have a pre-equilibrium, and eqn (8.23) may be simplified to give eqn (8.24).

$$\text{Rate of the pyridine-catalysed reaction} = \frac{k_1 k_2}{k_{-1}} \cdot \frac{[Pyr][H_2O][Ac_2O]}{[AcO^-]} \tag{8.24}$$

It was established experimentally[19] that the *pseudo* first-order rate constant, k_{obs}, is proportional to [Pyr] at constant $[AcO^-]$, as shown in Fig. 8.14, and that the rate of the pyridine-catalysed reaction is inversely proportional to $[AcO^-]$ as required by eqn (8.24). (Concurrent with the pyridine-catalysed reaction is the solvent-induced hydrolysis of acetic anhydride with first-order rate constant, k_0.)

Such congruence between experimental rate results and deductions

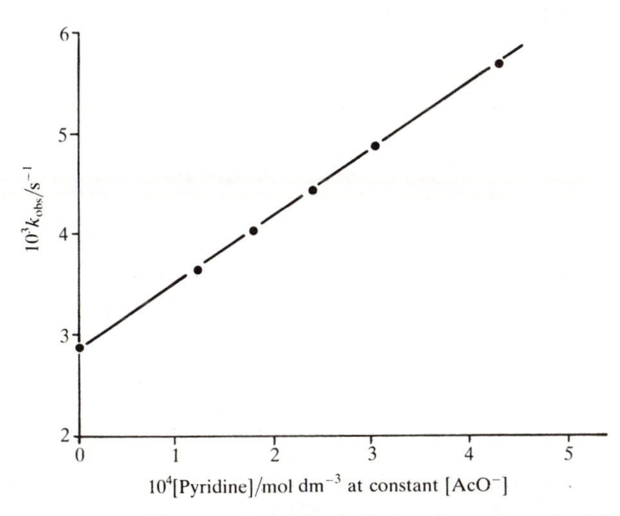

Fig. 8.14. Rate of pyridine-catalysed hydrolysis of acetic anhydride in water containing acetate buffer, 25 °C.[19]

based upon a hypothesis of how the reaction occurs adds credibility to the hypothesis. Moreover, the very short-lived N-acetylpyridinium cation (half-life *ca.* 0.1 s) has been detected spectroscopically during the reaction, and has been prepared under anhydrous conditions at low temperatures.

The catalytic effect of pyridine in reactions of this type is exploited in a well-established method of preparing esters from alcohols.[20]

8.6 Specific base catalysis[5,6,8]

The reaction of a compound S which has a rate law first order in [OH^-], but which does not include OH^- in the stoichiometric equation (eqn (8.25)) and which is not catalysed by any other bases, is said to show *specific base catalysis*.

$$\mathrm{S} \xrightarrow[k_{obs}]{OH^-,\ H_2O} \text{Products.} \tag{8.25}$$

If the reaction rate is first order in [S] and not equal to zero when [OH^-] = 0, the rate law at constant pH is given by eqns (8.9) and (8.26).

$$-\frac{d[S]}{dt} = k_{obs}[S] \tag{8.9}$$

where $k_{obs} = k_0 + k_{OH}[OH^-]$ (8.26)

= the experimental *pseudo* first-order rate constant,

k_0 = the first-order rate constant for the uncatalysed reaction, and

k_{OH} = the second-order catalytic constant for the reaction catalysed by OH^-.

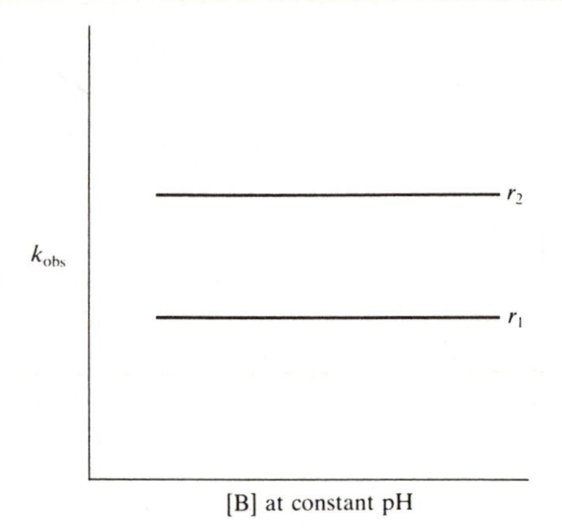

Fig. 8.15. Plot of k_{obs} against [B] at constant pH for a specific base-catalysed reaction.

As can be seen from eqn (8.26), a plot of k_{obs} against $[OH^-]$ in a specific base-catalysed reaction gives a straight line with gradient $= k_{OH}$ and intercept $= k_0$. This is exactly analogous to the determination of k_0 and k_H in a specific acid-catalysed reaction.

The experimental rate constant in a specific base-catalysed reaction depends only upon k_0, k_{OH}, and $[OH^-]$ (i.e. the pH) and not at all upon the source of OH^- or upon the concentration of any other base, B. Consequently, if k_{obs} is plotted against [B] for a series of reactions carried out at constant buffer ratio (constant pH), a graph such as Fig. 8.15 would be obtained if the reaction is specific base catalysed.

Figure 8.15 is analogous to Fig. 8.6 which was obtained for a specific acid-catalysed reaction, but whereas buffer ratio r_2 in Fig. 8.6 corresponds to the *lower* pH (higher $[H_3O^+]$) and *higher* rate constant, r_2 in Fig. 8.15 corresponds to the *higher* pH (higher $[OH^-]$) and *higher* rate constant.

Any reaction which involves rapid proton abstraction from the substrate followed by a slower step of the conjugate base of the substrate – either unimolecular or bimolecular but excluding proton transfer – will exhibit a specific base-catalysis rate law. (This is exactly analogous to the mechanisms in Fig. 8.7 and 8.9, pp. 324, 326 in which rapid and reversible protonation of a substrate followed by a slow reaction of the conjugate acid of the substrate leads to a specific acid-catalysis rate law.) The following mechanism for the self-condensation of acetaldehyde in dilute aqueous solution in the presence of hydroxide to give acetaldol (eqn (8.27)) is an example.

$$2CH_3CHO \xrightarrow[k_{obs}]{OH^-,\, H_2O} CH_3\overset{\overset{\displaystyle OH}{|}}{C}HCH_2CHO. \qquad (8.27)$$

Mechanism:

$$CH_3CHO + OH^- \overset{\text{fast}}{\rightleftharpoons} CH_2CHO^- + H_2O; \qquad K = \frac{[CH_2CHO^-]}{[CH_3CHO][OH^-]}$$

$$CH_3CHO + CH_2CHO^- \xrightarrow[\text{r.d.s.}]{k} CH_3\underset{}{\overset{O^-}{\overset{|}{C}}}HCH_2CHO \tag{8.28}$$

$$CH_3\overset{O^-}{\overset{|}{C}}HCH_2CHO + H_2O \overset{\text{fast}}{\rightleftharpoons} CH_3\overset{OH}{\overset{|}{C}}HCH_2CHO + OH^-.$$

$$\begin{aligned}\text{Rate of the catalysed reaction} &= k[CH_3CHO][CH_2CHO^-]\\ &= k[CH_3CHO]K[CH_3CHO][OH^-]\\ &= k_{OH}[OH^-][CH_3CHO]^2.\end{aligned} \tag{8.29}$$

This mechanism includes a rapid pre-equilibrium followed by a rate-determining bimolecular addition of the carbanion to the carbonyl group of a second acetaldehyde (eqn (8.28)). The rate law of eqn (8.29) may be confirmed by plotting Initial Rate/$[CH_3CHO]^2$ against $[OH^-]$. For different initial values of $[CH_3CHO]$, a common straight line passing through the origin is obtained whose gradient is k_{OH}, the third-order rate constant. One set of results is shown in Fig. 8.16; at constant pH and low initial concentrations of acetaldehyde, the rate constant is independent of the nature and concentration of other bases.[6,7]

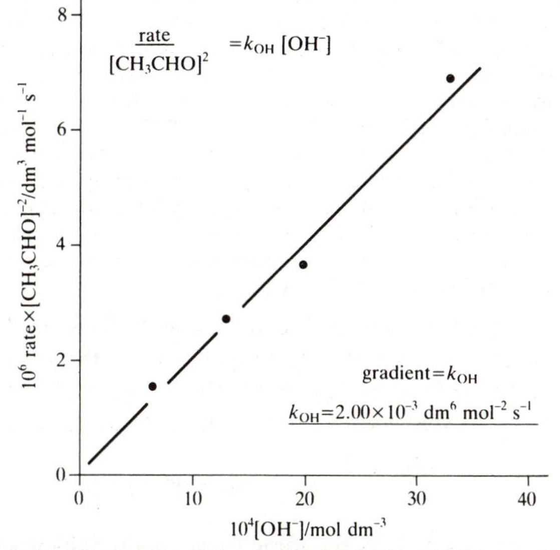

Fig. 8.16. Rate of self-condensation of acetaldehyde (Aldol reaction) in dilute aqueous solution, 25 °C. (Results taken from R. P. Bell and P. T. McTigue, *J. chem. Soc.* 2983 (1960). See also refs. 6 and 7.)

Also in accordance with this mechanism involving a pre-equilibrium, is the result that dilute acetaldehyde incorporates deuterium into its methyl group much faster than it undergoes the Aldol reaction in alkaline deuterium oxide.

Any reaction which involves nucleophilic interaction of OH^- with substrate in the rate-determining step, as opposed to proton abstraction, will also lead to a specific base-catalysis rate law. (Mechanistically, such a reaction is really nucleophile-induced but, because of the rate law, it is conveniently considered here.) There is strong evidence that the alkaline hydrolysis of simple esters involves nucleophilic attack of OH^- at the carbonyl;[10] this is illustrated for ethyl benzoate (eqn (8.30)) and a reaction profile is shown in Fig. 8.17.

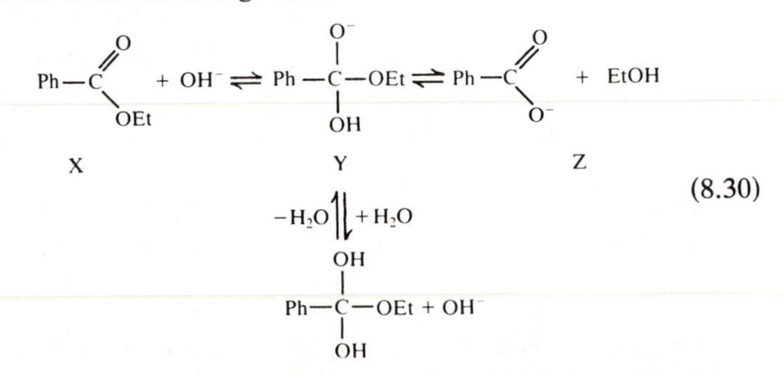

(8.30)

The thermodynamically-favourable proton transfer which accompanies the conversion of Y into Z will be very fast and concerted with the re-hybridization of the central carbon. Consequently, we do not include $PhCO_2H + EtO^-$ as an intermediate state between Y and Z. Comparison of the rate of exchange of ^{18}O from isotopically-labelled ester to solvent with the rate of the concurrent hydrolysis has established that, in this simple case, the hydrolysis is faster by a factor of 4.8 (25 °C).[11] So the elementary rate constant for the forward reaction from the intermediate is somewhat larger than that for the reverse. But, as in the specific acid-catalysed reaction (see p. 325), the factor is not large and depends upon the structure of the ester and the conditions. Although this isotope scrambling experiment implicates an uncharged tetrahedral intermediate (shown in eqn (8.30)), it is not included in Fig. 8.17.

Regardless of which activated complex is the less stable, that for the formation or the one for the decomposition of the negatively-charged tetrahedral intermediate in eqn (8.30) (Y in Fig. 8.17), this mechanism, together with a solvent-induced reaction, leads to the rate law:

$$-\frac{\mathrm{d[ester]}}{\mathrm{d}t} = (k_0 + k_{\mathrm{OH}}[\mathrm{OH}^-])[\mathrm{ester}].$$

In this rate expression, k_0 is the same (very small) first-order rate constant for the solvent-induced reaction that would be encountered in the specific acid-catalysed reaction under acidic reaction conditions.

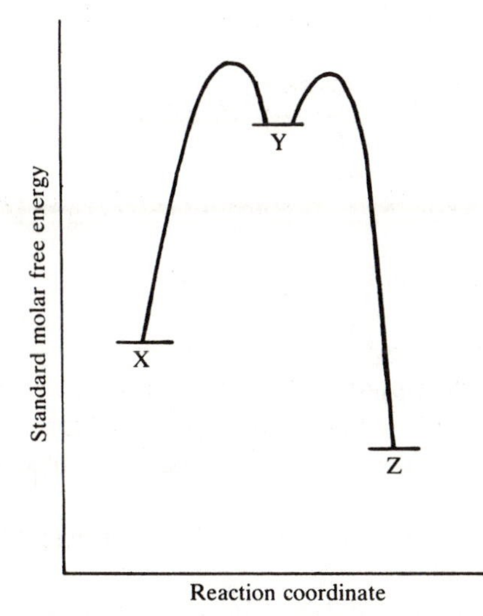

Fig. 8.17. Reaction profile for alkaline hydrolysis of ethyl benzoate (see eqn (8.30)).

The products of eqn (8.30) include benzoate, the conjugate base of an acid which is stronger than water ($pK_{AH} = 4.20$). There is, therefore, no subsequent step which regenerates OH^- because equilibrium in eqn (8.31) is far over to the left-hand side. Consequently, the overall hydrolysis, $X \rightarrow Z$, *consumes* hydroxide and is not an OH^--catalysed reaction; it is better described as a hydroxide-*induced* hydrolysis.

$$PhCO_2^- + H_2O \rightleftharpoons OH^- + PhCO_2H \tag{8.31}$$

Two important experimental consequences follow from the position of equilibrium in eqns (8.30) and (8.31). First, the alkaline hydrolysis of an ester is completely irreversible in contrast to the acid-catalysed process. Carboxylates do not react with alcohols, so alkaline conditions are usually preferred for the practical hydrolysis of simple esters.

Secondly, because OH^- is consumed, the rate of the reaction must be investigated either using second-order techniques or, using a *pseudo* first-order method, by having $[OH^-]$ buffered or very much larger than [ester] so that progress of the reaction does not cause a significant decrease in $[OH^-]$.

8.7 General base catalysis[5–7,17]

If a chemical reaction of compound S (eqn (8.32)) is catalysed not only by OH^- but also by any base B, the reaction will have a *general base-*

catalysis rate law. The experimental rate constant is given by eqn (8.33).

$$\mathrm{S} \xrightarrow[k_{\mathrm{obs}}]{\mathrm{B,\,OH^-,\,H_2O}} \text{Products} \tag{8.32}$$

$$-\frac{\mathrm{d[S]}}{\mathrm{d}t} = k_{\mathrm{obs}}[\mathrm{S}]^{\boldsymbol{n}} \quad \text{at constant } [\mathrm{OH^-}] \text{ and } [\mathrm{B}],$$

where $\boldsymbol{n}$ = reaction order with respect to [S],

$$k_{\mathrm{obs}} = k_0 + k_{\mathrm{OH}}[\mathrm{OH^-}] + k_{\mathrm{B}}[\mathrm{B}] \tag{8.33}$$

= the experimental *pseudo* $\boldsymbol{n}$-th order rate constant,
k_0 = the rate constant of the uncatalysed reaction,
k_{OH} = the catalytic constant of the hydroxide-catalysed reaction, and
k_{B} = the catalytic constant of the reaction catalysed by the general base, B.

As the reaction may be catalysed by any base, we may replace eqn (8.33) by eqn (8.34), the expanded rate law for a general base-catalysed reaction.

$$k_{\mathrm{obs}} = k_0 + k_{\mathrm{OH}}[\mathrm{OH^-}] + \sum k_{\mathrm{B}}[\mathrm{B}]. \tag{8.34}$$

General base-catalysed reactions can usually be distinguished from specific base-catalysed ones by investigating the relationship between k_{obs} and [B] at constant pH (the technique which is used to differentiate specific

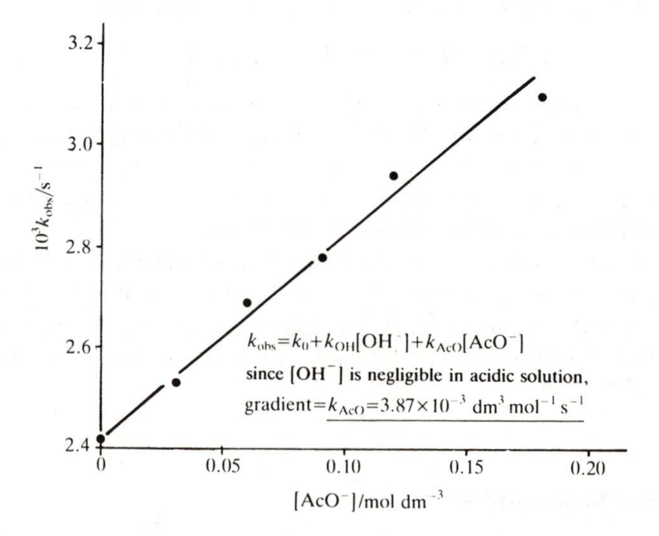

Fig. 8.18. Rate of the general base-catalysed hydrolysis of acetic anhydride in aqueous solution with acetic acid-sodium acetate buffers, 25 °C.[21]

and general acid catalysis). General base catalysis by a single general base B leads to a plot as in Fig. 8.18 (cf. Fig. 8.15 for a specific base-catalysed reaction). By eqn (8.33), the catalytic constant, k_B, is obtained from the gradient, and the intercept when $[B]=0$ gives $k_0+k_{OH}[OH^-]$. From results at two pHs, k_0 and k_{OH} may be obtained. Normally, however, they are better determined by investigating the rate constant dependence upon $[OH^-]$ in the absence of any general base.

We shall now consider two different mechanisms both of which lead to a general base-catalysis rate law.

8.7.1 Rate-determining proton abstraction from the substrate by any base (*Analogous to rate-determining protonation leading to general acid catalysis*, p. 329.)

If, in the Aldol reaction (eqn (8.35), see also p. 340), a base besides OH^- is present, we must consider two parallel reversible initial proton abstraction steps.

$$2CH_3CHO \xrightarrow{B,\ OH^-,\ H_2O} CH_3\overset{\displaystyle OH}{\overset{|}{C}}HCH_2CHO \qquad (8.35)$$

In our previous consideration of the Aldol reaction, we assumed that the initial proton abstraction is reversible and fast compared with the subsequent bimolecular step. We can now deal more comprehensively with this reaction by applying the steady-state approximation to the intermediate carbanion, and show that the specific base-catalysis rate law is found only

$$CH_3\text{—}CHO + OH^- \underset{k_{-1}}{\overset{k_1}{\rightleftharpoons}} CH_2CHO^- + H_2O$$

$$CH_3\text{—}CHO + B \underset{k_{-2}}{\overset{k_2}{\rightleftharpoons}} CH_2CHO^- + \overset{+}{B}H$$

$$CH_3CHO + CH_2CHO^- \xrightarrow{k_3} CH_3\overset{\displaystyle O^-}{\overset{|}{C}}HCH_2CHO \qquad (8.28)$$

Rapidly maintained equilibria:

$$CH_3\overset{\displaystyle O^-}{\overset{|}{C}}HCH_2CHO + H_2O \rightleftharpoons CH_3\overset{\displaystyle OH}{\overset{|}{C}}HCH_2CHO + OH^-$$

$$B + H_2O \rightleftharpoons \overset{+}{B}H + OH^-$$

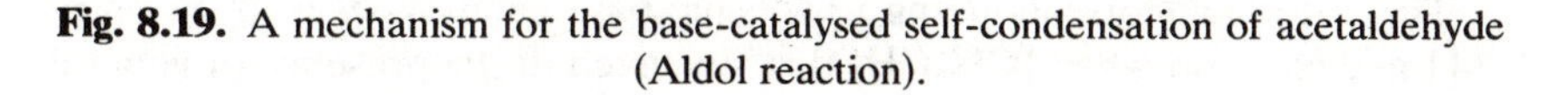

Fig. 8.19. A mechanism for the base-catalysed self-condensation of acetaldehyde (Aldol reaction).

under particular experimental conditions. Under other conditions, a general base-catalysis rate law is obtained.

$$\frac{d[CH_2CHO^-]}{dt} = k_1[CH_3CHO][OH^-] + k_2[CH_3CHO][B]$$
$$-(k_{-1}[CH_2CHO^-][H_2O] + k_{-2}[CH_2CHO^-][BH^+]$$
$$+ k_3[CH_2CHO^-][CH_3CHO])$$
$$= 0 \text{ (steady-state approximation)},$$

therefore

$$[CH_2CHO^-] = [CH_3CHO]\left\{\frac{k_1[OH^-] + k_2[B]}{k_{-1}[H_2O] + k_{-2}[BH^+] + k_3[CH_3CHO]}\right\},$$

and

$$\frac{d[aldol]}{dt} = k_3[CH_3CHO][CH_2CHO^-]$$
$$= k_3[CH_3CHO]^2\left\{\frac{k_1[OH^-] + k_2[B]}{k_{-1}[H_2O] + k_{-2}[BH^+] + k_3[CH_3CHO]}\right\}. \tag{8.36}$$

This more complete analysis shows that a simple overall rate law will not ensue except when certain inequalities obtain.

(i) When $k_{-1}[H_2O] + k_{-2}[BH^+] \gg k_3[CH_3CHO]$, i.e. when the carbanion is re-protonated faster than it adds to the carbonyl, eqn (8.36) becomes

$$\frac{d[aldol]}{dt} = k_3[CH_3CHO]^2\left\{\frac{k_1[OH^-] + k_2[B]}{k_{-1}[H_2O] + k_{-2}[BH^+]}\right\}.$$

But as the two initial steps are pre-equilibria by this mechanism,

$$\frac{[CH_2CHO^-]}{[CH_3CHO]} = \frac{k_1[OH^-]}{k_{-1}[H_2O]} = \frac{k_2[B]}{k_{-2}[BH^+]},$$

therefore

$$k_2[B] = \frac{k_1[OH^-] . k_{-2}[BH^+]}{k_{-1}[H_2O]},$$

and so

$$\frac{d[aldol]}{dt} = k_3[CH_3CHO]^2\left\{\frac{k_1[OH^-] + k_1[OH^-]k_{-2}[BH^+]/k_{-1}[H_2O]}{k_{-1}[H_2O] + k_{-2}[BH^+]}\right\}$$
$$= \frac{k_3[CH_3CHO]^2 k_1[OH^-]}{k_{-1}[H_2O]}$$
$$= k_{OH}[OH^-][CH_3CHO]^2. \tag{8.29}$$

This rate law, corresponding to specific base catalysis, was given on p. 341 and is found when $[CH_3CHO]$ is low, even in the presence of general bases.[6,7] The mechanism involves rate-determining attack of the carbanion from one aldehyde at the carbonyl of the second aldehyde with elementary rate constant k on p. 341, identical with k_3 above. In general, any reaction involving a rate-determining step not involving proton transfer of a deprotonated substrate generated in pre-equilibria will lead to a specific base-catalysis rate law even if general bases are directly involved in the initial pre-equilibria.

(ii) When $k_{-1}[H_2O]+k_{-2}[BH^+] \ll k_3[CH_3CHO]$, i.e. when generation of the carbanion is essentially irreversible and rate determining, eqn (8.36) becomes:

$$\frac{d[\text{aldol}]}{dt} = [CH_3CHO](k_1[OH^-]+k_2[B])$$

$$= k_{obs}[CH_3CHO],$$

where

$$k_{obs} = k_1[OH^-]+k_2[B], \tag{8.37}$$

which is the general base-catalysis rate law (eqn (8.33)) with $k_0 = 0$.

It has been confirmed experimentally[6,7] that at high concentrations of acetaldehyde (*ca.* 10 M), the rate of acetaldol formation is first order in $[CH_3CHO]$ and the reaction is general base catalysed according to eqn (8.37). Furthermore, under these conditions the reaction in deuterium oxide does not lead to any incorporation of deuterium into the methyl group of the aldol. This confirms that at high $[CH_3CHO]$ the carbanion in Fig. 8.19 reacts with another molecule of acetaldehyde faster than it is re-protonated.

8.7.2 Rate-determining base-assisted nucleophilic attack upon the substrate (*Analogous to rate-determining acid-assisted nucleophilic attack leading to general acid catalysis,* p. 333)

The hydrolysis of acetic anhydride is catalysed by acetate according to the general base-catalysis rate law:

$$-\frac{d[Ac_2O]}{dt} = k_0[Ac_2O]+k_{AcO}[AcO^-][Ac_2O].$$

This reaction was investigated in solution buffered with AcOH.[21] Consequently, any $k_{OH}[OH^-]$ term could be neglected and there was no significant acid catalysis, see Fig. 8.18. There cannot be nucleophile catalysis via a reactive intermediate (as in the pyridine-catalysed reaction, p. 337), because direct attack by acetate at the carbonyl of acetic anhydride will give a tetrahedral intermediate capable only of regenerating starting material. There must be some other mode of catalysis. The

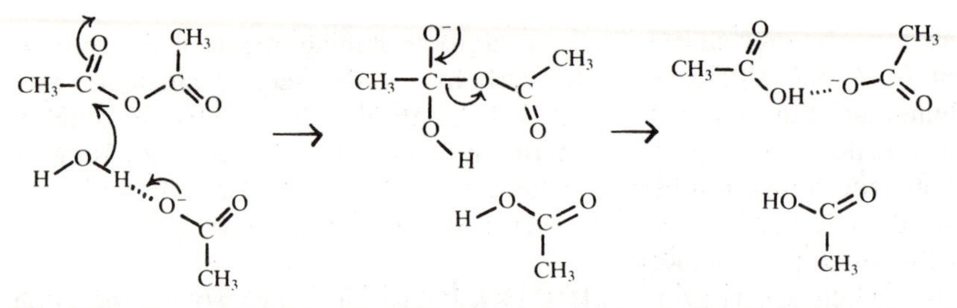

Fig. 8.20. Base-assisted nucleophilic attack by water in the hydrolysis of acetic anhydride – an example of mechanistic general base catalysis.

current view is that acetate enhances the nucleophilicity of water by accepting a proton (Fig. 8.20).[6,17,22]

Most simply, this reaction may be viewed as the substrate suffering nucleophilic attack by a water molecule hydrogen-bonded to the general base, as opposed to attack by water hydrogen-bonded to another water molecule in the solvent-induced reaction. This description allows us to see the solvent as a general base in the solvent-induced reaction which is complementary to its role as a general acid as described earlier (p. 334). The mechanism involving rate-determining base-assisted nucleophilic attack upon the substrate is sometimes referred to as general base catalysis. However, to avoid confusion with *kinetic* general base catalysis (which relates to a type of experimental rate law and is mechanistically ambiguous) we shall call the mechanism exemplified in Fig. 8.20 *mechanistic general base catalysis.* It corresponds to a single term in the kinetic general base-catalysis rate law. This distinction is exactly analogous to that between *kinetic* and *mechanistic* general acid catalysis (p. 334).

The hydrolysis of phenyl acetate is also catalysed by acetate,[23] eqn (8.38), and a credible mechanism can be drawn which involves acetate as a nucleophile catalyst and acetic anhydride as the reactive intermediate.

$$CH_3\overset{O}{\overset{\|}{C}}-O-C_6H_5 + H_2O \xrightarrow[H_2O]{Na^+\bar{O}Ac} CH_3CO_2H + HO-C_6H_5 \qquad (8.38)$$

$$k_{obs} = k_0 + k_{AcO}[AcO^-]$$

However, it was established that when the reaction is carried out in the presence of aniline, which is known to react with even a low concentration of aqueous acetic anhydride to give acetanilide, no acetanilide was detected. This careful product analysis excludes nucleophile catalysis, and effectively establishes that the reaction is mechanistic general base catalysed.

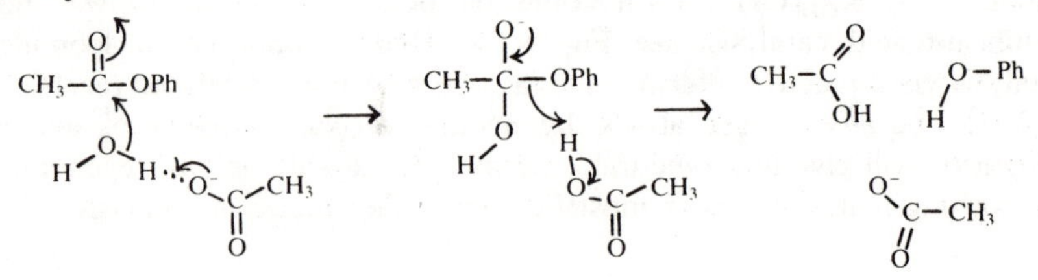

8.7.3 Nucleophile catalysis and mechanistic general base catalysis[22]

We have seen that at least three mechanisms (nucleophile catalysis, a mechanism with rate-determining proton abstraction, and mechanistic general base catalysis) can lead to the same rate law. When a general base-catalysis rate law is established for a reaction, a next step is to consider mechanism, and the two mechanisms usually most difficult to differentiate are nucleophile catalysis and mechanistic general base catalysis.

We have already encountered one approach: to attempt to detect spectroscopically or trap chemically the reactive intermediate, which, if successful, establishes the reaction as nucleophile catalysed. The hydrolysis of *p*-nitrophenyl acetate in neutral or slightly acidic aqueous solution is catalysed by imidazole (Im) and follows a general base-catalysis rate law:[24]

$$CH_3CO_2-C_6H_4-NO_2 \quad + \quad H_2O \xrightarrow[\text{Im}]{H_2O} CH_3CO_2H + HO-C_6H_4-NO_2$$

$$k_{obs} = k_0 + k_{Im}[\text{Im}].$$

When the reaction was carried out in the presence of aniline, acetanilide was formed which is strong evidence for the intermediacy of N-acetylimidazolium cation. Conclusive evidence was provided by the spectroscopic detection of the deprotonated intermediate, N-acetylimidazole, a known compound. At high concentration of imidazole, N-acetylimidazole was observed first to accumulate and then to decrease in concentration as N-acetylimidazolium is hydrolysed in a later product-forming step of the reaction. The mechanism is, therefore, nucleophile catalysis (Fig. 8.21).

A second general method is to investigate the solvent deuterium kinetic isotope effect,[25] a topic covered in more detail in Chapter 9. In mechanistic general base catalysis, e.g. Fig. 8.20, a water molecule suffers O–H bond weakening in the formation of the activated complex. Consequently, a substantial solvent kinetic isotope effect may be anticipated in deuterium oxide because O–D bonds are more difficult to weaken and break than O–H bonds. Indeed, in the mechanistic general base-catalysed hydrolysis of ethyl difluoroacetate by acetate, $k_{AcO}^{H_2O}/k_{AcO}^{D_2O} = 2.7$ (25 °C), and commonly results of $k_B^{H_2O}/k_B^{D_2O} \sim 2$–$4$ are obtained.[22]

In nucleophile catalysis in aqueous solution, the water is not directly involved in the formation of the reactive intermediate, so if this step were rate determining the rate should be much the same in D_2O as in H_2O. Even when the hydrolytic decomposition of the reactive intermediate is rate determining (rather than its formation), the solvent deuterium kinetic isotope effect tends to be smaller than when the reaction is mechanistic general base catalysed. For example, $k_{Im}^{H_2O}/k_{Im}^{D_2O} = 1.0$ for the imidazole-catalysed hydrolysis of *p*-nitrophenyl acetate, and $k_{HCO_2}^{H_2O}/k_{HCO_2}^{D_2O} = 1.07$ for

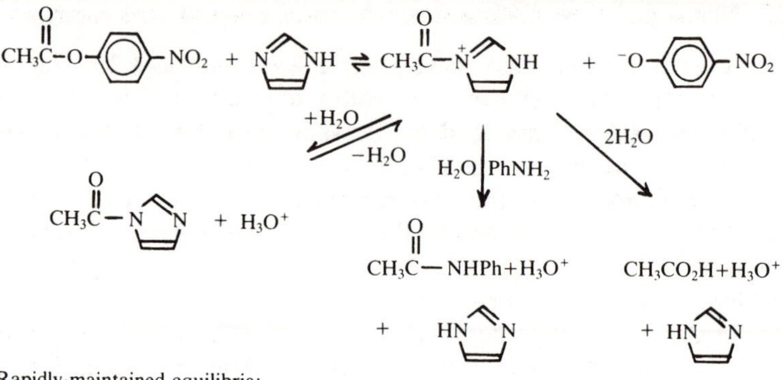

Rapidly-maintained equilibria:

Fig. 8.21. Nucleophile-catalysed hydrolysis of *p*-nitrophenyl acetate and interception of the reactive intermediate.[24]

the formate-catalysed hydrolysis of acetic anhydride. However, the result $k_{\mathrm{Im}}^{\mathrm{H_2O}}/k_{\mathrm{Im}}^{\mathrm{D_2O}} = 1.5$ (25 °C) for the nucleophile-catalysed hydrolysis of phenyl acetate with imidazole shows that the solvent deuterium kinetic isotope effect may be substantially greater than unity and values as high as 1.9 have been reported.

Since values of $k_{\mathrm{B}}^{\mathrm{H_2O}}/k_{\mathrm{B}}^{\mathrm{D_2O}} = 1.7\text{–}1.9$ are known for both nucleophile-catalysed and mechanistic general base-catalysed reactions, results in this range cannot alone lead to an assignment of mechanism. A further method for investigating the nucleophile catalysis – mechanistic general base catalysis ambiguity is based upon the Brönsted relationship and discussed in Chapter 10, p. 423.

8.8 Acid–base catalysis and pH-rate constant profiles

8.8.1 Specific acid–base-catalysed reactions[8]

We saw on p. 321 that the observed *pseudo* first-order rate constant for the hydrolysis of a simple ester in aqueous acidic solution is given by

$$k_{\mathrm{obs}} = k_0 + k_{\mathrm{H}}[\mathrm{H_3O^+}], \qquad (8.7)$$

and, for reaction in alkaline solution, the corresponding equation is

$$k_{\mathrm{obs}} = k_0 + k_{\mathrm{OH}}[\mathrm{OH^-}]. \qquad (8.26)$$

The *complete* rate law for this and any other reaction which is subject to both specific acid and specific base catalysis, based upon results of experiments carried out under quite different sets of conditions, is eqn (8.39).

$$k_{obs} = k_0 + k_H[H_3O^+] + k_{OH}[OH^-], \tag{8.39}$$

Equations (8.7) and (8.26) are approximations which are valid when either $[OH^-]$ or $[H_3O^+]$ is negligible.

Because $[H_3O^+]$ and $[OH^-]$ are not independent variables but related by

$[H_3O^+][OH^-] = K_{AP}$ (the autoprotolysis constant of water, see p. 163),

we may re-write eqn (8.39) as eqn (8.40):

$$k_{obs} = k_0 + k_H[H_3O^+] + k_{OH} \cdot \frac{K_{AP}}{[H_3O^+]}. \tag{8.40}$$

Consequently, by correlating the experimental results of k_{obs} with pH (the single parameter which describes $[H_3O^+]$ and $[OH^-]$ in dilute aqueous solution), we may express the same information that previously would have been included in two graphs more economically in a single graph. An enormous range in $[H_3O^+]$ (*ca.* 10^{14}) can also be included very conveniently by using the logarithmic pH scale in the *x*-axis. The *y*-axis may be either k_{obs} or, preferably, $\log(k_{obs})$.

The log–log correlation of a specific acid–base-catalysed reaction has the advantage that the plots obtained have linear sections and the gradients of the slopes in the high and low pH ranges indicate the order of the reaction with respect to $[OH^-]$ (gradient = 1 if first order in $[OH^-]$) and $[H_3O^+]$ (gradient = −1 if first order in $[H_3O^+]$), respectively.

We can demonstrate the interpretation of a plot of $\log(k_{obs})$ against pH by considering a hypothetical reaction for which numerically $k_H = k_{OH}$ and $k_0 \sim 0$.

(i) When $pH < 7$, $k_H[H_3O^+] \gg k_{OH}[OH^-]$

The first and third terms on the right-hand side of eqn (8.39) do not appreciably contribute to k_{obs} and a plot of $\log k_{obs}$ *vs.* pH is given by the line a–b in Fig. 8.22. Extrapolation along b–c to high pH shows the ever-decreasing contribution from the specific acid-catalysed mechanism in the increasingly alkaline medium.

(ii) When $pH > 7$, $k_{OH}[OH^-] \gg k_H[H_3O^+]$

In this case, only the third term on the right of eqn (8.39) is significant, and the line d–e represents the experimental correlation of $\log k_{obs}$ with pH. Extrapolation to low pH gives e–f.

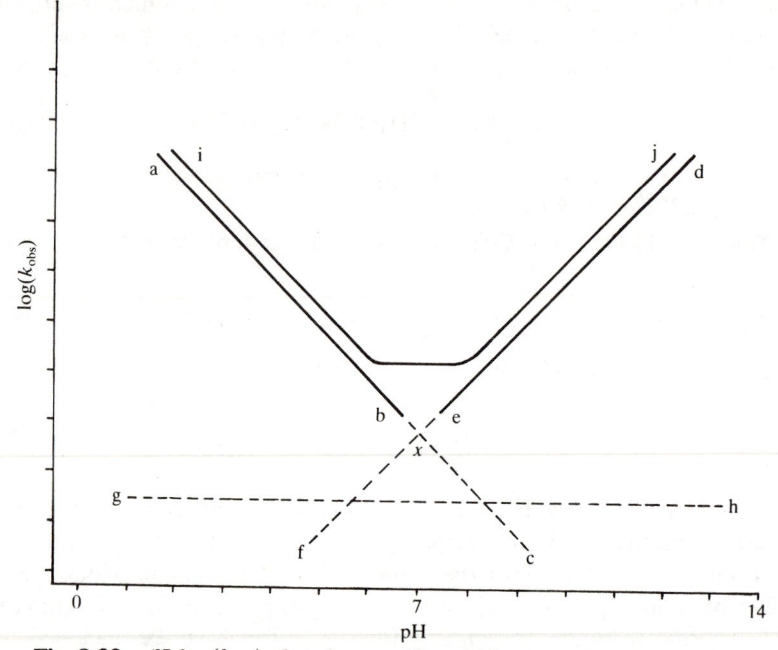

Fig. 8.22. pH-log(k_{obs}) plots for specific acid-base-catalysed reactions.

The point at which $k_H[H_3O^+] = k_{OH}[OH^-]$ is given by the intersection of the extended lines a–c and d–f. This point corresponds to the neutral point on the pH scale, pH=7 (25 °C) where $[H_3O^+]=[OH^-]$ since $k_{OH}=k_H$. Because $k_0 \sim 0$, i.e. there is no uncatalysed reaction, the minimum in log(k_{obs}) at x is quite sharp.

If $k_H > k_{OH}$, line a–c is displaced vertically upwards, and x moves to a higher pH. Correspondingly, if $k_{OH} > k_H$, line d–f is displaced upwards and x, the minimum in the profile, moves to a lower pH.

When the overall specific acid–base-catalysed reaction includes a significant uncatalysed mode ($k_0 \neq 0$), there is a pH-independent contribution to k_{obs}. (Such a reaction alone would correspond to g–h in Fig. 8.22.) Normally, such a contribution raises and, therefore, flattens the minimum in the log(k_{obs}) *vs.* pH profile but no major perturbation is noticed at high and low pH extremes. Line i–j represents a typical plot of log(k_{obs}) *vs.* pH for a reaction with an appreciable k_0 contribution and $k_H \sim k_{OH}$. The positions of the intersections of the linear portions which together constitute the overall pH-log(k_{obs}) profile of a specific acid–base-catalysed reaction, therefore, depend upon the relative magnitudes of the three rate constants k_0, k_H, and k_{OH}.

Results for the hydrolysis of ethyl acetate are shown in Fig. 8.23. In this, the minimum is at pH 5.5 because $k_{OH} > k_H$, and there is no appreciable horizontal section because the uncatalysed reaction is very slow.[10,26]

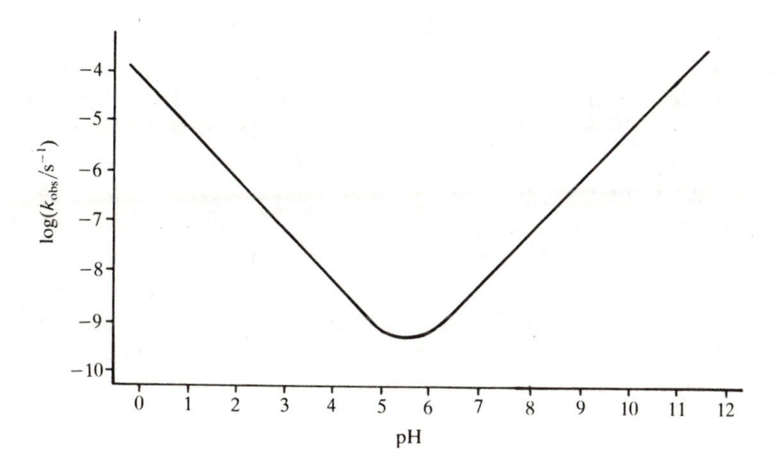

Fig. 8.23. pH-log(k_{obs}) profile for the hydrolysis of ethyl acetate in water, 25 °C.[26]

8.8.2 General acid–base-catalysed reactions[6,15]

It is not possible simply to plot log(k_{obs}) against pH for a general acid–base-catalysed reaction from results obtained in buffered solutions because, by definition, the observed rate constants depend not only upon the buffer *ratio*, *r* (i.e. the pH), but also upon the buffer *concentration.*

The enolization of acetone and other simple ketones is general acid catalysed (p. 331). It has also been shown to be general base catalysed, and the complete rate law of the forward reaction in aqueous solution buffered by general acid AH and base A^- is given in Fig. 8.24.

The investigation of a reaction in terms of a rate expression of this complexity is facilitated by making rate measurements under experimental conditions such that some of the components in eqn (8.41) are negligibly small. In this way it is possible to build up the complete equation a part at a time. For example, the reaction could be investigated

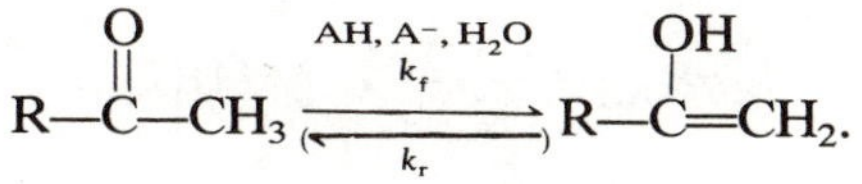

$$\mathrm{R{-}\overset{\overset{\displaystyle O}{\|}}{C}{-}CH_3} \underset{k_r}{\overset{\mathrm{AH,\,A^-,\,H_2O}}{\underset{(\longleftarrow)}{\xrightarrow{k_f}}}} \mathrm{R{-}\overset{\overset{\displaystyle OH}{|}}{C}{=}CH_2}.$$

Rate of forward reaction $= k_f[\mathrm{RCOCH_3}]$, where

k_f = the *pseudo* first-order rate constant for the forward reaction,

$$= k_0 + k_H[\mathrm{H_3O^+}] + k_{OH}[\mathrm{OH^-}] + k_{AH}[\mathrm{AH}] + k_A[\mathrm{A^-}]. \qquad (8.41)$$

Fig. 8.24. General acid–base-catalysed rate law for the enolization of a simple ketone.

first in the presence of only a strong acid, and then only a completely dissociated hydroxide. From these results k_0, k_H, and k_{OH} can be determined as described earlier under specific catalysis, and k_f at any pH, in the absence of general acids and bases, can be calculated. If a reaction has to be investigated under buffered conditions, k_f should be obtained for several buffer concentrations at each pH, then the value of k_f at this pH but zero buffer concentration may be obtained by extrapolation. This would be necessary, for example, if k_{OH} and k_H are very high, causing the OH^--and H_3O^+-catalysed reactions to be too fast for practical measurement at high and low pH. Experiments then have to be carried out not far from neutrality where buffering is necessary to control and maintain a constant pH, especially if the reaction consumes or produces H_3O^+ or OH^-.

The general acid and base terms of eqn (8.41) can be made more amenable to analysis if an equilibrium relationship can be exploited. For example, in the case of Fig. 8.24 the general base is the conjugate base of the general acid, therefore:

$$AH + H_2O \rightleftharpoons A^- + H_3O^+; \qquad K_{AH} = \frac{[A^-][H_3O^+]}{[AH]},$$

consequently, $[H_3O^+] = K_{AH}r$ where $r = [AH]/[A^-]$, the buffer ratio, and K_{AH} = the acidity constant of the catalytic acid. And since $[H_3O^+][OH^-] = K_{AP}$,

$$[OH^-] = \frac{K_{AP}}{K_{AH}r}.$$

Substitutions for $[H_3O^+]$, $[OH^-]$, and $[A^-]$ can now be made in eqn (8.41) to give an equation containing only a single concentration variable, [AH], at constant buffer ratio:

$$k_f = k_0 + k_H r K_{AH} + \frac{k_{OH} K_{AP}}{K_{AH} r} + k_{AH}[AH] + \frac{k_A}{r}[AH]$$

$$= k_0 + k_H r K_{AH} + \frac{k_{OH} K_{AP}}{K_{AH} r} + [AH]\left\{k_{AH} + \frac{k_A}{r}\right\}. \qquad (8.42)$$

The first three terms of the right-hand side of eqn (8.42) are constant at a given buffer ratio and constant ionic strength, therefore k_f may be plotted against [AH] at constant r and the gradient gives $k_{AH} + k_A/r$. From two such sets of results at different values of r (different pHs), the separate rate constants k_{AH} and k_A can be calculated,[6] and all the rate constants have, therefore, been determined. A value of k_f can now be calculated at any pH and known concentrations of AH and A^-.

Comparison of the plots described here for a reaction which is general acid *and* general base catalysed with the analogous plots for a reaction

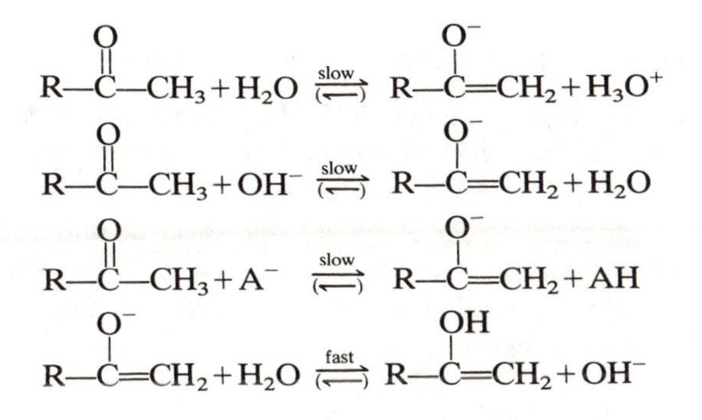

Fig. 8.25. A mechanism for the general base-catalysed enolization of a simple ketone.

which is *either* general acid *or* general base catalysed is helpful. Sets of results of k_{obs} versus [AH] for a general acid-catalysed reaction, obtained at different r-values and plotted using eqn (8.10) (p. 328), lead to parallel lines all of the same gradient, k_{AH}, with intercepts, $k_0 + k_H[H_3O^+]$. Analogously, results of k_{obs} versus $[A^-]$ for a reaction catalysed by general base A^- plotted using eqn (8.33) (p. 344) give parallel lines with gradient k_A and intercepts $k_0 + k_{OH}[OH^-]$. If a reaction is catalysed by *both* components of a buffer system AH and A^-, however, the gradient of a plot of k_{obs} against [AH] using eqn (8.42) includes r, so results obtained under different pH conditions (different r values) lead to straight lines not only with different intercepts, but also with different gradients, so the lines are not parallel.

A mechanism giving rise to the general acid-catalysis rate law for the enolization of carbonyl compounds was given earlier (p. 331). A mechanism leading to the kinetic general base catalysis is shown in Fig. 8.25 and involves rate-determining proton abstraction.

8.8.3 pH-rate constant profiles and changes in the rate-determining step

In the two preceding sections, the effect of pH upon the observed rate constant has been seen to be due to changes in the concentrations of four catalytic species H_3O^+, OH^-, HA, and A^- which, together with the solvent, account for up to five *concurrent* routes from reactants to products. Interesting and varied pH-log(rate constant) profiles also arise for complex reactions comprising *consecutive* steps if, for example, one of the reactants (or intermediates) in an uncatalysed step has weak acid–base properties, and another of the steps is acid–base catalysed. There is, of course, no reason why different steps of a sequential process should not be catalysed by different chemical species. But if a step is reversible, a catalyst for the forward direction will also catalyse the reverse.

8.8.3.1 *Reaction involving a weakly acidic compound*

Equation (8.43) involves a reactant YH which is weakly acidic (but more acidic than the solvent) and a second forward step, via a reactive intermediate C which is specific base catalysed. We shall use this simply to illustrate an approach to investigating the effect of buffered pH upon both the rate law and an observed overall rate constant for sequential complex reactions in general. One of many possible alternatives could equally well have been chosen.[5,17]

$$\mathrm{YH} + \mathrm{X} \underset{k_{-1}}{\overset{k_1}{\rightleftharpoons}} \mathrm{C} \underset{\mathrm{OH^-}}{\overset{k_2}{\longrightarrow}} \mathrm{D} \tag{8.43}$$

$$\mathrm{YH} + \mathrm{H_2O} \rightleftharpoons \mathrm{Y^-} + \mathrm{H_3O^+}; \qquad K_{\mathrm{AH}}(\mathrm{YH}) = \frac{[\mathrm{Y^-}][\mathrm{H_3O^+}]}{[\mathrm{YH}]},$$

the acidity constant of YH.

$$\frac{\mathrm{d[C]}}{\mathrm{d}t} = k_1[\mathrm{YH}][\mathrm{X}] - k_{-1}[\mathrm{C}] - k_2[\mathrm{C}][\mathrm{OH^-}]$$

$$= 0 \quad \text{(steady-state approximation)},$$

therefore

$$[\mathrm{C}] = \frac{k_1[\mathrm{YH}][\mathrm{X}]}{k_{-1} + k_2[\mathrm{OH^-}]}.$$

$$\text{Rate of formation of product} = \frac{\mathrm{d[D]}}{\mathrm{d}t}$$

$$= k_2[\mathrm{OH^-}][\mathrm{C}]$$

$$= \frac{k_2[\mathrm{OH^-}]k_1[\mathrm{YH}][\mathrm{X}]}{k_{-1} + k_2[\mathrm{OH^-}]},$$

or

$$\frac{\mathrm{d[D]}}{\mathrm{d}t} = k_{\mathrm{obs}}[\mathrm{YH}][\mathrm{X}] \text{ at constant pH},$$

where k_{obs} = the observed second-order rate constant based upon concentration variables [YH] and [X]

$$= \frac{k_2 k_1[\mathrm{OH^-}]}{k_{-1} + k_2[\mathrm{OH^-}]}, \quad \text{and} \tag{8.44}$$

[YH] = the concentration of the *un-dissociated* weakly acidic reactant. If $[\mathrm{YH}]_s$ = the total stoichiometric concentration of the weakly acidic reactant and its conjugate base,

$$[\mathrm{YH}]_s = [\mathrm{YH}] + [\mathrm{Y^-}], \tag{8.45}$$

and [YH] can be calculated from $[\mathrm{YH}]_s$ and its acidity constant if it is not directly measurable. Either way, the above rate law may be handled as described in the previous chapter (p. 290). It gives rise to a fairly simple relationship between k_{obs} and $[OH^-]$, i.e. pH. Rearrangement of eqn (8.44) yields

$$\frac{1}{k_{obs}} = \frac{k_{-1}}{k_1 k_2 [OH^-]} + \frac{1}{k_1},$$

so, by measuring k_{obs} at different pHs, a linear plot of $1/k_{obs}$ against $1/[OH^-]$ gives k_1 from the intercept and, knowing k_1, k_{-1}/k_2 from the gradient.

If the kinetics of the reaction of eqn (8.43) are investigated using

$$\frac{d[D]}{dt} = k'_{obs}[YH]_s[X] \quad \text{at constant pH,}$$

an alternative second-order rate constant is obtained which is based upon concentration variables $[YH]_s$ and $[X]$. k_{obs} and k'_{obs} are different except under acidic conditions when $[Y^-] = 0$ (so $[YH] = [YH]_s$), and the difference becomes greater with increasing pH as YH becomes more extensively deprotonated. The relationship between k'_{obs} and pH may also be rigorously established although it involves some fairly cumbersome equations.

$$\begin{aligned} [YH]_s &= [YH] + [Y^-] \\ &= [YH] + \frac{[YH]}{[H_3O^+]} \cdot K_{AH}(YH) \\ &= [YH](1 + K_{AH}(YH)/[H_3O^+]) \end{aligned}$$

therefore

$$[YH] = \frac{[YH]_s}{1 + K_{AH}(YH)/[H_3O^+]}$$

and

$$\begin{aligned} \frac{d[D]}{dt} &= \frac{k_1 k_2 [OH^-][X]}{(k_{-1} + k_2[OH^-])} \left\{ \frac{[YH]_s}{1 + K_{AH}(YH)/[H_3O^+]} \right\} \\ &= \frac{k_1 k_2 [X][YH]_s[OH^-][H_3O^+]}{(k_{-1} + k_2[OH^-])([H_3O^+] + K_{AH}(YH))} \\ &= \frac{k_1 k_2 K_{AP}[X][YH]_s}{(k_{-1}[H_3O^+] + k_{-1}K_{AH}(YH) + k_2 K_{AP} + k_2 K_{AH}(YH)[OH^-])}, \end{aligned}$$

where $K_{AP} = [H_3O^+][OH^-]$, or

$$\frac{d[D]}{dt} = k'_{obs}[X][YH]_s \quad \text{at constant pH,} \qquad (8.46)$$

where k'_{obs} = this alternative experimental second-order rate constant

$$= \frac{k_1 k_2 K_{AP}}{(k_{-1}[H_3O^+] + k_{-1}K_{AH}(YH) + k_2 K_{AP} + k_2 K_{AH}(YH)[OH^-])}. \tag{8.47}$$

Equation (8.47) shows that, as long as k_{-1} and $k_2 K_{AH}(YH)$ are broadly comparable, at high $[H_3O^+]$,

$$k'_{obs} \approx \frac{k_1 k_2 K_{AP}}{(k_{-1}[H_3O^+] + k_{-1}K_{AH}(YH) + k_2 K_{AP})}$$

and this decreases as $[H_3O^+]$ increases; and, at high $[OH^-]$,

$$k'_{obs} \approx \frac{k_1 k_2 K_{AP}}{(k_{-1}K_{AH}(YH) + k_2 K_{AP} + k_2 K_{AH}(YH)[OH^-])}$$

and this decreases as $[OH^-]$ increases. The experimental rate constant, k'_{obs}, will be maximal when $k_{-1}[H_3O^+] = k_2 K_{AH}(YH)[OH^-]$. Consequently, a plot of $\log(k'_{obs})$ against pH will show a maximum as illustrated in Fig. 8.26. The chemical interpretation of this dome-shaped $\log(k'_{obs})$-pH profile is as follows. At low pH (high $[H_3O^+]$) very little of the YH will be dissociated and the rate of the forward first step will be high. But because $[OH^-]$ is low, the specific base-catalysed second forward step is slow, and C reverts to YH+X faster than it proceeds to give products. The first step, therefore, is a pre-equilibrium and the second step is rate determining. In this pH range, $\frac{d(\log k'_{obs})}{d(pH)}$ is positive.

At high pH, however, YH is extensively dissociated, so the reactive [YH] within $[YH]_s$ is low which leads to a slow first forward step. But the second forward step from C, being specific base catalysed, has become faster at high $[OH^-]$ and effectively precludes reversal of the first step. At high pH, therefore, the first step is unidirectional and rate determining, and $\frac{d(\log k'_{obs})}{d(pH)}$ is negative. There will be an intermediate pH at which the second step is neither constrained by the first forward step nor slower

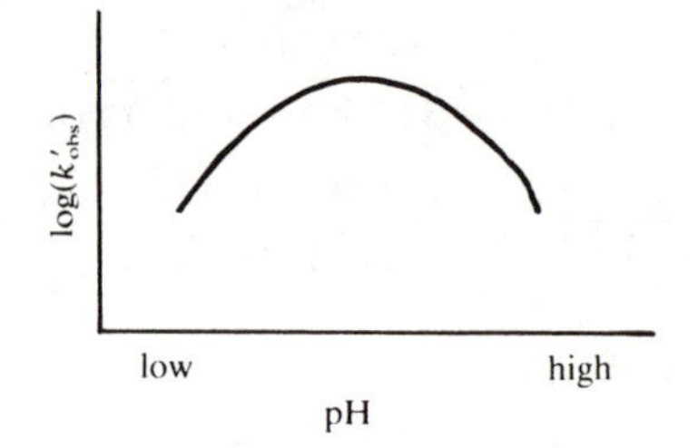

Fig. 8.26. pH-$\log(k'_{obs})$ profile for the reaction in eqn (8.43) with k'_{obs} defined by eqns (8.45) and (8.46).

than the reverse of the first step. The pH at which this occurs is determined by the relative magnitude of k_{-1}, k_2, and $K_{AH}(YH)$; and, at this maximal overall rate, $\dfrac{d(\log k'_{obs})}{d(pH)} = 0$.

8.8.3.2 *Reaction involving a weakly basic compound*

A comparable switch in rate-determining step with changing pH and a maximum in the pH-log(rate constant) profile can also be expected for a multistep reaction of a weakly basic compound if one of the later steps involves acid–base catalysis. The formation of oximes and semi-carbazones from aldehydes and ketones are examples. We shall consider qualitatively the reaction of acetone with hydroxylamine in water:

$$(CH_3)_2C{=}O + NH_2OH \underset{k_{-1}}{\overset{k_1}{\rightleftharpoons}} (CH_3)_2C(OH)NHOH \xrightarrow{k_0+k_H[H_3O^+]+k_{OH}[OH^-]} (CH_3)_2C{=}NOH + H_2O \quad (8.48)$$

$$\frac{d[\text{oxime}]}{dt} = k'_{obs}[(CH_3)_2CO][NH_2OH]_s \quad (8.49)$$

where $[NH_2OH]_s$ = the total stoichiometric concentration of hydroxylamine and its conjugate acid
$= [NH_2OH] + [\overset{+}{N}H_3OH]$.

It has been known for many years that the pH-log(k'_{obs}) profile for this reaction is dome-shaped between pH ~ 3–6, Fig. 8.27.[17,27] Between pH ~ 6.5–8, the correlation is approximately linear (gradient = −1), then, from pH ~ 8–10, k'_{obs} is independent of the pH. In the final section from

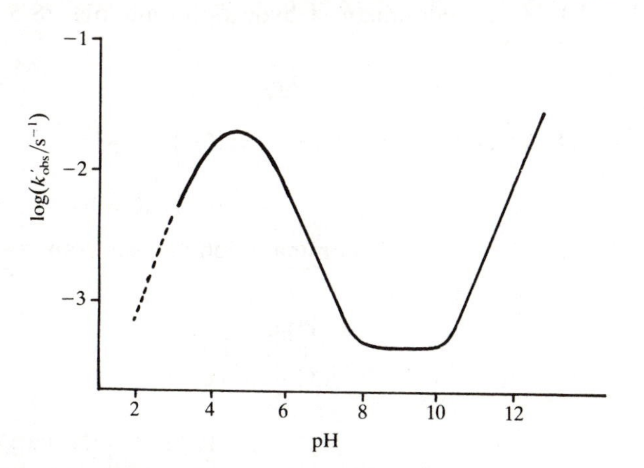

Fig. 8.27. pH-log(k'_{obs}) profile for the reaction of acetone and hydroxylamine in water, 25 °C, k'_{obs} being defined by eqn (8.49).

pH ~ 10–12.5, $\log(k'_{obs})$ increases linearly with pH (gradient = 1).[5,28] Equation (8.48) accounts quantitatively for these results. The reversible first step involves nucleophilic attack of the free amino compound at the carbonyl to give the hydroxyalkylhydroxylamine adduct. This intermediate then undergoes dehydration in the second step to give the product, and the pH determines whether the first forward step or the composite second forward step is rate determining.[28] In aqueous solutions of pH > ~7, hydroxylamine (pK_{BH} 5.96, 25 °C) is very largely unprotonated. Consequently, $[NH_2OH] \sim [NH_2OH]_s$ and $k'_{obs} \sim k_{obs}$ where k_{obs} is the conventional second-order rate constant defined by

$$\frac{d[\text{oxime}]}{dt} = k_{obs}[(CH_3)_2CO][NH_2OH]$$

where $[NH_2OH]$ = the concentration of free (unprotonated) hydroxylamine.

The addition step is fast in these alkaline solutions and the tetrahedral intermediate can be detected. The pH-$\log(k'_{obs})$ profile in Fig. 8.27 between pH ~ 7–12.5 indicates specific acid–base catalysis with an appreciable solvent-induced contribution in the composite rate-determining dehydration:

$$k'_{obs} \sim k_{obs} = \frac{k_1}{k_{-1}}(k_O + k_H[H_3O^+] + k_{OH}[OH^-])$$

$$(CH_3)_2C(OH){-}NHOH \xrightarrow{H_2O} (CH_3)_2C{=}NOH + H_2O$$

$k_O = 7.05 \times 10^{-3}\ s^{-1}$; predominant dehydration route pH ~ 8.5–9.5

$$(CH_3)_2C(OH){-}NHOH + H_3O^+ \underset{+H_2O}{\overset{-H_2O}{\rightleftharpoons}} (CH_3)_2C(\overset{+}{O}H_2){-}NHOH \longrightarrow$$

$$(CH_3)_2C{=}NOH + H_3O^+$$

$k_H = 1.52 \times 10^6\ dm^3\ mol^{-1}\ s^{-1}$; predominant dehydration route pH < ~8

$$(CH_3)_2C(OH){-}NHOH + OH^- \underset{+H_2O}{\overset{-H_2O}{\rightleftharpoons}} (CH_3)_2C(OH){-}\bar{N}OH \longrightarrow$$

$$(CH_3)_2C{=}NOH + OH^-$$

$k_{OH} = 1.25 \times 10^2\ dm^3\ mol^{-1}\ s^{-1}$; predominant dehydration route pH > ~10.

As the pH is decreased from about 7, the concentration of free hydroxylamine begins to decrease due to its basic nature:

$$NH_2OH + H_3O^+ \rightleftharpoons \overset{+}{N}H_3OH + H_2O;$$
$$[NH_2OH] < [NH_2OH]_s$$

and

$$k'_{obs} \neq k_{obs}.$$

Consequently, although the dehydration step continues to become faster with decreasing pH ($k_H \gg k_{OH}$), the second-order first forward step (rate constant k_1) which produces the intermediate begins to become slower as $[NH_2OH]$ (within $[NH_2OH]_s$) decreases. At pH ~ 4.5 (the maximum in the profile of Fig. 8.27), the intermediate partitions equally between the reverse of the first step and the forward composite second step ($k_{-1} = k_0 + k_H[H_3O^+]$; $k_{OH}[OH^-] \sim 0$) and the overall reaction has no single rate-determining step. Below pH ~ 4.5, the overall rate constant k'_{obs} decreases with decreasing pH as the second step is constrained by the first which is now unidirectional and rate determining, and becoming slower due to the ever-decreasing proportion of *free* hydroxylamine as $[H_3O^+]$ increases.

On the basis of this mechanism we can predict the nature of the pH-log(k'_{obs}) profile below pH ~ 3.

$$\frac{d[\text{oxime}]}{dt} = k_1[(CH_3)_2CO][NH_2OH] \tag{8.50}$$

(rate-determining adduct formation)

but $\overset{+}{N}H_3OH + H_2O \rightleftharpoons NH_2OH + H_3O^+$ with

$$K_{AH}(\overset{+}{N}H_3OH) = \frac{[NH_2OH][H_3O^+]}{[\overset{+}{N}H_3OH]}.$$

Substituting in eqn (8.50) for $[NH_2OH]$,

$$\frac{d[\text{oxime}]}{dt} = \frac{k_1[(CH_3)_2CO][\overset{+}{N}H_3OH]K_{AH}(\overset{+}{N}H_3OH)}{[H_3O^+]}.$$

But in practical terms, $[\overset{+}{N}H_3OH] \sim [NH_2OH]_s$ at pH $\leqslant$ 3, therefore

$$\frac{d[\text{oxime}]}{dt} \sim \frac{k_1[(CH_3)_2CO][NH_2OH]_s K_{AH}(\overset{+}{N}H_3OH)}{[H_3O^+]}$$
$$= k'_{obs}[(CH_3)_2CO][NH_2OH]_s \quad \text{at constant pH,}$$

where

$$k'_{obs} = \frac{k_1 K_{AH}(\overset{+}{N}H_3OH)}{[H_3O^+]}.$$

It follows, therefore, that below pH~3

$$\log(k'_{obs}) = \log k_1 + \log K_{AH}(\overset{+}{N}H_3OH) - \log[H_3O^+]$$
$$= \log k_1 - pK_{AH}(\overset{+}{N}H_3OH) + pH,$$

so the pH-log(k'_{obs}) correlation should be linear with gradient = 1 as indicated by the broken line in Fig. 8.27.

8.9 Problems

1. The following *pseudo* first-order rate constants were obtained for the hydrolysis of ethyl vinyl ether in aqueous perchloric acid at 26.7 °C.[14]

$10^2[HClO_4]$/mol dm^{-3}	$10^2\ k_{obs}$/s^{-1}
0.386	0.823
0.771	1.63
0.964	2.08
1.16	2.47
1.35	2.94
1.54	3.21
1.74	3.58

Calculate the second-order catalytic constant. Compare your result from this direct determination with the indirect value obtainable from the intercept in Fig. 8.10, p. 329, given that $pK_{AH}(CH_3CO_2H) = 4.64$ under the conditions of this experiment and the buffer ratio $r = [CH_3CO_2H]/[CH_3CO_2^-] = 2.00$; (assume $k_0 = 0$ in Fig. 8.10).

2. The dehydration of acetaldehyde hydrate in 92.5:7.5 acetone: water at 25 °C is catalysed by acetic acid

$$CH_3CH(OH)_2 \xrightarrow[CH_3CO_2H,\ k_{obs}]{(CH_3)_2CO+H_2O,\ 25\,°C} CH_3CHO + H_2O.$$

Calculate the second-order catalytic constant from the following observed *pseudo* first-order rate constants (reported by R. P. Bell and W. C. E. Higginson, *Proc. R. Soc.* A **197,** 141 (1949)).

$10^3[CH_3CO_2H]$/mol dm^{-3}	0.25	0.70	1.48	1.95	2.20	2.95	3.75	4.50
$10^2\ k_{obs}$/min^{-1}	0.748	1.59	3.08	3.71	4.35	5.65	7.08	8.55

3. For a rate law such as the one for the hydrolysis of ethyl acetate:

$$k_{obs} = k_0 + k_{OH}[OH^-] + k_H[H_3O^+],$$

show that the minimum in k_{obs} as a function of $[H_3O^+]$ occurs at $[H_3O^+] =$

$\left(\frac{k_{OH} \cdot K_{AP}}{k_H}\right)^{0.5}$ and that under these conditions

$$k_{obs} = k_0 + 2\sqrt{k_H \cdot k_{OH} \cdot K_{AP}},$$

where $K_{AP} = [H_3O^+][OH^-]$.

4. The rate law for the hydrolysis of ethyl acetate at 25 °C is

$$k_{obs} = k_0 + k_H[H_3O^+] + k_{OH}[OH^-],$$

with

$$k_H = 1.36 \times 10^{-4}\ dm^3\ mol^{-1}\ s^{-1}\ \text{(see p. 322)}$$

and

$$k_{OH} = 0.111\ dm^3\ mol^{-1}\ s^{-1}\ \text{(ref. 26)}.$$

Calculate (i) k_{obs} in pure water assuming $k_0 \sim 0$, (ii) the pH at which k_{obs} is a minimum (see Problem 3), and (iii) the minimum value of k_{obs} ($K_{AP} = 10^{-14}$ at 25 °C).

5. The following results were obtained for the pyridine-catalysed hydrolysis of *p*-nitrophenyl acetate in buffered aqueous solution at 25.5 °C (A. R. Butler and I. H. Robertson, *J. chem. Soc., Perkin* 2 660 (1975)).

$10^4\ k_{obs}/s^{-1}$	1.96	3.68	5.26	6.88	8.32
[Pyridine]/ mol dm^{-3}	0.08	0.16	0.24	0.32	0.40

Corresponding results in deuterium oxide are

$10^4\ k_{obs}/s^{-1}$	2.02	4.90	6.80	8.20
[Pyridine]/ mol dm^{-3}	0.11	0.26	0.36	0.46

(In both sets of experiments, [Pyridine] = concentration of the unprotonated base.) Calculate the catalytic rate constants and propose a mechanism for the catalysed reaction.

6. Demonstrate that the mechanism for hydrolysis of a vinyl alkyl ether given on p. 330 (rate-determining protonation of the substrate) leads to a general acid-catalysis rate law.

7. Show that a mechanism for tautomerism involving rapid deprotonation of a substrate followed by rate-determining protonation at a different site leads to a general base-catalysis rate law.

8. Calculate k_0, k_H, and k_{OH} in the rate law for the hydrolysis of β-butyrolactone in aqueous solution at 25 °C from the following results (reported by A. R. Olsen and R. J. Miller, *J. Am. chem. Soc.* **60,** 2687 (1938)).

$$k_{obs} = k_0 + k_H[H_3O^+] + k_{OH}[OH^-]$$

pH	−0.5	0	1	2	3–6	7	8	8.5	9	9.5	10	11	12
$10^4 k_{obs}/s^{-1}$	14.8	10.5	8.70	8.52	8.50	8.55	9.0	10.0	13.4	23.5	57.5	498.5	4908.5

9. Interpret the following pH-log(k_{obs}) profiles

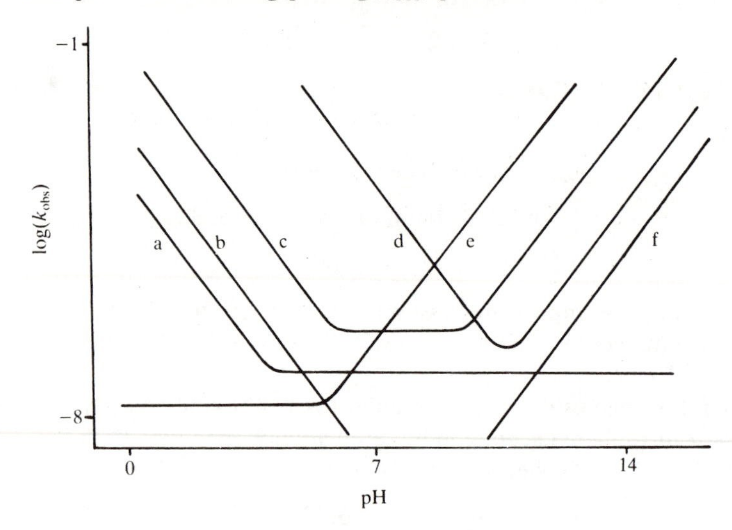

10. The oximation of benzaldehyde,

$$PhCHO + NH_2OH \underset{k_{-1}}{\overset{k_1}{\rightleftharpoons}} PhCH(OH)NHOH \xrightarrow{k_2} PhCH{=}NOH + H_2O,$$

where

$$k_2 = k_H[H_3O^+] + k_{OH}[OH^-],$$

is mechanistically similar to the corresponding reactions of acetone (p. 359) and pyruvate (Problem 10, p. 312).[17] The *pseudo* first-order rate law is given by:

$$-\frac{d[PhCHO]}{dt} = k_{obs}[PhCHO]$$

when $[PhCHO] \ll [NH_2OH]$, and the following results were obtained[28]

pH	6.0	6.5	7.05	7.7	8.55	9.0	9.5	10.2	11.08
$10^3 k_{obs}/s^{-1}$	40.3	15.1	4.17	0.892	0.467	1.21	2.56	14.6	71.0

Given that $k_{-1} \gg k_2$ and $k_1/k_{-1} = 17.5\ dm^3\ mol^{-1}$, calculate k_H and k_{OH}, the specific acid- and base-catalysis rate constants for the dehydration step.

8.10 References

1. F. Elsinger, J. Schreiber, and A. Eschenmoser, *Helv. chim. Acta* **43,** 113 (1960).
2. T-L. Ho, 'The hard and soft acids bases principle and organic chemistry', *Chem. Rev.* **75,** 1 (1975).

3. M. M. Khan and A. E. Martell, *Homogeneous catalysis by metal complexes*, Vol. 2, Academic Press, New York (1974).
4. R. Steinberger and F. H. Westheimer, *J. Am. chem. Soc.* **73,** 429 (1951); F. H. Westheimer, *Trans. N.Y. Acad. Sci.* **18,** 15 (1955).
5. M. L. Bender, *Mechanisms of homogeneous catalysis from protons to proteins*, Wiley-Interscience, New York (1971). This is a very readable book and provides a comprehensive coverage of homogeneous catalysis, including sections with inorganic and enzymic examples.
6. R. P. Bell, *The proton in chemistry*, 2nd edn, Chapman & Hall, London (1973). This authoritative specialist monograph contains chapters on all aspects of proton chemistry and includes one in which a few reactions catalysed by acids and bases are treated comprehensively.
7. J. W. Moore and R. G. Pearson, *Kinetics and mechanism*, 3rd edn, Chapt. 9, Wiley, New York (1981).
8. R. P. Bell, *Acid base catalysis*, Oxford University Press (1941). This seminal volume contains an invaluable account of early work. A modern treatment of those sections which are now out of date may be found in ref. 6, above.
9. P. A. Giguère, *J. chem. Educ.* **56,** 571 (1979).
10. M. L. Bender, *Chem. Rev.* **60,** 53 (1960).
11. M. L. Bender, *J. Am. chem. Soc.* **73,** 1626 (1951); D. Samuel and B. L. Silver, *Adv. phys. org. Chem.* **3,** 123 (1965).
12. E. H. Cordes, *Progr. phys. org. Chem.* **4,** 1 (1967); E. H. Cordes and H. G. Bull, *Chem. Rev.* **74,** 581 (1974).
13. C. K. Ingold, *Structure and mechanism in organic chemistry*, 2nd edn, (*a*) Chapt. 7, (*b*) Chapt. 13, Bell, London (1969).
14. A. J. Kresge and Y. Chiang, *J. chem. Soc.* B, 53 (1967).
15. P. Salomaa, A. Kankaanpera, and M. Lajunen, *Acta chem. Scand.* **20,** 1790 (1966); A. J. Kresge and H. J. Chen, *J. Am. chem. Soc.* **94,** 2818 (1972).
16. J. Hine, *Physical organic chemistry*, 2nd edn, Chapt. 10, McGraw-Hill, New York (1962).
17. W. P. Jencks, *Catalysis in chemistry and enzymology*, McGraw-Hill, New York (1969). This detailed book covers a substantial literature with special emphasis on the author's own contributions. It contains particularly useful chapters on reactions of carbonyl compounds, and practical kinetics.
18. E. S. Gould, *Mechanism and structure in organic chemistry*, Chapt. 10, Holt, Rinehart & Winston, New York (1959).
19. A. R. Butler and V. Gold, *J. chem. Soc.* 4362 (1961); A. R. Fersht and W. P. Jencks, *J. Am. chem. Soc.* **92,** 5432 (1970).
20. L. F. Fieser and M. Fieser, *Reagents for organic synthesis*, Vol. I, p. 958, Wiley, New York, (1967).
21. A. R. Butler and V. Gold, *J. chem. Soc.* 2305 (1961).
22. S. L. Johnson, *Adv. phys. org. Chem.* **5,** 237 (1967).
23. V. Gold, D. G. Oakenfull, and T. Riley, *J. chem. Soc.* B, 515 (1968).
24. M. L. Bender and B. W. Turnquest, *J. Am. chem. Soc.* **79,** 1652 (1957); T. C. Bruice and G. L. Schmir, ibid., p. 1663, and **80,** 148 (1958); D. G. Oakenfull, *J. chem. Soc.* B, 197 (1970).
25. V. Gold, *Adv. phys. org. Chem.* **7,** 259 (1969); R. L. Schowen, *Progr. phys. org. Chem.* **9,** 275 (1972).
26. E. K. Euranto, Ch. 11, 'Esterification and ester hydrolysis' in *The chemistry of carboxylic acids and esters* (ed. S. Patai), Wiley-Interscience, New York (1969).
27. L. P. Hammett, *Physical organic chemistry*, 2nd edn, McGraw-Hill, New York (1970).

28. A. Williams and M. Bender, *J. Am. chem. Soc.* **88,** 2508 (1966); see also W. P. Jencks, ibid. **81,** 475 (1959).

Supplementary references

M. L. Bender and L. J. Brubacher, *Catalysis and enzyme action*, McGraw-Hill, New York (1973).

T. C. Bruice and S. J. Benkovic, *Bio-organic mechanisms*, Vol. 1, Chapt. 1, W. A. Benjamin, New York (1966).

Techniques of chemistry, Vol. VI, *Investigation of rates and mechanisms of reactions*, Pt. I (ed. E. S. Lewis), Wiley-Interscience, New York (1974), Chapt. 11, 'Homogeneous solution catalysis by small molecules and by enzymes' by M. F. Dunn and S. A. Bernhard.

9

Isotope effects upon chemical reactions

9.1 Introduction

One of the principal aims of physical organic chemistry is to investigate how reactions of organic compounds take place. In earlier chapters, some methods and techniques were described; determination of the activation parameters from reaction rate measurements at different temperatures is an example. Another general method is to investigate how a modification to the structure of a reactant affects the rate (or equilibrium) under otherwise unchanged experimental conditions. If a chlorine substituent is replaced by bromine, or a methyl by an isopropyl group, the rate (or equilibrium) constant may change in a way which is consistent with some mechanisms but not with others. When such an approach is systematically extended and refined, it may narrow down very considerably the range of mechanistic possibilities for a series of related compounds. There is an inherent difficulty, however; initially, we want to know about the reaction of a single compound, and we finish up with results for a range of related but, nevertheless, different compounds. And in order to use information derived from the reaction of one compound as evidence in elucidating the mechanism of reaction of another, we assume a relationship between the two. The usual expression of this assumption is that the two compounds react by the same or by very closely related mechanisms.

This aspect of mechanistic organic chemistry will be considered more fully in the next chapter when we deal in general with the effect of substituents upon chemical reactions. In this chapter we consider the effect of a very special type of modification which does not cause the compound to be different in any qualitative sense. According to current theory, substitution of an atom within a molecule by an isotope does not affect the potential-energy hypersurface which describes the molecule either in isolation or when it is subjected to any sort of perturbation, e.g. the approach of another molecule. In other words, isotopic substitution does not qualitatively affect the mechanism of any reaction of a compound. It does, however, have a quantitative effect. The rate (or equilibrium) constant may become smaller or larger; but the rate (or equilibrium) constant of the isotopically-labelled compound is essentially for the same chemical compound reacting by the same mechanism.

Before we proceed with an account of how a change in atomic mass within a molecule can change a rate (or equilibrium) constant without changing the mechanism, and of how this assists in the elucidation of reaction mechanisms, there are a few matters of definition and terminology which must be explained.

In the acid–base reaction of eqn (9.1), the equilibrium constant is different according to whether L is ^{1}H (protium) or ^{2}H (deuterium, D).[1,2] This is an *equilibrium isotope effect.*

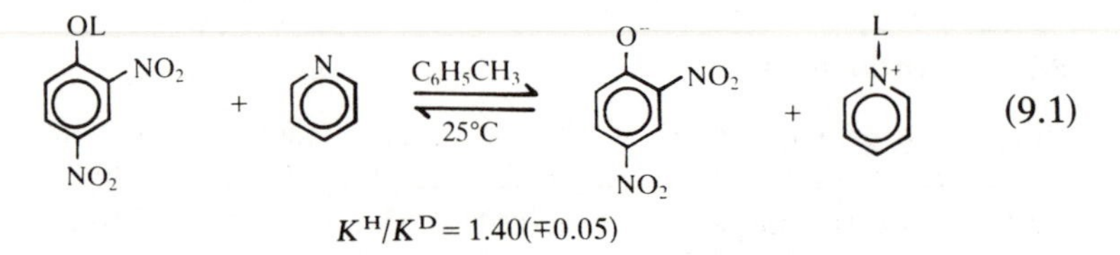

$$K^H/K^D = 1.40(\mp 0.05)$$

The value of K for the equation as written is larger for the lighter (protium) compound than for the heavier (deuterium) analogue; this is described as a *normal* isotope effect ($K^H > K^D$). There is an *inverse* equilibrium isotope effect ($K^H < K^D$ for the equation as written) in the reaction of eqn (9.2).[3]

$$C_6H_5C(=O)L + NH_2OH \underset{25\,^\circ C}{\overset{H_2O}{\rightleftharpoons}} C_6H_5C(OH)(L)\text{—}NHOH \qquad (9.2)$$

$$K^H/K^D = 0.735(\mp 0.020).$$

The rate constant in the Diels–Alder reaction of maleic anhydride with cyclopentadiene (eqn (9.3)) is larger when the two vinylic hydrogens of the dienophile are replaced by deuterium.[4]

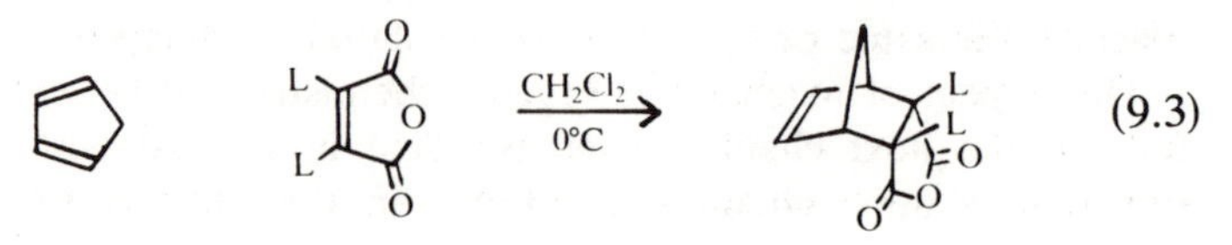

$k^H/k^D = 0.95$ (for both hydrogens substituted with deuterium; this is equivalent to 0.97 per deuterium). This, therefore, is also an inverse effect, and, since it relates to rate constants, it is an inverse *kinetic isotope effect.*

There is a normal kinetic isotope effect (k.i.e.) in the following substitution reaction[5]

$$\text{n-C}_4\text{H}_9\text{—}^*\text{Cl} + \text{C}_6\text{H}_5\text{S}^- \xrightarrow[20\,°\text{C}]{\text{CH}_3\text{OH}} \text{n-C}_4\text{H}_9\text{—SC}_6\text{H}_5 + {}^*\text{Cl}^-$$

$$k^{35}/k^{37} = 1.0089_5(\pm 0.00015),$$

and, because the isotopically-labelled atom, chlorine, is other than hydrogen, this is regarded as a *heavy atom isotope effect.*

There are also *solvent* kinetic isotope effects as we have already seen in the previous chapter, p. 349. Another example[6] is the proton abstraction from 2-acetylcyclohexanone by acetate in water (eqn (9.4)).

$$\text{2-acetylcyclohexanone} + \text{CH}_3\text{CO}_2^- \xrightarrow[25°\text{C}]{\text{L}_2\text{O}} \text{enolate} + \text{CH}_3\text{CO}_2\text{H} \qquad (9.4)$$

$$k^{\text{H}_2\text{O}}/k^{\text{D}_2\text{O}} = 1.20.$$

As used so far, kinetic, equilibrium, and solvent isotope effects refer to the outcome of isotopic substitution upon an experimentally-determined parameter – either a rate or an equilibrium constant. Further terms imply mechanism. When the bond to the isotopically-labelled atom is broken and reformed in a simple reaction (or during the rate-determining step of a more complex reaction) as, for example, in eqn (9.5),[1,7] we have a *primary* kinetic isotope effect.

$$\text{NO}_2\text{CL}_2\text{CO}_2\text{C}_2\text{H}_5 + \text{OH}^- \xrightarrow[25\,°\text{C}]{\text{H}_2\text{O}} \text{NO}_2\bar{\text{C}}\text{LCO}_2\text{C}_2\text{H}_5 + \text{LOH} \qquad (9.5)$$

$$k^H/k^D = 4.6.$$

Primary deuterium kinetic isotope effects of up to $k^H/k^D \sim 7$ are fairly common and some very much larger ones have been reported.

In contrast, we have a *secondary* kinetic isotope effect in the cycloaddition of maleic anhydride with cyclopentadiene (eqn (9.3)) since the bonds to the isotopically-labelled atoms are not broken in the reaction. Secondary kinetic deuterium isotope effects are seldom outside the range $k^H/k^D = 0.80–1.25$ per deuterium (25 °C).

There can also be secondary deuterium isotope effects upon equilibria, for example, the dissociation of formic and acetic acids:[8,9]

$$\text{LCO}_2\text{H} + \text{H}_2\text{O} \underset{25\,°\text{C}}{\overset{\text{H}_2\text{O}}{\rightleftharpoons}} \text{LCO}_2^- + \text{H}_3\text{O}^+: \qquad K_{AH}^H/K_{AH}^D = 1.08$$

$$\text{CL}_3\text{CO}_2\text{H} + \text{H}_2\text{O} \underset{25\,°\text{C}}{\overset{\text{H}_2\text{O}}{\rightleftharpoons}} \text{CL}_3\text{CO}_2^- + \text{H}_3\text{O}^+; \qquad K_{AH}^H/K_{AH}^D = 1.03.$$

Occasionally, it may not be possible to give a simple mechanistic assignment to an isotope effect. In the bromination of 3-methylpentan-2,4-dione (eqn (9.6)), water both abstracts a proton from the central carbon and is also the reaction medium, so the overall result is a composite of secondary and solvent kinetic isotope effects.[10]

$$CH_3\overset{O}{\overset{\|}{C}}CH(CH_3)\overset{O}{\overset{\|}{C}}CH_3 + Br_2 + L_2O \xrightarrow[25\,°C]{L_2O} CH_3\overset{O}{\overset{\|}{C}}CBr(CH_3)\overset{O}{\overset{\|}{C}}CH_3 + L_2\overset{+}{O}H + Br^- \tag{9.6}$$

$$k^{H_2O}/k^{D_2O} = 1.37.$$

For the autoprotolysis (dissociation) of water itself, the overall (normal) equilibrium effect is obviously a complex term.[1,11]

$$2L_2O \overset{25\,°C}{\rightleftharpoons} L_3O^+ + OL^-$$

$$K_{AP} = [L_3O^+][OL^-]$$

$K^H_{AP}/K^D_{AP} = 7.47(\pm 0.24)$ and $K^H_{AP}/K^T_{AP} = 16.4(\pm 0.2)$ where $T = {}^3H$, tritium, the unstable radioactive isotope of hydrogen.

The bond from oxygen to one of the isotopically-substituted atoms is broken, so there is a primary contribution; the bond from oxygen to the other hydrogen remains intact so substitution here contributes a secondary effect, and water is also the medium, so there is a solvent effect as well.

9.2 The molecular origin of isotope effects

9.2.1 Isotopic mass and vibrational properties

The bonding within a molecule, or indeed within any configuration of atoms, is determined entirely by the nuclear charges, the inner shell or core electrons of the individual atoms, and the outer shell or valence electrons. The core electrons surrounding each nucleus are held by the nuclear charge and in turn they screen the major electrostatic effect of the nucleus from the valence electrons. The addition of an extra neutron to the nucleus of a constituent atom in a molecule has no electrostatic effect and consequently does not affect the electronic distribution (the bonding) within the molecule.

We shall consider in detail first a simple diatomic molecule HX where X is some atom much heavier than hydrogen. The molecule has only one internal degree of freedom – a vibration – and its molecular potential-energy curve has already been discussed in Chapter 1 (p. 12, Fig. 1.3).

We saw that the potential-energy curve is approximately harmonic close to the minimum and the vibrational frequency, ν, is given by

$$\nu = \frac{1}{2\pi}\sqrt{\frac{\kappa}{\mu_{HX}}} \tag{9.7}$$

where κ = the force constant of the bond, and μ_{HX} = the reduced mass of the molecule given by

$$\frac{1}{\mu_{HX}} = \frac{1}{m_H} + \frac{1}{m_X}, \tag{9.8}$$

where m_H and m_X = the masses of H and X.

The force constant, κ, being a property of the bond, is a function of the nuclear charges and the total electronic distribution within the molecule. The addition of an extra neutron to the nucleus of either H or X will, therefore, have no effect upon κ. However, the reduced mass of the molecule could be affected according to whether the neutron is put into the hydrogen nucleus (to give deuterium, D) or into the nucleus of X (to give X′).

From eqn (9.8)

$$\mu_{HX} = \frac{m_H \,.\, m_X}{(m_H + m_X)}.$$

Assuming that $m_X \gg m_H$,

$$\mu_{HX} \sim m_H = 1\ \text{amu}.$$

If a neutron is put into the hydrogen nucleus,

$$\mu_{DX} \sim m_D = 2\ \text{amu},$$

so the reduced mass of the whole molecule DX is approximately twice that for HX. But if the neutron is put into the nucleus of X, it follows from eqn (9.8) that, as long as $m_X \gg m_H \ll m_{X'}$,

$$\mu_{HX'} \sim m_H = 1\ \text{amu},$$

and the reduced mass of HX′ is virtually the same as that for HX.

It also follows from these considerations and eqn (9.7) that the frequency of the molecular vibration of DX is lower than that of HX:

$$\nu_{HX} = \frac{1}{2\pi}\sqrt{\frac{\kappa}{\mu_{HX}}}$$

$$\nu_{DX} = \frac{1}{2\pi}\sqrt{\frac{\kappa}{\mu_{DX}}}$$

therefore

$$\frac{\nu_{HX}}{\nu_{DX}} = \sqrt{\frac{\mu_{DX}}{\mu_{HX}}} = \sqrt{2} = 1.41.$$

Example. The i.r. spectrum of gaseous HI shows an intense band at $\bar{\nu} = 2230\ \text{cm}^{-1}$.[12] Assuming the harmonic approximation, calculate the wave number of the corresponding band in the spectrum of the deuterium analogue ^{2}HI.
Since $\bar{\nu} = 1/\lambda$ and $\nu = c/\lambda$,

$$\nu = c \times \bar{\nu}$$

where c = the velocity of electromagnetic radiation, therefore

$$\frac{\bar{\nu}_{\text{HI}}}{\bar{\nu}_{\text{DI}}} = 1.41,$$

and

$$\bar{\nu}_{\text{DI}} = 2230\ \text{cm}^{-1}/1.41$$
$$= 1582\ \text{cm}^{-1}.$$

This is in fair agreement with the experimental result of $1600\ \text{cm}^{-1}$.[12]

We can expect agreement between values of $\bar{\omega}_e$ for ^{2}HX calculated from ^{1}HX results and values obtained from experiment to be better than for $\bar{\nu}$.
For ^{1}HI, $\bar{\omega}_e = 2309.1\ \text{cm}^{-1}$ (Chapter 1, p. 13) therefore

$$\bar{\omega}_e(^2\text{HI}) = 2309.1\ \text{cm}^{-1}/\sqrt{2}$$
$$= 1632.8\ \text{cm}^{-1}.$$

The experimental result is $1639.7\ \text{cm}^{-1}$.[12]

We saw in Chapter 1 that the molecular zero-point energy of a diatomic molecule, ε_0, is given by

$$\varepsilon_0 = 0.5\ h\nu,$$

where h = the Planck constant, and
ν = the vibrational frequency,
and the gap between successive energy levels = $h\nu$, assuming that the vibration remains harmonic.

Consequently,

(1) the zero-point energy is not the same for ^{1}HX as for ^{2}HX. (In the case of ^{1}HI and ^{2}HI the molar values are 13.3 and 9.6 kJ mol^{-1} respectively); and

(2) the difference between the ^{1}H and ^{2}H levels for a given quantum number in the harmonic region of a vibration gets larger as the vibrational quantum number increases. (The effect of anharmonicity, however, is to reduce this difference.)

We can now refine Fig. 1.3 (Chapter 1) by including energy levels for ^{1}HX and ^{2}HX to give Fig. 9.1 below.

Notice that the actual potential-energy curve itself, including D_e and r_e, is unchanged. But the sets of energy levels for molecular species HX and DX are different and, because the zero-point energy for ^{2}HX is lower than that for ^{1}HX, $D_0(^2\text{HX}) > D_0(^1\text{HX})$. This means that it requires more energy to dissociate ^{2}HX than ^{1}HX and the difference between the

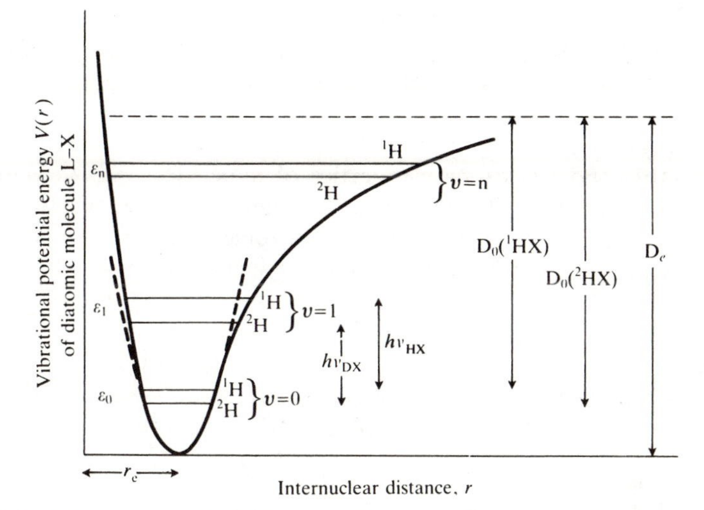

Fig. 9.1. Anharmonic molecular potential-energy diagram for diatomic molecules ^1HX and ^2HX.

spectroscopic bond dissociation energies, $D_0(^2HX) - D_0(^1HX)$, is equal to the difference in the zero-point energies: $0.5h(\nu_{HX} - \nu_{DX})$. But the cause of this is purely vibrational due, ultimately, to the different masses of ^1H and ^2H nuclei. It would be wrong to regard the bond in ^2HX as intrinsically stronger than the one in ^1HX; the bonds have a common potential-energy curve, the same D_e, and the same force constant because they are electronically identical.

The minimum in the curve of Fig. 9.1 corresponds to the equilibrium bond length in the hypothetical motionless state in the absence of a vibration for both ^1HX and ^2HX. As long as vibrations are harmonic (the curve symmetrical) the time-averaged internuclear distances in ^1HX and ^2HX remain identical. But when anharmonicity of the vibrations, i.e. the asymmetry of the potential-energy curve, is taken into account then the time-averaged bond lengths of ^1HX and ^2HX are seen to be unequal. The bond length is *shorter* in the *heavier* molecule ^2HX. But this apparently structural isotope effect is also vibrational in origin.

It also follows that, for the molecule ^3HX containing tritium ($^3H = T$),

(1) the vibrational frequency (ν_{TX}) is lower than that for the deuterium analogue: $\nu_{HX}/\nu_{TX} = \sqrt{\mu_{TX}/\mu_{HX}} \sim \sqrt{3} = 1.73$,

(2) the zero point energy $\varepsilon_0(TX) = 0.5h\nu_{TX}$ is lower than that of the deuterium analogue, and

(3) $D_0(TX)$ is greater than $D_0(DX)$.

9.2.2 Molecular vibrations, chemical equilibria, and rates of reactions

In Chapter 4, we dealt with equilibria from the viewpoint of classical thermodynamics and saw that the equilibrium condition is when the total free energy of a system is minimal. But there is nothing in the classical thermodynamic description of a reacting system which invokes the molecular nature of matter, consequently, it cannot help us to account for a reaction in molecular terms.

By applying elementary statistical thermodynamics and a knowledge of the structure and vibrational properties of molecules, we can achieve a molecular description of equilibrium in simple chemical reactions. This alternative perspective provides a very convenient basis for an account of the effect of isotopic substitution upon chemical reactions.

9.2.2.1 *Equilibria*

Figure 9.2 is the superposition of molecular potential-energy curves of two isomeric molecules X and Y which can interconvert.

$$X \rightleftharpoons Y.$$

The precise nature of the reaction coordinate does not matter. It could be a torsional vibration about a double bond as when X and Y are *trans* and *cis* alkenes.

Alternatively, it could be a pyramidal inversion if X and Y are a pair of aziridines.

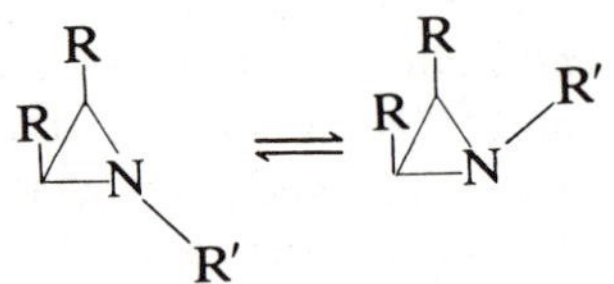

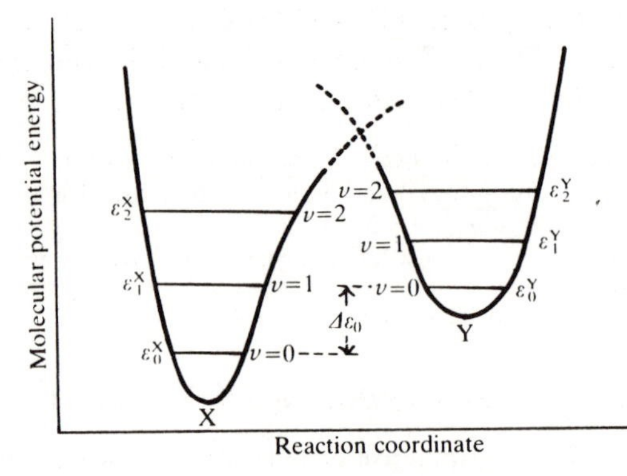

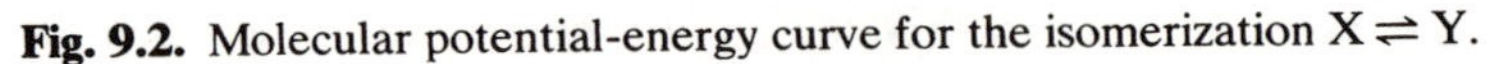

Fig. 9.2. Molecular potential-energy curve for the isomerization $X \rightleftharpoons Y$.

The first three quantized energy levels for X and Y within the reaction coordinate are included in Fig. 9.2.

We have already seen (Chapter 1, p. 10) that a large assembly of molecules in thermal equilibrium partition themselves among the available energy levels according to the Boltzmann Distribution Law. In the case of a single set of energy levels $0, 1, \ldots i \ldots$,

$$\frac{n_i}{n_0} = \mathrm{e}^{-(\varepsilon_i - \varepsilon_0)/k_B T},$$

where n_0 and n_i = the numbers of molecules in levels 0 and i,
ε_0 and ε_i = the energy of levels 0 and i,
k_B = the Boltzmann constant, and
T = the absolute temperature.

If N = the total number of molecules, then

$$N = \sum n_i = n_0 \sum \mathrm{e}^{-(\varepsilon_i - \varepsilon_0)/k_B T} = n_0 Q, \tag{9.9}$$

where $Q = \sum \mathrm{e}^{-(\varepsilon_i - \varepsilon_0)/k_B T}$ = the molecular partition function.

In Fig. 9.2 we have a pair of intersecting potential-energy curves and initially we treat them separately using eqn (9.9):

$$N_X = n_0^X \,.\, Q_X(\mathrm{rc})$$

and

$$N_Y = n_0^Y \,.\, Q_Y(\mathrm{rc})$$

where N_X and N_Y = the total numbers of molecules of X and of Y at equilibrium,
n_0^X and n_0^Y = the numbers of molecules of X and of Y in their respective zero energy levels within the vibrational mode which is the reaction coordinate, and
$Q_X(\mathrm{rc})$ and $Q_Y(\mathrm{rc})$ = the molecular partition functions of X and Y in this same reaction coordinate vibrational mode.

However, if compounds X and Y are in chemical equilibrium at temperature T, then the numbers of molecules of X and Y in their respective zero energy levels must also be given by the Boltzmann Distribution Law,

$$\frac{n_0^Y}{n_0^X} = \mathrm{e}^{-(\varepsilon_0^Y - \varepsilon_0^X)/k_B T} \tag{9.10}$$

where $(\varepsilon_0^Y - \varepsilon_0^X) = \Delta\varepsilon_0$, the molecular potential-energy difference between the zero levels of X and Y in the reaction coordinate mode as indicated in Fig. 9.2. Therefore

$$\frac{N_Y}{N_X} = \frac{Q_Y(\mathrm{rc})}{Q_X(\mathrm{rc})} \,.\, \mathrm{e}^{-\Delta\varepsilon_0/k_B T},$$

but $N_Y/N_X = K$, the equilibrium constant for this isomerization, therefore

$$K = \frac{Q_Y(\text{rc})}{Q_X(\text{rc})} \cdot e^{-\Delta\varepsilon_0/k_BT}. \tag{9.11}$$

As introduced in this derivation of eqns (9.10) and (9.11), $\Delta\varepsilon_0$ and the molecular partition functions $Q(\text{rc})$ relate only to the degree of freedom which is included in Fig. 9.2 – the reaction coordinate. This is an over-simplification.

Molecules X and Y have other degrees of freedom according to their molecular complexity, each one of X transforming into a corresponding one of Y as molecule X changes into molecule Y. Each degree of freedom has a partition function at temperature T and, in principle, a zero-point energy all of which must be considered for inclusion in an extended version of eqn (9.11). This is because the statistical thermodynamic equilibrium extends to all degrees of freedom of X and Y, not just the one which is the reaction coordinate. Consider general points first.

If the partition function for a degree of freedom of Y is equal to one for X, these two partition functions do not affect the equilibrium since *their ratio* in the pre-exponential part of an extended eqn (9.11) is unity.

Correspondingly, if the zero-point energy of a degree of freedom of X is equal to one for Y, they also have no net contribution since *their difference* in the exponential term of an extended eqn (9.11) is zero.

We shall now consider these matters in a little more detail, the partition functions first.

The expression for a translational molecular partition function is

$$Q(\text{trans}) = \left(\frac{2\pi m k_B T}{h^2}\right)^{3/2} \cdot V,$$

where m = the mass of the molecule, and
V = the available volume.

Consequently, X and Y, being isomers, have identical translational partition functions at the same temperature and these will cancel.

The precise expression for the molecular partition function of a rotational degree of freedom of a molecule about an axis depends upon its molecular symmetry and moment of inertia. Values for isomers are likely to be very similar indeed, if not identical. So although in principle they should be considered, rotational partition functions are unlikely to make any appreciable contribution in an extended version of eqn (9.11).

This leaves vibrational degrees of freedom other than the reaction coordinate which we have already considered. Ratios of partition functions for molecular vibrations have an intermediate effect – greater than those for translational and rotational modes, but not so great as that for the vibration in the reaction coordinate – and must be included.

The second aspect concerns the zero-point energy terms and is more easily dealt with since the total zero-point energy of all translational and rotational degrees of freedom is zero for reactant and product. So, to summarize, we need really consider for inclusion in an extended version of eqn (9.11) only molecular partition functions and zero-point energy terms for vibrations, and even then only those which are appreciably different for reactant and product.

We now extend eqn (9.11) for the reaction of Fig. 9.2 to give eqn (9.12) which includes *overall* molecular partition functions (Q) and zero-point energy terms (these now on a molar scale, $\Delta E_0 = N_A \,.\, \Delta\varepsilon_0$ and $R = N_A \,.\, k_B$) for all contributing degrees of freedom.

$$K = \frac{Q_Y}{Q_X} \,.\, e^{-\Delta E_0/RT} \tag{9.12}$$

This expression is in fact for a specific case – an isomerization – developed here for the purpose of illustration. The general expression for any chemical equilibrium

$$\nu_A \,.\, A + \nu_B \,.\, B + \ldots \overset{T}{\rightleftharpoons} \nu_X \,.\, X + \nu_Y \,.\, Y \ldots$$

is

$$K = \frac{(Q_X)^{\nu_X} \,.\, (Q_Y)^{\nu_Y} \ldots}{(Q_A)^{\nu_A} \,.\, (Q_B)^{\nu_B} \ldots} \,.\, e^{-\Delta E_0/RT}, \tag{9.13}$$

where $\Delta E_0 = \{\nu_X \,.\, E_0(X) + \nu_Y \,.\, E_0(Y) + \ldots\} - \{\nu_A \,.\, E_0(A) + \nu_B \,.\, E_0(B) + \ldots\}$,
Q_i = the overall molecular partition function of component i, and
$E_0(i)$ = the total molar vibrational zero-point energy of component i.

9.2.2.2 *The effect of isotopes*

We now acknowledge that if an atom in X and Y of Fig. 9.2 is substituted by a heavier isotope, molecules of the new compounds X′ and Y′ will occupy different sets of energy levels within the same potential-energy hypersurface. This is illustrated in Fig. 9.3 with only the zero levels

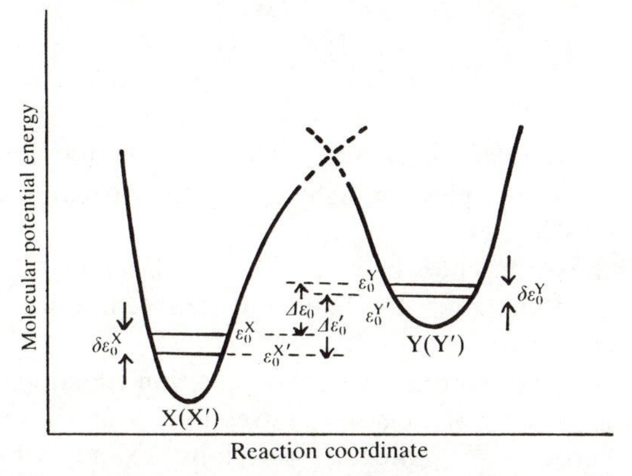

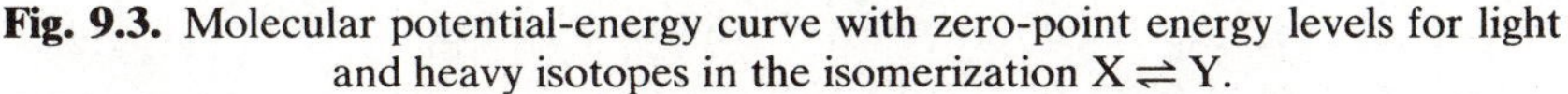

Fig. 9.3. Molecular potential-energy curve with zero-point energy levels for light and heavy isotopes in the isomerization X ⇌ Y.

included in a single dimension – the reaction coordinate. Each level for the heavier compounds X′ and Y′ is lower than the corresponding level with the same quantum number for the lighter analogues X and Y. The equilibrium constant K' for the new reaction

$$\mathrm{X}' \rightleftharpoons \mathrm{Y}'; \qquad K' = \frac{\mathrm{Y}'}{\mathrm{X}'},$$

is given by

$$K' = \frac{Q_{\mathrm{Y}'}}{Q_{\mathrm{X}'}} \,.\, \mathrm{e}^{-\Delta E_0'/RT},$$

and the isotope effect upon the equilibrium is:

$$K/K' = \frac{Q_{\mathrm{Y}}}{Q_{\mathrm{Y}'}} \cdot \frac{Q_{\mathrm{X}'}}{Q_{\mathrm{X}}} \,.\, \mathrm{e}^{-(\Delta E_0 - \Delta E_0')/RT}, \tag{9.14}$$

where the symbols have the same meanings as earlier, the prime indicating terms for the compounds labelled with the heavier isotope.

As seen in Fig. 9.3, for the one degree of freedom which is included,

$$(\Delta\varepsilon_0 - \Delta\varepsilon_0') = (\delta\varepsilon_0^{\mathrm{Y}} - \delta\varepsilon_0^{\mathrm{X}}),$$

where $\delta\varepsilon_0^{\mathrm{X}}$ = the difference between zero-point energies for light and heavy reactant molecules X and X′, and
$\delta\varepsilon_0^{\mathrm{Y}}$ = the difference between zero-point energies for light and heavy product molecules Y and Y′.

On a molar scale, therefore, eqn (9.14) for the equilibrium isotope effect may be replaced by eqn (9.15)

$$K/K' = \frac{Q_{\mathrm{Y}}}{Q_{\mathrm{Y}'}} \cdot \frac{Q_{\mathrm{X}'}}{Q_{\mathrm{X}}} \,.\, \mathrm{e}^{-(\delta E_0^{\mathrm{Y}} - \delta E_0^{\mathrm{X}})/RT}, \tag{9.15}$$

where δE_0^{X} = the difference between the total molar vibrational zero-point energies for light and heavy reactant compounds (X and X′),
δE_0^{Y} = the difference between the total molar vibrational zero-point energies for light and heavy product compounds (Y and Y′),

and, in principle, the overall molecular partition functions Q are for all vibrational and rotational degrees of freedom. But those which hardly change as X becomes Y (molecular rotations and any vibrations largely unaffected by the reaction) will contribute a factor close to unity in the pre-exponential term of eqn (9.15).

9.2.2.3 Reaction rates

From Transition State Theory (Chapter 6, p. 239), we saw that the rate constant k (with proper units) of an elementary reaction is given by

$$k = \frac{k_B T}{h} . K^{\ddagger} (1 \text{ mol dm}^{-3})^{1-n},$$

where $K^{\ddagger}$ = the equilibrium constant for the postulated equilibrium between reactant(s) and the activated complex, and
n = the molecularity (order) of the elementary reaction (1 or 2).
It follows, therefore, that

$$k/k' = K^{\ddagger}/K'^{\ddagger}, \tag{9.16}$$

where k' and $K'^{\ddagger}$ are the rate and activation equilibrium constants for the reaction with the heavier isotope, and exactly the same principles are involved in discussing kinetic isotope effects as in the previous section on equilibrium effects.

As mentioned in the previous section, compounds have a number of degrees of freedom determined by their size and molecular complexity. Depending upon the reaction type and the nature of these degrees of freedom, some or all of them have to be taken into account, not just the one which is the reaction coordinate. To acknowledge this point, we have included in Fig. 9.4 two potential-energy curves as well as the reaction coordinate for the reaction

$$A \underset{(\leftarrow --)}{\rightleftarrows} A^{\ddagger} \longrightarrow \text{products}$$

(but for the sake of clarity, we have left out the quantized energy levels within the reaction coordinate). One is superimposed upon the reaction coordinate where the latter corresponds to reactant, and the other at the transition state. These describe a representative vibration of the reactant which turns into a corresponding vibration of the activated complex. According to the diagram, the force field of this vibration becomes weaker as the molecular transformation takes place, but this will not always be the case. Note also that Fig. 9.4, in contrast to Fig. 9.3, includes energy on the molar scale.

The following general expression for a kinetic isotope effect is obtained by treating the *pseudo*-equilibrium between reactant A and activated complex $A^{\ddagger}$ exactly as in the earlier treatment of a real equilibrium between reactant and product.

$$k/k' = \frac{Q^{\ddagger}}{Q'^{\ddagger}} \cdot \frac{Q_{A'}}{Q_A} . e^{-(\Delta E_0^{\ddagger} - \Delta E_0'^{\ddagger})/RT} \tag{9.17}$$

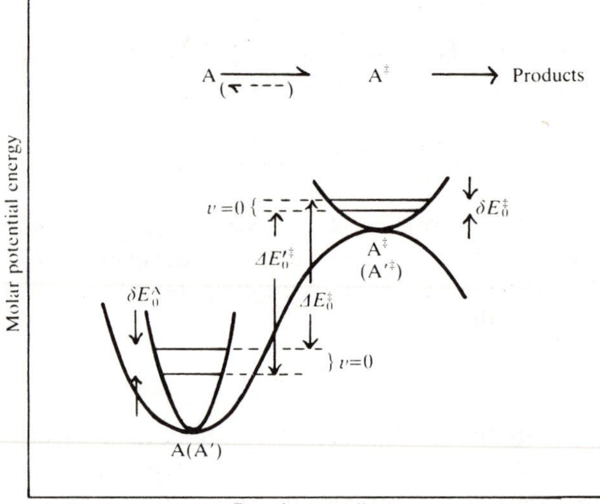

Fig. 9.4. Vibrational potential-energy curves with zero-point energy levels for light and heavy isotopes superimposed upon the reaction coordinate for the formation of the activated complex in a unimolecular reaction.

where Q = the overall molecular partition function (superscript ‡ indicates activated complex $A^{\ddagger}$ and subscript $_A$ indicates reactant),

$\Delta E_0^{\ddagger}$ = the total molar vibrational zero-point energy of activation for the reaction with the light isotope indicated in Fig. 9.4,

and the primed symbols refer to the terms for the reaction with the heavy isotope.

As in the expression for the equilibrium isotope effect, we can transform eqn (9.17) to give eqn (9.18).

$$k/k' = \frac{Q^{\ddagger}}{Q'^{\ddagger}} \cdot \frac{Q_{A'}}{Q_A} \cdot e^{-(\delta E_0^{\ddagger} - \delta E_0^{A})/RT} \tag{9.18}$$

where δE_0^{A} = the difference between the total molar vibrational zero-point energies for light and heavy reactants (A and A′), and

$\delta E_0^{\ddagger}$ = the difference between the total molar vibrational zero-point energies for light and heavy activated complexes ($A^{\ddagger}$ and $A'^{\ddagger}$).

Since the difference between the levels due to light and heavy compounds in the transition state can be much less than in either reactant or product, kinetic isotope effects are frequently much larger than equilibrium isotope effects.

We shall not pursue the mathematical analysis of these equilibrium and kinetic isotope effect expressions any further in general, but we shall return to the overall conclusions. At a molecular level, we can identify two causes for isotopic substitution in a compound bringing about a change in an equilibrium (or rate) constant without changing at all the nature of the reaction. Isotopic substitution changes

(1) the *difference* in total vibrational zero-point energy between reactant and product (or activated complex), and

(2) the *ratio* of the overall molecular partition functions for product (or activated complex) and reactant.

It turns out, however, that, quantitatively, the exponential zero-point energy terms in eqns (9.14), (9.15), (9.17), and (9.18) are invariably dominant and usually account for almost the whole of any observed isotope effect.

9.3 Primary deuterium kinetic isotope effects

9.3.1 Hydrogen transfer between single atoms

We are now in a position to draw upon the above general treatment to obtain an approximate theory for the kinetic isotope effect in a hydrogen-transfer reaction, eqn (9.19).

$$\text{X–L} + \text{Y} \rightleftharpoons \text{X} + \text{L–Y}. \tag{9.19}$$

This reaction is bimolecular and reaction profiles for other bimolecular reactions usually include a translational degree of freedom as the reaction coordinate (see Chapter 6, Fig. 6.5, p. 239). For the hydrogen transfer of eqn (9.19), molecules XL and Y approach each other and, when they are sufficiently close and in the right relative configuration, the hydrogen moves across from X to Y. X and LY then separate to complete the reaction. We shall presume first that the lowest energy path of this reaction maintains the three atoms X···L···Y in a straight line (the right relative configuration) and secondly that, during the actual transfer of L from X to Y, the X···Y internuclear distance does not change significantly. In this second approximation, we are uncoupling the motion of light and heavy atoms and assuming that the former is fast compared with the latter.

The coming together of XL and Y and, if the reaction is in solution, any associated changes in solvation naturally affect the rate of the overall reaction in a major way. But we shall demonstrate that (according to current theory) these translational aspects of the overall process do not affect the primary deuterium kinetic isotope effect. In anticipation, therefore, we represent the hydrogen transfer in Fig. 9.5 by superimposed molecular potential energy curves for the X–L and L–Y stretching vibrations and ignore those energy changes which accompany the initial coming together of XL and Y, re-solvation, and any other essentially translational (as opposed to vibrational) steps.

Figure 9.5, therefore, represents a very simple process: the migration of

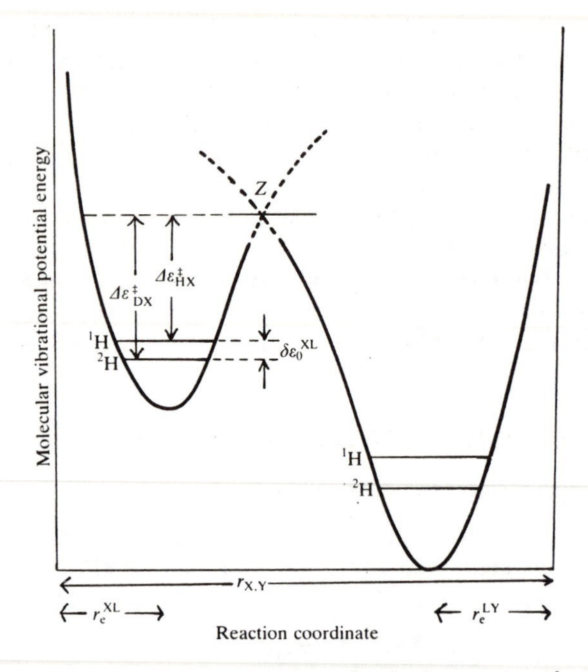

Fig. 9.5. Molecular potential-energy curve for hydrogen transfer between X and Y.

hydrogen from X to Y at a fixed internuclear separation of X and Y ($r_{X,Y}$), so a single structural variable – the position of L within the straight line joining X to Y – determines the potential energy of the system. Since this process is reversible, it can also be seen as the vibration of L within the linear arrangement of atoms X···L···Y. And it does not matter whether a cation L^+, an atom $L^{\cdot}$, or an anion L^- is transferred, so we have left out electrical charges.

At the intersection of the two curves, *Z* in Fig. 9.5, the hydrogen is partially bonded to X and to Y so it never becomes free during the transfer at this particular internuclear separation of X and Y. If we were to consider the hydrogen transfer at a smaller X···Y separation, the heavy atoms are moved closer together (smaller $r_{X,Y}$), the potential energy at the transition state becomes lower (relative to the outer minima), and L becomes less free during its transfer than previously. Conversely, for hydrogen transfer at larger separations of X and Y, the potential energy of the system at the transition state for hydrogen transfer, *Z*, is higher (relative to the outer minima). At very large separations of the heavy atoms, L becomes effectively free during its transfer, the potential-energy barrier being D_0(XL) in the forward direction and D_0(LY) for the reverse. We can, therefore, take this one translational degree of freedom, $r_{X,Y}$, into account by using a series of diagrams such as Fig. 9.5 (but this analysis would break down at very small X···Y distances since then the shapes of the individual potential-energy curves would be grossly affected by small changes in $r_{X,Y}$).

The separation of ^{1}H and ^{2}H vibrational levels will not usually be the same in XL as in LY as indicated for the zero levels in Fig. 9.5 (higher levels have been left out for the sake of clarity). However, for a very small but finite element of the reaction coordinate at *Z*, the transition state for the transfer, the reaction coordinate is really a translation (rather than a vibration) so there can be no difference between ^{1}H and ^{2}H levels. Consequently, there is only the single common (zero-point) energy level for $X\cdots{}^1H\cdots Y$ and $X\cdots{}^2H\cdots Y$ in the reaction coordinate at *Z* as shown. The significance of this will be clearer if we establish the expression for the deuterium kinetic isotope effect, k^H/k^D, for the forward reaction of eqn (9.19).

By eqn (9.16) (p. 379)

$$k^H/k^D = K_H^\ddagger/K_D^\ddagger$$

where $K_L^\ddagger$ = the *quasi*-equilibrium constant for the formation of the activated complex $(X\cdots L\cdots Y)^\ddagger$ from XL and Y, and, by eqn (9.13) (p. 377),

$$K_L^\ddagger = \frac{Q_{XLY^\ddagger}}{Q_{XL}\cdot Q_Y}\cdot e^{-\Delta E_0^\ddagger/RT},$$

where Q_i = the total molecular partition function for species *i*, and
$\Delta E_0^\ddagger$ = the molar zero-point energy of activation (the total molar zero-point energy of the activated complex minus that for XL; Y, being monatomic, has none).

Therefore,

$$k^H/k^D = \frac{Q_{XHY^\ddagger}}{Q_{XDY^\ddagger}}\cdot\frac{Q_{XD}}{Q_{XH}}\cdot e^{-(\Delta E_0^\ddagger - \Delta E_0'^\ddagger)/RT},$$

where the prime indicates the energy term for the heavier isotope.

This equation may be transformed in the same way that eqns (9.14) and (9.17) gave eqns (9.15) and (9.18), respectively, to give

$$k^H/k^D = \frac{Q_{XHY^\ddagger}}{Q_{XDY^\ddagger}}\cdot\frac{Q_{XD}}{Q_{XH}}\cdot e^{-(\delta E_0^\ddagger - \delta E_0^{XL})/RT}$$

where $\delta E_0^\ddagger$ = the change in total molar zero-point energy of the activated complex due to isotopic substitution, and
δE_0^{XL} = the change in total molar zero-point energy of the reactant XL due to isotopic substitution.

If we now factorize the exponential energy term by separating out the component of the activated complex due to the reaction coordinate which alone is included in Fig. 9.5 (reactant XL has only the vibration in the reaction coordinate) we obtain:

$$e^{-(\delta E_0^\ddagger - \delta E_0^{XL})/RT} = e^{-(\delta E_0^\ddagger(rc) - \delta E_0^{XL})/RT}\cdot e^{-\delta E_0^\ddagger(nrc)/RT},$$

where (rc) indicates the zero-point energy in the reaction coordinate (included in Fig. 9.5) and (nrc) indicates zero-point energy terms due to all the other degrees of freedom of the activated complex (not included in Fig. 9.5).

But we have just seen that $X\cdots H\cdots Y^{\ddagger}$ and $X\cdots D\cdots Y^{\ddagger}$ have a common zero-point energy (of zero) in the reaction coordinate, i.e. $\delta E_0^{\ddagger}(\mathrm{rc}) = 0$, therefore

$$k^{\mathrm{H}}/k^{\mathrm{D}} = \frac{Q_{\mathrm{XHY}^{\ddagger}}}{Q_{\mathrm{XDY}^{\ddagger}}} \cdot \frac{Q_{\mathrm{XD}}}{Q_{\mathrm{XH}}} \cdot \mathrm{e}^{\delta E_0^{\mathrm{XL}}/RT} \cdot \mathrm{e}^{-\delta E_0^{\ddagger}(\mathrm{nrc})/RT}. \tag{9.20}$$

So, as far as the energetics of the reaction coordinate are concerned, the kinetic isotope effect is determined by $\delta E_0^{\mathrm{XL}}(= N_{\mathrm{A}} \,.\, \delta\varepsilon_0^{\mathrm{XL}})$ and the actual barrier height is irrelevant. In other words, although the heavy atom separation $r_{\mathrm{X,Y}}$ fixes $\Delta\varepsilon_{\mathrm{HX}}^{\ddagger}$ and $\Delta\varepsilon_{\mathrm{DX}}^{\ddagger}$ (Fig. 9.5), i.e. the actual rates of the atom-transfer reactions, it does not affect $\delta\varepsilon_0^{\mathrm{XL}}$ or, therefore, the reaction coordinate's contribution towards $k^{\mathrm{H}}/k^{\mathrm{D}}$ – the kinetic isotope effect.

The second exponential energy term in eqn (9.20) relates to the vibrational degrees of freedom of the activated complex other than the in-line vibration of L at fixed $X\cdots Y$ separation (the reaction coordinate) and these, so far, we have not considered.

A triatomic system has a total of nine degrees of freedom. There are three translational and, if the system is linear, two rotational modes the energetics of which we may ignore as far as the kinetic isotope effect is concerned since they have no zero-point energy. This leaves four vibrational modes one of which – the asymmetric stretch – we have already dealt with in Fig. 9.5 since this is the reaction coordinate. We are left with a doubly-degenerate out-of-line (bending) vibration:

$$\left(\mathrm{X}\cdots\overset{\uparrow\nearrow}{\underset{\swarrow\downarrow}{\mathrm{L}}}\cdots\mathrm{Y}\right)^{\ddagger}$$

and an in-line symmetric stretching vibration of the linear triatomic system:

$$\left(\underset{\rightarrow}{\overset{\leftarrow}{\mathrm{X}}}\cdots\mathrm{L}\cdots\underset{\leftarrow}{\overset{\rightarrow}{\mathrm{Y}}}\right)^{\ddagger}.$$

A more complete theory of primary deuterium kinetic isotope effects must take these into account but they have the effect of refining the description given here rather than of drastically altering it. Quantitatively, as can be inferred from eqn (9.20), they reduce somewhat the effect due to the vibration of the reaction coordinate. A simple means of acknowledging their existence would be to introduce into Fig. 9.5 representative potential energy curves superimposed at the initial state and the transition state as in Fig. 9.4.

9.3.2 Hydrogen transfer between complex molecules

We can now extend our qualitative account of the primary deuterium kinetic isotope effect to more complex molecules. This corresponds to X and Y in eqn (9.19) and Fig. 9.5 being polyatomic groups. As the number of atoms in XH increases, so does the number of internal degrees of freedom. Initially, we assume that the reaction coordinate in a diagram corresponding to Fig. 9.5 is *essentially* one of the normal vibrational coordinates of XL. The early stages of the transfer of L from XL to Y corresponds to a stretching vibration of the X–L bond (admittedly perturbed by the proximate molecule Y).

But we also have to take into account the other molecular vibrations of XL as the atomic configuration changes to that of the activated complex. There are several stages in our consideration. First, there is the question of whether a vibration changes significantly as reactant becomes activated complex. A localized vibration in a large group X remote from the seat of the isotopic substitution (the reaction site) may not. The potential-energy curve for this type of vibration and the sets of energy levels for ^{1}H and ^{2}H analogues will be much the same in X–L as in $(X\cdots L\cdots Y)^{\ddagger}$. Such a vibration, therefore, makes no significant contribution to a primary deuterium kinetic isotope effect for the reasons discussed in general terms above.

If a vibration does change as reactant becomes activated complex, we have a second point to consider. Is there a decrease in the force field for the vibration, as in Fig. 9.4, or an increase? If the former (force constant, κ, decreases), the differences between X–^{1}H and X–^{2}H levels in this particular vibration are smaller in the activated complex than in the reactant as shown in Fig. 9.4. This vibrational mode, therefore, produces an enhancing contribution towards the overall normal deuterium kinetic isotope effect.

Conversely, if the vibration undergoes a change in force field between reactant and activated complex which leads to an increase in the gaps between ^{1}H and ^{2}H levels (force constant, κ, increases), it will tend to decrease the overall normal isotope effect.

To put these matters into perspective, however, it appears that the overwhelming contribution towards a primary deuterium kinetic isotope effect in a reaction like that of eqn (9.19) is due to the vibration which effectively disappears in the activated complex, i.e. the hydrogen stretching vibration itself which we have taken as the reaction coordinate.

Example. Calculate the primary deuterium kinetic isotope effect in a reaction, the rate-determining step of which involves breaking a carbon–hydrogen bond.

We make two approximations in using eqn (9.20), the general expression for a primary kinetic isotope effect. We shall (1) neglect the pre-exponential terms involving partition functions and (2) assume that the contributions of vibrations other than the C—H/D stretch are negligible. Our simplified version of eqn (9.20) becomes

$$k^{\mathrm{H}}/k^{\mathrm{D}} \sim e^{\delta E_0^{\mathrm{XL}}/RT},$$

where δE_0^{XL} = the difference in the zero-point energy for C—H and C—D stretching vibrations of the reactant. From infra-red spectroscopy,[13] $\bar{\nu} = 2914$ and $2085\ \mathrm{cm}^{-1}$ for the symmetric stretches of CH_4 and CD_4 respectively.

The zero-point energies of C—H and C—D stretching modes, on a molar scale, are easily calculated.

$$\begin{aligned} \varepsilon_0 &= \tfrac{1}{2}h\nu \\ &= \tfrac{1}{2}hc\bar{\nu} \end{aligned}$$

therefore,

$$\begin{aligned} \delta E_0^{\mathrm{XL}} &= \tfrac{1}{2}hc \times N_{\mathrm{A}}(\bar{\nu}_{\mathrm{CH}} - \bar{\nu}_{\mathrm{CD}}) \\ &= 0.5 \times 6.6262 \times 10^{-34}\ \mathrm{J\,.\,s} \times 3.00 \times 10^{10}\ \mathrm{cm\,.\,s^{-1}} \times 6.02 \\ &\qquad \times 10^{23}\ \mathrm{mol^{-1}}\ (2914\text{–}2085)\ \mathrm{cm}^{-1} \\ &= 4.96\ \mathrm{kJ\ mol^{-1}}, \end{aligned}$$

therefore

$$k^{\mathrm{H}}/k^{\mathrm{D}} \sim e^{4.96\ \mathrm{kJ\ mol^{-1}}/RT}$$

$$k^{\mathrm{H}}/k^{\mathrm{D}} \sim 7.5 \text{ at } 25\ ^{\circ}\mathrm{C}.$$

Some experimental results are shown in Table 9.1. In all cases they confirm that the rate-determining step of the reaction involves cleavage of a bond to hydrogen. Note that in all cases except one, the hydrogen is transferred from carbon to either oxygen or nitrogen.

9.4 Secondary deuterium kinetic isotope effects

When a reaction involves no rebonding at the isotopically-labelled hydrogen atom, the gaps between ^{1}H and ^{2}H levels in the vibrational mode which approximates the reaction coordinate undergo only very small or no changes as reactant becomes activated complex. Depending upon the nature of the reaction, however, there may be appreciable changes in the gaps between ^{1}H and ^{2}H levels in other vibrations of the molecule which more directly involve the isotopic atom. These changes will be large or small depending upon how distant the isotopic atom is from the reaction site, and upon the nature of the reaction. We can use Fig. 9.4 (p. 380) to represent a reaction in which virtually the whole isotope effect resides in

---

Notes (*continued from Table* 9.1)

[f] DMSO = dimethyl sulphoxide, $(CH_3)_2SO$; S. B. Kaldor and W. H. Saunders, *J. chem. Phys.* **68,** 2509 (1978).

[g] W.-B. Chiao and W. H. Saunders, *J. Am. chem. Soc.* **100,** 2802 (1978).

[h] R. P. Bell, *The proton in chemistry*, 1st edn, p. 201, Cornell University Press (1959).

[i] F. H. Westheimer and N. Nicolaides, *J. Am. chem. Soc.* **71,** 25 (1949).

TABLE 9.1
Primary deuterium kinetic isotope effects[a]

Reaction	k^H/k^D	*Note*
2-methylpyridine + $NO_2CL_2CO_2C_2H_5$ $\xrightarrow[25\,°C]{H_2O}$ 2-methylpyridinium-$\overset{+}{N}$L + $NO_2\bar{C}LCO_2C_2H_5$	9.6	*b*
$H_2O + CH_3\overset{O}{\overset{\|}{C}}CL_2SO_3^- Na^+ \xrightarrow[25\,°C]{H_2O} H_2\overset{+}{O}L + CH_3\overset{O}{\overset{\|}{C}}\bar{C}LSO_3^- Na^+$	2.5	*b*
$CH_3CO_2^- + CH_3\overset{O}{\overset{\|}{C}}CL(CH_3)CO_2C_2H_5 \xrightarrow[25\,°C]{H_2O} CH_3CO_2L + CH_3\overset{O}{\overset{\|}{C}}\bar{C}(CH_3)CO_2C_2H_5$	5.92	*c*
$ArCH(O^-)\text{—}\overset{+}{N}H_2OCH_3 + CN(CH_2)_2\overset{+}{N}L_3 \xrightarrow[Ar = pCH_3OC_6H_4]{L_2O} ArCH(OL)\text{—}\overset{+}{N}H_2OCH_3 + CN(CH_2)_2NL_2$	2.8	*d*
$C_2H_5O^- + C_6H_5CL(CH_3)CH_2Br \xrightarrow[25\,°C]{C_2H_5OH} C_2H_5OL + C_6H_5C(CH_3){=}CH_2 + Br^-$	7.51	*e*
$OH^- + C_6H_5CL_2CH_2\overset{+}{N}(CH_3)_3 \xrightarrow[50\,°C]{H_2O:DMSO(1:1)} C_6H_5CL{=}CH_2 + LOH + N(CH_3)_3$	3.89	*f*
$C_6H_5CL(L)\text{—}CH_2\text{—}\overset{+}{N}(CH_3)_2\text{—}O^-$ (cyclic) $\xrightarrow[59.8\,°C]{DMSO} C_6H_5CL{=}CH_2 + (CH_3)_2NOL$	2.87	g
$CL_3\overset{O}{\overset{\|}{C}}CL_3 + OH^- \xrightarrow[25\,°C]{H_2O} CL_3\overset{O}{\overset{\|}{C}}\bar{C}L_2 + HOL$	10.2	*h*
$(CH_3)_2CLOH \xrightarrow[(oxid.);\ 40\,°C]{Cr(VI),\ H_3O^+} (CH_3)_2C{=}O$	5.9	*i*

Notes

[a] See also eqns (9.4) and (9.5), p. 369.
[b] Ref. 7.
[c] R. P. Bell and J. E. Crooks, *Proc. R. Soc.* A **286,** 285 (1965).
[d] N.-A. Bergman, Y. Chiang, and A. J. Kresge, *J. Am. chem. Soc.* **100,** 5954 (1978).
[e] V. J. Shiner and M. L. Smith, ibid. **83,** 593 (1961).

molecular vibrations whose potential energies are represented by curves superimposed upon the reaction coordinate. Because of dimensional restrictions, we have in Fig. 9.4 only one representative vibration of the reactant and the one which this becomes in the transition state. In each curve, the two horizontal lines represent the zero energy levels for ^{2}H and ^{1}H analogues.

As there is a 'loosening' of the vibration involving the isotopic atom in Fig. 9.4, the gap between ^{1}H and ^{2}H levels decreases and this vibration contributes towards a normal isotope effect ($k^H/k^D > 1$). However, the decrease in the gap between ^{1}H and ^{2}H levels for such a vibration will never be as large as in the hydrogen stretching vibration for a hydrogen-transfer reaction when that degree of freedom is the reaction coordinate (Fig. 9.5). Consequently, secondary deuterium kinetic isotope effects are smaller than primary ones and are seldom larger than $k^H/k^D \sim 1.25$ (per deuterium).

On the other hand, if the force field for a vibration involving the isotopic atom increases with progress along the reaction coordinate, the vibrational potential-energy curve becomes steeper (as in Fig. 9.6), the quantized levels become further apart, and the gaps between ^{1}H and ^{2}H levels increase. Such a vibrational change would contribute towards an inverse kinetic isotope effect ($k^H/k^D < 1$). Results of $k^H/k^D < 0.80$ are rare.

If there is no change at all in the gaps between ^{1}H and ^{2}H levels as reactant becomes activated complex, i.e. if there is no interaction at all

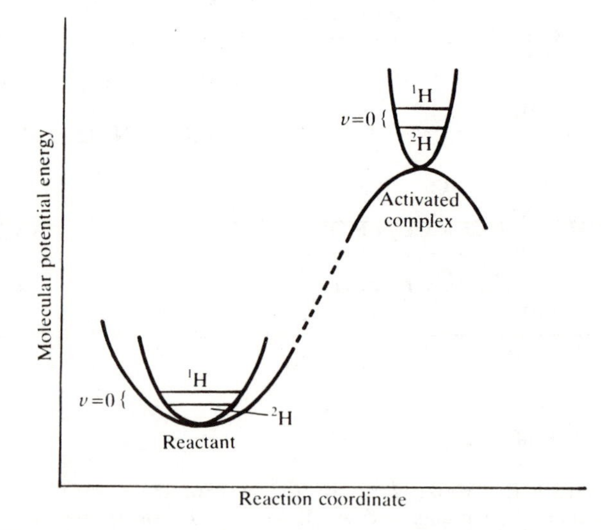

Fig. 9.6. Potential-energy curves for the formation of an activated complex with an increase in the force field for a vibration involving the labelled hydrogen.

between vibrations involving the isotopic label and the reaction coordinate, then there will be no kinetic isotope effect and $k^{H} = k^{D}$. Some results for a variety of reaction types are given in Table 9.2.

Although it is possible that a single vibrational mode may dominate a secondary deuterium kinetic isotope effect, the overall effect in principle will always be a composite of several vibrational contributions perhaps with opposing tendencies. The α-deuterium effects for the bimolecular substitution reactions in Table 9.3 range between $k^{H}/k^{D} \sim 0.96$–1.18 (per deuterium) according to the nature of the nucleophile, the leaving group, the alkyl residue, and the temperature. When the leaving group and nucleophile both bond through oxygen, the result is usually very close to 1.00 (25 °C), and this has been used as an empirical criterion of an S_N2 mechanism.

In contrast, a larger effect of $k^{H}/k^{D} \sim 1.19$–1.25 (25 °C) is found for the α-deuterium k.i.e. in an S_N1 reaction of a secondary alkyl compound with oxygen-bonded leaving group and nucleophile (Table 9.4). Here the methine C–H/D bending vibration of the saturated carbon which becomes a 'looser' out-of-plane bending vibration at the developing trigonal carbon in the incipient carbonium ion appears to dominate the overall effect when ionization is rate determining.[18,19] If the rate-determining step of the solvolysis involves a fully developed carbonium ion, i.e. when separation of a reversibly formed intimate ion-pair is rate determining, the α-k.i.e. is maximal at $k^{H}/k^{D} \sim 1.25$ (25 °C) since the out-of-plane C–H/D bending vibration at the α-carbon is now uninhibited by a vestigially-bonded leaving group in the activated complex.[20,21]

This clear distinction between an α-k.i.e. (25 °C) of *ca.* 1.00 for S_N2 reactions and $\sim$1.19–1.25 for S_N1 reactions of secondary alkyl substrates with oxygen-bonded leaving groups and nucleophiles has been established for reactions whose mechanisms are otherwise well-founded.[18] A low α-k.i.e. has subsequently been used as a criterion of nucleophilic assistance in substitution reactions, in particular to detect neighbouring group participation (intramolecular catalysis).[15,18]

The *cis*-2-acetoxycyclohexyl sulphonate (1) undergoes a slow solvolysis in 97:3, trifluoroethanol:water with an α-deuterium k.i.e. = 1.20 (93 °C).[22]

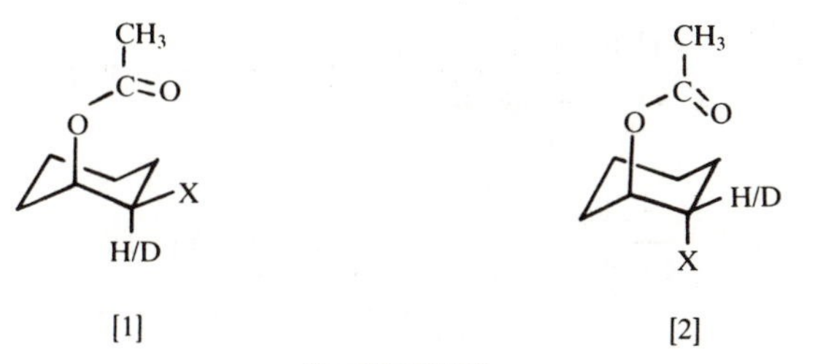

TABLE 9.2
Secondary deuterium kinetic isotope effects[a]

Reaction	k^H/k^D	*Note*
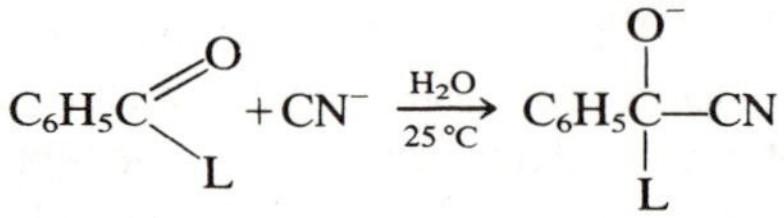 $C_6H_5C(=O)L + CN^- \xrightarrow[25\,°C]{H_2O} C_6H_5C(O^-)(L){-}CN$	0.833	*b*
$LCO_2R + H_2O \xrightarrow[R=CH_3,\ C_2H_5;\ 25°C]{H_3O^+} LCO_2H + ROH$	0.81	*c*
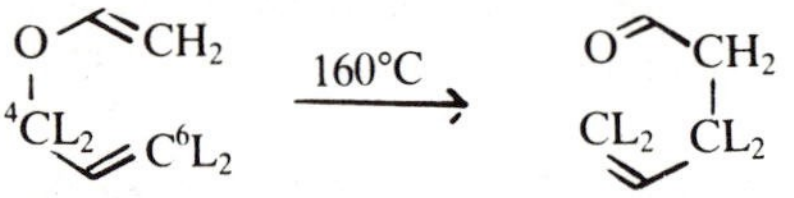 $O{-}CH{=}CH_2,\ {}^4CL_2{-}CH{=}C^6L_2 \xrightarrow{160°C} O{=}CH{-}CH_2{-}CL_2{-}CL_2{-}CH{=}$	At C-4, 1.092 at C-6, 0.976	*d*
Racemization of 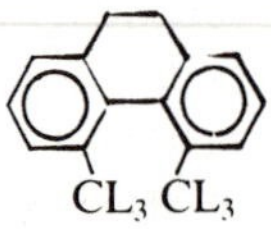 [bridged biphenyl with CL_3 CL_3], benzene, 42°C,	0.88	*e*
Hydrolysis of $CL_3CH_2CH_2Br$ (80 °C)	0.921	*f*
Hydrolysis of $CH_3CL_2CH_2Br$ (80 °C)	1.054	*f*
Solvolysis of $(CL_3)_3CCl$ (60E, 25 °C)	2.327(1.098)	g
Solvolysis of  [norbornyl: L, H, OBs] (80E, 54.5°C)	1.000	*h*
Solvolysis of  [norbornyl: L, H, OBs] (80E, 25.5°C)	1.081	*i*
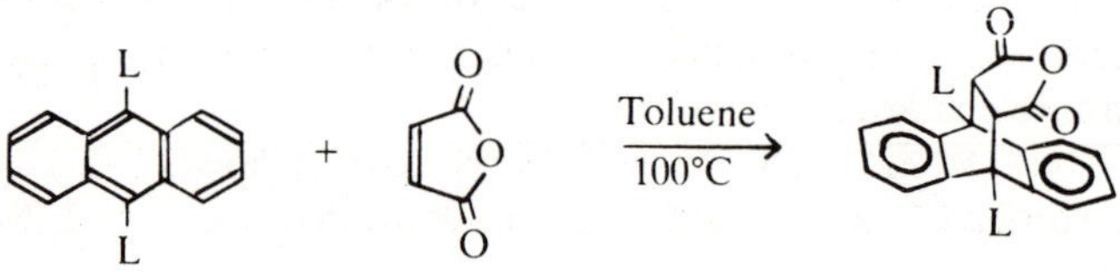 [9,10-L-anthracene] + [maleic anhydride] $\xrightarrow[100°C]{Toluene}$ [adduct with L, L]	0.91(0.95)	*j*
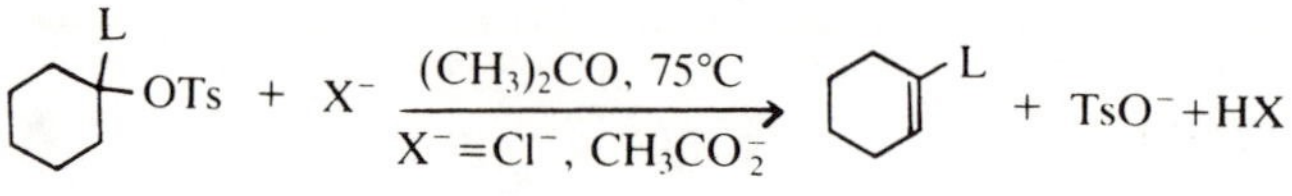 [1-L-cyclohexyl OTs] + $X^- \xrightarrow[X^-=Cl^-,\ CH_3CO_2^-]{(CH_3)_2CO,\ 75°C}$ [1-L-cyclohexene] + $TsO^- + HX$	1.14	*k*
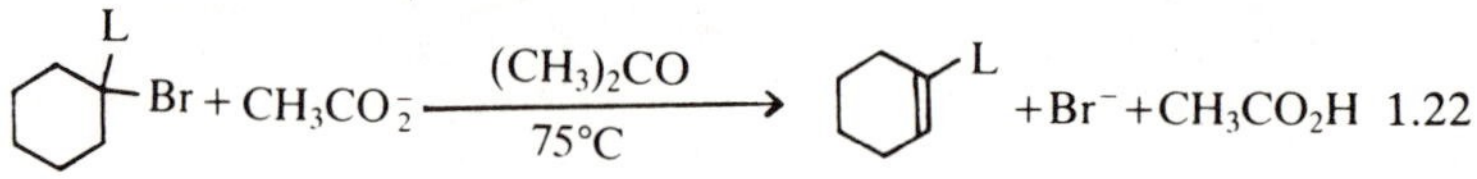 [1-L-cyclohexyl Br] + $CH_3CO_2^- \xrightarrow[75°C]{(CH_3)_2CO}$ [1-L-cyclohexene] + $Br^- + CH_3CO_2H$	1.22	*k*

This is the value expected of a reaction by an S_N1 mechanism with no nucleophilic involvement by the solvent or the acetoxy group (which is above the same face of the molecule as the leaving group X).

The α-deuterium k.i.e. for the much faster solvolysis of the *trans*-isomer (2), $k^H/k^D = 1.03$ (55 °C), is considerably lower and not at all characteristic of a mechanism involving a simple secondary carbonium ion intermediate; it is symptomatic of a mechanism in which the leaving group is displaced by some nucleophile. This suggests that the rate-determining step in the solvolysis of (2) is a rear-side displacement of the sulphonate by the acetoxy group to give an intermediate (3) which then suffers nucleophilic capture by water in a very easy subsequent step. In

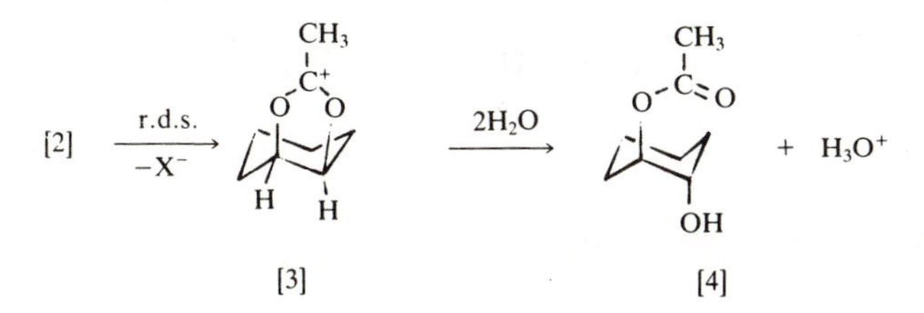

accordance with these interpretations, (2) cleanly gives a single product (4) corresponding to overall retention of relative configuration (but presumably with racemization if (2) were chiral since the intermediate (3) has a mirror plane). In contrast, (1) gives a number of products characteristic of a reaction through a simple secondary carbonium ion, the principal ones corresponding to substitution with overall inversion of relative configuration and, presumably, without racemization since no achiral intermediates are involved.

We close this section on secondary deuterium kinetic isotope effects by emphasizing two matters. The first is that the effects are always small and k^H/k^D values between 0.95 and 1.05 are commonly quoted as being

---

Notes

[a] See also eqn (9.3), p. 368. In all cases, the effect is for the number of deuteriums shown; the result calculated per deuterium atom is shown in parentheses.
[b] V. Okano, L. do Amaral, and E. H. Cordes, *J. Am. chem. Soc.* **98,** 4201 (1976).
[c] Z. Bilkadi, R. de Lorimier, and J. F. Kirsch, ibid. **97,** 4317 (1975).
[d] J. J. Gajewski and N. D. Conrad, ibid. **101,** 2747 (1979).
[e] K. Mislow, R. Graeve, A. J. Gordon, and G. H. Wahl, ibid. **85,** 1199 (1963).
[f] Ref. 14.
[g] 60E = ethanol : water, 60 : 40 by volume; V. J. Shiner, B. L. Murr, and G. Heinemann, *J. Am. chem. Soc.* **85,** 2413 (1963).
[h] OBs = *p*-bromobenzenesulphonate, 80E = ethanol : water, 4 : 1 by volume; H. Maskill, *J. chem. Soc., Perkin 2* 1889 (1976).
[i] Ref. 15.
[j] Ref. 4.
[k] Ref. 16.

TABLE 9.3

α-Deuterium kinetic isotope effects in S_N2 reactions[a]

Reaction	k^H/k^D	Note
$CL_3I + N_3^- \xrightarrow[40\,°C]{H_2O} CL_3N_3 + I^-$	0.907	*b*
$CL_3I + AcO^- \xrightarrow[40\,°C]{H_2O} CL_3OAc + I^-$	0.882	*b*
$CL_3OTs + 2H_2O \xrightarrow[70\,°C]{H_2O} CL_3OH + TsO^-H_3O^+$	0.96(0.984)	c
$CH_3CL_2OTs + 2H_2O \xrightarrow[54.3\,°C]{H_2O} CH_3CL_2OH + TsO^-H_3O^+$	1.038(1.020)	*d*
$CH_3CH_2CL_2Br + 2H_2O \xrightarrow[80\,°C]{H_2O} CH_3CH_2CL_2OH + Br^-H_3O^+$	0.980	*e*
$PhCL_2Cl + N_3^- \xrightarrow[60\,°C]{\text{aqueous acetone}} PhCL_2N_3 + Cl^-$	(1.033)	*f*
$PhCL_2OBs + 2HY \xrightarrow[25\,°C]{HY(95E)} PhCL_2Y + BsO^-H_2Y^+$	(1.053)	g
cyclohexyl(L)OTs + $AcO^- \xrightarrow[75°C]{\text{acetone}}$ cyclohexyl(L)OAc + TsO^-	1.05	*h*
cyclohexyl(L)Br + $Cl^- \xrightarrow[75°C]{\text{acetone}}$ cyclohexyl(L)Cl + Br^-	1.08	*h*
$CH_3O-C_6H_4-CL_2Br + X^- \xrightarrow[25°C]{\text{aqueous dioxan}} CH_3O-C_6H_4-CL_2X + Br^-$		*i*
$X = NEt_3$	(1.014)	
$X^- = OH^-$	(1.022)	
$X^{2-} = S_2O_3^{\ 2-}$	(1.045)	
$X^- = N_3^-$	(1.050)	
$X^- = SCN^-$	(1.069)	
$PhCL_2\overset{+}{N}(CH_3)_2Ph + PhS^- \xrightarrow[0\,°C]{DMF} PhCL_2SPh + PhN(CH_3)_2$	1.179(1.086)	*j*
$CH_3O{-}CL_2{-}\overset{+}{N}(Me)_2Ar + X^- \xrightarrow[H_2O,\ 25\,°C]{Ar = m\text{-}NO_2C_6H_4} CH_3O{-}CL_2{-}X + ArNMe_2$		*k*
$X^- = F^-$	(0.99)	
CN^-	(1.08)	
I^-	(1.18)	

Notes

[a] The effect is for the number of deuteriums shown except for those in parentheses which are results per deuterium.

[b] C. M. Won and A. V. Willi, *J. phys. Chem.* **76,** 427 (1972).

mechanistically significant. In order to rely upon such results, a high degree of precision is required in the rate constant determinations. If the ^1H and ^2H compounds are investigated on different occasions, it is very unlikely that the experimental conditions could be reproduced sufficiently well for us to have any confidence in the comparison. k^H and k^D have to be measured at the same time using identical conditions to give a ratio k^H/k^D. This should be repeated several times so that a mean value can be quoted along with some indication of the precision of the result.

The second point is that the theory of secondary deuterium kinetic isotope effects is less well developed than that for primary effects and the range of reactions in which they are measured much more diverse. Consequently, the use of secondary deuterium kinetic isotope effect measurements in mechanistic investigations tends to be rather empirical. Results are measured first for reactions whose mechanisms ostensibly are otherwise understood. These results are then used as criteria of those mechanisms and taken in the next stage as standards against which new results may be compared. Although an empirical approach should not be scorned (frequently it may be the only way forward), secondary deuterium kinetic isotope effect measurements should be considered only when they are known to be precise and then always with circumspection and in the light of as much other evidence as is available.

9.5 Heavy atom kinetic isotope effects[18,23]

We have already seen that any kinetic (or equilibrium) isotope effect is due ultimately to the different masses of isotopes, and the same principles are involved regardless of whether the isotopic substitution is of hydrogen or a much heavier atom, e.g. a halogen. Heavy atom effects are also designated either primary or secondary according to whether or not bonds to the heavy atom are broken and made in the rate-determining step.

Notes (*continued from Table* 9.3)

[c] J. A. Llewellyn, R. E. Robertson, and J. M. W. Scott, *Canad. J. Chem.* **38,** 222 (1960); calculated value for 1D corrected to 25 °C.[18]

[d] K. T. Leffek, J. A. Llewellyn, and R. E. Robertson, ibid., p. 1505; calculated value for 1D corrected to 25 °C.[18]

[e] Ref. 14.

[f] Ref. 17.

[g] 95E = ethanol : water, 95 : 5 by volume; OBs = *p*-bromobenzenesulphonate; V. J. Shiner, M. W. Rapp, and H. R. Pinnick, *J. Am. chem. Soc.* **92,** 232 (1970).

[h] Reference 16.

[i] V. P. Vitullo, J. Grabowski, and S. Sridharan, *J. Am. chem. Soc.* **102,** 6463 (1980).

[j] K. C. Westaway and S. F. Ali, *Canad. J. Chem.* **57,** 1089 (1979).

[k] B. L. Knier and W. P. Jencks, *J. Am. chem. Soc.* **102,** 6789 (1980).

TABLE 9.4
α-Deuterium kinetic isotope effects in S_N1 solvolytic reactions of secondary alkyl sulphonates

Compound[a]	*Solvent*[b] *(Temp.)*	k^H/k^D	*Note*
$(CH_3)_2CLOTs$	TFA (25 °C)	1.22	*c*
L, OBs	70E (25 °C)	1.187	*d*
	90HFIP (25 °C)	1.23	*e*
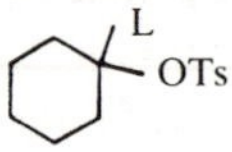 L, OTs	CH_3CO_2H (50 °C)	1.22	*f*
OTs, L	50E (44.8 °C)	1.200	g
	97HFIP (40 °C)	1.232	
L, OTs	50E (44.8 °C)	1.16	g
	97HFIP (56.2 °C)	1.175	
OTs, L	50E (24.8 °C)	1.214	g
	97HFIP (25.2 °C)	1.246	
L, OTs	50E (36 °C)	1.198	g
	97HFIP (25.2 °C)	1.234	
 L, OBs	80E (55 °C)	1.193	*h*
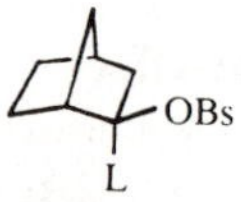 OBs, L	80E (25 °C)	1.124	*h*
OTr, L	50E (25 °C)	1.225	*i*
	97TFE (25 °C)	1.228	

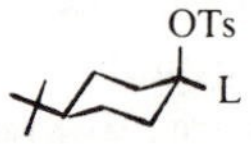

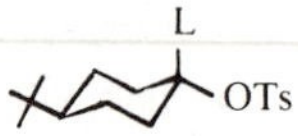

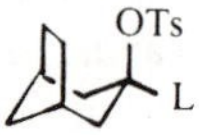

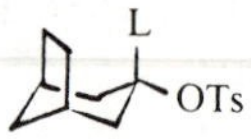

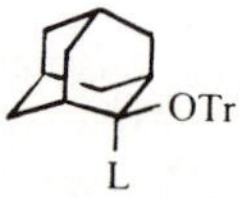

Notes

[a] OTs = *p*-Toluenesulphonate; OBs = *p*-Bromobenzenesulphonate; OTr = 2,2,2-Trifluoroethanesulphonate.
[b] TFA = Trifluoroacetic acid; 70E = Ethanol : water, 7 : 3 by volume; 90 HFIP = Hexafluoropropan-2-ol : water, 9 : 1 by weight; 50E = Ethanol : water, 1 : 1 by volume; 97 HFIP = Hexafluoropropan-2-ol : water, 97 : 3 by weight; 80E = Ethanol : water, 4 : 1 by volume.
[c] A. Streitwieser and G. A. Dafforn, *Tetrahedron Lett.* 1263 (1969).

Adequate precision (±0.1 and ±0.01 in the rate ratio k/k' for primary and for secondary effects, respectively) can usually be obtained for deuterium kinetic isotope effect measurements by direct comparisons of individually-determined rate constants from separate (but ideally simultaneous) experiments using protium and deuterium analogues, the latter of high isotopic purity. Quantitatively, much smaller effects are to be expected for heavy atoms since their isotopic mass ratios are much closer to unity, and it is this *ratio*, not the *difference*, which matters. This means that greater experimental precision is necessary (∓0.0005 or better for k/k').

Furthermore, compounds with a high incorporation of an uncommon heavy isotope are very expensive, so it is seldom feasible to experiment with such isotopically-pure materials. An entirely different technique is necessary. A compound containing an accurately known ratio of two isotopes is reacted with an unlabelled material in a competition experiment. The ratio of the two isotopes is then accurately measured again in residual starting compound or in the product. From the difference in the isotopic ratio before and after a known extent of reaction, the kinetic isotope effect can be calculated.[23] Since the reactions of the isotopically-different compounds take place in the same reaction mixture, there can be no uncertainty due to non-identical reaction conditions.

Mass spectrometry is normally used to measure isotopic ratios of stable isotopes, e.g. $^{12}C/^{13}C$, $^{14}N/^{15}N$, or $^{35}Cl/^{37}Cl$. For non-natural isotopes, e.g. ^{14}C, a radioactivity-measuring device such as a scintillation counter is used. In both methods, there must be many checks and control experiments to establish high precision for the method in the particular application.

There is also a difficulty of another kind. No theory of heavy atom isotope effects is sufficiently well developed yet to allow reliable quantitative *ab initio* predictions on the basis of reaction mechanisms. Nevertheless, to establish the existence (or otherwise) of an effect, or whether it is normal or inverse, is often sufficient in itself to narrow down a range of mechanistic possibilities for a particular reaction.

Only a few heavy elements have actually been used in isotope effect measurements, and in most cases the choice of isotopes is dictated by

Notes (*continued from Table* 9.4)

[d] J. O. Stoffer and J. D. Christen, *J. Am. chem. Soc.* **92,** 3190 (1970).

[e] R. C. Seib, V. J. Shiner, V. Sendijarevic, and K. Humski, ibid. **100,** 8133 (1978).

[f] W. H. Saunders and K. T. Finley, ibid. **87,** 1384 (1965).

[g] Ref. 20.

[h] B. L. Murr and J. A. Conkling, *J. Am. chem. Soc.* **92,** 3462 (1970); see also ref. 15.

[i] Ref. 21.

availability, e.g. chlorine, ^{35}Cl and ^{37}Cl, bromine, ^{79}Br and ^{81}Br. In the case of carbon, however, as in the case of hydrogen, there is a choice. ^{13}C may be used and the ratio $^{12}C/^{13}C$ can be measured accurately at natural abundance levels by mass spectrometry. ^{14}C has also been used and has the advantage that it leads to a larger isotope effect because of the greater $^{12}C/^{14}C$ mass ratio. There are also disadvantages, however. ^{14}C is a radioactive, man-made isotope and consequently its use requires synthesis of the substrate with the isotopic label incorporated at some specific site within the molecule. A radioactivity counter is then required to measure $^{12}C/^{14}C$ isotopic content. We shall not deal any further with the relative merits of ^{13}C and ^{14}C in isotope effect measurements. It is important, however, to note which carbon isotopes have been used when comparing results from different experiments.

Kinetic isotope effects in the thermal decarboxylation of malonic acid, eqn (9.21), have been measured for both carbons.[24]

$$\text{*CO}_2\text{H}-{}^{\blacktriangle}\text{CH}_2-\text{CO}_2\text{H} \xrightarrow{154°\text{C}} {}^{\blacktriangle}\text{CH}_3\text{CO}_2\text{H} + \text{*CO}_2 \qquad (9.21)$$

At * $k^{12}/k^{14} = 1.065$ and at ▲ $k^{12}/k^{14} = 1.076$. These results establish that cleavage of the carbon–carbon bond is involved in the rate-determining step implicating a cyclic mechanism.

$$\text{HO}-\text{C}(=\text{O})-\text{CH}_2-\text{C}(=\text{O})\text{OH} \xrightarrow[\text{r.d.s.}]{} \left[\text{O}{=}\text{C}\cdots\text{CH}_2\cdots\text{C(OH)}{=}\text{O}\cdots\text{H}\cdots\text{O (cyclic)}\right]^{\ddagger} \longrightarrow \text{CO}_2 + \left[\text{CH}_2{=}\text{C(OH)}-\text{OH}\right] \longrightarrow \text{CH}_3\text{CO}_2\text{H}$$

In the Mn^{2+}-catalysed decarboxylation of oxaloacetic acid (see p. 319), there is also an effect so, here again, carbon–carbon bond cleavage is involved in the rate-determining step.[25]

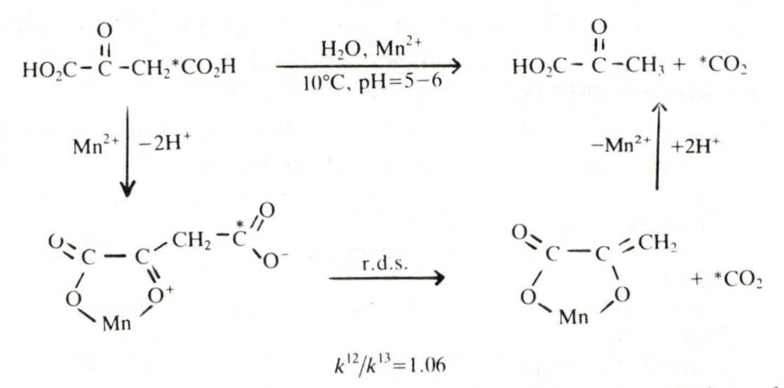

However, in the corresponding enzyme-catalysed decarboxylation, there is no carbon isotope effect so, in this case, a different step must be rate determining.[25]

Some ^{13}C and ^{14}C isotope effects are given below for various substitution reactions.

$$^{\blacktriangle}CH_3{-}I + {}^{*}CN^- \xrightarrow[11.4\,°C]{H_2O^{26}} {}^{\blacktriangle}CH_3{-}{}^{*}CN + I^- \qquad (9.22)$$

$$k^{12}/k^{13} = 1.0733(\mp 0.0056) \text{ at } \blacktriangle, \text{ and}$$
$$k^{12}/k^{13} = 1.0149(\mp 0.0020) \text{ at } *.$$

$$C_6H_5{-}{}^{*}CL_2Cl + Y^- \xrightarrow[60\,°C]{\text{aqueous acetone}^{17}} C_6H_5{-}{}^{*}CL_2Y + Cl^-$$

$$k^{12}/k^{14} = 1.130(\pm 0.007), \text{ and}$$
$$k^H/k^D = 1.033 \text{ per deuterium for } Y^- = N_3^-;$$
$$k^{12}/k^{14} = 1.085(\pm 0.007), \text{ and}$$
$$k^H/k^D = 1.003 \text{ per deuterium for } HY = H_2O.$$

$$C_6H_5{-}{}^{*}CH_2O_3SC_6H_5 + C_6H_5N(CH_3)_2 \xrightarrow[45\,°C]{(CH_3)_2CO^{27}}$$
$$C_6H_5{-}{}^{*}CH_2\overset{+}{N}(CH_3)_2C_6H_5 \;\; \overset{-}{O}_3SC_6H_5$$

$$k^{12}/k^{14} = 1.135(\pm 0.003).$$

These results are compatible with the established S_N2 mechanism for substitution as also are the secondary deuterium kinetic isotope effects in those reactions with doubly-labelled substrates. Accordingly, there should be appreciable heavy atom kinetic isotope effects in the nucleophile and the leaving group. This has been established for $^{*}CN^-$ in eqn (9.22), and for chlorine, bromine, and nitrogen leaving groups in the following examples.

$$NO_2{-}C_6H_4{-}CH_2{-}{}^{*}Cl + Y^- \xrightarrow[30°C]{\text{aqueous dioxan}^{28}} NO_2{-}C_6H_4{-}CH_2{-}Y + {}^{*}Cl^-$$

$k^{35}/k^{37}(\mp 0.0002)$	$Y^-(YH)$
1.0057	CN^-
1.0058	$S_2O_3^{2-}$
1.0076	H_2O .

$$n\text{-}C_4H_9{-}{}^{*}X + C_6H_5S^- \xrightarrow[20\,°C]{CH_3OH^{5.29}} n\text{-}C_4H_9SC_6H_5 + {}^{*}X^-$$

$$k^{35}/k^{37} = 1.0089_5(\pm 0.00015) \quad \text{for} \quad X^- = Cl^-$$
$$k^{79}/k^{81} = 1.00169(\pm 0.00003) \quad \text{for} \quad X^- = Br^-.$$

$$ArCL_2{}^{*}\overset{+}{N}(CH_3)_2C_6H_5 + C_6H_5S^- \xrightarrow[0\,°C]{\text{dimethylformamide}^{30}} ArCL_2SC_6H_5$$
$$+ C_6H_5{}^{*}N(CH_3)_2 \qquad (9.23)$$

Ar	k^{H}/k^{D} (per deuterium)	k^{14}/k^{15}
p-$CH_3OC_6H_4$	1.099	1.01974(±0.00034)
C_6H_5	1.086	1.0200(±0.0007)
p-ClC_6H_4	1.073	1.0202(±0.0009).

Somewhat smaller kinetic isotope effects at carbon but larger ones in the leaving group are found for S_N1 reactions:

methanolysis of C_6H_5—*$CH(CH_3)$—Br, 25 °C[31],

$$k^{12}/k^{13} = 1.0065(\pm 0.0007);$$

hydrolysis (aqueous dioxan) of $(CH_3)_3$*CCl, 25 °C[32],

$$k^{12}/k^{14} = 1.027(\pm 0.015);$$

methanolysis of $(CH_3)_3C$—*X, 20 °C[5,29],

$$k^{35}/k^{37} = 1.0105_8(\pm 0.00015) \quad \text{for} \quad X = Cl,$$

$$k^{79}/k^{81} = 1.00310(\pm 0.00004) \quad \text{for} \quad X = Br.$$

Small carbon isotope effects have also been detected in substitution reactions at unsaturated carbon (mechanistically, addition–elimination):[33]

$$C_6H_5\text{—}{}^{*}C(=O^{\blacktriangle})\dot{O}CH_3 + OH^- \xrightarrow[25°C]{H_2O} C_6H_5CO_2^- + CH_3OH$$

$$k^{12}/k^{13} = 1.0426(\pm 0.0026) \quad \text{at} \quad *,$$

$$k^{16}/k^{18} = 1.0046(\pm 0.0020) \quad \text{at} \quad \blacktriangle,$$

$$k^{16}/k^{18} = 1.0062(\pm 0.0006) \quad \text{at} \quad \bullet.$$

As indicated, oxygen isotope effects were also measured in this investigation and shown to be quite small, but present.

Primary nitrogen kinetic isotope effects have been measured in substitution reactions[30] (see eqn (9.23) above) and elimination reactions of substituted alkylammonium cations.[18,23] In the elimination reactions where there is other evidence for an E2 mechanism, there is also a substantial primary deuterium kinetic isotope effect:[34]

$$C_2H_5O^- + C_6H_5CL_2\text{—}CH_2\text{—}{}^{*}\overset{+}{N}(CH_3)_3 \xrightarrow[40\ °C]{C_2H_5OH} C_6H_5CL{=}CH_2 + C_2H_5OL + {}^{*}N(CH_3)_3$$

$$k^{14}/k^{15} = 1.0133(\pm 0.0002)$$

$$k^{H}/k^{D} = 3.23(\pm 0.06).$$

9.6 Solvent kinetic isotope effects[23,35]

We saw in the previous chapter (p. 349) that an aqueous solvent deuterium kinetic isotope effect may sometimes identify the mechanism of a reaction which shows a general base-catalysis rate law. If the mechanism is nucleophile catalysis, the result is usually rather small, $k^{H_2O}/k^{D_2O} \sim 1.0$–$1.6$, and due entirely to secondary effects such as solvation. In contrast, mechanistic general base catalysis in aqueous solution involves cleavage of an O—H bond in a water molecule during the formation of the activated complex. Consequently, there is a primary contribution to the solvent deuterium kinetic isotope effect and overall results of $k^{H_2O}/k^{D_2O} \sim 2$–4 are commonly obtained.

Substantial solvent kinetic isotope effects are also found in solvent-induced (pH independent) reactions without the addition of a base catalyst. Examples include the hydrolysis of acetylimidazolium cation in aqueous acid,[36] and the acid hydrolysis of bis(p-nitrophenyl) carbonate,[37]

$$CH_3\overset{O}{\overset{\|}{C}}\text{-N}\langle\text{imidazolium}\rangle N^+\text{-L} + L_2O \xrightarrow[25°C]{pL<3} CH_3CO_2L + LN\langle\text{imidazolium}\rangle N^+L$$

$$k^{H_2O}/k^{D_2O} = 2.58.$$

$$(ArO)_2CO + L_2O \xrightarrow[50\,°C]{pL<5} 2ArOL + CO_2$$

$$Ar = p\text{-}NO_2C_6H_4; \qquad k^{H_2O}/k^{D_2O} = 2.24.$$

Very similar results had been reported earlier for the hydrolysis and methanolysis of acetic anhydride:[38]

$$Ac_2O + ROL \xrightarrow[25\,°C]{ROL} AcOL + ROAc$$

$$R = L, \qquad k^{H_2O}/k^{D_2O} = 2.89;$$

$$R = CH_3, \qquad k^{CH_3OH}/k^{CH_3OD} = 2.82.$$

Such reactions may also be regarded as mechanistic general base catalysed (see p. 348), one solvent molecule acting as the base and another as the nucleophile. In the aqueous reactions, two water molecules are, therefore, implicated in the activated complex:

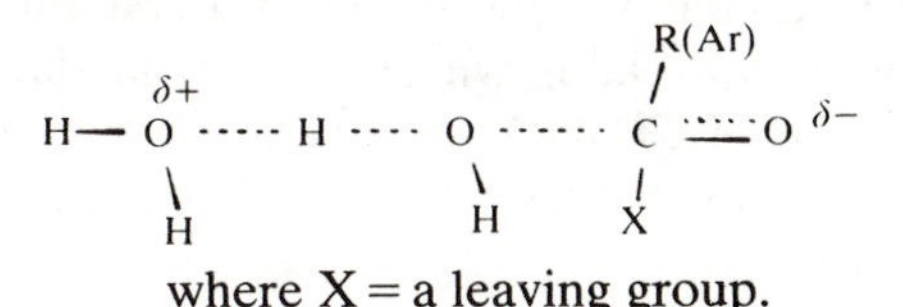

where X = a leaving group.

The magnitude of the experimental deuterium solvent isotope effect is not always a reliable indicator of whether there is a primary contribution. If the rate-determining step of a reaction involves protonation of the substrate by an acid which rapidly exchanges protons with the solvent, we should expect a solvent isotope effect with a substantial primary contribution.

This is evidently the case for the hydroloysis of ethyl vinyl ether (eqn (9.24)) catalysed by formic acid, the rate-determining step of which is initial protonation of the ether by the catalytic acid:[39]

$$CH_2{=}CH{-}OC_2H_5 + L_2O \xrightarrow[25°C]{HCO_2L} LCH_2CHO + C_2H_5OL \qquad (9.24)$$

$$CH_2{=}CH{-}OC_2H_5 \xrightarrow[\text{r.d.s.}]{HCO_2L} LCH_2{-}CH{=}\overset{+}{O}C_2H_5 + HCO_2^- \xrightarrow{L_2O} LCH_2{-}CH(\overset{+}{O}L_2){-}OC_2H_5 \xrightarrow{-L^+} LCH_2CH(OL){-}OC_2H_5 \longrightarrow LCH_2CHO + C_2H_5OL$$

$$k^{H_2O}/k^{D_2O} = 6.8$$

In the corresponding reaction catalysed by L_3O^+ rather than HCO_2L, however, the solvent kinetic isotope effect is only $k^{H_2O}/k^{D_2O} = 2.95$. It appears that the modest *overall* rate ratio for the hydronium-catalysed reaction is a composite of a normal primary effect and a substantial *inverse* secondary effect.

The solvent isotope effect in the acid-catalysed hydration of eqn (9.25) is even smaller.[40]

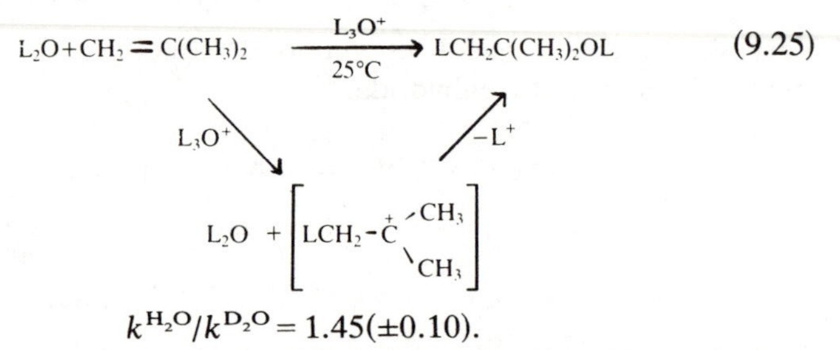

$$k^{H_2O}/k^{D_2O} = 1.45(\pm 0.10).$$

But the initial protonation step must still be irreversible and, therefore, rate determining since the product was found to contain only one deuterium atom per molecule bonded to carbon when the reaction was carried out in D_2O. In contrast, the solvent isotope effect in the proton abstraction reaction of eqn (9.26), which can only be due to the differential solvating properties of CH_3OH and CH_3OD between the initial state and the activated complex, is relatively large.[41]

$$(CH_3)_2CHNO_2 + CH_3O^- \xrightarrow[25\,°C]{CH_3OL} (CH_3)_2C{=}NO_2^- + CH_3OH$$

$$k^{CH_3OH}/k^{CH_3OD} = 2.28. \qquad (9.26)$$

Solvent isotope effects upon the solvolysis of alkyl arenesulphonates and halides have also been investigated but in very few solvents.[23,42,43] Representative results are included in Table 9.5.

Results for the hydrolysis of halides, regardless of mechanism, are

TABLE 9.5
Solvent kinetic isotope effects in solvolysis reactions[a]

Compound[b]	Reaction (Temp./°C)	Solvent k.i.e.[c]
$CH_3O_3SC_6H_5$	H (50)	1.11
CH_3Cl	H (80)	1.28
CH_3OTs	A (95)	1.03
$C_2H_5O_3SC_6H_5$	H (50)	1.10
C_2H_5Br	H (80)	1.23
$(CH_3)_2CHO_3SC_6H_5$	H (30)	1.09
$(CH_3)_2CHBr$	H (50)	1.28
$(CH_3)_3CCl$	H (4)	1.43
$C_6H_5CH_2OTs$	A (70)	1.10
$C_6H_5CH_2Cl$	H (60)	1.27
$CH_2{=}CH{-}CH_2Br$	H (60)	1.30
1-Adamantyl OTs	A (30)	1.05
p-$CH_3OC_6H_4C(CH_3)_2CH_2OTs$	A (60)	1.10
2-Adamantyl azoxytosylate(5)	F (51)	1.02_5

[a] Results for hydrolysis (H), acetolysis (A), and reaction in hexafluoropropan-2-ol (F) from refs 42, 43, and 44, respectively.
[b] OTs = p-toluenesulphonate.
[c] Solvent k.i.e. = k^{H_2O}/k^{D_2O}(H), $k^{CH_3CO_2H}/k^{CH_3CO_2D}$(A), or $k^{(CF_3)_2CHOH}/k^{(CF_3)_2CHOD}$(F).

systematically larger than those for the oxygenated substrates. Most of these reactions are, on the basis of other evidence, solvent-induced S_N2 whereas 1-adamantyl tosylate, for example, solvolyses by an S_N1 mechanism. Even so, for all the oxygenated substrates in two solvents, the results are uniformly low and not at all indicative of solvolytic mechanism. A result close to unity ($1.02_5 \pm 0.02$, 51 °C) is also found for the solvolysis of 2-adamantyl azoxytosylate (5) in hexafluoropropan-2-ol, a reaction whose mechanism involves an initial rate-determining fragmentation and a nitrogen-bonded leaving group.[44]

$$\text{2-Ad}{-}\overset{+}{N}(O^-){=}N{-}OTs \;\; (5) \xrightarrow[\text{r.d.s.}]{(CF_3)_2CHOL} \text{2-Ad}^+N_2OTsO^- \xrightarrow[-N_2O]{2(CF_3)_2CHOL} \text{2-Ad}{-}OCH(CF_3)_2 + (CF_3)_2CHOL_2^+TsO^-$$

9.7 Problems

1. Calculate $\bar{\nu}$ and $\bar{\omega}_e$ for the deuteriated analogues of the following compounds (data taken from Table 1.2, Chapter 1, p. 13).

Compound	$\bar{\nu}/cm^{-1}$	$\bar{\omega}_e/cm^{-1}$
HF	3961.6	4138.3
HCl	2885.9	2989.7
HBr	2559.3	2649.2

2. Calculate $\bar{\omega}_e$ for the deuteriated analogues of the following unstable species (data taken from *JANEF Thermochemical Tables*, 2nd edn (ed. D. R. Stull and H. Prophet), NSRDS-NBS37, National Bureau of Standards, Washington, DC, 1970).

Compound	$\bar{\omega}_e/\text{cm}^{-1}$
CH	2861.4
NH	3315
OH	3735.2
SH	2702
SiH	2041.8

3. Calculate the approximate primary deuterium kinetic isotope effect for a chemical reaction at 298 K, the rate-determining step of which involves cleavage of (i) O—H and (ii) N—H bonds. $\bar{\nu}$ of the O—H stretch of methanol = 3645 cm^{-1} (L. J. Bellamy, *The infrared spectra of complex molecules*, Vol. 1, 3rd edn, Chapman and Hall, London (1975)) and of the N—H stretch in diethylamine = 3334 cm^{-1} (R. A. Russell and H. W. Thompson, *J. chem. Soc.* 483 (1955)).

4. Bands at 3570 and 3312 cm^{-1} in the gas-phase infra-red spectra of HCO_2H and HCN have been ascribed to O—H and C—H stretching vibrations respectively. Corresponding bands for HCO_2D and DCN are found at 2666 and 2629 cm^{-1}. Compare these experimental results for the deuteriated analogues with values calculated from the protium compounds assuming that the molecules may be treated as diatomics X—H. Compare the agreement here with that for the real diatomic molecules ^{1}HI and ^{2}HI given on p. 372. (Data taken from G. Herzberg, *Infrared and Raman spectra of polyatomic molecules*, Van Nostrand, New York (1945).)

9.8 References

1. R. P. Bell, *The proton in chemistry*, 2nd edn, Chapman & Hall, London (1973). This important book contains chapters on kinetic and equilibrium isotope effects in proton-transfer reactions and includes theory and experimental results.
2. R. P. Bell and J. E. Crooks, *J. chem. Soc.* 3513 (1962).
3. L. do Amaral, M. P. Bastos, H. G. Bull, and E. H. Cordes, *J. Am. chem. Soc.* **95,** 7369 (1973).
4. D. E. Van Sickle and J. O. Rodin, *J. Am. chem. Soc.* **86,** 3091 (1964).
5. C. R. Turnquist, J. W. Taylor, E. P. Grimsrud, and R. C. Williams, *J. Am. chem. Soc.* **95,** 4133 (1973).
6. T. Riley and F. A. Long, *J. Am. chem. Soc.* **84,** 522 (1962).
7. D. J. Barnes and R. P. Bell, *Proc. R. Soc.* A **318,** 421 (1970).
8. R. P. Bell and W. B. T. Miller, *Trans. Faraday Soc.* **59,** 1147 (1963).
9. A. Streitwieser and H. S. Klein, *J. Am. chem. Soc.* **85,** 2759 (1963).
10. F. A. Long and D. Watson, *J. chem. Soc.* 2019 (1958).

11. P. Salomaa, *Acta chem. Scand.* **25,** 367 (1971); M. Goldblatt and W. M. Jones, *J. chem. Phys.* **51,** 1881 (1969).
12. S. C. Hurlock, R. M. Alexander, K. N. Rao, and N. Dreska, *J. molec. Spectrosc.* **37,** 373 (1971).
13. E. B. Wilson, J. C. Decius, and P. C. Cross, *Molecular vibrations,* p. 183. McGraw-Hill, New York (1955).
14. K. T. Leffek, J. A. Llewellyn, and R. E. Robertson, *J. Am. chem. Soc.* **82,** 6315 (1960).
15. H. Maskill, *J. Am. chem. Soc.* **98,** 8482 (1976).
16. D. Cook, *J. org. Chem.* **41,** 2173 (1976).
17. V. F. Raaen, T. Juhlke, F. J. Brown, and C. J. Collins, *J. Am. chem. Soc.* **96,** 5928 (1974).
18. *Isotope effects in chemical reactions* (eds. C. J. Collins and N. S. Bowman), A.C.S. Monograph 167, Van Nostrand-Reinhold, New York (1970). This includes two chapters on the theory and origin of isotope effects (by W. A. Van Hook, and E. K. Thornton and E. R. Thornton), two on deuterium effects on solvolysis reactions including neighbouring group participation (by V. J. Shiner, and D. E. Sunko and S. Borcic), and a final chapter on heavy atom kinetic isotope effects (by A. Fry). In all, this is a very useful book written primarily from the point of view of the organic chemist.
19. A. Streitwieser, R. H. Jagow, R. C. Fahey and S. Suzuki, *J. Am. chem. Soc.* **80,** 2326 (1958).
20. R. M. Banks, H. Maskill, R. Natarajan, and A. A. Wilson, *J. chem. Soc., Perkin Trans. 2,* 427 (1980); see also R. M. Banks and H. Maskill, ibid. 1991 (1977).
21. V. J. Shiner and R. D. Fisher, *J. Am. chem. Soc.* **93,** 2553 (1971).
22. S. Richter, I. Bregovec, and D. E. Sunko, *J. org. Chem.* **41,** 785 (1976).
23. L. Melander and W. H. Saunders, *Reaction rates of isotopic molecules,* Wiley–Interscience, New York (1980). This recent authoritative book by two eminent workers in the field of kinetic isotope effects provides the best and widest coverage of the subject presently available. The ten chapters include sections on theory, experimental methods and the derivation of results, primary and secondary deuterium k.i.e., solvent k.i.e., and heavy atom k.i.e.
24. G. A. Ropp and V. F. Raaen, *J. Am. chem. Soc.* **74,** 4992 (1952).
25. S. Seltzer, G. A. Hamilton, and F. H. Westheimer, *J. Am. chem. Soc.* **81,** 4018 (1959).
26. K. R. Lynn and P. E. Yankwich, *J. Am. chem. Soc.* **83,** 53 and 790 (1961).
27. H. Yamataka and T. Ando, *J. Am. chem. Soc.* **101,** 266 (1979).
28. J. W. Hill and A. Fry, *J. Am. chem. Soc.* **84,** 2763 (1962).
29. J. F. Willey and J. W. Taylor, *J. Am. chem. Soc.* **102,** 2387 (1980).
30. K. C. Westaway and S. F. Ali, *Canad. J. Chem.* **57,** 1354 (1979).
31. J. B. Stothers and A. N. Bourns, *Canad. J. Chem.* **38,** 923 (1960).
32. M. L. Bender and G. J. Buist, *J. Am. chem. Soc.* **80,** 4304 (1958).
33. M. H. O'Leary and J. F. Marlier, *J. Am. chem. Soc.* **101,** 3300 (1979).
34. P. J. Smith and A. N. Bourns, *Canad. J. Chem.* **48,** 125 (1970); **52,** 749 (1974).
35. R. L. Schowen, 'Mechanistic deductions from solvent isotope effects' in *Progr. phys. org. Chem.* **9,** 275 (1972).
36. J. L. Hogg, M. K. Phillips, and D. E. Jergens, *J. org. Chem.* **42,** 2459 (1977).
37. F. M. Menger and K. S. Venkatasubban, *J. org. Chem.* **41,** 1868 (1976).
38. B. D. Batts and V. Gold, *J. chem. Soc.* A 984 (1969); V. Gold and S. Grist, *J. chem. Soc.* B 2285 (1971).
39. A. J. Kresge and Y. Chiang, *J. chem. Soc.* B 58 (1967).

40. V. Gold and M. A. Kessick, *J. chem. Soc.* 6718 (1965).
41. V. Gold and S. Grist, *J. chem. Soc.* B 2282 (1971).
42. R. E. Robertson, 'Solvolysis in water' in *Progr. phys. org. Chem.* **4,** 213 (1967).
43. T. S. C. C. Huang and E. R. Thornton, *J. Am. chem. Soc.* **98,** 1542 (1976).
44. H. Maskill, J. T. Thompson, and A. A. Wilson, *J. Chem. Soc., Perkin Trans. 2* 1693 (1984).

Supplementary references

Isotope effects on enzyme-catalysed reactions (eds. W. W. Cleland, M. H. O'Leary, and D. B. Northrop), University Park Press, Baltimore (1977). This book contains chapters directly relevant to several of the topics covered in the present chapter, in particular 'Magnitude of primary hydrogen isotope effects' by A. J. Kresge, 'Solvent isotope effects on enzymic reactions' by R. L. Schowen, and 'Secondary kinetic isotope effects' by J. F. Kirsch.

Proton transfer reactions (eds. E. Caldin and V. Gold), Chapman & Hall, London (1975). Chapt. 8 'Substrate isotope effects' by R. A. More O'Ferrall; Chapt. 9 'Solvent isotope effects', by W. J. Albery.

Isotopes in organic chemistry (eds. E. Buncel and C. C. Lee).

Vol. 1: Chapt. 2, 'Isotope effects in pericyclic reactions' by W. R. Dolbier.

Vol. 2: Chapt. 1, 'The effect of structure on isotope effects in proton transfer reactions' by M. M. Kreevoy; Chapt. 3 'Proton transfer in nitro compounds' by K. T. Leffek; Chapt. 4, 'Isotope effects in hydrogen transfer reactions' by E. S. Lewis; Chapt. 6, 'Isotope effects in elimination reactions' by P. J. Smith; Chapt. 7, 'Isotopes in oxidation reactions' by R. Stewart.

Vol. 3: Chapt. 1, 'Carbon-13 effects in decarboxylation reactions' by G. E. Dunn; Chapt. 5, 'Kinetic carbon-13 and other isotope effects in cleavage and formation of bonds to carbon' by A. V. Willi.

Vol. 4: Chapt. 2, 'Isotope effects requiring the use of tritium' by A. J. Kresge.

R. P. Bell, *Chem. Soc. Rev.* **3,** 513 (1974), 'Recent advances in the study of kinetic hydrogen isotope effects'.

V. Gold, *Adv. phys. org. Chem.* **7,** 259 (1969), 'Protolytic processes in H_2O–D_2O mixtures'.

Solute–solvent interactions Vol. 1 (eds. J. F. Coetzee and C. D. Ritchie), Marcel Dekker, New York (1969), Chapt. 6, 'Solvent isotope effect on thermodynamics of nonreacting solutes' by E. M. Arnett and D. R. McKelvey; Chapt. 7 'Solvent isotope effects for equilibria and reactions' by P. M. Laughton and R. E. Robertson.

H. Simon and D. Palm, 'Isotope effects in organic chemistry and biochemistry', *Angew. Chem., int. edn* **5,** 920 (1966).

10

Molecular structure and chemical reactivity

10.1 Introduction

In Chapter 5 we considered the relationship between molecular structure and some equilibrium constants; later in the present chapter we consider how a structural modification to an organic compound affects its reactivity, i.e. a rate constant in a chemical reaction. We shall also discuss how the chemical nature of the *catalyst* affects the catalytic rate constant of a reaction which is susceptible to general acid or base catalysis.

One method of investigating a reaction's mechanism is to consider the effect of substituents, or some minor change to the reaction conditions, upon the rate (or equilibrium) constant. A powerful qualitative method of interpreting such results, and for deciding what sort of modification to the reactants or conditions should next be investigated, is to use an extended form of a reaction profile.

As presented in Chapter 6, a reaction profile is normally a plot of either molecular potential energy or standard molar free energy against a reaction coordinate. Molecular potential-energy profiles were also encountered in Chapter 1 and relate directly to properties of individual molecules such as bond energies, bond lengths, and bond angles, and correspond to real physical processes. Standard molar free-energy profiles use the same reaction coordinate, but the x-axis is not a simple molecular property, nor can the process be real. As a hypothetical macroscopic measure of overall stability, the standard molar free energy of a whole system under particular experimental conditions is shown as a function of molecular configuration specified by the reaction coordinate (see p. 233).

A notional standard molar free-energy barrier, $\Delta G^{\ominus\ddagger}$, between two chemical states at absolute temperature T is related to the rate constant for their interconversion, k_c, by the expression

$$\ln k_c = \ln\left\{\frac{k_B T}{h}\right\} - \frac{\Delta G^{\ominus\ddagger}}{RT},$$

in which all symbols have their usual meanings (see Chapter 6).

The standard molar free-energy difference between the two states at absolute temperature T, $\Delta G^{\ominus}$, expresses their relative thermodynamic stability; sometimes this is given as an equilibrium constant $K^{\ominus}$ (see Chapter 4):

$$\Delta G^{\ominus} = -RT \ln K^{\ominus}.$$

10.2 Reaction maps

10.2.1 Molecular potential-energy reaction maps[1]

Figure 6.4 (Chapter 6, p. 237) is a three-dimensional molecular potential-energy surface which describes the interconversion of a pair of isomers A and B. It is part of a multidimensional potential-energy hypersurface describing the complete relationship between total molecular potential energy and the configuration of the assembly of atoms which constitute A (or B). There are two mutually perpendicular axes which represent molecular structure coordinates, and the third axis represents molecular potential energy. Alternatively, we can describe part of a three-dimensional potential-energy surface by a contour diagram, Fig. 10.1, in which the lines join points representing configurations of equal potential energy. We shall call this a *molecular potential-energy reaction map.*

Different lines across the map from the bottom left to the top right correspond to different routes across the potential-energy surface and different reaction profiles for the conversion of A into B. One route represented in the diagram by the slightly curved diagonal is of lower energy than any other. This corresponds to the trough between the minima in the potential-energy surface of Fig. 6.4 (Chapter 6, p. 237).

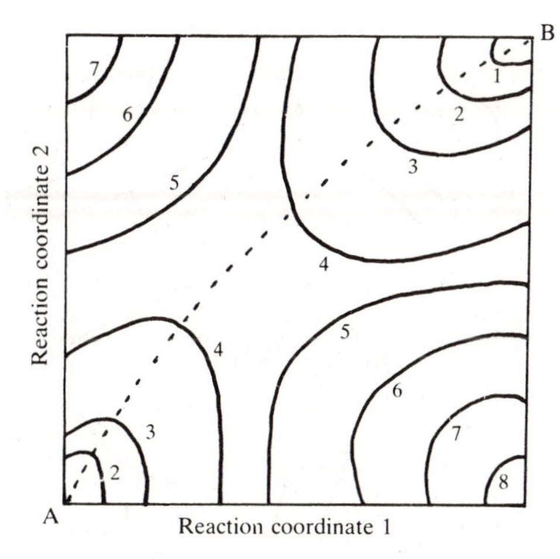

Fig. 10.1. Molecular potential-energy reaction map for the isomerization A$\rightleftharpoons$B. (Increasing numbers indicate contours of increasing potential energy.)

The transformation is a single-step reaction involving concerted bond reorganization and no intermediates since there are no local minima *en route.*

One merit of a molecular potential-energy reaction map is that, in principle, it includes the exact real correspondence between molecular configuration and molecular potential energy. This is true even when, in an isomerization of a complex molecule, the two reaction coordinates selected to describe the molecular configuration do not correspond to simple single structural parameters but are composite terms. Furthermore, barrier heights are a fair indication of the rate of an isomerization (the higher the potential-energy barrier, the smaller the rate constant), and the molecular potential-energy difference between minima reflects well the corresponding equilibrium constant. For isomerizations, therefore, molecular potential-energy reaction maps are usually adequate for interpreting or predicting structural effects upon rate and equilibrium constants.

Figure 10.2, for example, may be used to discuss the effect of R upon the isomerization of a bicyclo[2.2.0]hexane to the corresponding hexa-1,5-diene.[2]

The overall reaction corresponds to a $\rightarrow$ d in Fig. 10.2 and experimental results[2] suggest that, in the parent system (R = H), this involves initial

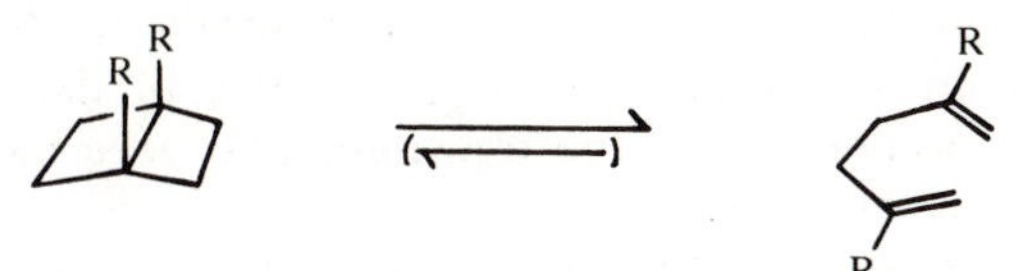

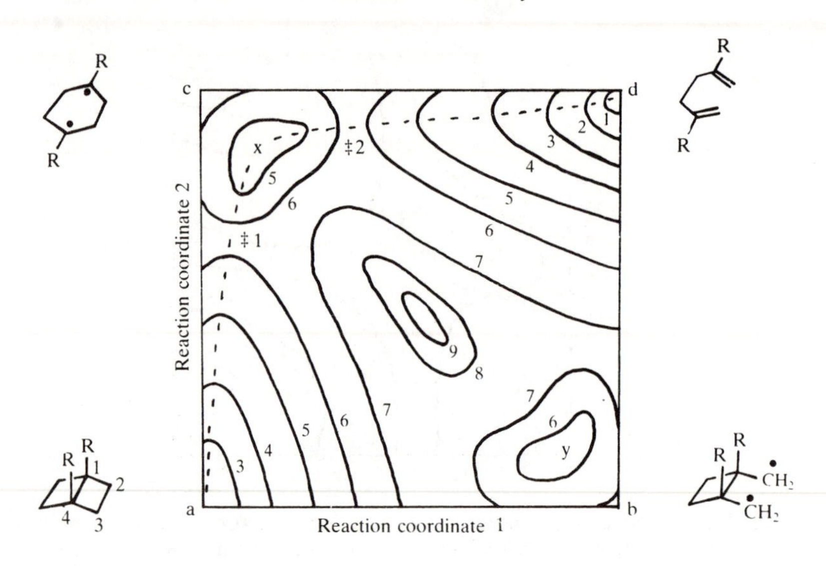

Fig. 10.2. Molecular potential-energy reaction map for the isomerization of a bicyclo[2.2.0]hexane into a hexa-1,5-diene. (Increasing numbers indicate contours of increasing potential energy.)

homolysis of the C1–C4 bond (a → x). The diradical intermediate corresponding to the potential-energy minimum represented by the loops in the energy contours at x then either reverts to starting material (x → a) or proceeds to give the acyclic diene product by cleavage of the C2–C3 bond with the associated electronic reorganization (x → d). The two activated complexes corresponding to ‡1 and ‡2 are probably of similar energy, so $k_{-1} \sim k_2$:

$$\underset{k_{-1}}{\overset{k_1}{\rightleftharpoons}} \qquad \xrightarrow{k_2}$$

In principle, some substituents R could have the effect of destabilizing the diradical intermediate at x by raising c relative to b. In this event, the contours would be re-drawn to indicate that the new reaction involves either the alternative sequential bond homolyses via the isomeric diradical intermediate at y, or a concerted single-step bond reorganization bypassing any diradical intermediates. The reaction path for the former possibility would be a → y close to the bottom of the map, then y → d close to the right-hand edge. The second possibility would involve a merging of ‡1 and ‡2 somewhere in the centre of the map to give a saddle-point corresponding to the transition state in a more direct path from a to d.

The effect of the nature of R upon the equilibrium constant can also be described by Fig. 10.2. Conjugating or bulky alkyl groups, for example, would favour the acyclic diene so d would be lowered relative to a.

If a reaction is not an isomerization, however, or if we wish to describe the effect of solvent upon a reaction in solution, or if there is a strong entropy contribution to either the rate or equilibrium constant, then a molecular potential-energy map is not usually adequate.

10.2.2 Standard molar free-energy reaction maps[1]

Figure 10.3 is a second type of extended reaction profile which includes in the plane of the paper two orthogonal molecular coordinates just as in a molecular potential-energy reaction map. But this time the third dimension (out of the plane of the paper) is the standard molar free energy of the system. Figure 10.3 includes contour lines which join points representing states of equal standard molar free energy; it is, therefore, a *standard molar free-energy reaction map*, and includes all contributions to the standard molar free energy of the system: enthalpy (energy) and entropy aspects of both intramolecular and intermolecular effects including those due to the solvent for reactions in solution.

Figure 10.3 actually represents a familiar reaction mechanism: a simple S_N2 reaction of nucleophile Y^- with a substrate comprising an alkyl Lewis acid residue R^+ and leaving group (nucleofuge) X^-. The overall reaction is

$$Y^- + R{-}X \xrightarrow{S_N2} Y{-}R + X^-$$

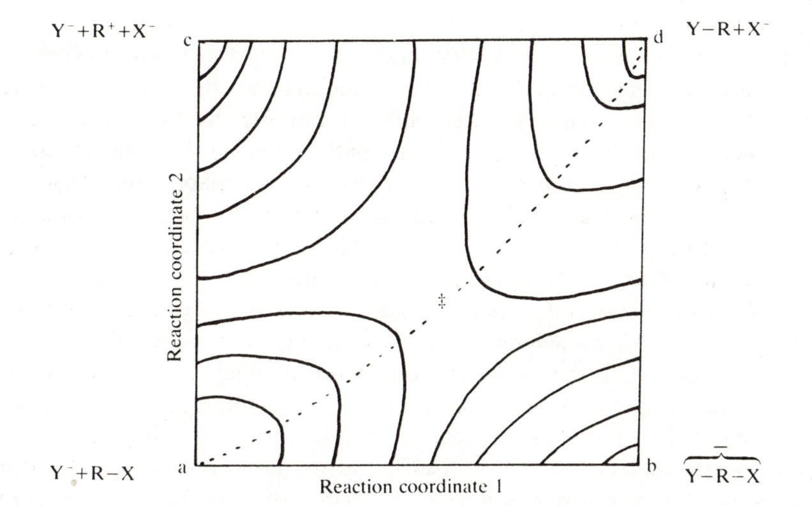

Fig. 10.3. Standard molar free-energy reaction map for a simple S_N2 mechanism.

and is represented in Fig. 10.3 by the progress from the bottom left of the reaction map, a, to the top right, d.

Movement across the map from left to right (reaction coordinate 1) corresponds to shortening the $Y^- \cdots R$ distance, and movement from bottom to top (reaction coordinate 2) corresponds to lengthening the R–X bond. If the left–right movement is parallel with the top and bottom of the map, the $Y^- \cdots R$ distance is altered *with no change in the R–X bond length.* Analogously, movement up or down the map parallel to the sides corresponds to altering the R–X bond length *at constant* $Y^- \cdots R$ *distance.* Thus, movement from a to b corresponds to Y^- approaching R–X, with no lengthening of the R–X bond, to produce a complex anion $(Y—R—X)^-$ which must include a pentacoordinate carbon in R. This process cuts across the free-energy contours and corresponds to increasing steeply the standard molar free energy of the system, i.e. the hypothetical formation of one mole of an increasingly unstable species.

Movement from a to c corresponds to ionization of R—X with no approach of Y^-. This, as drawn, is also an increasingly unfavourable process.

The contours of the map indicate that *concerted* approach of Y^- to R as the R–X bond lengthens, i.e. *coupled* progress along *both* reaction coordinates, follows a less steep standard free-energy increase than progression along either coordinate alone. Indeed, although the route is initially up a standard free-energy gradient, a saddle-point is reached at ‡ from which the path turns downwards. The broken line from a to d via ‡, then, is the lowest standard molar free-energy route for this particular bimolecular substitution, and the saddle-point at ‡ is the transition state. We describe the molecular species corresponding to the thermodynamic state at ‡ as the activated complex (see p. 237).

There are several important general features of this illustration.

1. The contours indicate no local minima, so there are no intermediates by any route in this map. According to this very simple hypothetical mechanism, therefore, all paths from the bottom left to the top right are single-step reactions. Although this map is unrealistically simple, it does indicate that, in principle, concerted bimolecular substitution reactions may involve a considerable degree of charge separation without the intervention of ionic intermediates.

2. The two most unfavourable routes are $a \rightarrow b \rightarrow d$ and $a \rightarrow c \rightarrow d$. These correspond to *sequential* progress along the two reaction coordinates in contrast to the most favourable route which crosses the central part of the map and involves *concerted, concurrent* progress along *both* coordinates. But note that successive progression along component reaction coordinates does not of itself require the involvement of intermediates; the overall reaction $a \rightarrow d$ by either of these highly unfavourable routes is still a single-step mechanism. A reaction tending towards the

a → c → d extreme is becoming an *uncoupled* S_N2 reaction – a single-step process in which the incoming nucleophile provides little or no assistance to the departure of the leaving group.

3. The broken line from a to d representing the least unfavourable reaction route is not straight. When ‡ is to the lower right of the straight diagonal from a to d, the transition state is described as *tight.* As far as *structure* is concerned, the formation of the Y–R bond up to the transition state in such a reaction is running ahead of, but still concerted with, the cleavage of the R–X bond. If, in contrast, the reaction path of another example is curved towards the upper left of the straight diagonal, we describe the transition state as *loose* since, *structurally*, heterolysis of R—X up to the transition state is just ahead of the concerted formation of the Y–R bond. Whether an S_N2 reaction has a tight or a loose transition state depends upon the nature of the reactants Y^- and R—X; in particular, it depends upon the force constants and anharmonicities of the respective R–X and R–Y bonds. For a reaction in solution, it will also be affected by the nature of the solvent. We can regard the tightness or looseness of the transition state in terms of the quality of the symmetry of the free-energy map about the a–d diagonal. And the effect of a modification to the structure of a reactant, or of a change in the reaction conditions, upon the reaction can be predicted on this basis. For example, changing to a solvent with improved Lewis acid–base properties which can better stabilize charge separation will lower the standard free energy of states to the top left of the map more than of those to the bottom right. Consequently, the middle part of the reaction trough is displaced towards the top left of the map, perhaps to the extent that the broken line becomes curved towards c rather than b, and the transition state is loosened. In contrast, the transition state of such an S_N2 reaction which is transferred from solution to the gas phase, with loss of all charge separation stabilizing solvation, will become very tight; the reaction path a → ‡ → d will then curve much more strongly towards b.

4. The transition state ‡ is not configurationally half-way between a and d except when X = Y. The effect of raising or lowering the molar free energy of the top left (c) or the bottom right (b) of the reaction map of Fig. 10.3 when the overall reaction is from bottom left (a) to top right (d) is to move the transition state (‡) in a direction *perpendicular* to the composite reaction coordinate at the transition state. This is in contrast to an effect *in* (or *parallel to*) the composite reaction coordinate caused by raising or lowering the energy of the reactant (a) or the product (d). These latter effects were discussed earlier under the Hammond Postulate (p. 261) and are known as Hammond effects. A reaction profile along the overall reaction coordinate is, of course, convex at the transition state, and Hammond effects tend to move the transition state within the reaction coordinate *away from* the side whose energy is lowered (or

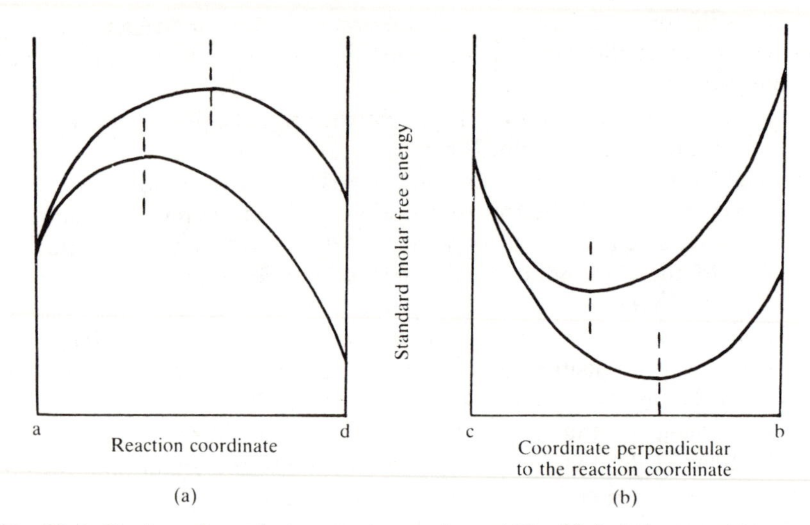

Fig. 10.4. Sections through the reaction surface of Fig. 10.3 at the transition state (a) along the reaction coordinate illustrating the Hammond effect of selectively raising the energy of d, and (b) perpendicular to the reaction coordinate illustrating the non-Hammond effect of selectively raising the energy of b.

towards the side whose energy is raised). This is illustrated in Fig. 10.4(a). The surface of the reaction map of Fig. 10.3 *perpendicular to* the reaction coordinate at the transition state ‡ is concave and a non-Hammond (or perpendicular effect) moves the transition state *towards* the side whose energy is lowered (or away from the side whose energy is raised), Fig. 10.4(b).

A simple model for an S_N1 reaction proceeding via a single intermediate ion-pair is shown in Fig. 10.5 in competition with a concurrent S_N2 reaction.

This reaction map includes the following mechanistic features.

1. There are two quite distinct routes from a to d which follow troughs (i) a → ‡1 → *l* → ‡2 → d: the simple S_N1 mechanism, and (ii) a → ‡3 → d: the simple S_N2 mechanism.

2. The barrier ‡1 separating the intermediate *l* from the reactants at a is higher than barrier ‡2 which separates *l* from the products at d. The initial ionization of this S_N1 mechanism, therefore, is rate determining.

3. The barrier in the S_N2 mechanism at ‡3 is broadly comparable with ‡1 in the S_N1 route, so the two routes from reactants to products have similar rate constants. Which, in practice, would be the faster reaction is determined by $[Y^-]$ (see pp. 276, 305).

4. The intermediate at *l* is a (solvated) ion-pair R^+X^- with no covalent involvement by the nucleophile Y^-.

5. There is a standard free-energy ridge with a maximum at *k* which

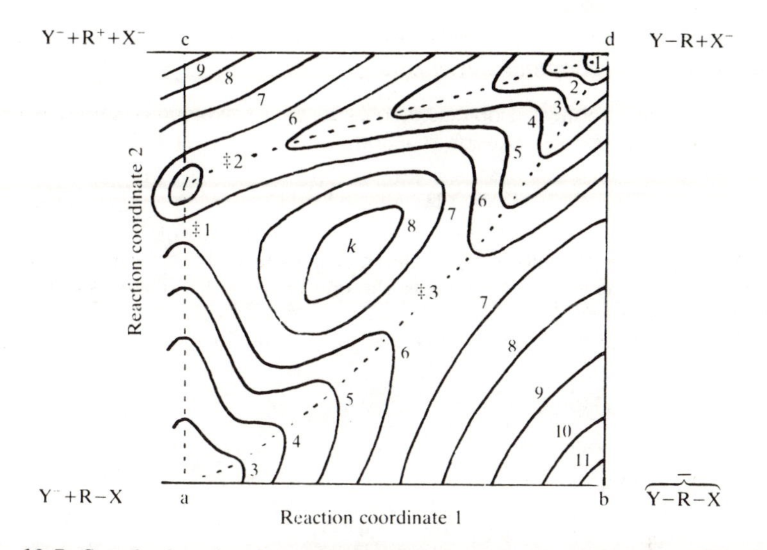

Fig. 10.5. Standard molar free-energy reaction map for competing simple S_N1 and S_N2 mechanisms. (Increasing numbers indicate contours of increasing standard molar free energy.)

separates the troughs corresponding to the S_N1 and S_N2 mechanisms. These two mechanisms, therefore, are quite distinct parallel processes and no central part of the free-energy map is common to both routes.

Prediction of structural or medium effects upon this complex overall reaction is exactly analogous to that for the simple S_N2 reaction.

If $(Y—R—X)^-$ is selectively destabilized, e.g. by making R a very bulky group so that nucleophilic approach is sterically hindered and the penta-coordinated activated complex severely strained, then b is raised relative to c. Or if any other change in the nature of the reactants or the experimental conditions selectively stabilizes $Y^- + R^+ + X^-$, such as the use of a more ionizing solvent, then c is lowered relative to b. The structural effect upon ‡3 of either of these changes is to move it towards c – a non-Hammond effect. Either change also facilitates the S_N1 route *vis-a-vis* the S_N2. If the effect is gross, then the col between b and *k* at ‡3 will not only move towards c, but may disappear altogether and, with it, the trough corresponding to the S_N2 route.

Alternatively, selective destabilization of R^+ will raise c relative to b and disfavour the S_N1 route compared with the S_N2. Or stabilization of $(Y—R—X)^-$, e.g. by making R less bulky or better able, somehow, to accommodate negative charge, will lower b relative to c. If, for whatever reason, the barrier ‡1 becomes much higher than ‡3 and the minimum at *l* very shallow, then the reaction will proceed virtually entirely by the S_N2 route via transition state ‡3 which will have been displaced perpendicular to the reaction coordinate towards b in the map of Fig. 10.5. A minimum in a reaction map which does not lie on a feasible reaction route corresponds to a *phantom intermediate.*

One of the commonest problems associated with solvolytic reactions is to establish whether a substitution reaction,

$$RX + YH \xrightarrow{YH} RY + HX,$$

is S_N1 or solvent-induced S_N2. The mechanistic dichotomy may be described by a reaction map similar to the one in Fig. 10.5. There is a major difficulty, however.

The relative proportions of S_N1 and S_N2 routes in the reaction of Fig. 10.5

$$RX + Y^- \xrightarrow{\text{solvent}} RY + X^-$$

are easily established.

$$S_N1: \text{rate} = k_1[RX]$$
$$S_N2: \text{rate} = k_2[RX][Y^-]$$
$$\text{overall rate} = k_1[RX] + k_2[RX][Y^-]$$
$$= k_{exp}[RX]$$

where $k_{exp} = k_1 + k_2[Y^-]$ and both k_1 and k_2 can in principle be determined from measurements of k_{exp} at different $[Y^-]$ values (see Chapter 7).

But in the solvolytic reaction, the solvent-induced S_N2 mechanism inevitably leads to a *pseudo* first-order term,

$$S_N1: \text{rate} = k_1[RX]$$
$$\text{solvent-induced } S_N2: \text{rate} = k_2[RX][HY]$$
$$= k_1'[RX] \text{ since [HY] is not variable.}$$

The overall rate, therefore, is given by

$$\text{overall rate} = k_1[RX] + k_1'[RX]$$
$$= k_{exp}[RX],$$

where $k_{exp} = k_1 + k_1'$, and it is not possible to resolve k_{exp} into k_1 and k_1'. A reaction map which describes such a mechanistic dichtomy must indicate that both uni- and bi-molecular routes lead to first-order rate constants.

Example. Interpret the following standard molar free-energy reaction map and predict the effect upon the reaction of (1) electron-withdrawing substituents in the aryl group, and (2) replacement of the H at C-1 by a more electron-donating substituent. Increasing numbers indicate contours of increasing standard molar free energy.

1. The overall reaction a → d is an elimination.
2. Reaction coordinate 1 corresponds to proton abstraction from the substrate by the base B.
3. Reaction coordinate 2 corresponds to heterolysis of the C–X bond.
4. The reaction with the largest rate constant a → ‡2 → d involves proton

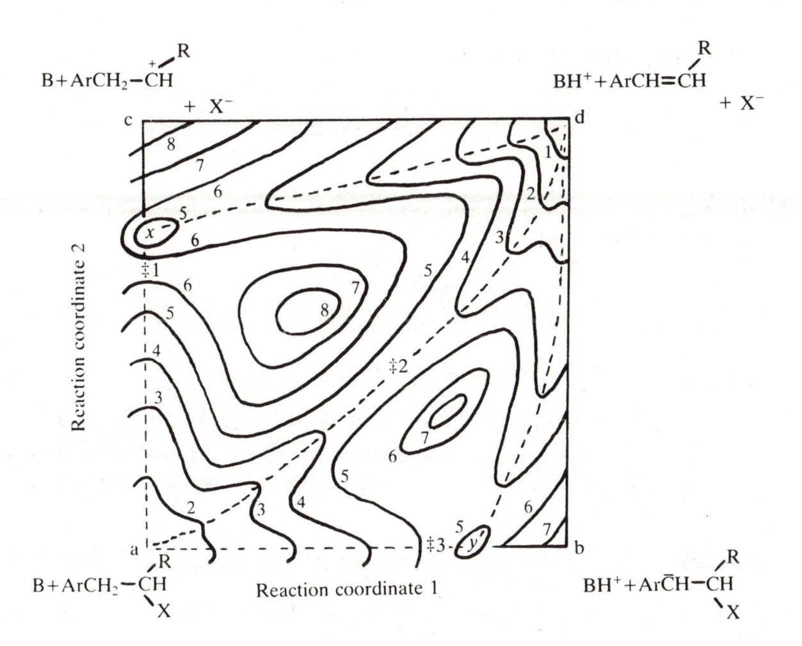

abstraction concerted with departure of the nucleofuge. This is the E2 mechanism.

5. The map includes minima at x and y corresponding to carbonium ion and carbanion intermediates, respectively, in E1 and $E1_{cb}$ mechanisms with smaller rate constants than the E2 reaction.
6. Electron-withdrawing groups in the aryl residue will facilitate carbanion formation and lower the standard free energy of b relative to c possibly causing the rate constant of the $E1_{cb}$ mechanism to become larger than that of the E2 (‡3 lower than ‡2). Even if the principal mechanism does not become stepwise via the intermediate carbanion at y, the E2 reaction path will still curve more strongly towards b and the overall rate constant will become larger.
7. A better electron-donating group than H at C-1 will facilitate carbonium ion formation and lower the standard molar free energy of c relative to b. Such substituents will move ‡2 towards c, cause the overall rate constant to become larger, and may be sufficient to cause ‡1 to become lower than ‡2, i.e. the rate constant of the E1 mechanism to become larger than that of the E2.

10.3 The Brönsted Catalysis Law

10.3.1 General acid-catalysed reactions

We saw in Chapter 8 that some reactions are catalysed by undissociated acids, e.g. the hydrolysis of vinyl ethers[3] (eqn (10.1))

$$CH_2{=}CH{-}OR + H_2O \xrightarrow[H_2O]{H_3O^+,\ AH} CH_3CHO + ROH. \qquad (10.1)$$

We may generalize this equation as

$$X \xrightarrow[H_2O]{H_3O^+,\ AH} \text{Products,}$$

where X = the substrate. If the rate is first order in substrate concentration, the rate law is given by

$$-\frac{d[X]}{dt} = k_{exp}[X],$$

where $k_{exp} = k_0 + k_H[H_3O^+] + k_{AH}[AH]$ (10.2)

= the experimental *pseudo* first-order rate constant,

k_0 = the first-order rate constant for the water-induced reaction,

k_H = the second-order catalytic constant for the hydronium-ion-catalysed reaction, and

k_{AH} = the second-order catalytic constant for the reaction catalysed by the undissociated acid AH.

If there are several undissociated acids in the solution, then we write eqn (10.2) as

$$k_{exp} = k_0 + k_H[H_3O^+] + \sum k_{AH}[AH]. \qquad (10.3)$$

Some reactions, such as the hydrolysis of a vinyl ether, do not occur in the absence of an acid catalyst so $k_0 = 0$ in eqns (10.2) and (10.3). Usually $k_H \gg k_{AH}$ (e.g. hydrolysis of vinyl ethers, p. 329) but if the reaction is carried out in a solution buffered at a pH such that $[H_3O^+]$ is negligibly small, catalysis by hydronium ions is suppressed, and the reaction is catalysed only by the undissociated weak acid of the buffer system.

It was discovered experimentally in the 1920s that, for some reactions, e.g. the conversion of α (or β) glucose into the equilibrium mixture (the mutarotation of glucose),[4]

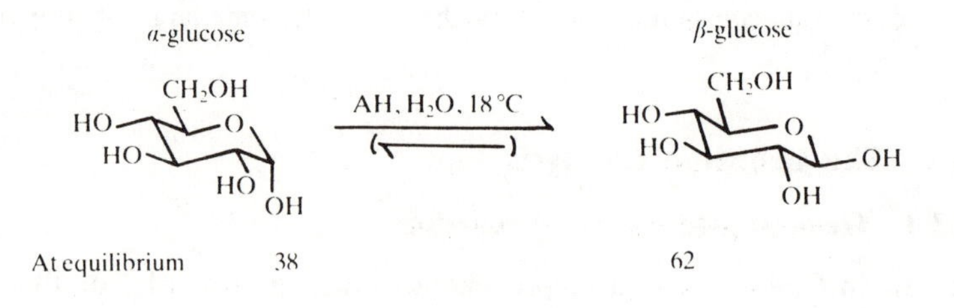

the stronger a catalyst is *as an acid*, the more effective it is *as a catalyst.* The *Brönsted relationship* (or *Catalysis Law*) between the acidity constant K_{AH} of a range of catalytic acids and their catalytic constants k_{AH} in the reaction is logarithmic:

$$\log k_{AH} = \alpha \,.\, \log K_{AH} + \text{constant}, \qquad (10.4)$$

or, in its differential form,

$$\delta \log k_{AH} = \alpha \,.\, \delta \log K_{AH}, \qquad (10.5)$$

where δ represents an incremental change in the function indicated and α is a parameter associated with the particular reaction being catalysed. $\alpha = 0.84$ for the hydrolysis of phenyl vinyl ether[3] (eqn (10.1) $R = C_6H_5$) and 0.3 for the mutarotation of glucose.[4]

Since $pK_{AH} = -\log K_{AH}$, eqns (10.4) and (10.5) may also be written

$$\log k_{AH} = -\alpha \,.\, pK_{AH} + \text{constant}, \qquad (10.6)$$

and

$$\delta \log k_{AH} = -\alpha \,.\, \delta pK_{AH}. \qquad (10.7)$$

We saw in Chapter 8 that a catalytic constant k_{AH} for a single acid AH in a general acid-catalysed reaction is determined by measuring the experimental rate constant k_{exp} at constant pH (constant buffer *ratio* $[AH]/[A^-]$) but at different buffer concentrations, i.e. at different values of [AH]. From eqn (10.2), a plot of k_{exp} against [AH] at constant pH gives a straight line of gradient $= k_{AH}$.

The value of α for a general acid-catalysed reaction is obtained by plotting $\log k_{AH}$ against $\log K_{AH}$ or, as shown in Fig. 10.6, pK_{AH} using a range of catalytic acids.

Example. Calculate the Brönsted α for the hydrolysis of ethyl vinyl ether from the experimental results given below.[3]

$$\underset{X}{C_2H_5OCH{=}CH_2} + H_2O \xrightarrow[AH,\ H_3O^+]{H_2O,\ 25\,°C} C_2H_5OH + CH_3CHO$$

$$-\frac{d[X]}{dt} = k_{exp}[X]$$

$$k_{exp} = k_H[H_3O^+] + k_{AH}[AH]$$

Catalyst, AH	pK_{AH}	$k_{AH}/dm^3\,mol^{-1}\,s^{-1}$	$\log(k_{AH}/dm^3\,mol^{-1}\,s^{-1})$
$CNCH_2CO_2H$	2.47	0.0442	−1.35
$ClCH_2CO_2H$	2.76	0.0324	−1.49
$CH_3OCH_2CO_2H$	3.57	0.0111	−1.95
HCO_2H	3.75	0.00672	−2.17
$HOCH_2CO_2H$	3.83	0.00553	−2.26
CH_3CO_2H	4.76	0.00138	−2.86
$CH_3CH_2CO_2H$	4.88	0.00104	−2.98

The result $\alpha = 0.70$ is obtained from the graph in Fig. 10.6 on p. 418.

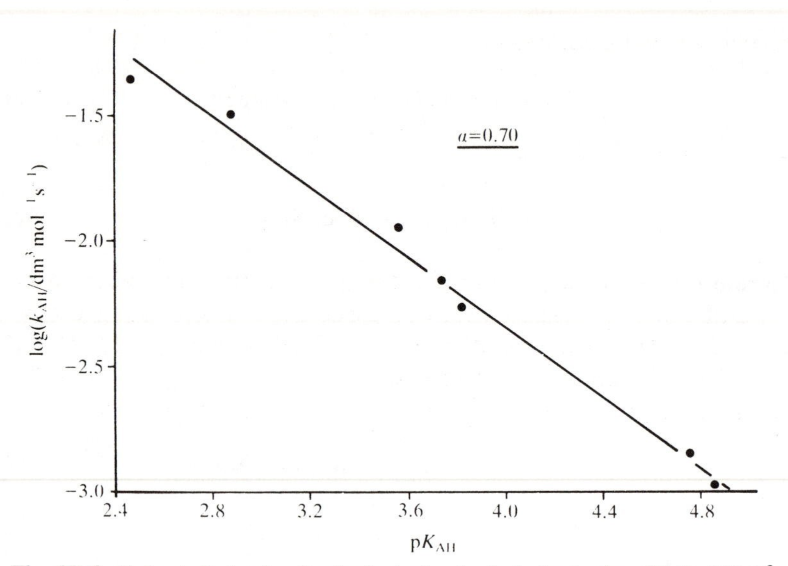

Fig. 10.6. Brönsted plot for the hydrolysis of ethyl vinyl ether, H_2O, 25°C.[3]

Values of α for different reactions in aqueous solution normally range between 0 and 1.0 and some results are given in Table 10.1.

Values of α close to 0 or 1.0 can be difficult to establish experimentally; the reason is evident from eqn (10.5) (or (10.7)):

$$\delta \log k_{AH} = \alpha \,.\, \delta \log K_{AH}. \qquad (10.5)$$

If α is close to zero, then replacing a catalytic acid by one considerably stronger, e.g. phenol by 2,4-dinitrophenol, will not give an appreciably better catalyst. In other words, the rate constant of the catalysed reaction with a very small α is not very sensitive to the acid strength of the

TABLE 10.1

Brönsted coefficients α for some general acid-catalysed reactions in water 25 °C

Reaction	α	Ref. or Note
$CH_3CH(OH)_2 \longrightarrow CH_3CHO + H_2O$	0.54	ref. 5
$PhCH(OCMe_3)_2 + H_2O \longrightarrow PhCHO + 2Me_3COH$	0.6	*a*
$CH_2{=}CH{-}OC_2H_5 + H_2O \longrightarrow CH_3CHO + C_2H_5OH$	0.70	ref. 3
$CH_2{=}CH{-}OPh + H_2O \longrightarrow CH_3CHO + PhOH$	0.84	ref. 3
$pClC_6H_4CHO + PhNH_2 \longrightarrow pClC_6H_4CH{=}NPh + H_2O$	0.25	*b*
α-Glucose $\xrightarrow[(-)]{18\,°C}$ β-Glucose	0.3	ref. 4

Notes

[a] E. Anderson and T. H. Fife, *J. Am. chem. Soc.* **93,** 1701 (1971).
[b] E. H. Cordes and W. P. Jencks, *J. Am. chem. Soc.* **84,** 832 (1962).

catalyst. Consequently, the relative proportions of parallel routes of an overall reaction catalysed by different acids present will be determined largely by their relative concentrations. Since the *pseudo* first-order rate constant k_0 in eqns (10.2) and (10.3) is equal to $k_{H_2O}[H_2O]$, k_{H_2O} being the real second-order rate constant for the water-induced process and broadly comparable in magnitude with k_{AH} etc. if α is very small, then, because $[H_2O] \gg [H_3O^+]$, [AH] etc., the reaction as a whole will be overwhelmingly solvent-induced. It is, therefore, quite difficult, experimentally, to detect acid catalysis at all in such reactions; they appear to be uncatalysed.

At the other extreme ($\alpha \sim 1$), the reaction is very sensitive indeed to the acid strength of the catalyst. Consequently, if two acids which differ in acid strength by $2pK_{AH}$ units are present at comparable concentrations, then, by eqn (10.7), reaction via the stronger will be about 100 times faster than that via the weaker. So in aqueous solution at $pH<7$ the overwhelming proportion of an overall reaction with $\alpha \sim 1$ will be via the single strongest proton donor present: H_3O^+. In acidic aqueous solutions, therefore, reactions with $\alpha \sim 1$ appear to be specific acid catalysed.

We see from this way of regarding acid-catalysed reactions in aqueous solution that experimental distinctions between solvent-induced reactions, general acid-catalysed reactions, and specific acid-catalysed reactions are not always clear-cut.

10.3.2 General base-catalysed reactions

The empirical relationship between the thermodynamic acid–base properties of a compound and its effectiveness as a catalyst was established first for a general base-catalysed reaction: the decomposition of nitramide.[6,7]

$$NH_2NO_2 \xrightarrow[H_2O]{OH^-,\ B} N_2O + H_2O$$

$$-\frac{d[NH_2NO_2]}{dt} = k_{exp}[NH_2NO_2],$$

$$\text{where } k_{exp} = k_0 + k_{OH}[OH^-] + k_B[B] \qquad (10.8)$$

= the experimental *pseudo* first-order rate constant,

k_0 = the first-order rate constant for the water-induced reaction,

k_{OH} = the second-order catalytic constant for the hydroxide-catalysed reaction, and

k_B = the second-order catalytic constant for the reaction catalysed by the general base B.

If several bases are present, then eqn (10.8) may be expanded to give

$$k_{exp} = k_0 + k_{OH}[OH^-] + \sum k_B[B],$$

which represents the rate law of any general base-catalysed reaction.

The nitramide decomposition was initially investigated in aqueous solutions buffered on the acidic side of neutrality to suppress catalysis by OH^- ($[OH^-]$ negligible at pH<7). Under these conditions Brönsted and Pedersen[6] found that k_{exp} for a given base was independent of the precise pH, i.e. the buffer ratio, but linearly dependent upon the buffer concentration, i.e. upon the concentration of the anion of the weak acid which, with the weak acid, constituted the buffer system. By measuring values of k_B for a range of buffer systems, they established that the effectiveness of a carboxylate anion as a catalyst was related to its base strength; the more basic the anion (i.e. the weaker its conjugate acid), the better it is as a catalyst.

The relationship is logarithmic:

$$\log k_B = -\beta \,.\, \log K_{BH} + \text{constant},$$

where k_B = the second-order catalytic constant for the reaction catalysed by general base B,

β = the Brönsted parameter associated with the particular reaction being catalysed by general bases,

K_{BH} = the *acidity* constant of the *conjugate acid* of the catalytic base B;

or, since $pK_{BH} = -\log K_{BH}$,

$$\log k_B = \beta \,.\, pK_{BH} + \text{constant}, \qquad (10.9)$$

which may be expressed in its differential form

$$\delta \log k_B = \beta \,.\, \delta pK_{BH}. \qquad (10.10)$$

The value of β for a particular general base-catalysed reaction is determined by first measuring k_B for each of a series of base catalysts (see Chapter 8) and then plotting the results logarithmically against the pK_{BH} values of the catalysts as illustrated in Fig. 10.7 from the results given below.[6]

Catalytic base	pK_{BH}[a]	$k_B/dm^3\ mol^{-1}\ min^{-1}$	$\log(k_B/dm^3\ mol^{-1}\ min^{-1})$
$CH_3CH_2CO_2^-$	4.87	0.65	−0.19
$CH_3CO_2^-$	4.76	0.50	−0.30
$C_6H_5CH_2CO_2^-$	4.28	0.23	−0.64
$C_6H_5CO_2^-$	4.20	0.19	−0.72
HCO_2^-	3.75	0.082	−1.09
o-$HOC_6H_4CO_2^-$	3.0	0.021	−1.68
$Cl_2CHCO_2^-$	1.26	0.0007	−3.2

[a] pK_{BH} of the carboxylate base = pK_{AH} of the corresponding conjugate carboxylic acid, see Chapter 5.

Values for β are usually between 0 and 1.0 and some results are shown in Table 10.2.

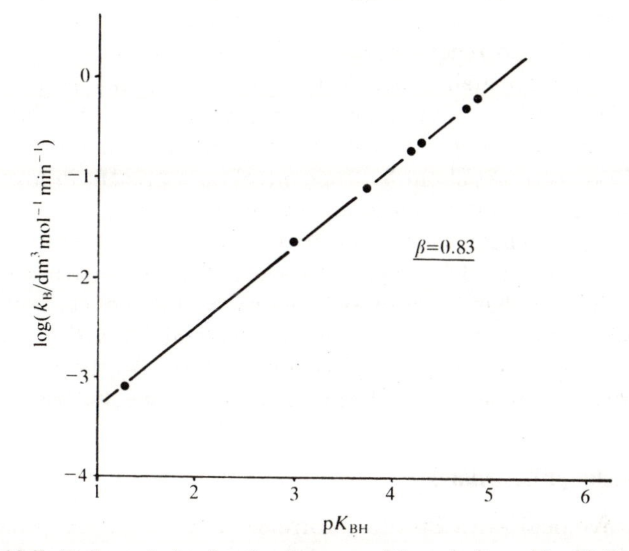

Fig. 10.7. Brönsted plot for the decomposition of nitramide, H_2O, 15°C.[6]

TABLE 10.2

Brönsted coefficients β for some general base-catalysed reactions in water

Reaction	β	Ref. or Note
α-Glucose $\xrightarrow[(\leftarrow)]{18\,°C}$ β-Glucose	0.4	ref. 4
$Cl_2CHCO_2Et + H_2O \xrightarrow{25\,°C} Cl_2CHCO_2H + EtOH$	0.47	*a*
$CH_3COCH_3 \xrightarrow[(\leftarrow)]{25\,°C} CH_3C(OH){=}CH_2$	0.88	*b*
$CH_3COCH_2COCH_3 \xrightarrow[(\leftarrow)]{25\,°C} CH_3C(OH){=}CHCOCH_3$	0.48	*c*
$NO_2NH_2 \xrightarrow{15\,°C} N_2O + H_2O$	0.83	ref. 6
$ArOCH_2CH_2C({=}O)CH_3 \xrightarrow{25\,°C} ArOH + CH_2{=}CHC({=}O)CH_3$ ($Ar = 4\text{-}NO_2C_6H_4$)	0.75	*d*

Notes

[a] W. P. Jencks and J. Carriuolo, *J. Am. chem. Soc.* **83,** 1743 (1961).
[b] R. P. Bell and O. M. Lidwell, *Proc. R. Soc. A* **176,** 88 (1940).
[c] R. P. Bell, E. Gelles, and E. Möller, *Proc. R. Soc. A* **198,** 308 (1949).
[d] D. J. Hupe and D. Wu, *J. Am. chem. Soc.* **99,** 7653 (1977).

For reasons exactly analogous to those given when we considered general acid-catalysed reactions, values of $\beta \sim 0$ and ~ 1 may be difficult to establish experimentally. The former indicates a reaction for which bases of widely different strengths are broadly comparable in catalytic effectiveness. Consequently, the major part of the reaction in aqueous solution will be brought about by the general base present in overwhelming preponderance – water – and base catalysis by dilute solutes will be difficult to detect. The reaction will, therefore, appear uncatalysed.

A reaction with $\beta \sim 1$ is very sensitive to the base strength of the catalyst. If different bases are present at comparable concentrations, the strongest will be most effective and provide the major reaction route. The strongest possible base in aqueous solution is OH^-, so the rate law of a reaction with $\beta \sim 1$ in alkaline solution will indicate specific base catalysis.

10.3.3 Nucleophile catalysis

So far we have dealt with Brönsted correlations for reactions which show general acid- or base-catalysis rate laws with no mention of reaction mechanism. The overwhelming majority of reactions which have been subjects of Brönsted-type investigations appear to have rate-determining steps involving proton transfer. But we saw in Chapter 8 that a general acid- or base-catalysis rate law is in itself mechanistically ambiguous. Several mechanisms can be devised which do not involve proton transfer in the rate-determining step but which would still lead to a general acid- or base-catalysis rate law. The most important of these is nucleophile catalysis which leads to kinetic general base catalysis, (p. 349).

The hydrolysis of acetic anhydride in aqueous solution is catalysed by tertiary aromatic amines.

$$(CH_3CO)_2O + H_2O \xrightarrow[k_{exp},\ 0\,°C]{H_2O,\ B} 2CH_3CO_2H$$

$$k_{exp} = k_0 + k_{cat}[B].$$

The mechanism involves nucleophile catalysis (see Chapter 8):[8]

$$(CH_3CO)_2O + B \underset{k_{-1}}{\overset{k_1}{\rightleftharpoons}} CH_3COB^+ + CH_3CO_2^-$$

$$CH_3COB^+ + H_2O \xrightarrow{k_2} CH_3CO_2H + BH^+$$

$$BH^+ + CH_3CO_2^- \overset{fast}{\rightleftharpoons} B + CH_3CO_2H.$$

k_1 and k_{-1} are second-order rate constants for forward and reverse directions of the first step and k_2 = the *pseudo* first-order rate constant for the capture of the reactive intermediate CH_3COB^+ by water in the second step, but k_1 is rate determining[8] i.e. $k_2 \gg k_{-1}[CH_3CO_2^-]$.

A Brönsted-type plot of $\log k_{cat}$ against pK_{BH} gives a reasonable straight line for a limited range of bases and the correlation is given by

$$\log k_{cat} = 0.92_5 pK_{BH} + \text{constant}.$$

Frequently, the slope of the correlation for a nucleophile-catalysed reaction is higher than would be expected if the reaction were mechanistic general base catalysed. This may be used as another criterion for distinguishing between the two kinetically equivalent mechanisms (see p. 349).

10.3.4 Limitations to linear Brönsted correlations

Some Brönsted plots, e.g. for the dehydration of acetaldehyde hydrate:[5]

$$CH_3CH(OH)_2 \xrightarrow{HA,\ H_2O} CH_3CHO + H_2O,$$

have been shown to be linear for catalysts whose pK_{AH} values span a very wide range (~8 pK_{AH} units). There are, however, several experimental and mechanistic factors which for other reactions limit this linear range. Obviously, one cannot attempt to construct a Brönsted plot at all unless the reaction is general acid or base catalysed, and a reaction may be subject to general catalysis in only one pH range. There may be specific catalysis in another range, and, in yet another, the reaction could be uncatalysed. If general catalysis of a reaction is restricted to a narrow pH range, then general catalysts of only a relatively narrow pK_{AH} (or pK_{BH}) range may be used to construct the Brönsted plot.

We shall consider three simple factors any of which could cause the acid AH in eqn (10.11) (or base B in eqn (10.12)) to cease becoming more effective as a catalyst as it becomes stronger as an acid (or a base).

$$X + AH \underset{(k_{-1})}{\overset{k_1\ (\text{r.d.s.})}{\rightleftharpoons}} XH^+ + A^- \xrightarrow{k_2\ (\text{fast})} \text{Products.} \qquad (10.11)$$

1. The first may be regarded as a kinetic limitation. If the second forward step in the mechanism of eqn (10.11) is much faster than the reverse of the first, initial protonation of the substrate X is rate determining. This mechanism yields a general acid-catalysis rate law (see p. 329); use of a stronger catalytic acid, A′H, leads to a higher catalytic rate constant and we have the beginnings of a Brönsted correlation.

But this cannot continue indefinitely. If nothing else occurs, the rate at some stage will become limited not by the rate of the protonation of X by the catalytic acid, but by the rate at which catalyst and substrate encounter each other in solution prior to the proton transfer. In other words, the reaction will become *diffusion-controlled* and further increase in

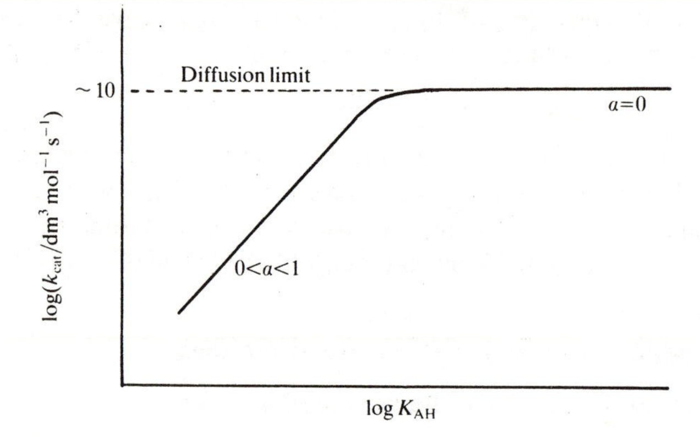

Fig. 10.8. Brönsted plot for a general acid-catalysed reaction as k_{cat} attains the diffusion-controlled limit.

acid strength of the catalyst will not lead to a higher catalytic constant. This causes a discontinuity in the Brönsted plot, and α becomes equal to 0 when k_{cat} attains the diffusion limit, about 10^{10} dm^3 mol^{-1} s^{-1}, Fig. 10.8. The reaction is still general acid catalysed, however, since increasing the concentration of the catalyst still increases the rate of the reaction.

Analogously, eqn (10.12) leads to a general base-catalysis rate law when the second forward step is much faster than the reverse of the first.

$$\mathrm{YH+B} \underset{(k_{-1})}{\xrightarrow{k_1\ \text{(r.d.s.)}}} \mathrm{Y^- + BH^+} \xrightarrow{k_2\ \text{(fast)}} \text{Products.} \qquad (10.12)$$

Initially, a more basic catalyst gives rise to a larger catalytic constant, but at some stage (if nothing else occurs) the reaction rate becomes limited not by the actual proton transfer from substrate to base catalyst, but by the prior diffusion together of HY and B. At this point, there is a discontinuity in the Brönsted plot corresponding to the diffusion-controlled value of k_{cat} (Fig. 10.9) and any further increase in the base strength of the catalyst can cause no further increase in k_{cat}; at this stage, therefore, β has become equal to 0. The reaction remains, however, general base catalysed.

2. Secondly, there may be a *thermodynamic* restriction to linear Brönsted plots following from the mechanisms of eqns (10.11) and (10.12). As the acid strength of AH in eqn (10.11) is increased, it may become at some stage almost fully deprotonated by the aqueous medium before k_{cat} has attained the diffusion limit:

$$\mathrm{AH + H_2O \rightleftharpoons A^- + H_3O^+}.$$

Consequently, there is a practical restriction to the acid strength of

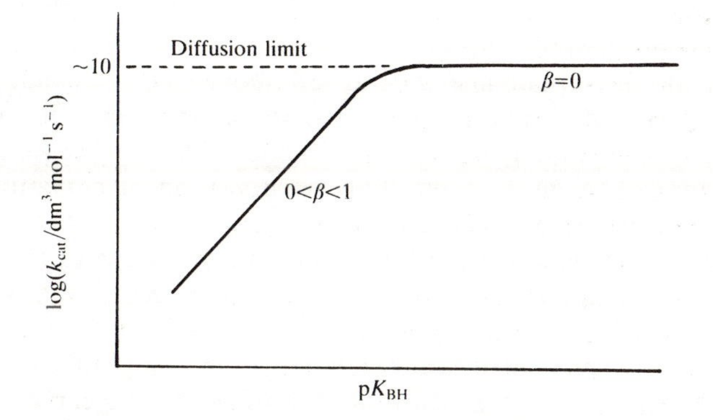

Fig. 10.9. Brönsted plot for a general base-catalysed reaction as k_{cat} attains the diffusion-controlled limit.

general catalysts which may be used, and this is determined by the base strength of the solvent. This type of limitation will be encountered with unreactive substrates which are less basic than the solvent, and leads to specific acid catalysis.

A corresponding thermodynamic restriction may be encountered in general base-catalysed reactions if an unreactive substrate YH is less acidic than the solvent. Increasing the base strength of the general catalyst B in eqn (10.12) could lead it to become almost completely protonated by water before k_{cat} has reached the diffusion limit:

$$B + H_2O \rightleftharpoons BH^+ + OH^-.$$

At this stage, there is no point in using catalysts of higher base strength since the reaction will have become specific base catalysed.

3. A third possible cause for the reactions of eqns (10.11) and (10.12) showing sudden discontinuities in Brönsted plots with $\alpha(\beta) \rightarrow 0$ corresponds to a mechanistic change within the reactions themselves. This may occur when the second forward step (with rate constant k_2) is not hugely faster than the reverse of the first (rate constant k_{-1}) at the low-reactivity end of the Brönsted correlation.

As k_1 in eqn (10.11) increases due to AH becoming a stronger acid, a stage may be reached at which the first step ceases to be rate determining. In other words, the reverse of the first step begins to compete with the second forward step. If the first step becomes a pre-equilibrium, the second step is rate determining and it is easily shown that this mechanism now leads to specific acid catalysis (as long as the second forward step does not involve deprotonation).

Analogously, increasing the base strength of the catalyst in eqn (10.12) could cause the first step to become a pre-equilibrium (rather than irreversible and hence rate determining). This would effect a changeover from general to specific base catalysis if the second forward step does not involve re-protonation.

Any of the three phenomena discussed above causes a sudden discontinuity in a Brönsted plot and the slope to become zero. (In two of the three, the reason is really because the reaction ceases being general acid–base catalysed.) A more interesting phenomenon is curvature of the Brönsted plot whilst the reaction remains general acid or base catalysed. Good theoretical reasons have long been known for such curvature, but unambiguous examples were relatively late in being established.[7,9(a),10] Principally, this was because some feature of the reaction, such as those discussed above, prevented the correlation from being sufficiently extensive for the curvature to be detected. Before we discuss curved Brönsted plots, however, we need to deal with proton transfer itself in a little more detail.

10.4 Proton transfer and the Brönsted Catalysis Law

The algebraic representation of the Brönsted Catalysis Law given so far (eqns (10.4)–(10.7) and (10.9)–(10.10)) includes k_{cat} and K_{AH} (or K_{BH}) where K_{AH} (or K_{BH}) is the conventional acidity constant of a Brönsted acid in water, i.e. it is a measure of the equilibrium extent of protonation of the aqueous solvent by AH (or BH). These equations, therefore, relate the rate constant of one process to the equilibrium constant of a *different* reaction.

It is more convenient for our present purposes if we are able to relate k_{cat}, the rate constant for an elementary reaction which involves proton transfer, to the equilibrium constant K for that same proton-transfer reaction. If we write out the equilibria for the protonation of water and a base B^- by two acids of closely related structure AH and A′H,

$$\mathrm{AH + H_2O \rightleftharpoons A^- + H_3O^+}; \qquad K_{AH} = \frac{[\mathrm{A^-}][\mathrm{H_3O^+}]}{[\mathrm{AH}]}$$

$$\mathrm{AH + B^- \rightleftharpoons A^- + BH}; \qquad K = \frac{[\mathrm{A^-}][\mathrm{BH}]}{[\mathrm{AH}][\mathrm{B^-}]}$$

$$\mathrm{A'H + H_2O \rightleftharpoons A'^- + H_3O^+}; \qquad K'_{AH} = \frac{[\mathrm{A'^-}][\mathrm{H_3O^+}]}{[\mathrm{A'H}]}$$

$$\mathrm{A'H + B^- \rightleftharpoons A'^- + BH}; \qquad K' = \frac{[\mathrm{A'^-}][\mathrm{BH}]}{[\mathrm{A'H}][\mathrm{B^-}]},$$

we see that

$$\frac{K_{AH}}{K'_{AH}} = \frac{[\mathrm{A^-}][\mathrm{A'H}]}{[\mathrm{AH}][\mathrm{A'^-}]} = \frac{K}{K'}$$

therefore

$$\log K_{AH} - \log K'_{AH} = \log K - \log K'$$

or

$$\delta \log K_{AH} = \delta \log K.$$

This establishes that a structural modification to AH which causes a change in $\log K_{AH}$ (for the equilibrium protonation of water) also causes an *identical* change in $\log K$ for its equilibrium protonation of any other base B^-. (The same would be true if we had used an uncharged base B.) From this it follows that the Brönsted relationship for a perfectly general proton-transfer reaction

$$AH + B^- \underset{(\longleftarrow)}{\overset{k}{\longrightarrow}} A^- + HB; \quad K \tag{10.13}$$

may be written

$$\delta \log k = \alpha \,.\, \delta \log K$$

where k = the rate constant for the forward reaction and K = the equilibrium constant for the overall reaction as written.

In this reaction we need not include a distinction between catalytic acid AH and substrate B^- on the one hand, and catalytic base B^- plus substrate AH on the other. The reaction of eqn (10.13) serves as a paradigm for any type of general acid- or general base-catalysed reaction in which the rate-determining step involves proton transfer. Consequently, it does not matter whether the modification which causes a change in K is in AH or B^-. If it is in B^-, we would normally write the Brönsted relationship for the forward reaction of eqn (10.13) as

$$\delta \log k = \beta \,.\, \delta \log K. \tag{10.14}$$

If we do not wish to specify whether the changes in K and k of the reaction of eqn 10.13 are caused by alterations to AH or to B^-, we could use either α or β for the Brönsted coefficient; we shall use β and eqn 10.14.[9(a)]

There are different routes which may at this stage prove helpful. In one, we translate eqn (10.14) into standard molar free energy terms and construct a standard molar free-energy reaction profile for the reaction of eqn (10.13). This quasi-thermodynamic approach is perfectly general and illustrates the obvious parallel with the kinetic form of the Hammett equation to be considered in a later section of this chapter. Additionally, we proceed to the molecular level using the methods exploited at length in the previous chapter on hydrogen isotope effects (to which proton transfer is so obviously related).

10.4.1 The Brönsted Catalysis Law as a linear standard molar free-energy relationship

From the conclusions of Chapter 6 we can write the relationship between the rate constant k and the standard molar free energy of activation $\Delta G^{\ominus\ddagger}$ of the forward reaction of eqn (10.13) as

$$k = \frac{k_B T}{h} \, . \, e^{-\Delta G^{\ominus\ddagger}/RT} \, . \, (1 \text{ mol dm}^{-3})^{-1},$$

where k = the second-order rate constant with units $\text{dm}^3 \text{ mol}^{-1} \text{ s}^{-1}$. Since all units cancel, this may also be written

$$\ln k = -\frac{\Delta G^{\ominus\ddagger}}{RT} + \text{constant},$$

where k now represents only the numerical value of the rate constant.

We also have from Chapter 4 the more rigorously based relationship between the standard molar free energy $\Delta G^{\ominus}$ and the equilibrium constant $K^{\ominus}$ of the reaction of eqn (10.13):

$$\Delta G^{\ominus} = -RT \ln K^{\ominus}.$$

It follows from eqn (10.14) and these relationships that the Brönsted Catalysis Law for the reaction of eqn (10.13) may be expressed as

$$\underline{\delta \, \Delta G^{\ominus\ddagger} = \beta \, . \, \delta \, \Delta G^{\ominus}.}$$

The linear logarithmic relationship between rate and equilibrium constants, therefore, translates into a linear relationship between the standard free energy of activation and the standard free-energy change of the overall reaction. This is illustrated in Fig. 10.10, the upper line representing the molar transformation of eqn (10.13) when AH is a weaker acid than BH (the reaction thermodynamically unfavourable from left to right).

If, without altering AH, we modify B^- to give a stronger base B'^-, then we have a slightly different overall reaction, eqn (10.15), which is thermodynamically less unfavourable from left to right, $K' > K$.

$$AH + B'^- \underset{(\longleftarrow)}{\overset{k'}{\longrightarrow}} A^- + HB'; \qquad K'. \qquad (10.15)$$

We superimpose the reaction profile for eqn (10.15) upon that for eqn (10.13) in Fig. 10.10 in such a way that the initial states coincide in the free-energy coordinate and the reaction coordinates are scaled to match. This gives the lower line of Fig. 10.10 and we see that $\Delta G'^{\ominus\ddagger}$ is less than $\Delta G^{\ominus\ddagger}$, so the forward rate constant of the new reaction is larger, $k' > k$.

The version of the Brönsted relationship in eqn (10.14) for a simple

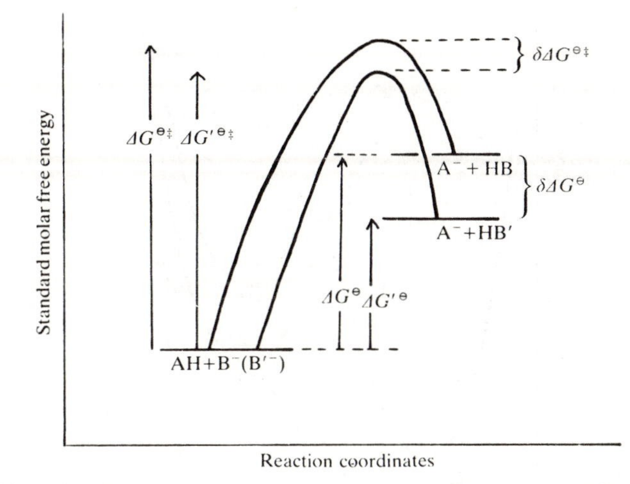

Fig. 10.10. Standard molar free-energy reaction profile for the reaction $AH+B^- \rightleftharpoons A^- + HB$; $pK_{AH}(BH) < pK_{AH}(AH)$.

proton-transfer reaction such as in eqn (10.13) in conjunction with Fig. 10.10 shows that any structural change in the base B^- which causes a change $\delta\,\Delta G^{\ominus}$ in the standard free energy of the proton-transfer reaction causes a smaller but linearly related change $\delta\,\Delta G^{\ominus\ddagger}$ in the standard free energy of activation, and the constant of proportionality between $\delta\,\Delta G^{\ominus}$ and $\delta\,\Delta G^{\ominus\ddagger}$ is the Brönsted parameter β. Alternatively, we could have developed this graphical representation of the Brönsted Catalysis Law using a thermodynamically favourable proton transfer or by considering the structural change in B^- to be base-weakening. Equally well, we could have considered structural changes in AH rather than B^- (and then used α rather than β). Note, however, that standard free-energy profiles such as Fig. 10.10 are simply representations of experimental findings. Expressing the Brönsted Catalysis Law as a linear standard free-energy relationship tells us nothing more about the mechanism of proton transfer or acid–base catalysis. In order to speculate about mechanism, we must consider the reaction at a molecular level.

10.4.2 A simple molecular description of the Brönsted relationship

A mechanism for the reaction of eqn (10.13) in terms of the familiar intersection of molecular potential-energy curves is shown in Fig. 10.11, again with BH a stronger acid than AH.

Initially we have one curve for AH and one for HB. The point of intersection, X, corresponds to the activated complex for the proton transfer between A^- and B^-. In addition, we have also included the curve

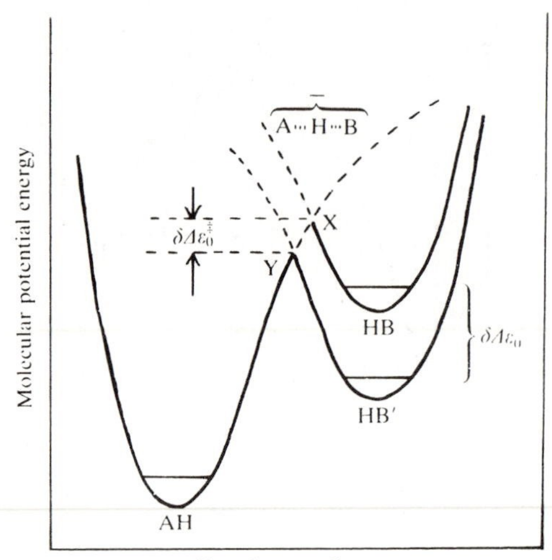

Fig. 10.11. Molecular potential-energy reaction profile for the reaction of eqn (10.13) when BH is a stronger acid than AH.

for a modified compound HB′ which is a weaker acid than HB (the HB′ curve is lower than that for HB). The transition state for proton transfer from AH to B'^- at Y is both lower in energy and earlier in its reaction coordinate than that for proton transfer to B^- at X.

Figures 10.10 and 10.11 give complementary descriptions of a simple proton transfer. The former, being a standard free-energy profile, describes the empirical logarithmic relationship between the rate and equilibrium constants. The latter includes molecular potential energy and so, strictly, does not itself represent the Brönsted relationship. It does provide, however, at a molecular level, a qualitative relationship between the extent of proton transfer in the activated complex and the degree of endo- (or exo-)thermicity of the proton transfer. This is another example of a Hammond effect (see p. 261, and p. 411).

10.4.3 The Eigen mechanism of proton transfer[10]

The simplistic theoretical model which is the basis of (Fig. 10.11) cannot account satisfactorily for the Brönsted coefficient in a simple proton-transfer reaction being close to 0 or to 1.0, yet such values are known. We shall now describe a model for proton transfer which does explain such values.

First we re-write the proton-transfer reaction of eqn (10.13) as eqn (10.16) by including the overall second-order rate constants for the forward and reverse directions:

$$\mathrm{AH + B^-} \underset{k_r}{\overset{k_f}{\rightleftharpoons}} \mathrm{A^- + HB}; \qquad K. \tag{10.16}$$

The equilibrium constant for the reaction as written is given by

$$K = \frac{[\mathrm{A^-}][\mathrm{HB}]}{[\mathrm{AH}][\mathrm{B^-}]} = \frac{k_f}{k_r}.$$

But, since the acidity constants for each of AH and BH separately are given by

$$\mathrm{AH + H_2O \rightleftharpoons H_3O^+ + A^-}; \qquad K_{\mathrm{AH}}(\mathrm{AH}) = \frac{[\mathrm{H_3O^+}][\mathrm{A^-}]}{[\mathrm{AH}]}$$

$$\mathrm{BH + H_2O \rightleftharpoons H_3O^+ + B^-}; \qquad K_{\mathrm{AH}}(\mathrm{BH}) = \frac{[\mathrm{H_3O^+}][\mathrm{B^-}]}{[\mathrm{BH}]},$$

it follows that

$$K = \frac{k_f}{k_r} = \frac{K_{\mathrm{AH}}(\mathrm{AH})}{K_{\mathrm{AH}}(\mathrm{BH})},$$

so

$$\begin{aligned} \log k_f - \log k_r &= \log K_{\mathrm{AH}}(\mathrm{AH}) - \log K_{\mathrm{AH}}(\mathrm{BH}) \\ &= \mathrm{p}K_{\mathrm{AH}}(\mathrm{BH}) - \mathrm{p}K_{\mathrm{AH}}(\mathrm{AH}), \end{aligned}$$

or

$$\log k_f - \log k_r = \Delta \mathrm{p}K, \tag{10.17}$$

where

$$\Delta \mathrm{p}K = \mathrm{p}K_{\mathrm{AH}}(\mathrm{BH}) - \mathrm{p}K_{\mathrm{AH}}(\mathrm{AH}). \tag{10.18}$$

The effect upon the rates of forward and reverse reactions in eqn (10.16) of a change in the acid strength of AH or BH is given by the differential of eqn (10.17) with respect to $\Delta \mathrm{p}K$:

$$\frac{\partial(\log k_f)}{\partial(\Delta \mathrm{p}K)} - \frac{\partial(\log k_r)}{\partial(\Delta \mathrm{p}K)} = 1. \tag{10.19}$$

In order to understand *how* at a molecular level a change in thermodynamic driving force of the overall proton transfer in eqn (10.16) (the magnitude of K) affects the forward and reverse reaction rates, we need to specify mechanism in a little more detail. Equation (10.20) is one model with elementary rate constants included. The next step is to predict the reactivity changes caused by alterations to the acid–base properties of the reactants based upon this mechanism, and then finally we have to compare these predictions quantitatively with experimental results. The quality of the agreement will be a measure of our confidence in the mechanism of eqn (10.20).

$$AH + B^- \underset{k_{-1}}{\overset{k_1}{\rightleftharpoons}} AH.B^- \underset{k_{-2}}{\overset{k_2}{\rightleftharpoons}} A^-.HB \underset{k_{-3}}{\overset{k_3}{\leftrightharpoons}} A^- + HB. \quad (10.20)$$

This mechanism of the reaction from left to right comprises three distinct steps:

(1) initial diffusion together of acid and base to give an encounter complex $AH.B^-$,

(2) intramolecular proton transfer to give an isomeric complex $A^-.HB$, and

(3) dissociation of the second complex to give conjugate base and acid.

We may take k_1 and k_3, the second-order rate constants for diffusion together of two solutes, to be broadly comparable (both involve neutral and anionic species) with maximal values of about $10^{10}\,dm^3\,mol^{-1}\,s^{-1}$. Eigen[10] then postulated that if the overall reaction is thermodynamically favourable ($pK_{AH}(AH) < pK_{AH}(BH)$ or ΔpK = positive) and the basic sites of A^- and B^- are oxygen or nitrogen, then k_2 will be exceedingly high and will not restrict the rate of the forward reaction. Since the acidic hydrogen of an O—H or an N—H will be hydrogen bonded within the encounter complex to the basic site which is to receive the proton, the trajectory of the intramolecular proton transfer will be along the hydrogen bond and involve very little heavy atom motion:

$$\text{e.g.} \quad -\ddot{O}-H\text{-----}:\bar{N}\!< \;\underset{k_{-2}}{\overset{k_2}{\rightleftharpoons}}\; -\ddot{O}:^-\text{----}H-\overset{+}{N}\!<$$

This postulate allowed the ready interpretation of experimental results for the rates of some very fast simple proton-transfer reactions in solution and is the basis of the Eigen treatment of proton transfer.

A consequence of the mechanism of eqn (10.20) for the reaction of eqn (10.16) is that the forward rate is controlled solely by the diffusion together of the reactants in the initial step when AH is appreciably stronger as an acid than BH (when $pK_{AH}(BH) > pK_{AH}(AH)$, ΔpK = positive, and the reaction is thermodynamically favourable in the forward direction). As long as this condition prevails, any structural change in AH or B^- which makes ΔpK more or less positive will not affect the rate of the forward reaction.

Consequently, $\dfrac{\partial(\log k_f)}{\partial(\Delta pK)} = 0$ which corresponds to a Brönsted $\beta = 0$ for the forward reaction under diffusion control; but it then follows from eqn (10.19) that

$$\frac{\partial(\log k_r)}{\partial(\Delta pK)} = -1.$$

This corresponds to the rate constant for the *reverse* reaction *decreasing* if the reaction in the *forward* direction becomes thermodynamically *more* favourable (i.e. the *reverse* reaction becomes *less* favourable) through some structural change in A^- or BH. For the *reverse* reaction of eqn (10.16), therefore, $\beta = +1$ as long as ΔpK (defined in eqn (10.18)) remains positive.

Exactly the same reasoning shows that when the overall reaction of eqn (10.16) is thermodynamically favourable in the reverse direction (BH a stronger acid than AH so, by eqn (10.18), ΔpK = negative), the rate constant of the reaction from right to left by the mechanism of eqn (10.20) is determined solely by k_3; and any structural modifications to A^- or BH which alter the magnitude of the negative ΔpK (eqn (10.18)) will not affect this rate constant.

Consequently, $\dfrac{\partial(\log k_r)}{\partial(\Delta pK)} = 0$ corresponding to Brönsted $\beta = 0$ for the diffusion-controlled reaction from right to left and, by eqn (10.19) therefore,

$$\frac{\partial(\log k_f)}{\partial(\Delta pK)} = 1$$

which corresponds to a Brönsted β value = 1 for the forward reaction. So any structural modification to either AH or B^- in eqn (10.16) which causes the reaction to be less unfavourable from left to right increases the rate constant in this forward direction.

The model for the reaction of eqn (10.16) which is embodied in eqn (10.20) with the postulate of an unrestricting intramolecular proton-transfer step when ΔpK (eqn (10.18)) is substantially positive or negative does, therefore, account for Brönsted coefficients of 0 and 1.0. Predicted values for the Brönsted parameters at different values of ΔpK are shown in Fig. 10.12.[10]

The line abcd is for the forward direction of eqn (10.16) with rate constant $k = k_f$, and efgh is for the reverse direction with rate constant $k = k_r$; k_m = the maximum value for both k_f and k_r (equal to k_1 and k_3 in eqn (10.20) which, in turn, are equal to diffusion-controlled second-order rate constants for neutral and anionic solutes encountering each other).

As the broken lines indicate, the diagram may be divided into three zones.

(1) When ΔpK = negative, the favourable reverse direction is diffusion-controlled with $k_r = k_m$,

$$\frac{\partial(\log k_r)}{\partial(\Delta pK)} = 0,$$

and $k_f < k_m$ but

$$\frac{\partial(\log k_f)}{\partial(\Delta pK)} = \beta = 1.$$

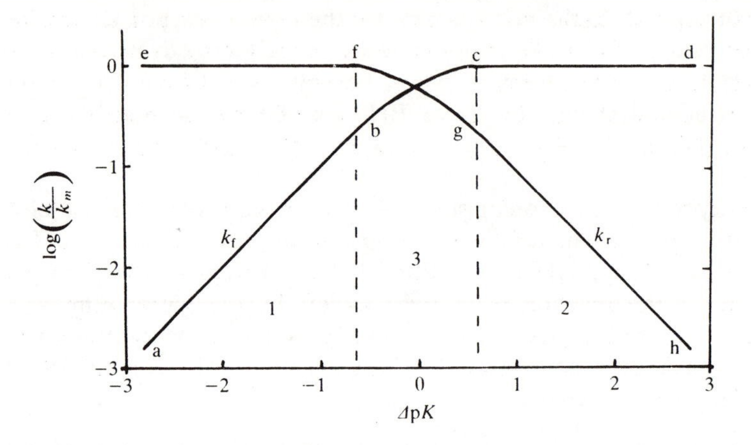

Fig. 10.12. Eigen diagram relating $\log k_f$ and $\log k_r$ to ΔpK for the reaction

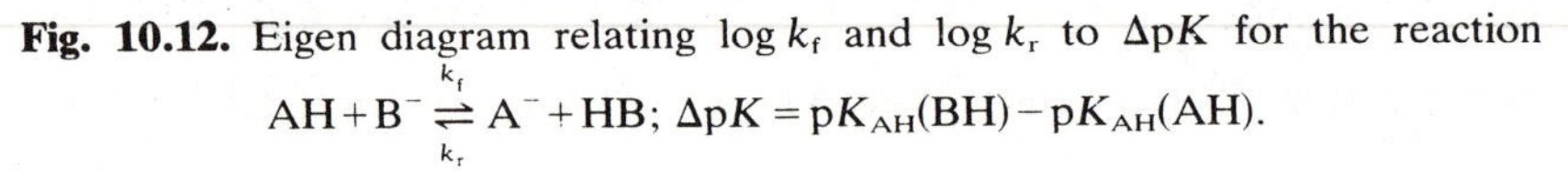

$AH + B^- \underset{k_r}{\overset{k_f}{\rightleftharpoons}} A^- + HB$; $\Delta pK = pK_{AH}(BH) - pK_{AH}(AH)$.

(2) When ΔpK = positive, the favourable forward direction is diffusion-controlled with $k_f = k_m$,

$$\frac{\partial(\log k_f)}{\partial(\Delta pK)} = 0,$$

and now $k_r < k_m$ with

$$\frac{\partial(\log k_r)}{\partial(\Delta pK)} = -\beta = -1 \quad (\beta = 1).$$

(3) In the central zone, as the curved sections indicate, there is a narrow range of ΔpK around $\Delta pK \sim 0$ over which $k_f < k_m$ and $k_r < k_m$ and the gradients $\partial(\log k)/\partial(\Delta pK)$ change from 1.0 to 0 for the forward reaction and 0 to -1.0 for the reverse reaction. The mechanism of eqn (10.20) fully describes the curved lines in this region, the overall forward and reverse rate constants k_f and k_r being complex functions of the individual elementary rate constants k_1, k_{-1}, k_2, k_{-2}, k_3, and k_{-3}.

The high symmetry of Fig. 10.12 is due to the symmetry of the charge type of the reaction in eqn (10.16). If the proton transfer were of the charge type

$$AH + B \underset{k_r}{\overset{k_f}{\rightleftharpoons}} A^- + HB^+, \qquad (10.21)$$

the Eigen diagram would be qualitatively similar but unsymmetrical. This is because the rate constant for a diffusion-controlled reaction between neutral molecules is not the same as that for one between oppositely-charged ions. For the reaction of eqn (10.21), $k_f(\max) < k_r(\max)$ and an Eigen diagram[10] is shown in Fig. 10.13 for a specific example in which

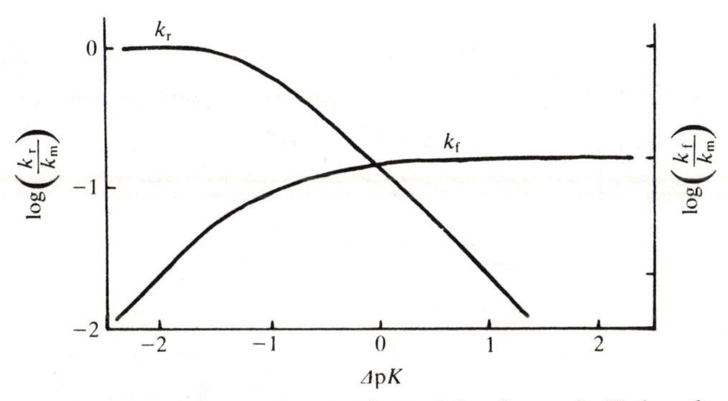

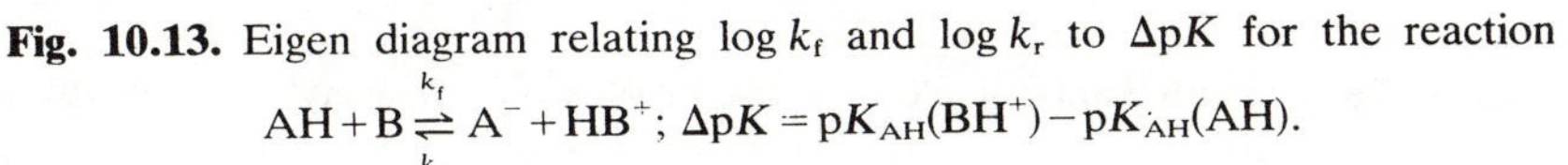

Fig. 10.13. Eigen diagram relating $\log k_f$ and $\log k_r$ to ΔpK for the reaction

$$AH + B \underset{k_r}{\overset{k_f}{\rightleftharpoons}} A^- + HB^+;\ \Delta pK = pK_{AH}(BH^+) - pK_{AH}(AH).$$

$k_f(\max) = 0.1k_r(\max)$. Again, the correlations for the rate constants of forward and reverse directions with ΔpK have 3 phases, the gradients corresponding to Brönsted β values of 0 or 1 where ΔpK is large, and a narrow range around $\Delta pK \sim 0$ where $0 < \beta < 1$.

Having established comprehensive quantitative predictions for the acid–base strength versus reactivity relationships for the reaction of eqn (10.16) on the basis of the mechanism in eqn (10.20), it now remains to compare these predictions with experimental results and thereby test the model.

Results for proton transfer from several oxygen acids to ammonia in aqueous solution (eqn (10.22)) are shown in Fig. 10.14.

$$ROH + NH_3 \underset{(\leftarrow)}{\overset{k_f}{\rightleftharpoons}} RO^- + NH_4^+ \tag{10.22}$$

$$\Delta pK = pK_{AH}(NH_4^+) - pK_{AH}(ROH)$$

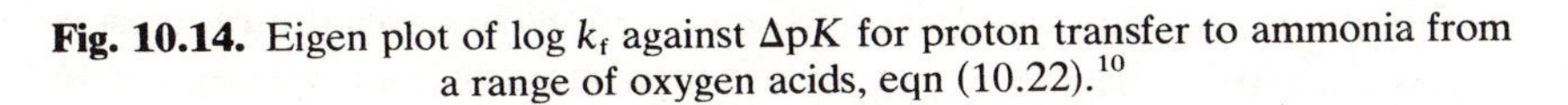

Fig. 10.14. Eigen plot of $\log k_f$ against ΔpK for proton transfer to ammonia from a range of oxygen acids, eqn (10.22).[10]

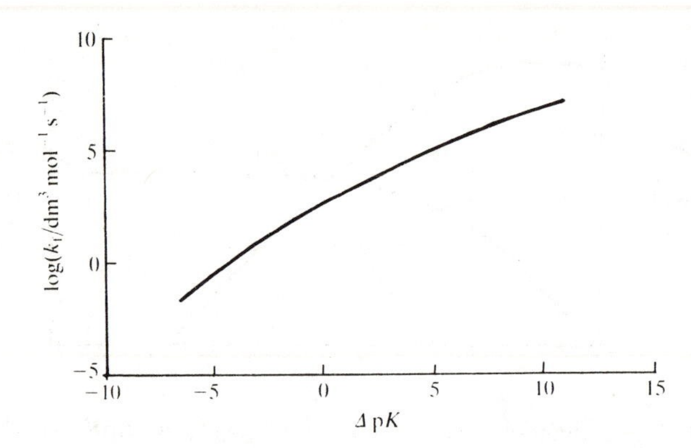

Fig. 10.15. Curved Brönsted plot for the protonation of acetylacetonate by a range of oxygen acids, eqn (10.23).[10]

Clearly, the agreement between Eigen theory and experimental results is very satisfactory in this example as it is for many proton-transfer reactions between oxygen and nitrogen bases (Eigen bases).

There are, however, other reactions involving proton transfer which do not conform to the Eigen model. They include not only reactions with conventional Brönsted correlations, linear over quite a wide range of ΔpK with gradients between about 0.3 and 0.8 as described earlier, but also those such as the reaction of eqn (10.23) for which extended curvature has been demonstrated as shown in Fig. 10.15.[9(a),10]

$$CH_3\overset{\overset{O}{\|}}{C}\bar{C}H\overset{\overset{O}{\|}}{C}CH_3 + ROH \xrightarrow{k_f} CH_3\overset{\overset{O}{\|}}{C}CH_2\overset{\overset{O}{\|}}{C}CH_3 + RO^- \qquad (10.23)$$

$$\Delta pK = pK_{AH}(CH_3COCH_2COCH_3) - pK_{AH}(ROH)$$

Characteristic features of such curves include

(1) a gradually changing β over a wide ΔpK range, the correlation often seeming linear over a limited ΔpK range with $0 < \beta < 1$, and

(2) maximum values of k which are much lower than the diffusion-controlled limiting value.

The failure of the Eigen model to describe accurately these reactions indicates that the mechanism of eqn (10.20) is too simple (or defective in some other respect).

The overall process clearly involves proton transfer, but the detailed mechanism must involve complicating features not included in eqn (10.20). The prevailing current view is that formation of the encounter complex requires more than just a coming together of the reactants; extensive reorganization of solvating molecules is necessary. Furthermore, the rate-determining step of such reactions involves proton transfer coupled with some heavy atom reorganization within the encounter complex.[7,9(a),11] In other words, motion of the proton is not the only component of the reaction coordinate in the region of the transition state of the reaction. Two obvious types of heavy atom reorganization are due to

(1) further changes in the solvation of the encounter complex to facilitate proton transfer, and
(2) rehybridization of the carbon upon deprotonation of a carbon acid.

Both are almost certainly involved in reactions such as the protonation of acetylacetonate (eqn (10.23)) and are accommodated by an extension to the Eigen mechanism which originated in what is generally known as Marcus theory.

10.4.4 Marcus theory

The equation which bears his name was presented by Marcus initially to describe rates of electron transfer between metal ions in solution, but it was subsequently recognized as being also applicable to proton-transfer reactions.[7,9(a),11,12] Unfortunately, the equation appears in different contexts in a bewildering variety of forms, and sometimes the complexities of the expressions obscure their underlying unity. The form we shall use for a proton-transfer reaction in solution, e.g. eqn (10.16),

$$\mathrm{AH} + \mathrm{B}^- \underset{(\leftarrow)}{\xrightarrow{k_f}} \mathrm{A}^- + \mathrm{HB}; \qquad \Delta G^{\ominus} \tag{10.16}$$

is eqn (10.24)

$$\Delta G^{\ominus\ddagger} = w^{\mathrm{r}} + \Delta G_0^{\ominus\ddagger}\left(1 + \frac{\Delta G_{\mathrm{e}}^{\ominus}}{4\Delta G_0^{\ominus\ddagger}}\right)^2 \tag{10.24}$$

where $\Delta G^{\ominus\ddagger}$ = the standard molar free energy of activation of the forward reaction (rate constant $= k_f$),

$\Delta G_{\mathrm{e}}^{\ominus}$ = the standard molar free-energy difference between reactant and product encounter complexes (Fig. 10.16),

w^{r} = the *work term* for the forward reaction (the standard molar free-energy change involved in bringing AH and B^- together in the right configuration and properly solvated for proton transfer then to take place),

$\Delta G_0^{\ominus\ddagger}$ = the *intrinsic barrier* (equal to $\Delta G^{\ominus\ddagger} - w^{\mathrm{r}}$ when $\Delta G^{\ominus} = 0$), and

$|\Delta G^{\ominus}| \not> 4\Delta G_0^{\ominus\ddagger}$. (This restriction follows from Marcus's description of proton transfer within an encounter complex using intersecting potential-energy curves as in Fig. 10.11, p. 430, but these being unperturbed parabolas of equal curvature).

The intrinsic barrier, $\Delta G_0^{\ominus\ddagger}$, is the overall standard free-energy barrier, $\Delta G^{\ominus\ddagger}$, *less the work term*, w^{r}, *when* $B = A$ i.e. in the degenerate (or identity)

reaction. $\Delta G_0^{\ominus\ddagger}$ for the reaction of eqn (10.16) when AH cannot in practice be the same as BH is normally taken to be the mean of the standard free-energy barriers $\Delta G_0^{\ominus\ddagger}$(A) and $\Delta G_0^{\ominus\ddagger}$(B) between the properly configured encounter complexes in the two separate symmetrical proton-transfer steps:

$$\mathrm{AH\,.\,A^-} \xrightarrow{\Delta G^{\ominus\ddagger}(\mathrm{A})} \mathrm{A^-\,.\,HA}$$

$$\mathrm{BH\,.\,B^-} \xrightarrow{\Delta G_0^{\ominus\ddagger}(\mathrm{B})} \mathrm{B^-\,.\,HB}$$

$$\Delta G_0^{\ominus\ddagger}\,(\text{eqn }(10.16)) = 0.5(\Delta G_0^{\ominus\ddagger}(\mathrm{A}) + \Delta G_0^{\ominus\ddagger}(\mathrm{B})).$$

Proton-transfer reactions between Eigen bases have $\Delta G_0^{\ominus\ddagger} \sim 17$–25 kJ mol^{-1} but values in excess of 60 kJ mol^{-1} are known for reactions involving carbon acids.[12]

If we return to the Eigen mechanism for the overall proton transfer of eqn (10.16), i.e. eqn (10.20),

$$\mathrm{AH + B^-} \underset{(\leftarrow)}{\xrightarrow{k_1}} \mathrm{AH\,.\,B^-} \underset{(\leftarrow)}{\xrightarrow{k_2}} \mathrm{A^-\,.\,HB} \underset{(\leftarrow)}{\longrightarrow} \mathrm{A^- + HB}, \quad (10.20)$$

and construct for it a standard molar free-energy profile (Fig. 10.16), the terms of the Marcus equation are readily identified along with $\Delta G^{\ominus}$, the

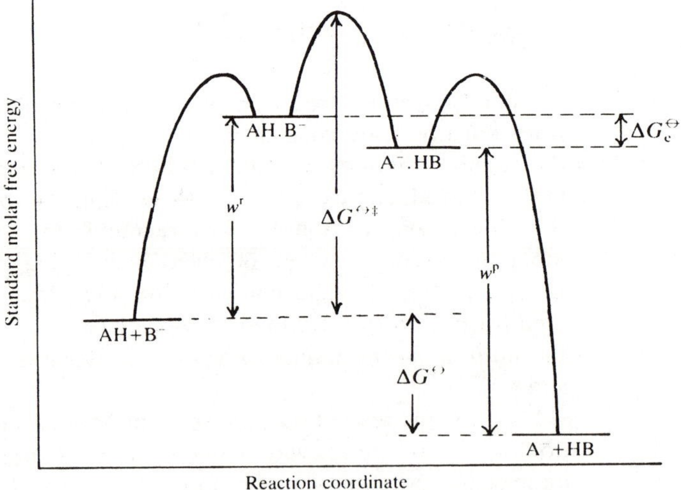

Fig. 10.16. Standard free-energy profile for the mechanism of eqn (10.20), illustrating the terms of the Marcus equation. The intrinsic barrier, $\Delta G_0^{\ominus\ddagger}$, is equal to $\Delta G^{\ominus\ddagger} - w^r$ when A = B ($w^r = w^p$ and $\Delta G^{\ominus} = \Delta G_e^{\ominus} = 0$) and the profile is symmetrical.

overall standard free-energy change, and w^p, the *work term* for the reverse reaction.

In proton-transfer reactions between simple Eigen bases, resolvation during the formation of the encounter complex will not involve much work. w^r, therefore, includes little more than the free-energy change involved in bringing the reactants together, so, if the actual proton transfer itself is not restrictive as when $\Delta pK > \sim 1$, the reaction rate approaches the diffusion-controlled limit. The state of solvation of the encounter complex in the proton transfer from a carbon acid to an Eigen base, however, will be very different from that of the solvated carbon acid and Eigen base apart. Consequently, w^r in such a reaction will be appreciably greater than just the diffusional component. For this reason at least, therefore, maximal reaction rates of deprotonation of carbon acids would be much lower than the diffusion-controlled limit even if the actual proton-transfer step within the encounter complex involved no serious barrier. As mentioned above, however, these reactions also have substantial intrinsic barriers, consequently rates of deprotonation of even quite strong carbon acids are invariably much slower than the diffusion-controlled limit.

Like the Brönsted relationship, eqn (10.24) relates a rate constant of a reversible process (k_f in eqn (10.16)) via the $\Delta G^{\ominus\ddagger}$ term to the reaction's equilibrium constant via $\Delta G_e^{\ominus}$ and hence $\Delta G^{\ominus}$. The variation in k_f, expressed as changes in $\Delta G^{\ominus\ddagger}$, caused by changes in $\Delta G^{\ominus}$ (the overall standard free-energy change) is the partial differential $(\partial \Delta G^{\ominus\ddagger}/\partial \Delta G^{\ominus})$. Expansion of eqn (10.24) followed by replacement of $\Delta G_e^{\ominus}$ by $\Delta G^{\ominus} + w^r - w^p$ and differentiation (assuming $\Delta G_0^{\ominus\ddagger}$, w^r, and w^p are constants for a series of reactions between a single acid AH and a family of bases B'^-, B''^-, etc.) gives, upon re-insertion of $\Delta G_e^{\ominus}$,

$$\left(\frac{\partial\, \Delta G^{\ominus\ddagger}}{\partial\, \Delta G^{\ominus}}\right) = 0.5\left(1 + \frac{\Delta G_e^{\ominus}}{4\, \Delta G_0^{\ominus\ddagger}}\right).$$

But $(\partial\, \Delta G^{\ominus\ddagger}/\partial\, \Delta G^{\ominus})$ for a reaction series of this sort is the Brönsted parameter, β, therefore

$$\beta = 0.5\left(1 + \frac{\Delta G_e^{\ominus}}{4\, \Delta G_0^{\ominus\ddagger}}\right). \tag{10.25}$$

This important theoretically-derived equation indicates that β should be a function not only of the standard molar free-energy change of the overall proton transfer (as is implicit in the Brönsted relationship), but also of the nature of the activated complex in the elementary step of the actual proton transfer. For the degenerate reaction (A = B, $\Delta G_e^{\ominus} = 0$), Marcus theory leads to the calculated result $\beta = 0.5$. If structural changes to the base cause $\Delta G_e^{\ominus}$ to become modestly negative or positive, without at the same time causing any gross change in the structure of the activated complex, then by eqn (10.25), β becomes less than or greater than 0.5 but still well within the range 0–1. If the Hammond effect of these structural changes in B^- is taken into account, we see that β is a measure of transition state structure, i.e. of the extent of proton transfer in the activated complex:

(i) $\Delta G_e^{\ominus}$ = negative, $0<\beta<0.5$, early transition state, low extent of proton transfer in the activated complex;

(ii) $\Delta G_e^{\ominus}$ = positive, $0.5<\beta<1$, late transition state, proton largely transferred in the activated complex.

Had we considered reactions of the single base B^- with a family of acids A′H, A″H, etc., then we would have come to exactly the same conclusions but, following convention, we would have called the Brönsted parameter α rather than β. Either way, the variation in $\beta(\alpha)$ along a reaction series corresponds to gradual curvature in a Brönsted plot and Marcus theory accounts quantitatively for such curves in a way that the simpler Eigen mechanism does not.

If modifications to B^- (or AH, eqn (10.16) with both A^- and B^- Eigen bases) cause proton transfer to become increasingly favourable, but still cause no *qualitative* change to the nature of the activated complex, then β (or α) tends towards the limiting value of 0. In the opposite (unfavourable) direction, of course, α (or β) tends towards 1 for this reversible process since $\beta(k_f)+\alpha(k_r)=1$. When ΔpK for the overall transfer becomes greater than 1–2 units, then the reaction rate becomes controlled by the diffusion together of the reactants rather than by resolvation or the actual proton transfer within the encounter complex, the Marcus equation is no longer applicable, and we have the Eigen mechanism with $\beta=0$ in the favourable direction and 1 in the unfavourable.

Structural modifications to a carbon acid may have a greater effect (but in the same sense) upon the nature and stability of the activated complex for proton transfer than upon the overall standard molar free-energy change. In such cases, a Brönsted α greater than unity is expected, and a *negative* β would necessarily be observed for the reverse reaction since $\alpha(k_f)+\beta(k_r)=1$. Such results have been reported for the deprotonation of arylnitromethanes by several bases including hydroxide:[13]

$$\mathrm{ArCH_2NO_2 + OH^-} \xrightarrow[(\leftarrow)\;k_r]{\mathrm{H_2O,\ 26\,^\circ C}\;\; k_f} \mathrm{ArCH{=}NO_2^- + H_2O}$$

$$\alpha(k_f)=1.54, \qquad \beta(k_r)=-0.54.$$

Interestingly, when the effect of the base strength of a series of amines B upon the rate of deprotonation of any one of several 1-arylnitro-ethanes was investigated,

$$Ar(CH_3)CHNO_2 + B \xrightarrow[H_2O,\ 25\,°C]{k_f} Ar(CH_3)C{=}NO_2^- + BH^+,$$

the unexceptional result of $\beta(k_f) = 0.55$ was obtained.[13]

In contrast, a negative Brönsted $\alpha(k_f)$ was observed for the deprotonation of another series of nitro-alkanes by hydroxide:[14]

$$R^1R^2CHNO_2 + OH^- \underset{(\leftarrow)\ k_r}{\xrightarrow{H_2O,\ 25\,°C\ \ k_f}} R^1R^2C{=}NO_2^- + H_2O$$

$$R^1, R^2 = H, CH_3; \qquad \alpha(k_f) = -0.5, \qquad \beta(k_r) = 1.5.$$

Empirically, a negative α indicates that a structural modification to an acid which causes it to become a stronger proton donor (increases the equilibrium constant for its deprotonation) actually slows down the *rate* of the deprotonation. At the molecular level, such a finding requires that the structural modification to the acid has *opposite* effects upon the stabilities of (i) the activated complex for deprotonation, and (ii) the fully deprotonated form.

Whilst 'anomalous' Brönsted α-values for nitro-alkanes (either negative or greater than unity) are now reasonably well understood within the context of Marcus theory in terms of different electronic effects in (i) the activated complex, and (ii) the nitronate anion,[14,15] they do indicate that Brönsted coefficients may not be precise or reliable indicators of transition state structure in the deprotonation of carbon acids.[11(a),13]

Over the past decade or so, there has been a much wider recognition of the applicability of a Marcus-type approach to reactions other than electron and proton transfer.[11(b),16] It has been applied successfully, for example, to S_N2 reactions such as methyl transfer:

$$X^- + CH_3{-}Y \underset{(\leftarrow)}{\xrightarrow{k}} X{-}CH_3 + Y^-; \qquad \Delta G^{\ominus}.$$

Whether or not the approach can be extended legitimately to additions, e.g.

$$X^- + \rangle C{=}O \underset{(\longleftarrow)}{\xrightarrow{k}} X{-}\overset{|}{\underset{|}{C}}{-}O^-; \ \Delta G^{\ominus},$$

or electrophile (E^+)-nucleophile (N^-) combinations

$$E^+ + N^- \underset{(\leftarrow)}{\xrightarrow{k}} E{-}N; \qquad \Delta G^{\ominus},$$

for which the identity reactions, and hence the intrinsic barriers, are less easily formulated, is presently a controversial question.[11(b),17] This and the development of other theoretically-based quantitative equations of correlation, especially when they can be linked to the use of reaction maps,[11(a),18] are major areas of current research in physical organic chemistry.

10.5 The Hammett equation and chemical reactivity

The Hammett equation (eqn (10.26)) applied to a series of chemical equilibria involving a family of aromatic compounds (1) with different substituents X but a common reaction site Y was introduced in Chapter 5.

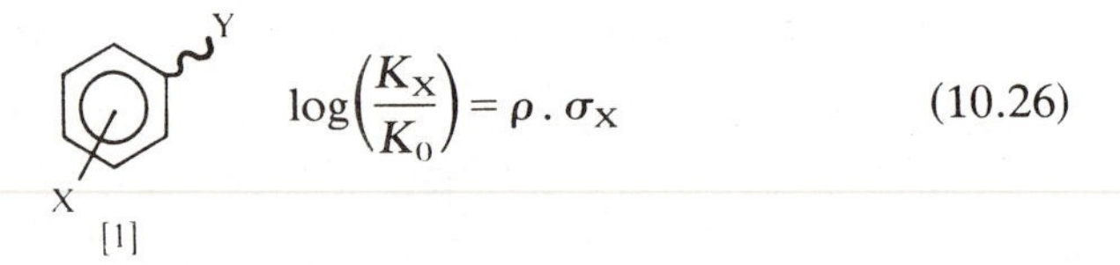

$$\log\left(\frac{K_X}{K_0}\right) = \rho \, . \, \sigma_X \qquad (10.26)$$

In eqn (10.26), K_X and K_0 = the equilibrium constants for the reactions of the X-substituted aromatic compound and the compound with X = H, ρ = a reaction parameter for the particular reaction series, and σ_X = a substituent parameter defined by the effect of X upon the acidity constant of benzoic acid in water at 25 °C:

$$XC_6H_4CO_2H + H_2O \underset{25\,°C}{\overset{H_2O}{\rightleftharpoons}} XC_6H_4CO_2^- + H_3O^+ \qquad (10.27)$$

$$K_{AH}(XC_6H_4CO_2H) = \frac{[H_3O^+][XC_6H_4CO_2^-]}{[XC_6H_4CO_2H]}$$

$$\sigma_X = \log K_{AH}(XC_6H_4CO_2H) - \log K_{AH}(C_6H_5CO_2H),$$

or

$$\sigma_X = pK_{AH}(C_6H_5CO_2H) - pK_{AH}(XC_6H_4CO_2H). \qquad (10.28)$$

A selected list of σ-values is given in Table 10.3.

The normal practice in establishing a Hammett correlation for equilibrium constants in a reaction series involves plotting $\log K_X$ for the differently substituted members of the family of compounds (or $\log(K_X/K_0)$) against σ_X for the substituents, and the gradient is equal to ρ, e.g. Fig. 5.3 (Chapter 5, p. 204).

It was also found[19,20(a)] that, for some reactions, linear plots are obtained when the *rate constants* k_X of differently substituted aromatic compounds (1) are plotted logarithmically against σ_X. Results for an early example, the base-induced hydrolysis of a family of benzoates (eqn (10.29)),[21] are shown in Fig. 10.17.

$$XC_6H_4CO_2C_2H_5 + OH^- \xrightarrow[\text{85\% aqueous ethanol}]{k_X,\ 25\,°C} XC_6H_4CO_2^- + C_2H_5OH. \qquad (10.29)$$

TABLE 10.3
Hammett substituent constants[9(b)]

Substituent	σ_m	σ_p
NO_2	0.71	0.78
CN	0.61	0.70
CF_3	0.43	0.54
CH_3CO_2	0.39	0.31
Br	0.39	0.23
CH_3CO	0.38	0.48
$CO_2C_2H_5$	0.37	0.45
Cl	0.37	0.22
CHO	0.36	0.44
CO_2H	0.35	0.44
I	0.35	0.28
F	0.34	0.06
C≡CH	0.20	0.23
SCH_3	0.15	~0
OH	0.13	−0.38
OCH_3	0.11	−0.28
C_6H_5	0.05	~0
H	0	0
CH_3	−0.06	−0.17
C_2H_5	−0.07	−0.15
$CH(CH_3)_2$	−0.07	−0.15
$C(CH_3)_3$	−0.10	−0.20
$N(CH_3)_2$	−0.15	−0.63
NH_2	−0.16	−0.57

The mathematical equation which describes correlations exemplified in Fig. 10.17, regardless of the kinetic order of the reaction, is

$$\log\left(\frac{k_X}{k_0}\right)=\rho\,.\,\sigma_X \tag{10.30}$$

where k_X and k_0 = rate constants for the reactions of the X-substituted aromatic compound and the compound with X = H, ρ = the reaction parameter associated with the reaction series (e.g. $\rho = 2.54$ from Fig. 10.17 for the reaction of eqn (10.29)), and σ_X = the substituent parameter defined in eqn (10.28).

Equation (10.30), which is the usual expression of the Hammett equation applied to reaction rates, may also be written

$$\log k_X=\rho\,.\,\log K_{AH}(XC_6H_4CO_2H)+\text{constant},$$

or, in its differential form

$$\delta\log k_X=\rho\,.\,\delta\log K_{AH}(XC_6H_4CO_2H), \tag{10.31}$$

where δ indicates incremental changes in the functions shown, k_X now

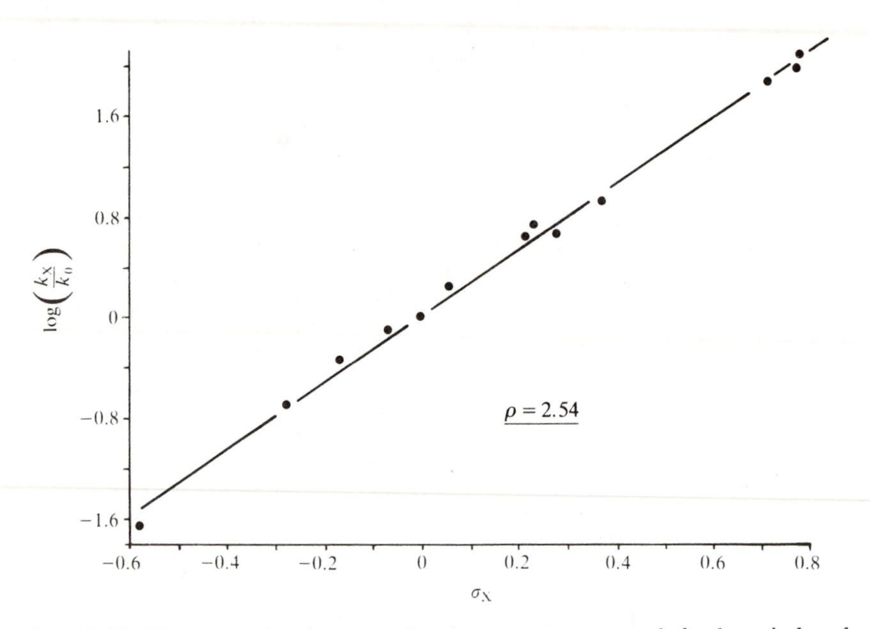

Fig. 10.17. Hammett plot for second-order rate constants of the base-induced hydrolysis of ethyl benzoates, 85 per cent aqueous ethanol, 25 °C.[21]

represents only the numerical value of the rate constant (the units having cancelled in taking the ratio k_X/k_0 in eqn (10.30), and ρ is seen to be the constant of proportionality between the incremental changes in these functions, $\log k_X$ and $\log K_{AH}(XC_6H_4CO_2H)$, brought about by the same change in substituent X.

A positive ρ-value indicates that the reaction as written is facilitated (the rate constant increased) by introducing a substituent which increases the acidity of benzoic acid (one which is electron-withdrawing). Conversely, if the introduction of electron-withdrawing (acid-enhancing) substituents decreases the reactivity of the aromatic substrate in a particular reaction, then that reaction series will have a negative ρ-value. And the magnitude of ρ expresses the sensitivity of the reaction series, in either sense, to the introduction of electron-withdrawing substituents. A reaction series for which $\rho = 0$ indicates that electron-withdrawing substituents have no effect upon the rate constant of the reaction.

The ρ-value used to correlate the equilibrium constants of a family of compounds (Chapter 5) will not usually have the same value as that for the corresponding rate constants, but it is easily shown that, for a reversible reaction,

$$\rho(\text{equil.const.}) = \rho(\text{forward rate const.}) - \rho(\text{reverse rate const.})$$

Example. From the σ-values and rate constants $k_2(X)$ given below for the second-order forward reaction as written,[22]

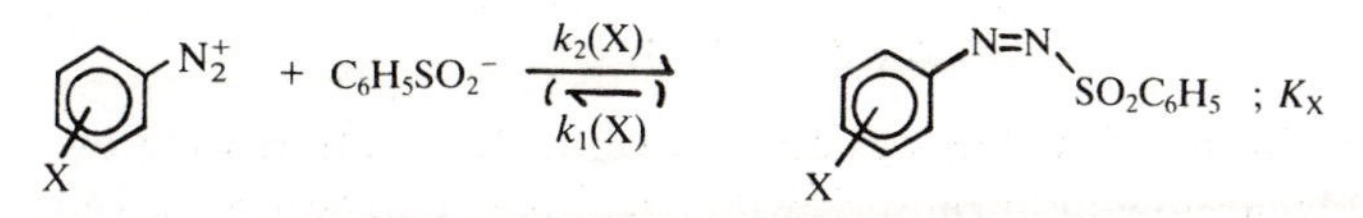

and the result $\rho = 3.76$ for the equilibrium constants K_X (given in Chapter 5, p. 211), calculate ρ for the first-order reverse rate constants $k_1(X)$.

Substituent X	σ_X	$10^{-3}\, k_2(X)/dm^3\, mol^{-1}\, s^{-1}$	$\log\{k_2(X)/k_2(H)\}$
p-CH_3	−0.17	0.158	−0.31
H	0	0.324	0
p-Cl	0.22	0.983	0.48
p-Br	0.23	0.715	0.34
m-Cl	0.37	3.64	1.05
m-CF_3	0.43	4.72	1.16
p-CN	0.70	14.7	1.66
p-NO_2	0.78	19.2	1.77

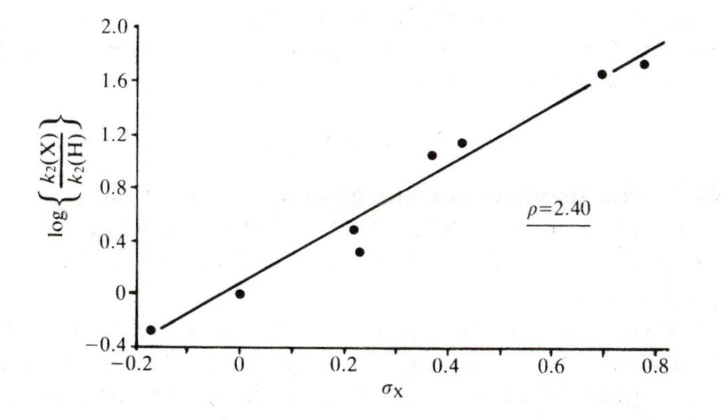

From these results, the ρ-value for the dissociation of the diazo-compound, $k_1(X)$, is given by

$$3.76 = 2.40 - \rho(k_1(X)),$$

therefore

$$\underline{\rho(k_1(X)) = -1.36.}$$

10.5.1 The Hammett equation as a linear standard molar free-energy relationship

We saw in Chapter 4 that the thermodynamic equilibrium constant $K_a^\ominus$ is simply an alternative to the standard molar free-energy change $\Delta G^\ominus$ for expressing the direction and thermodynamic driving force of a reversible reaction. For the dissociation of a substituted benzoic acid, eqn (10.27),

$$\Delta_{AH}G^{\ominus} = -RT \ln K_a^{\ominus}.$$

To the extent that the acidity constant K_{AH} approximates the true thermodynamic equilibrium constant $K_a^{\ominus}$ (see Chapter 5, p. 164),

$$\Delta_{AH}G^{\ominus} = -RT \ln K_{AH}. \tag{10.32}$$

The relationship between the rate constant k (with proper units) for a reaction of compound (1) of molecularity (kinetic order) $\boldsymbol{n}$ and the standard free energy of activation $\Delta G^{\ominus\ddagger}$ is obtained from Chapter 6:

$$k = \frac{k_B T}{h} \,.\, e^{-\Delta G^{\ominus\ddagger}/RT} \,.\, (1 \text{ mol dm}^{-3})^{1-\boldsymbol{n}},$$

or, since the units of k cancel with those of the right-hand side,

$$\ln k = -\frac{\Delta G^{\ominus\ddagger}}{RT} + \text{constant} \tag{10.33}$$

where k now represents only the numerical value of the rate constant.

Equations (10.32) and (10.33) allow the differential form of the Hammett equation applied to reaction rates (eqn (10.31)) to be transformed into

$$\delta\, \Delta G^{\ominus\ddagger} = \rho \,.\, \delta\, \Delta_{AH}G^{\ominus} \tag{10.34}$$

where $\Delta G^{\ominus\ddagger}$ = the standard free energy of activation of the (substituted) aromatic compound (1) in a reaction involving the functional group Y, and $\Delta_{AH}G^{\ominus}$ = the standard free energy of the dissociation of the correspondingly substituted benzoic acid.

Equation (10.34) tells us that if a reaction series is correlated by the Hammett equation, the incremental change in $\Delta G^{\ominus\ddagger}$ for the reaction involving the group Y of (1) caused by a change in substituent X is proportional to the incremental change in $\Delta_{AH}G^{\ominus}$ for the dissociation of the benzoic acid caused by the same change in X. So the Hammett equation for reaction rates is another linear standard free-energy relationship. It is like the Brönsted Catalysis Law in that it correlates kinetic and thermodynamic parameters; but, in contrast to the Brönsted relationship, the Hammett equation correlates the effects of aromatic substituents upon the rate constants of a wide variety of reaction types with their effects upon a single standard chemical equilibrium – the dissociation of benzoic acid in water at 25 °C.

10.5.2 The Hammett equation and reaction mechanisms

We shall now consider a *molecular* interpretation of the correlation between the effect of the substituent upon the acidity constant of benzoic acid (H_2O, 25 °C) and the effect upon the rate constant for the reaction of some benzene derivative using a specific example. Good Hammett correlations have been obtained for the reactions of diphenyldiazomethane with several types of aromatic acids under a range of experimental conditions; one such reaction series is represented by eqn (10.35).[23]

$$\mathrm{ArCH_2CO_2H + Ph_2CN_2 \xrightarrow[C_2H_5OH,\ 30\,°C]{k_{exp}} ArCH_2CO_2CHPh_2 + N_2} \quad (10.35)$$

$$\rho = 0.40.$$

A reasonable mechanism with the initial proton transfer of eqn (10.36) as the rate-determining step is shown below:

$$\mathrm{ArCH_2CO_2H + Ph_2CN_2 \xrightarrow[k_1]{r.d.s.} ArCH_2CO_2^- + Ph_2CHN_2^+} \quad (10.36)$$

$$\mathrm{Ph_2CHN_2^+ \xrightarrow{fast} Ph_2CH^+ + N_2}$$

$$\mathrm{ArCH_2CO_2^- + Ph_2CH^+ \xrightarrow{fast} ArCH_2CO_2CHPh_2.}$$

One resonance canonical of the activated complex which intervenes between reactants and products in the bimolecular elementary step of eqn (10.36), the rate-determining step of the reaction of eqn (10.35), is shown in (2).

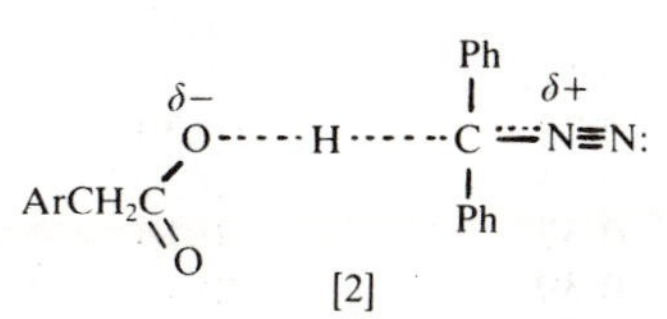

[2]

It involves *partial* proton transfer from $ArCH_2CO_2H$ to the diazo-compound. And a change in the substituent in the aryl residue which affects the rate constant k_1 in eqn (10.36) (and hence k_{exp} in eqn (10.35)) alters $\Delta G^{\ominus\ddagger}$, the standard molar free energy for the formation of (2) from reactants, by an amount $\delta\,\Delta G^{\ominus\ddagger}$.

Substituents which affect K_{AH} for the dissociation of benzoic acid (eqn (10.27)) alter the standard free energy $\Delta_{AH}G^{\ominus}$ for the *complete* proton transfer from $PhCO_2H$ to H_2O by an amount $\delta\,\Delta_{AH}G^{\ominus}$. It is, therefore, at a molecular level perfectly reasonable that the effect of a substituent upon the *partial* removal of a proton from $ArCH_2CO_2H$ by the diazo-compound in the elementary reaction of eqn (10.36) should be proportional to the effect induced by the same substituent upon the *complete* removal of a proton from the *related* carboxylic acid $ArCO_2H$, by H_2O (eqn (10.27)). The magnitude of the constant of proportionality between these two effects, ρ in eqn (10.34) for the reaction of eqn (10.35), will, of course, be affected by the nature of the solvent, the temperature, and the extent of the proton transfer in (2). Each different family of aromatic acids,

e.g. $\mathrm{ArC{\equiv}CCO_2H}$, $\mathrm{ArCH_2CH_2CO_2H}$, and $\mathrm{ArOCH_2CO_2H}$, has its own ρ-value for given experimental conditions in the reaction with diphenyldiazomethane; $\rho = 0.31$, 0.22, and 0.25, respectively, for these three acids in ethanol at 30 °C.[23] The value for substituted benzoic acids themselves, $ArCO_2H$ in this reaction in ethanol, $\rho = 0.94$, compares with $\rho = 1.65$ for the equilibrium dissociation of benzoic acids in ethanol.[24] As expected, therefore, a substitution in the aromatic ring of benzoic acid has a greater effect upon the ease of complete proton removal by solvent than upon partial deprotonation *of the same acid* by diphenyldiazomethane under the same reaction conditions.

Example. Construct a Hammett plot from the results below[25] and comment upon the ρ-value.

$$XC_6H_4\ddot{N}(CH_3)_2 + CH_3I \xrightarrow[\text{acetone}]{k_X,\ 35\,°C} XC_6H_4\overset{+}{N}(CH_3)_3I^- \qquad (10.37)$$

Substituent, X	σ_X	$10^2 k_X/dm^3\,mol^{-1}\,min^{-1}$	$\log\left\{\frac{k_X}{k_0}\right\}$
p-CH_3O	−0.28	9.0	0.89
p-CH_3	−0.17	3.40	0.46
m-CH_3	−0.06	1.85	0.20
H	0	1.17	0
m-CH_3O	0.11	1.03	−0.06
p-Cl	0.22	0.39	−0.48
p-Br	0.23	0.37	−0.50
m-Cl	0.37	0.16	−0.86
m-Br	0.39	0.14	−0.92

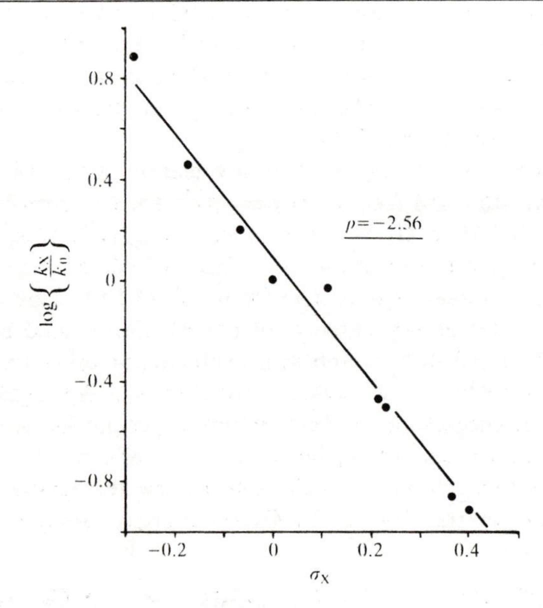

Empirically, $\rho = -2.56$ indicates that electron-withdrawing groups reduce considerably the reactivity of the arylamine. This is entirely compatible with an S_N2 mechanism involving an activated complex (3) which is destabilized by electron-withdrawing groups:

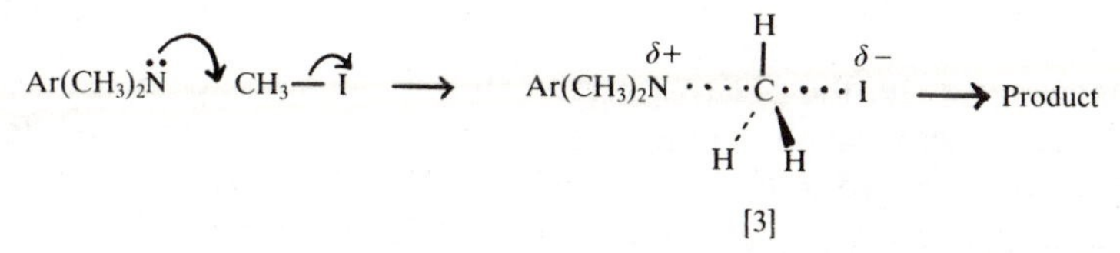

The nitrogen atom must, therefore, bear a substantial partial positive charge at the transition state.

The reactivity of the nucleophilic amine in the addition–elimination reaction of eqn (10.38) is also reduced by substituents with positive (acid-enhancing) σ-values:[26]

$$ArNH_2 + C_6H_5CO.Cl \xrightarrow[25\,°C]{C_6H_6} ArNHCO.C_6H_5 + HCl \qquad (10.38)$$

$$\rho = -3.21.$$

If the aromatic residue is in the leaving group of a nucleophilic substitution (or addition–elimination) reaction rather than the nucleophile, a positive ρ-value is always expected regardless of mechanistic detail. This has been established for the following solvolytic reactions of arenesulphonates:[28]

solvent-induced S_N2,

$$CH_3O_3SAr \xrightarrow[70\,°C]{C_2H_5OH} CH_3OC_2H_5 + ArSO_3H$$

$$\rho = 1.3;$$

S_N1,

$$\text{2-adamantyl-}O_3SAr \xrightarrow[70\,°C]{C_2H_5OH} \text{2-adamantyl-}OC_2H_5 + ArSO_3H$$

$$\rho = 1.8.$$

The effect of substituents in an aromatic group of the electrophilic residue in nucleophilic substitution reactions has also been investigated. Here, the sign and magnitude of the ρ-value depend subtly upon the nature of the reactants and the mechanism.

Equation (10.39) includes an alkyl halide whose reactivity is modestly increased by electron-withdrawing substituents in the aromatic residue,[29] and the activated complex, assuming an S_N2 mechanism, is shown in (4).

$$ArCH_2Cl + I^- \xrightarrow[\text{acetone}]{k_X,\ 20\,°C} ArCH_2I + Cl^- \qquad (10.39)$$

$$\rho = 0.79$$

$$\left(\overbrace{\mathrm{I}\cdots\cdots\underset{\underset{\displaystyle \mathrm{Ar}}{|}}{\mathrm{CH_2}}\cdots\cdots\mathrm{Cl}}^{-}\right)^{\ddagger}$$

(4)

Mechanistically, the positive ρ indicates an activated complex which is stabilized to some small extent by electron-withdrawing substituents. This suggests an increase in electron density at the CH_2 of the reaction site as the substrate is transformed into activated complex by the approach of the nucleophile. We conclude, therefore, that bonding by the I^- runs just ahead of the unbonding of the Cl^-, and a standard free-energy reaction map of this type of reaction would be like Fig. 10.3, p. 409.

If the breaking of the bond to the anionic leaving group runs ahead of the formation of the bond to an incoming anionic nucleophile, the reaction path in the standard free-energy reaction map curves in the opposite sense to that of Fig. 10.3. This represents the formation of an activated complex with a decrease in electron density at the carbon of the reaction site. A negative ρ-value would be one indication of such a mechanism, and such results have been reported although they are usually based upon rate constants of only a few compounds. With a more extended family of benzylic compounds whose substituents X span a wider range in σ, negative ρ-values have been obtained but a curved Hammett plot often indicates that the ρ-value changes along the series. It may even be U-shaped with ρ changing from negative, through zero, to positive as exemplified in Fig. 10.18 for the reaction series of eqn (10.40).[31]

$$XC_6H_4CH_2Br + p\text{-}CH_3OC_6H_4S^- \xrightarrow[CH_3OH]{k_X,\ 20\,^\circ C} XC_6H_4CH_2SC_6H_4OCH_3 + Br^-. \tag{10.40}$$

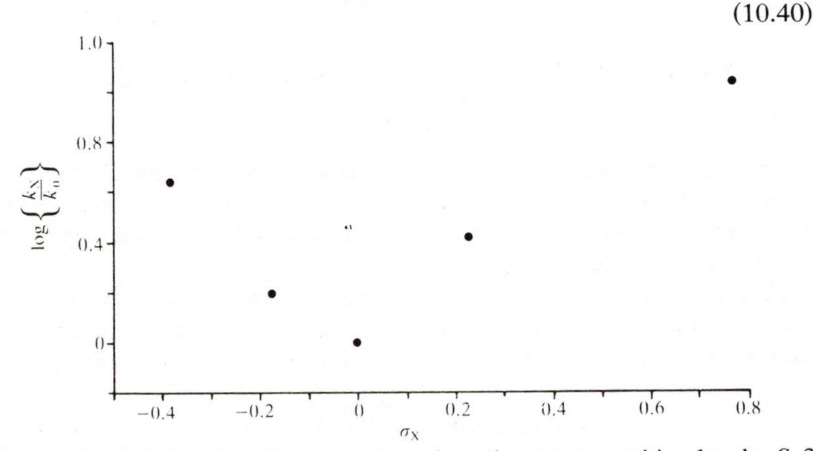

Fig. 10.18. Variation in ρ from negative, through zero, to positive for the S_N2 reaction of eqn (10.40).[31]

The ρ-value of the addition-elimination reaction of eqn (10.41) shows that electron-withdrawing substituents in the acid chloride have a modest rate-enhancing effect.[26]

$$C_6H_5NH_2 + ArCO.Cl \xrightarrow[25\,°C]{C_6H_6} C_6H_5NHCO.Ar + HCl \qquad (10.41)$$

$$\rho = 1.18.$$

This result should be compared with the opposite effect of the same substituents in the arylamine, eqn (10.38). Clearly, any statement of the ρ-value of a reaction with more than one aromatic residue must indicate the substitution site.

10.5.3 Correlations with σ^-

We saw in Chapter 5 that, when a reaction site involves π-electron density directly conjugated to a *para*-substituent with a mesomeric electron-withdrawing capability, a modified substituent constant, σ^-, is used. The σ^--value of such a substituent is determined by fitting the pK_{AH} of the *p*-substituted phenol to the Hammett plot established using other phenols (*meta*-substituted and those with *para*-substituents which are not conjugatively electron-withdrawing); see Fig. 5.4 (Chapter 5, p. 208). σ^--values so obtained then give much better correlations for other comparable equilibria, such as the dissociation of anilinium cations (Fig. 5.5, Chapter 5, p. 210), than do the simple σ-values. Some σ^--values are given in Table 10.4.

TABLE 10.4
σ_p and σ_p^- substituent constants[9(b)]

p-Substituent	σ_p	σ_p^-
NO_2	0.78	1.27
CN	0.70	0.88
CH_3CO	0.48	0.84
$CO_2C_2H_5$	0.45	0.74
CHO	0.44	1.04
CO_2H	0.44	0.78
$C{\equiv}CH$	0.23	0.52
C_6H_5	0	0.08

For exactly the same reasons, a *p*-nitro group has a much larger deactivating effect upon the nucleophilicity of the phenoxide in the S_N2 reaction of eqn (10.42) than is anticipated on the basis of its ordinary σ-value.[25]

$$XC_6H_4O^- + C_2H_5I \xrightarrow[C_2H_5OH,\ 35\,°C]{k_X} XC_6H_4OC_2H_5 + I^-. \qquad (10.42)$$

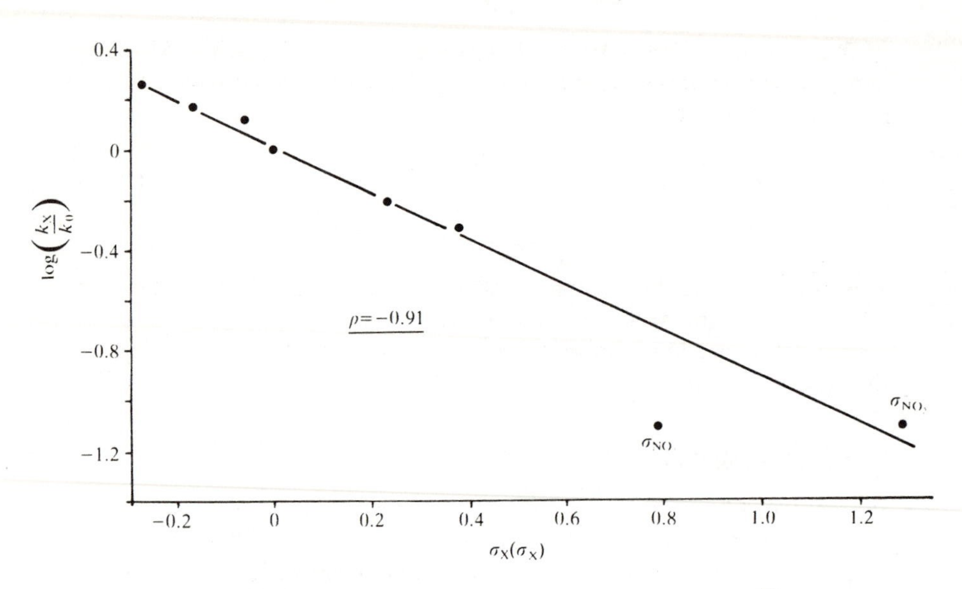

Fig. 10.19. Hammett plot for the reaction of eqn (10.42).[25]

This is due to conjugation between the nucleophilic oxygen and the *p*-nitro group:

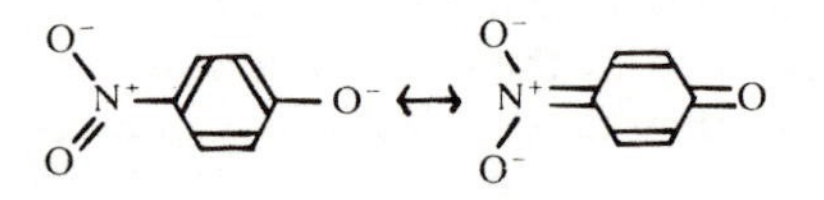

The effect is illustrated in Fig. 10.19 which includes the correlation between $\log\{k_X/k_0\}$ and σ_X for the reaction of eqn (10.42); clearly, σ^- rather than σ for *p*-nitro fits the correlation.

$\text{Log}(k_X/k_0)$ of arylamines with *para*-substituents such as CH_3CO, CHO, or NO_2 in the reactions of eqns (10.37) and (10.38) would also be expected to correlate with σ^- rather than σ.

Perhaps the most important class of reactions for which σ^--values are used is nucleophilic aromatic substitution, eqn (10.43).

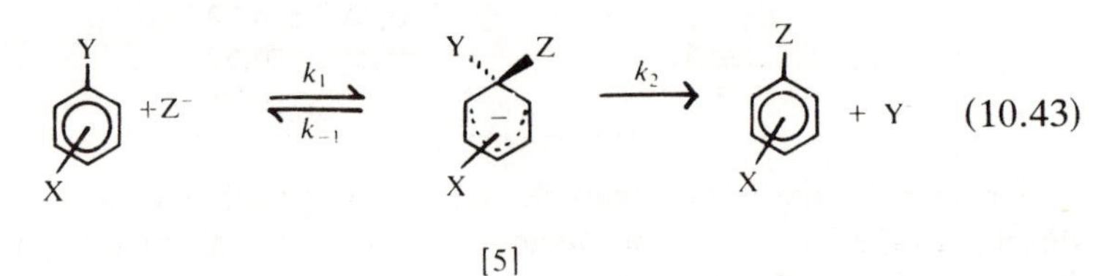

(10.43)

[5]

In contrast to the other reactions considered so far in this chapter, the one of eqn (10.43) is at a carbon of the aromatic ring. Consequently, the increase in negative charge in the six-membered ring of the activated complex is much more effectively delocalized by *para*-substituents which have a conjugative electron-withdrawing capability. This is true regardless of whether the formation of the Meisenheimer intermediate (5) or (less commonly) its subsequent decomposition is rate determining. We can illustrate the effect for a *p*-acetyl group in the intermediate (5),

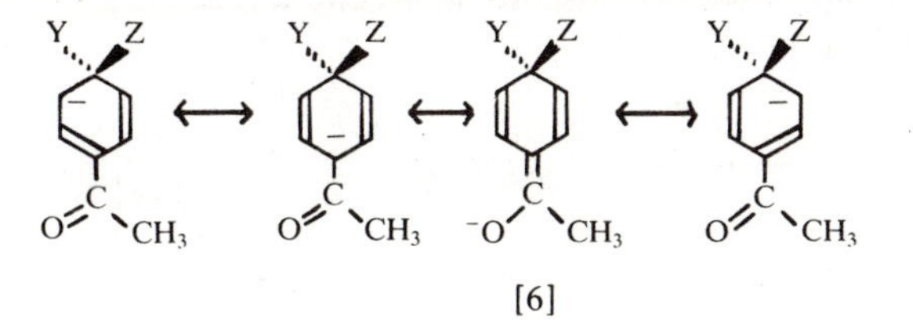

[6]

but, of course, the kinetic effect is really due to the resonance interaction at the activated complex which precedes (or succeeds) the formation of (5). If the acetyl group is *meta* to the reaction site, or if another substituent with no electron-withdrawing conjugative effect is in the *para*-position, then a resonance canonical corresponding to (6) cannot be drawn and the species is that much less stable.

Some results are shown below:[32]

$$\text{(Y, X, } NO_2\text{-substituted ring)} + Z^- \xrightarrow[50\,^\circ C]{CH_3OH} \text{(Z, X, } NO_2\text{-substituted ring)} + Y^-$$

Y^-	Z^-	$\rho(\sigma^-)$
I^-	CH_3O^-	3.87
I^-	N_3^-	3.12
I^-	SCN^-	5.05
Cl^-	CH_3O^-	3.90

10.5.4 Correlations with σ^+

We have seen that substituents which are conjugatively electron-withdrawing have an effect beyond what is expected on the basis of their σ-values upon reactions involving π-electron density at a reaction site conjugated with the aromatic ring, e.g. eqn (10.42).

Correspondingly, a substituent which can mesomerically supply π electron density, such as *p*-$CH_3\ddot{O}$– or *p*-$\ddot{N}H_2$–, has an extra facilitating effect upon a reaction which generates an electron-deficient site when the substituent and the incipient vacant *p*-orbital on the sp^2-hybridized carbon of the reaction centre are conjugated.

Poor Hammett correlations are obtained in such reactions, of which there are many, using simple σ-values defined by eqn (10.28) relating to the standard reaction of eqn (10.27). But when alternative substituent constants, σ^+, are used, determined by a different standard reaction, much better correlations ensue.

The new standard reaction is the hydrolysis of t-cumyl chlorides (2-arylprop-2-yl chlorides, (7)) in 90 per cent aqueous acetone (90A) at 25 °C, eqn (10.44).[33,34]

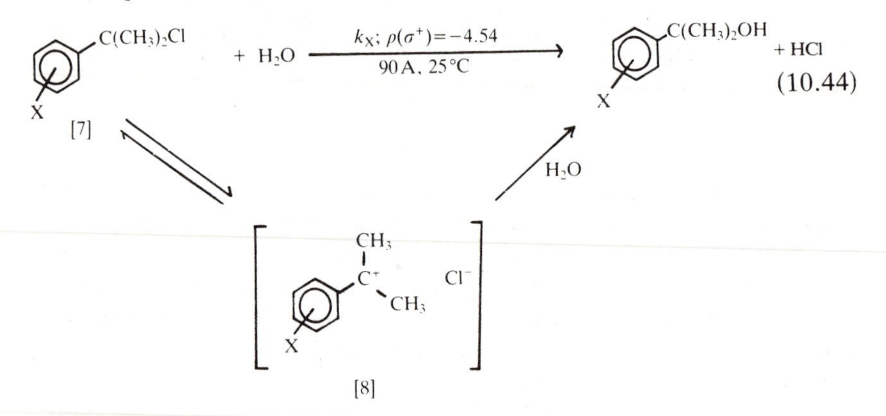

Regardless of whether initial ionization, conversion of one type of ion-pair into another, or subsequent nucleophilic capture of the carbonium ion is rate determining, the activated complex in this reaction involves an electron-deficient carbon atom bonded to the aromatic ring. If we approximate the activated complex by the intermediate carbonium ion-pair (8), the resonance form (9) indicates the effect of conjugatively electron-supplying *para*-substituents beyond what is anticipated on the basis of their simple σ-values.

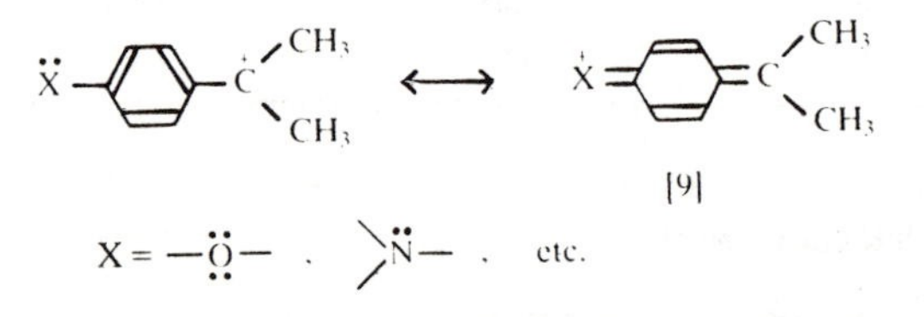

The ρ-value for the reaction of eqn (10.44) was defined as -4.54 on the basis of some *meta*-substituents and *para*-substituents which are unable conjugatively to assist. σ^+-values were then allocated to all other substituents X from experimental results of k_X for the solvolysis of (7) in the reaction of eqn (10.44) using the equation

$$\log\left\{\frac{k_X}{k_0}\right\} = -4.54 \,.\, \sigma_X^+,$$

where k_0 is the value for the parent compound ($X = H$).[33,34] Note that

TABLE 10.5
σ_p and σ_p^+ substituent constants[9(b)]

p-Substituent	σ_p	σ_p^+
I	0.28	0.13
Br	0.23	0.15
C≡CH	0.23	0.18
Cl	0.22	0.11
F	0.06	−0.07
SCH_3	~0	−0.6
C_6H_5	~0	−0.21
CH_3	−0.17	−0.31
OCH_3	−0.28	−0.78
OH	−0.38	−0.92
NH_2	−0.57	−1.3
$N(CH_3)_2$	−0.63	−1.7

this, unlike the standard reactions which define σ and σ^-, is based upon rate and not equilibrium constants. Differences between σ and σ^+ are appreciable only for those *para*-substituents with electron-donating mesomeric properties, although there are minor differences for other substituents due to the regressive manner in which the σ^+-values are determined. Table 10.5 includes some results; notice that a few substituents such as C≡CH and C_6H_5 have *both σ_p^+- and σ_p^-*-values (Table 10.4) significantly different from their ordinary σ_p parameters.

The following solvolysis reactions give good correlations with σ^+ and, as expected, have large negative ρ-values.

$$ArCH(C_6H_5)Cl + C_2H_5OH \xrightarrow[C_2H_5OH]{k_X,\ 25\,°C^{33}} ArCH(C_6H_5)OC_2H_5 + HCl \tag{10.45}$$

$$\rho(\sigma^+) = -4.05.$$

$$ArC(CH_3)_2Cl + ROH \xrightarrow[ROH]{k_X,\ 25\,°C^{34}} \text{Products} \tag{10.46}$$

$$\rho(\sigma^+) = -4.82, \quad ROH = CH_3OH,$$
$$= -4.67, \quad ROH = C_2H_5OH,$$
$$= -4.43, \quad ROH = (CH_3)_2CHOH.$$

Good ρ–σ^+ correlations are not surprising for these reactions since they are very closely related to the standard reaction of eqn (10.44) which is used to establish the σ^+ parameters.

A benzylic compound which solvolyses by a solvent-induced S_N2 mechanism with a loose activated complex involving partial positive charge development at the CH_2 group (eqn (10.47)) should also correlate with σ^+.

$$\mathrm{R\ddot{O}H} \quad \underset{\displaystyle \mathrm{Ar}}{\mathrm{CH_2}}\!-\!\mathrm{X} \longrightarrow \left(\underset{\displaystyle \mathrm{H}}{\overset{\delta\delta+}{\mathrm{R{-}O}}} \cdots\cdots \underset{\displaystyle \mathrm{Ar}}{\overset{\delta\delta+}{\mathrm{CH_2}}} \cdots\cdots \overset{\delta-}{\mathrm{X}} \right)^{\ddagger} \longrightarrow \mathrm{ROCH_2Ar + HX} \qquad (10.47)$$

activated complex

But since this mechanism involves a smaller extent of positive charge development at the carbon bonded to the aromatic residue than the reactions of eqns (10.44)–(10.46), it should lead to smaller negative ρ-values.

Figure 10.20 illustrates results for the acetolysis of substituted benzyl tosylates. For the more reactive substrates (X = H, *p*-F, *m*-CH_3, *p*-CH_3), there is an excellent linear correlation with a large negative ρ-value (−5.58) corresponding to an S_N1 mechanism. For the less reactive compounds (X = *m*-Cl, *m*-CF_3, *p*-CF_3), a second correlation is evident with $\rho = -2.81$. This smaller ρ-value corresponds to a solvent-induced S_N2 mechanism.

Electrophilic aromatic substitutions are a very important class of reactions whose rates correlate with σ^+:[33,36,37]

$$ArY + Z^+ \rightarrow ArZ + Y^+, \qquad (10.48)$$

where Z^+ is an electrophile usually generated in the reaction, and most commonly Y = H.

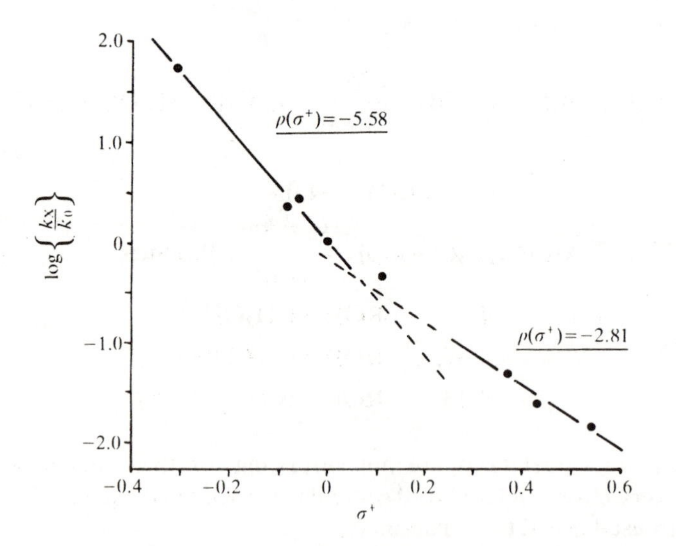

Fig. 10.20. Hammett $\rho - \sigma^+$ correlation for the acetolysis of substituted benzyl tosylates, 40 °C.[35]

There is a complication in these reactions which is absent from the nucleophilic aromatic substitutions considered in the previous section and, indeed, from all the other reaction types considered so far in this chapter. When $Ar = XC_6H_4$ and $Y = H$ in eqn (10.48), reaction of the monosubstituted benzene may occur at more than one site:

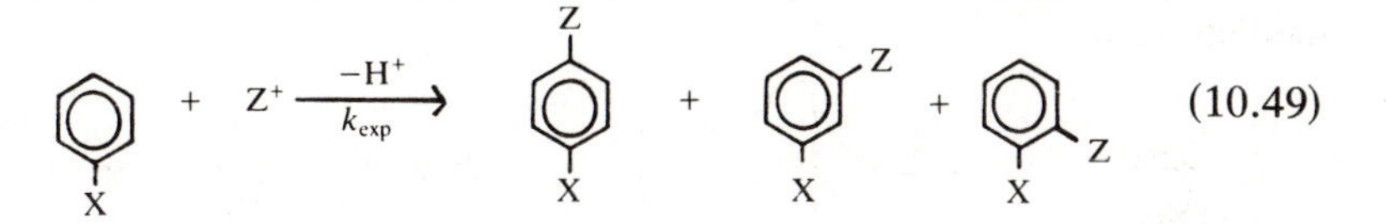

(10.49)

Equation (10.49) is different from one representing, for example, an $E1/S_N1$ reaction which gives a mixture of products via a common intermediate; it is a short-hand representation of a complex reaction comprising three *independent* parallel routes:

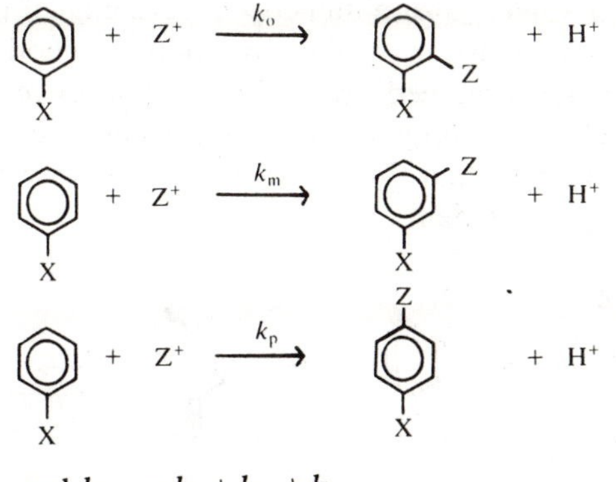

and $k_{exp} = k_o + k_m + k_p$.

We saw in Chapter 7 that the individual elementary rate constants for the (kinetically controlled) complex reaction represented by eqn (10.49) may be obtained from the overall reaction rate constants k_{exp} and a product analysis (see the example on p. 273). If sets of elementary rate constants $k_m(X)$ and $k_p(X)$ for the formation of *meta*- and *para*-disubstituted products from a series of reactants C_6H_5X are obtained, then the construction of a Hammett plot may be investigated.

The elementary rate constants k_m, k_p, and k_o as determined in the example of Chapter 7 refer to the formation of *meta*-, *para*-, and *ortho*-disubstituted products. For the Hammett plot, we really need rate constants for reactions at the *single* positions *meta* and *para* to the substituent in the reactant. Since there are two *meta*-positions but only one *para*-position, k_m as determined above has to be divided by 2 in order to get the parameter for inclusion in the Hammett plot. In practice, absolute elementary rate constants are not usually measured; *partial rate factors*, *f*,[38] are obtained by competition methods,

$$f = \frac{\text{rate constant for one position in the aromatic compound}}{\text{rate constant for one position in benzene}},$$

and f_m and f_p may legitimately be used in the Hammett plot.[36]

The mechanism of electrophilic aromatic substitution involves an intermediate (10):

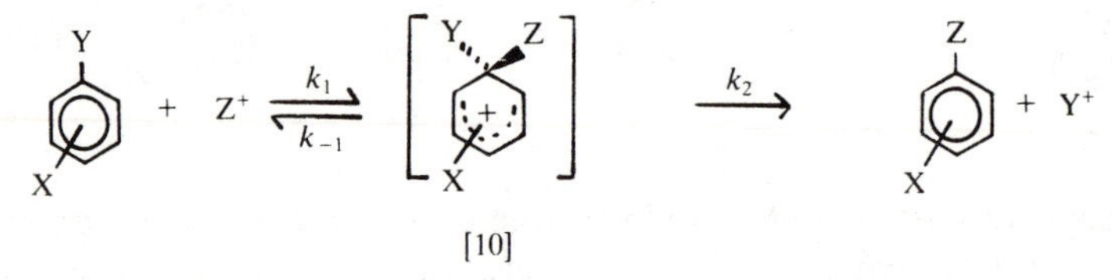

[10]

Normally, the formation of this so-called Wheland intermediate is rate-determining ($k_2 > k_{-1}$). An electron-withdrawing substituent X will obviously destabilize the activated complex leading to (10), consequently, large negative ρ-values are obtained in reasonable plots for *meta*-substitution using ordinary σ_m values. In contrast, good correlations for *para*-substitution are obtained only with σ_p^+ parameters. In particular, substituents which can supply π-electron density give rise to rate constants which are higher than their ordinary σ_p-values would lead one to expect. The reason is evident:

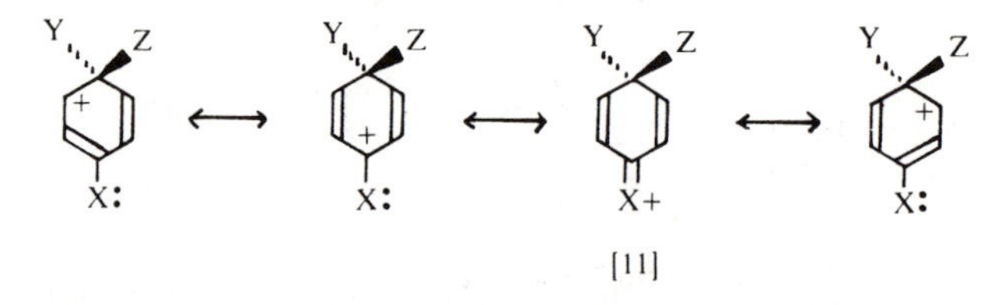

[11]

An extra resonance canonical (11) is provided for the intermediate (10) by such substituents, and some of the associated stabilization will be available at the preceding transition state. Some results are shown below:

$$\text{ArH} + \text{Br}_2 \xrightarrow[\text{CH}_3\text{CO}_2\text{H}]{k_X,\ 25\,^\circ\text{C}^{36}} \text{ArBr} + \text{HBr},$$

$$\rho(\sigma^+) = -12.1;$$

$$\text{ArH} + \text{Cl}_2 \xrightarrow[\text{CH}_3\text{CO}_2\text{H}]{k_X,\ 25\,^\circ\text{C}^{36}} \text{ArCl} + \text{HCl},$$

$$\rho(\sigma^+) = -10.0;$$

$$\text{ArH} + \text{CH}_3\text{CO.Cl} \xrightarrow[\text{C}_2\text{H}_4\text{Cl}_2,\ \text{AlCl}_3]{k_X,\ 25\,^\circ\text{C}^{36}} \text{ArCO.CH}_3 + \text{HCl},$$

$$\rho(\sigma^+) = -9.1;$$

$$ArH + HNO_3 \xrightarrow[(CH_3CO)_2O]{k_X,\ 25\ °C^{37}} ArNO_2 + H_2O,$$

$$\rho(\sigma^+) = -6.2;$$

$$ArH + (CH_3CO_2)_2Hg \xrightarrow[CH_3CO_2H]{k_X,\ 25\ °C^{36}} ArHgOAc + CH_3CO_2H,$$

$$\rho(\sigma^+) = -4.0.$$

There are many reactions of the type

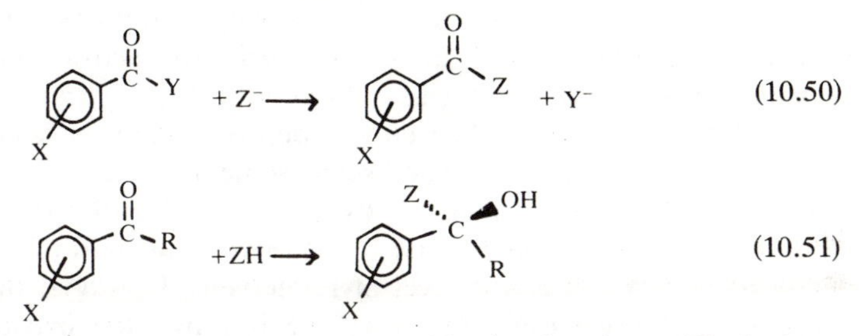

which may be expected to give correlations with σ^+ rather than σ. The basis for this expectation is that resonance canonical (12) should attenuate the effect of canonical (13) in the overall hybrid, thereby rendering the reactant less susceptible to nucleophilic attack than otherwise would be anticipated.

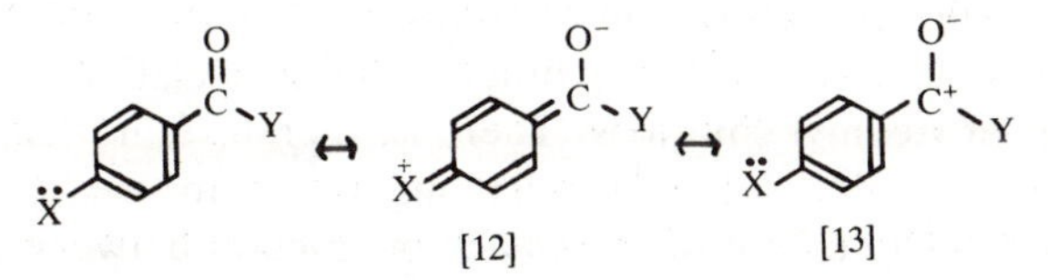

Reactions represented by eqns (10.50) and (10.51), therefore, should give ρ–σ^+ correlations with *positive* ρ-values. In practice, such reactions usually give very adequate correlations with the ordinary σ parameters. The reason is probably that the standard reaction which defines σ, the ionization–dissociation of benzoic acids, already includes some measure of the resonance effect identified above since a contribution from canonical (14)

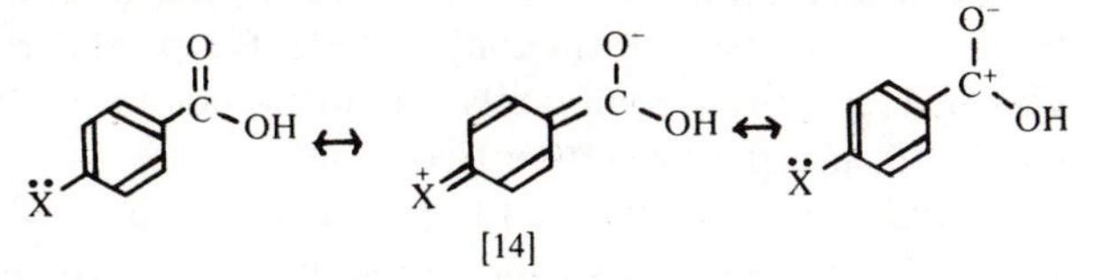

will be included in the overall resonance hybrid of a benzoic acid substituted with a group such as $CH_3\ddot{O}$–. In other words, the ordinary σ-value of such a *para*-substituent already includes a (small) resonance contribution.

10.5.5 The Yukawa–Tsuno equation

We saw at the end of the last section that it is not always clear whether a Hammett correlation should be with σ or σ^+. In other cases, the question is whether a correlation should be with σ or σ^-. For example, the dissociation of thiophenols[39] gives a good correlation with σ even though direct conjugation between the electron-rich reaction site and electron-withdrawing *para*-substituents seems possible.

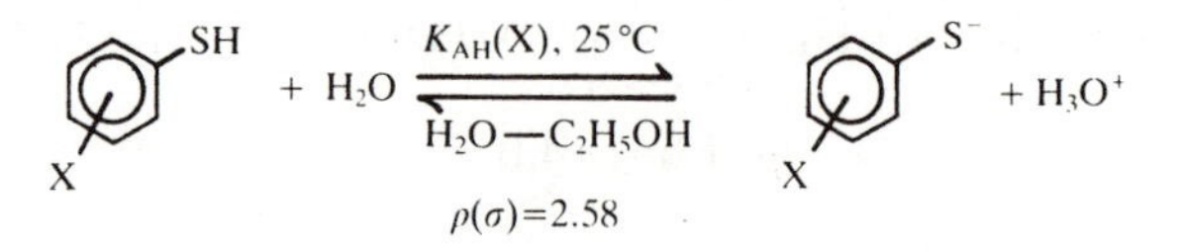

The *simple* Hammett equation uses only two parameters. One, ρ, is a measure of the sensitivity of the reaction series to electronic properties of substituents, and the other, σ, is a measure of the (electronic) effect of a substituent upon a standard reaction – the dissociation of benzoic acid in water at 25 °C. But, as we have seen, some reactions are sensitive to electronic properties of substituents in a way that the dissociation of benzoic acid is not, hence the use of *alternative* standard reactions which define σ^+ and σ^- as a first level of refinement. However, the modified Hammett equations using either σ^+ or σ^-, like the original, remain two-parameter correlations. And, in correlating rate (or equilibrium) constants against only one of σ, σ^+, or σ^-, we impose upon our reaction a comparison with a standard involving *either* no significant conjugation between reaction site and substituent *or* conjugation of the same type and to the same extent as exists in one or other of the dissociation of phenols or the solvolysis of t-cumyl chlorides. There is, therefore, no scope using a two-parameter equation for acknowledging that some reactions may involve an intermediate extent of resonance interaction between reaction site and substituent.

One suggestion has been to allocate to each substituent a range of σ-values for use in different types of reactions.[27] This, in a sense, is an extension of what we already do in selecting either σ or σ^+ (or σ^-) according to reaction type; but instead of selecting either the parameter for no significant conjugation or the one for full resonance interaction, we would also be able to choose intermediate values for partial resonance interaction. This would remove completely the independence between the reaction parameter ρ and the substituent constant.

An alternative and more widely accepted means of refining the Hammett approach which accommodates the possibility of intermediate degrees of resonance interaction is embodied in the Yukawa–Tsuno equation.[9(c),40]

Each substituent retains σ, σ^+, and σ^- parameters (not usually both σ^+ and σ^-) as described previously, and the extent to which σ^+ (or σ^-) is required, in addition to σ, for a particular set of rate (or equilibrium) constants to give a good linear correlation with gradient (reaction constant) ρ is reflected in the magnitude of a second reaction parameter, *r*. In eqn (10.52), therefore, r^+ measures the extent of conjugative interaction between a substituent with an electron-donating mesomeric effect and an electrophilic reaction centre.

$$\log\left\{\frac{k_X}{k_0}\right\} \quad or \quad \log\left\{\frac{K_X}{K_0}\right\} = \rho\{\sigma + r^+(\sigma^+ - \sigma)\}. \qquad (10.52)$$

In principle, the method for determining r^+ is just an extension to that used for the determination of ρ. First, ρ is determined using only *meta*- and *para*-substituents with no resonance capabilities (for which $\sigma = \sigma^+$, therefore the r^+ term in eqn (10.52) disappears). Then, using the ρ-value so obtained and *para*-substituents for which $\sigma \neq \sigma^+$, the value of r^+ is determined from the gradient of a plot of

$$\left(\frac{1}{\rho} . \log\left\{\frac{k_X}{k_0}\right\} - \sigma\right) \quad or \quad \left(\frac{1}{\rho} . \log\left\{\frac{K_X}{K_0}\right\} - \sigma\right) \quad \text{against} \quad (\sigma^+ - \sigma).$$

The modern practice, however, would be to use a computer-based multiple regression analysis to obtain values of ρ and r^+ which best fit the experimental results to eqn (10.52).[20(a)]

Clearly, if $r^+ = 0$, then eqn (10.52) simplifies to the original Hammett equation and mechanistically this indicates a reaction series with no resonance interaction between reaction site and substituents. If $r^+ = 1$ we have a direct correlation with σ^+ since eqn (10.52) becomes equal to the modified two-parameter equation of the previous section:

$$\log\left\{\frac{k_X}{k_0}\right\} \quad or \quad \log\left\{\frac{K_X}{K_0}\right\} = \rho . \sigma^+.$$

In other words, the reaction of eqn (10.44) is the standard for which $r^+ = 1$ (as well as $\rho = -4.54$). Some results are given below.[40]

$$XC_6H_4\overset{\overset{+}{O}H}{\overset{\|}{C}}CH_3 + H_2O \underset{H_2SO_4\text{–}H_2O}{\overset{K_X,\ 25\ ^\circ C}{\rightleftharpoons}} XC_6H_4\overset{O}{\overset{\|}{C}}CH_3 + H_3O^+,$$

$$\rho = 2.20, \qquad r^+ = 0.76;$$

$$XC_6H_4C(C_6H_5)_2Cl + CH_3OH \xrightarrow[CH_3OH]{k_X,\ 25\ ^\circ C} XC_6H_4C(C_6H_5)_2OCH_3 + HCl,$$

$$\rho = -4.02, \qquad r^+ = 1.23;$$

$$XC_6H_4CH(OH)CH{=}CHCH_3 \xrightarrow[H_2O\text{-dioxan}]{k_X,\ 30\ ^\circ C} XC_6H_4CH{=}CHCH(OH)CH_3,$$

$$\rho = -4.06, \qquad r^+ = 0.40;$$

$$C_6H_5X + HOBr \xrightarrow[HClO_4,\ H_2O\text{-dioxan}]{k_X,\ 25\ ^\circ C} XC_6H_4Br + H_2O,$$

$$\rho = -5.28, \qquad r^+ = 1.15;$$

$$C_6H_5X + HNO_3 \xrightarrow[CH_3CO_2H]{k_X,\ 25\ ^\circ C} XC_6H_4NO_2 + H_2O,$$

$$\rho = -6.38, \qquad r^+ = 0.90.$$

As some of the above show, values of $r^+ > 1$ are known; this indicates that resonance involvement in the reaction is greater than in the hydrolysis of t-cumyl chlorides, eqn (10.44).

A version of the Yukawa–Tsuno equation applicable to reactions which involve a nucleophilic reaction site has also been proposed:[41]

$$\log \frac{k_X}{k_0} \left(\text{or } \log\left\{\frac{K_X}{K_0}\right\}\right) = \rho\{\sigma + r^-(\sigma^- - \sigma)\}. \qquad (10.53)$$

Its use exactly parallels that of eqn (10.52) and the magnitude of r^- reflects the extent of resonance between the electron-rich reaction site and a substituent with a potential electron-withdrawing mesomeric effect.

Results for the reaction of eqn (10.38) (p. 449) are shown in Fig. 10.21.[26]

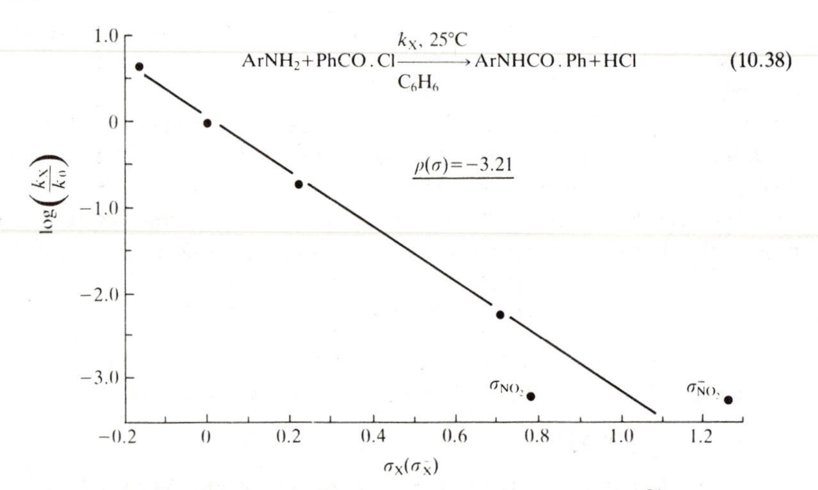

Fig. 10.21. Hammett plot for the reaction of eqn (10.38).[26]

Neither σ nor σ^- for *p*-nitro fits the experimental rate result for *p*-nitroaniline to the correlation based upon the other results, so eqn (10.53) should be used for this and other such reactions. Some results are shown below:

$$XC_6H_4NHSO_2C_6H_4NH_2 + H_2O \underset{H_2O}{\overset{K_X,\ 27\,°C^{41}}{\rightleftharpoons}} XC_6H_4\bar{N}SO_2C_6H_4NH_2 + H_3O^+,$$

$$\rho = 1.88, \qquad r^- = 0.54;$$

$$CH_3CO_2C_6H_4X + OH^- \xrightarrow[H_2O\text{-acetone}]{k_X,\ 1\,°C^{42}} CH_3CO_2^- + XC_6H_4OH,$$

$$\rho = 1.69, \qquad r^- = 0.20;$$

$$(NO_2)_3C_6H_2Cl + XC_6H_4O^- \xrightarrow[H_2O\text{-}C_2H_5OH]{k_X,\ 30\,°C^{42}} (NO_2)_3C_6H_2OC_6H_4X + Cl^-$$

$$\rho = -2.27, \qquad r^- = 0.94.$$

10.5.6 The Taft equation

Restrictions upon the use of the simple Hammett equation have already been discussed and two modifications (use of σ^+ and σ^-) and an extension (Yukawa–Tsuno equation) which have widened its range of applicability have been presented. But we have not been able to use even these refined versions for reactions other than those of *meta*- and *para*-substituted benzene derivatives. Other modified forms and extensions of the basic equation have allowed correlations to include *ortho*-substituted benzene derivatives,[43] non-aromatic unsaturated systems,[44] and polycyclic arenes and hetero-aromatic compounds.[9(d)]

Several equations of correlation for reactions of aliphatic and alicyclic compounds have been proposed, used, and subsequently modified.[9,20] In a saturated system, resonance contributions are absent so a substituent's electronic effect is purely inductive (polar), but there may also be a (non-electronic) steric effect. An early hope was to define inductive and steric substituent constants for use in an equation with the same general form as the Hammett equation. A fairly substantial literature describes what was not to be a simple quest.[20(b),45]

In one approach to defining a purely inductive substituent parameter σ_I (originally given the symbol σ'), the acidity constants $K_{AH}(X)$ of a family of 4-X-bicyclo[2.2.2]octane-1-carboxylic acids (15) in 50 per cent aqueous ethanol at 25 °C were determined.[46]

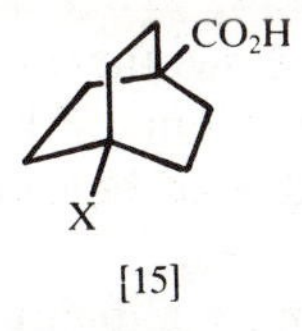

[15]

The rigid molecular structure of these compounds ensures a constant spatial relationship between substituents and reaction site but no steric interaction.

Instead of defining $\sigma_I(X)$ as the direct effect of X upon the logarithm of the acidity constant (or the pK_{AH}) of (15), eqn (10.54) is used:[47]

$$\log\left\{\frac{K_{AH}(X)}{K_{AH}(H)}\right\} = 1.65\sigma_I(X), \tag{10.54}$$

where $K_{AH}(H)$ = the acidity constant of (15) with X = H. This corresponds to defining $\rho_I = 1.65$ for the dissociation of (15) in 50 per cent aqueous ethanol (by weight) and was chosen because $\rho = 1.65$ for the dissociation of substituted benzoic acids in this medium and leads to the very convenient result, $\sigma_I \sim \sigma_m$.

A range of other mutually consistent methods have also been used to determine σ_I values partly because compounds (15) with a wide range of

TABLE 10.6
σ_m and σ_I substituent constants[9(b)]

Substituent	σ_m	σ_I
NO_2	0.71	0.72
CN	0.61	0.58
CF_3	0.43	0.43
CH_3CO_2	0.39	0.41
Br	0.39	0.46
$CO_2C_2H_5$	0.37	0.34
Cl	0.37	0.47
I	0.35	0.39
F	0.34	0.50
C≡CH	0.20	0.30
SCH_3	0.15	0.25
C_6H_5	0.05	0.10
H	0	0
CH_3	−0.06	−0.05
C_2H_5	−0.07	−0.05
$CH(CH_3)_2$	−0.07	−0.06
$C(CH_3)_3$	−0.10	−0.07

substituents X are not easily available. These other methods include the effect of substituents X upon the acidity constant of acetic acids XCH_2CO_2H (see below) and upon various spectroscopic properties of compounds such as chemical shifts of ^{19}F n.m.r. signals.[9(b)].

As the results in Table 10.6 show, there is good agreement between σ_I and σ_m for a range of substituents. The origin of the effect of a *meta*-substituent upon the acidity of benzoic acid appears, therefore, to be almost entirely polar (inductive).

The reaction parameter ρ_I for an aliphatic or alicyclic reaction series is determined from rate (or equilibrium) constants and substituent constants σ_I using eqn (10.55).

$$\log\left\{\frac{k_X}{k_0}\right\}\left(\text{or } \log\left\{\frac{K_X}{K_0}\right\}\right) = \rho_I \,.\, \sigma_I. \qquad (10.55)$$

The interpretation of ρ_I is as for ρ in the Hammett equation itself. It is a measure of the sensitivity of the chemical reaction as written to the introduction of substituents whose electronic-inductive effect is defined by eqn (10.54): positive and negative ρ_I parameters correspond, respectively, to reactions which are facilitated or inhibited by electron-withdrawing substituents. Some results are given below.[48]

$$XCH_2CO_2H + H_2O \xrightleftharpoons[H_2O]{K_X,\ 25\,^{\circ}C} XCH_2CO_2^- + H_3O^+,$$

$$\rho_I = 3.82;$$

$$XCH_2CH(CH_3)OBs \xrightarrow[\text{solvolysis, } CH_3CO_2H]{k_X,\ 75\,°C} \text{Products},$$

$$\rho_I = -7.8;$$

$$XCH_2CH_2OTs + C_2H_5OH \xrightarrow[\text{solvolysis, } C_2H_5OH]{k_X,\ 100\,°C} XCH_2CH_2OC_2H_5 + TsOH,$$

$$\rho_I = -1.65;$$

$$XCH_2CH_2Br + C_6H_5S^- \xrightarrow[CH_3OH]{k_X,\ 20\,°C} XCH_2CH_2SC_6H_5 + Br^-,$$

$$\rho_I = -1.35;$$

$$XCH_2CH(OC_2H_5)_2 + H_2O \xrightarrow[H_3O^+,\ H_2O\text{-dioxan}]{k_X,\ 25\,°C} XCH_2CHO + 2C_2H_5OH,$$

$$\rho_I = -8.12;$$

$$XCH_2CO_2H + Ph_2CN_2 \xrightarrow[C_2H_5OH]{k_X,\ 25\,°C} XCH_2CO_2CHPh_2 + N_2,$$

$$\rho_I = 2.61.$$

Equation (10.55) is a modification of the Taft equation which was developed in the early 1950s and originally presented as eqn (10.56).[45,20(b)]

$$\log\left\{\frac{k_X}{k_0}\right\}\left(\text{or } \log\left\{\frac{K_X}{K_0}\right\}\right) = \rho^* . \sigma^* \qquad (10.56)$$

σ^* was defined by a complex expression involving the rate constants for acid- and base-induced hydrolysis of esters (16) in aqueous ethanol with $Z = CH_3$ as the parent compound of this standard reaction series (rather than $Z = H$), so $\sigma^*(CH_3) = 0$. Correspondingly, k_0 (or K_0) in eqn (10.56) is the value for the methyl-substituted parent compound in the reaction series under investigation.

ZCO_2R ($R = CH_3$ or C_2H_5) (16) $\qquad XCH_2CO_2H$ (17)

It was found that the acidity constants of substituted acetic acids (17) in water, 25 °C, gave a good correlation with $\sigma^*(XCH_2)$ with a gradient $\rho^* = 1.72$.[45] Measurements of $K_{AH}(XCH_2CO_2H)$ for further substituted acetic acids followed by interpolation using the known ρ^*-value provided a convenient means of determining σ^*-values of further XCH_2 groups. It was also found that the function $\sigma_I(X)$ defined by eqn (10.54) using acidity constants of acids (15) gave a good correlation with the σ^*-values of the groups XCH_2 with gradient $\rho^* = 0.45$, eqn (10.57).[9(b)]

$$\sigma_I(X) = 0.45\sigma^*(CH_2X) \qquad (10.57)$$

This allows σ_I-values of substituents X to be determined (i) from older compilations of $\sigma^*(CH_2X)$ results, and (ii) via acidity constant measurements on substituted acetic acids (17) which are much more accessible than compounds (15). We can demonstrate the method as follows:

since $\rho^* = 1.72$ for the dissociation of ZCO_2H,[45]

$$\log\left\{\frac{K_{AH}(XCH_2CO_2H)}{K_{AH}(CH_3CO_2H)}\right\} = 1.72\sigma^*(CH_2X);$$

by eqn (10.57),

$$\log\left\{\frac{K_{AH}(XCH_2CO_2H)}{K_{AH}(CH_3CO_2H)}\right\} = \frac{1.72}{0.45}\cdot\sigma_I(X),$$
$$= 3.82\sigma_I(X)$$

or

$$\sigma_I(X) = 0.262\log\left\{\frac{K_{AH}(XCH_2CO_2H)}{K_{AH}(CH_3CO_2H)}\right\}.$$

It also follows from eqn (10.57) that the ρ^* of eqn (10.56) for a reaction series correlated against $\sigma^*(Z)$ is related to the ρ_I of eqn (10.55) for the same reaction series correlated against $\sigma_I(X)$ by

$$\rho^* = 0.45\rho_I, \tag{10.58}$$

when $Z = CH_2X$.

This simple relationship facilitates comparisons of some reactions in the older literature using ρ^* with other results involving ρ_I.

As implied earlier, there are many equations of correlation besides the ones discussed above. Some are multiparameter equations, some have a rather esoteric and limited range of applicability, but none involve new matters of chemical principle.

10.6 Problems

1. Sketch molecular potential-energy reaction maps for the following gas-phase reactions showing the possibility of either single- or two-step mechanisms.

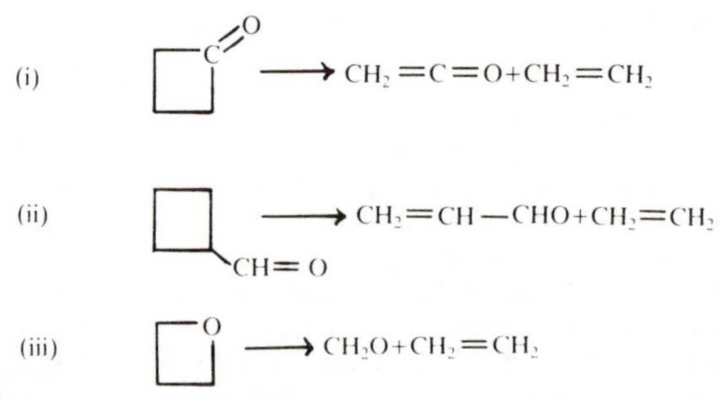

2. A mechanism for the solvolysis of some alkyl arenesulphonates involving rate-determining formation of an intermediate with partial covalent bonding to the incoming nucleophile has been proposed. Construct a standard molar free-energy map for such a mechanism. (T. W. Bentley and P. von R. Schleyer, *J. Am. chem. Soc.* **98,** 7658 (1976).)

3. The following results have been obtained for the general acid-catalysed hydrolysis of ethyl cyclopentenyl ether in water at 25 °C.[3]

$$\text{(cyclopentenyl)}-OC_2H_5 + H_2O \xrightarrow[k_{obs}]{HA} \text{(cyclopentanone)}=O + C_2H_5OH$$

$$k_{obs} = k_H[H_3O^+] + k_{AH}[AH]$$

AH	pK_{AH}	$k_{AH}/dm^3\ mol^{-1}\ s^{-1}$
$CNCH_2CO_2H$	2.47	20.2
$ClCH_2CO_2H$	2.87	14.5
$CH_3OCH_2CO_2H$	3.57	5.51
HCO_2H	3.75	3.19
$HOCH_2CO_2H$	3.83	2.94
CH_3CO_2H	4.76	0.758
$C_2H_5CO_2H$	4.88	0.710

Calculate the Brönsted coefficient α for this reaction and estimate the catalytic constant for phenylacetic acid ($pK_{AH} = 4.31$).

4. The following catalytic constants were obtained for the general base-catalysed bromination of ethyl acetoacetate in water at 25 °C.

$$CH_3COCH_2CO_2C_2H_5 + Br_2 \xrightarrow[-H^+]{B,\ k_{obs}} CH_3COCHBrCO_2C_2H_5 + Br^-$$

$$k_{obs} = k_{OH}[OH^-] + k_B[B]$$

Base	pK_{BH}	$k_B/dm^3\ mol^{-1}\ min^{-1}$
Monochloroacetate	2.87	2.28
m-Nitrobenzoate	3.49	6.5
Glycollate	3.83	8.0
Benzoate	4.18	15
Phenylacetate	4.31	20
Isobutyrate	4.86	39

The reactions are all zero order in bromine concentration. Propose a mechanism for this reaction type (see Chapter 8) and determine the Brönsted coefficient β. (Kinetics results from R. P. Bell, E. Gelles, and E. Möller, *Proc. R. Soc.* A **198,** 308 (1949); pK_{BH} values from A. Albert and E. P. Serjeant, *Ionization constants of acids and bases,* Methuen, London (1962).)

5. Calculate the Brönsted coefficient for the substituted-pyridine-

induced cleavage of 2,4-dinitrophenyl methyl carbonate in aqueous solution at 25 °C from the following results (taken from E. A. Castro and F. J. Gil, *J. Am. chem. Soc.* **99,** 7611 (1977)).

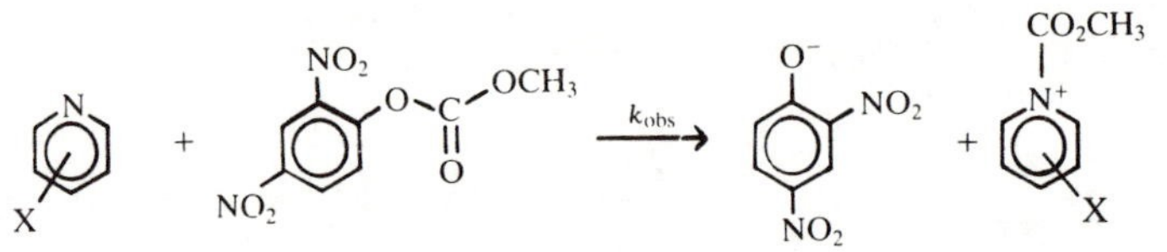

$$k_{obs} = k_0 + k_{Pyr}[Pyr]$$

X	pK_{BH}	$k_{Pyr}/dm^3\ mol^{-1}\ s^{-1}$
4-CN	1.98	6.9×10^{-4}
3-Cl	2.98	5.6×10^{-3}
3-$CONH_2$	3.46	1.2×10^{-2}
H	5.30	0.7
3-CH_3	5.81	1.6
4-CH_3	6.15	3.9

For pyridines of $pK_{BH} > \sim 7$, the Brönsted plot curves and β decreases towards $\beta \sim 0.2$. Account for this change in β for the above reaction on the basis of a step-wise mechanism.

6. The acid-catalysed hydrolysis of 2-aryl-2-methoxy-1,3-dioxolanes in water at 25 °C has been investigated and the following catalytic constants determined:

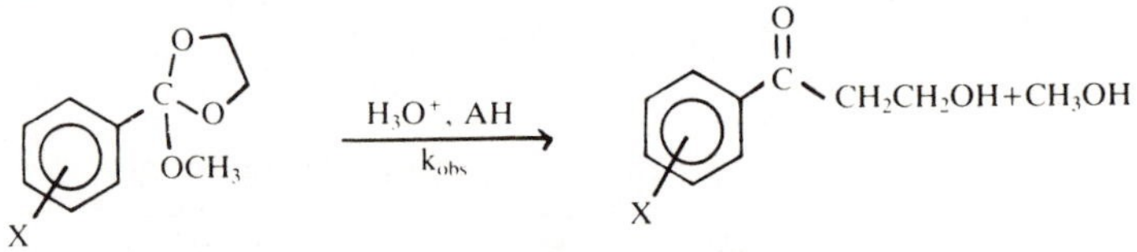

$$k_{obs} = k_0 + k_H[H_3O^+] + k_{AH}[AH]$$

X	σ_X	$10^{-2}\ k_H/dm^3\ mol^{-1}\ s^{-1}$
4-CH_3O	−0.28	19.3
4-CH_3	−0.17	12.8
H	0	7.32
4-F	0.06	5.02
4-Cl	0.22	3.29
4-NO_2	0.78	0.39

Calculate ρ for H_3O^+ catalysis in this reaction series, interpret the result, and estimate k_H for the H_3O^+-catalysed hydrolysis of the 4-CF_3-substituted analogue. (Rate results taken from Y. Chiang, A. J. Kresge, and C. I. Young, *Finn. chem. Lett.* 13 (1978); σ_X values from Table 10.3, p. 443, $\sigma(4\text{-}CF_3) = 0.54$.)

7. The following kinetics results were obtained for the alkaline hydrolysis of the esters shown:

$$X\text{-}C_6H_4\text{-}C_6H_4\text{-}CO_2C_2H_5 + OH^- \xrightarrow[k,\ 40^\circ C]{88.7\%\ C_2H_5OH\text{—}H_2O} X\text{-}C_6H_4\text{-}C_6H_4\text{-}CO_2^- + C_2H_5OH$$

X	σ_X	$10^3\ k/dm^3\ mol^{-1}\ s^{-1}$
p-CH_3O	−0.28	1.56
p-CH_3	−0.17	1.90
H	0	2.32
p-Cl	0.22	3.13
p-Br	0.23	3.18
m-Br	0.39	3.88
m-NO_2	0.71	5.95
p-NO_2	0.78	6.58

Calculate ρ for this reaction series (rate results taken from E. Berliner and L. H. Liu, *J. Am. chem. Soc.* **75,** 2417 (1953); σ_X values from Table 10.3, p. 443). Account for the difference between this result and that of $\rho = 2.54$ for the hydroxide-induced hydrolysis of ethyl benzoates (p. 442).

8. Rate constants given below were obtained for the second-order displacement reaction between benzenesulphonyl chloride and substituted benzoate anions in methanol at 25 °C (taken from O. Banjoko, and R. Okwuiwe, *J. Org. Chem.* **45,** 4966 (1980)).

$$XC_6H_4CO_2^- + C_6H_5SO_2Cl \xrightarrow{k_2} XC_6H_4C(=O)\text{—}O\text{—}S(=O)_2C_6H_5 + Cl^-$$

X	σ_X	$10^3\ k_2/dm^3\ mol^{-1}\ s^{-1}$
p-CH_3O	−0.28	6.58
p-CH_3	−0.17	5.75
m-CH_3	−0.06	5.33
H	0	5.18
m-CH_3O	0.11	4.56
p-Cl	0.22	4.27
p-Br	0.23	4.04
m-Br	0.39	3.80
m-NO_2	0.71	2.91
p-NO_2	0.78	2.78

Calculate the Hammett ρ for this reaction series and comment upon its magnitude. How would the result have been quoted and interpreted if the same data had been used for a Brönsted-type plot?

9. The H_3O^+-catalysed hydrolysis of substituted benzaldehyde diethyl acetals in 50 per cent aqueous dioxan at 30 °C gave the following results

$$XC_6H_4CH(OC_2H_5)_2 + H_2O \xrightarrow[k_H]{H_3O^+} XC_6H_4CHO + 2C_2H_5OH$$

X	$\sigma_X(\sigma_X^+)$	k_H/dm^3 mol^{-1} min^{-1}
p-CH$_3$O	−0.28(−0.78)	29 200
p-CH$_3$	−0.17	3 937
m-CH$_3$	−0.06	1 170
H	0	723.3
m-CH$_3$O	0.11	453.1
p-Cl	0.22(0.11)	181.3
m-NO$_2$	0.71	2.48
p-NO$_2$	0.78	1.84

Determine the ρ-value for this reaction series using the Yukawa–Tsuno equation with $r^+ = 0.5$ (T. H. Fife and L. K. Jao, *J. org. Chem.* **30,** 1492 (1965); see also E. H. Cordes, *Progr. phys. org. Chem.* **4,** 1 (1967)). Account for the difference between this result and that for the reaction in Problem 6 above.

10. The results given below were obtained for the reaction between picryl chloride and substituted anilines in 75 per cent ethanol-water at 30 °C.[42]

$$2,4,6\text{-}(NO_2)_3C_6H_2Cl + NH_2C_6H_4X \xrightarrow[k_2]{-H^+} 2,4,6\text{-}(NO_2)_3C_6H_2\text{-}NH\text{-}C_6H_4X + Cl^-$$

X	$\sigma_X(\sigma_X^-)$	k_2/dm^3 mol^{-1} s^{-1}
m-CH$_3$	−0.06	6.5
H	0	3.2
m-CH$_3$O	0.11	1.8
m-Cl	0.37	0.136
m-CH$_3$CO	0.38	0.197
m-NO$_2$	0.71	8.05×10^{-3}
p-C$_6$H$_5$N$_2$	0.29(0.69)	4.91×10^{-2}
p-C$_6$H$_5$CO	0.36(0.88)	1.07×10^{-2}
p-CO$_2$C$_2$H$_5$	0.45(0.74)	1.51×10^{-2}
p-CH$_3$CO	0.48(0.84)	8.62×10^{-3}
p-NO$_2$	0.78(1.27)	2.52×10^{-4}

Construct a simple Hammett plot using the results for the m-substituted anilines and determine ρ. Discuss the most appropriate method to use the results for the p-substituted anilines. (σ- and σ^{-}-values from Tables 10.3 and 10.4, pp. 443, 451, and ref. 9(*b*).)

11. The following rate results were obtained for the solvolysis of the tertiary chlorides:

$$RCH_2CH_2C(CH_3)_2Cl \xrightarrow[60\,^{\circ}C]{80\%\ \text{ethanol–water}} \text{Products.}$$

R	σ_I	$10^6\,k/s^{-1}$
CN	0.58	8.38
Cl	0.47	22.0
$CO_2C_2H_5$	0.34	84.1
CH_2Cl	0.16	216
CH_3	−0.05	820
H	0	902

Calculate ρ_I for this reaction series and compare the result with those given on p. 465 for other solvolysis reactions. (Rate results taken from C. A. Grob and A. Waldner, *Helv. chim. Acta* **62,** 1736 (1979); σ_I values from Table 10.6, p. 464, and ref. 9(*b*).)

10.7 References

1. Since an important paper by R. A. More O'Ferrall, *J. chem. Soc.* B 274 (1970), which drew upon ideas formulated by E. R. Thornton, *J. Am. chem. Soc.* **89,** 2915 (1967), there have been many demonstrations of the interpretative and predictive uses of what here we have called reaction maps, especially by W. P. Jencks, *Chem. Rev.* **72,** 705 (1972), *Acc. chem. Res.* **13,** 161 (1980), and *Chem. Soc. Rev.* **10,** 345 (1981). Occasionally, however, it is unclear whether authors intend the reaction maps to include molecular potential energy or standard molar free energy.
2. C. Steel, R. Zand, P. Hurwitz, and S. G. Cohen, *J. Am. chem. Soc.* **86,** 679 (1964). See also S. W. Benson and H. E. O'Neal, *Kinetic data on gas phase unimolecular reactions* NSRDS-NBS21, National Bureau of Standards, Washington, 1970, p. 328.
3. A. J. Kresge, H. L. Chen, Y. Chiang, E. Murrill, M. A. Payne, and D. S. Sagatys, *J. Am. chem. Soc.* **93,** 413 (1971).
4. J. N. Brönsted and E. A. Guggenheim, *J. Am. chem. Soc.* **49,** 2554 (1927). See also R. P. Bell, *Acid–base catalysis*, Oxford University Press, (1941).
5. R. P. Bell, *Adv. phys. org. Chem.* **4,** 1 (1966).
6. J. N. Brönsted and K. Pedersen, *Z. phys. Chem.* **108,** 185 (1923).
7. R. P. Bell, *The proton in chemistry*, 2nd edn, Chapman & Hall, London (1973).
8. V. Gold and E. G. Jefferson, *J. chem. Soc.* 1409 (1953).
9. *Correlation analysis in chemistry*: *Recent advances*, (eds. N. B. Chapman and

J. Shorter), Plenum Press, New York (1978): (*a*) Chapt. 2, 'The Brönsted equation – its first half-century', by R. P. Bell; (*b*) Chapt. 10, 'A Critical compilation of substituent constants' by O. Exner; (*c*) Chapt. 4, 'Multiparameter extensions of the Hammett equation', by J. Shorter; (*d*) Chapt. 5, 'Applications of linear free-energy relationships to polycyclic arenes and to heterocyclic compounds' by M. Charton.

10. M. Eigen, *Angew. Chem., int. edn* **3,** 1 (1964).
11. (*a*) J. R. Murdoch, *J. Am. chem. Soc.* **94,** 4410 (1972); (*b*) W. J. Albery, *A. Rev. phys. Chem.* **31,** 227 (1980).
12. R. A. Marcus, *J. phys. Chem.* **72,** 891 (1968); A. O. Cohen and R. A. Marcus, ibid., p. 4249.
13. F. G. Bordwell and W. J. Boyle, *J. Am. chem. Soc.* **94,** 3907 (1972).
14. A. J. Kresge, *Canad. J. Chem.* **52,** 1897 (1974).
15. R. A. Marcus, *J. Am. chem. Soc.* **91,** 7224 (1969); *Proton transfer reactions* (eds. E. Caldin and V. Gold), Chapman & Hall, London (1975), Chapt. 7 by A. J. Kresge.
16. W. J. Albery and M. M. Kreevoy, *Adv. phys. org. Chem.* **16,** 87 (1978); W. J. Albery, *Pure appl. Chem.* **51,** 949 (1979); J. R. Murdoch, *J. Am. chem. Soc.* **105,** 2159 (1983); M. J. Pellerite and J. I. Brauman, ibid., p. 2672.
17. C. D. Ritchie, C. Kubisty, and G. Y. Ting, *J. Am. chem. Soc.* **105,** 279 (1983).
18. J. R. Murdoch, *J. Am. chem. Soc.* **105,** 2660 (1983).
19. L. P. Hammett, *Physical organic chemistry*, McGraw-Hill, New York, 1st edn (1940), 2nd edn (1970).
20. *Advances in linear free-energy relationships* (eds. N. B. Chapman and J. Shorter), Plenum Press, London (1972). (*a*) Chapt. 1 'The Hammett equation—the present position' by O. Exner; (*b*) Chapt. 2 'The separation of polar, steric, and resonance effects by the use of linear free-energy relationships' by J. Shorter.
21. C. K. Ingold and W. S. Nathan, *J. chem. Soc.* 222 (1936); D. P. Evans, J. J. Gordon, and H. B. Watson, ibid. 1430 (1937).
22. C. D. Ritchie, J. D. Saltiel, and E. S. Lewis, *J. Am. chem. Soc.* **83,** 4601 (1961).
23. K. Bowden, N. B. Chapman, and J. Shorter, *Canad. J. Chem.* **42,** 1979 (1964).
24. C. D. Ritchie and R. E. Uschold, *J. Am. chem. Soc.* **90,** 2821 (1968).
25. R. A. Benkeser, C. E. DeBoer, R. E. Robinson, and D. M. Sauve, *J. Am. chem. Soc.* **78,** 682 (1956).
26. ρ-value calculated[27] from experimental results reported by E. G. Williams and C. N. Hinshelwood, *J. chem. Soc.* 1079 (1934).
27. H. van Bekkum, P. E. Verkade, and B. M. Wepster, *Rec. Trav. Chim.* **78,** 815 (1959).
28. D. N. Kevill, K. C. Kolwyck, D. M. Shold, and C-B. Kim., *J. Am. chem. Soc.* **95,** 6022 (1973).
29. ρ-value calcuated[30] from experimental results reported by G. M. Bennett and B. Jones, *J. chem. Soc.* 1815 (1935).
30. H. H. Jaffé, *Chem. Rev.* **53,** 191 (1953).
31. R. F. Hudson and G. Klopman, *J. chem. Soc.* 1062 (1962). See also P. R. Young and W. P. Jencks, *J. Am. chem. Soc.* **101,** 3288 (1979), and V. P. Vitullo, J. Grabowski, and S. Sridharan, ibid. **102,** 6463 (1980).
32. J. Miller, A. J. Parker, and B. A. Bolto, *J. Am. chem. Soc.* **79,** 93 (1957).
33. Y. Okamoto and H. C. Brown, *J. org. Chem.* **22,** 485 (1957).

34. Y. Okamoto, T. Inukai, and H. C. Brown, *J. Am. chem. Soc.* **80,** 4972 (1958).
35. ρ-values calculated using $\sigma^+(\sigma)$ values from ref. 9(*b*) and rate constants reported by A. Streitwieser, H. A. Hammond, R. H. Jagow, R. M. Williams, R. G. Jesaitis, C. J. Chang, and R. Wolf, *J. Am. chem. Soc.* **92,** 5141 (1970).
36. L. M. Stock and H. C. Brown, *Adv. phys. org. Chem.* **1,** 35 (1963).
37. H. C. Brown and Y. Okamoto, *J. Am. chem. Soc.* **80,** 4979 (1958).
38. T. H. Lowry and K. S. Richardson, *Mechanism and theory in organic chemistry*, 2nd edn, Harper & Row, New York (1981).
39. F. G. Bordwell and H. M. Andersen, *J. Am. chem. Soc.* **75,** 6019 (1953).
40. Y. Yukawa and Y. Tsuno, *Bull. chem. Soc. Japan* **32,** 971 (1959); Y. Yukawa, Y. Tsuno, and M. Sawada, ibid. **39,** 2274 (1966).
41. M. Yoshioka, K. Hamamoto, and T. Kubota, *Bull. chem. Soc. Japan* **35,** 1723 (1962).
42. J. J. Ryan and A. A. Humffray, *J. chem. Soc.* B 842 (1966); 1300 (1967).
43. M. Charton, *Progr. phys. org. Chem.* **8,** 235 (1971); T. Fujita and T. Nishioka, ibid. **12,** 49 (1976).
44. M. Charton, *Progr. phys. org. Chem.* **10,** 81 (1973).
45. R. W. Taft, Chapt. 13 'Separation of polar, steric, and resonance effects in reactivity' in *Steric effects in organic chemistry* (ed. M. S. Newman) Wiley, New York (1956).
46. J. D. Roberts and W. T. Moreland, *J. Am. chem. Soc.* **75,** 2167 (1953).
47. H. D. Holtz and L. M. Stock, *J. Am. chem. Soc.* **86,** 5188 (1964).
48. ρ_I values calculated using eqn (10.58) from ρ^* results reported in ref. 45.

Supplementary references

The Hammett equation by C. D. Johnson, Cambridge University Press, (1973).

Correlation analysis in organic chemistry, by J. Shorter, Oxford University Press, (1973).

Rates and equilibria of organic reactions, by J. E. Leffler and E. Grunwald, Wiley, New York (1963).

C. D. Johnson, 'Linear free-energy relationships and the reactivity-selectivity principle', *Chem. Rev.* **75,** 755 (1975).

A. Pross, 'The reactivity–selectivity principle and its mechanistic applications', *Adv. phys. org. Chem.* **14,** 69 (1977).

Answers to numerical problems

Chapter 1

1. i: a, 1.24×10^{-18}; **b,** 8.64×10^{-19}; **c,** 3.97×10^{-19}; **d,** 1.79×10^{-20}; **e,** 1.99×10^{-23}; **f,** 5.96×10^{-26} J.
ii: a, 748; **b,** 520; **c,** 239; **d,** 10.8 kJ mol^{-1}; **e,** 12.0; **f,** 3.59×10^{-2} J mol^{-1}.
3. i, 467, **ii,** 383; **iii,** 316; **iv,** 254 kJ mol^{-1}.
4. i, 43.6; **ii,** 16.1; **iii,** 12.4 kJ mol^{-1}.
5. i: a, 5.1×10^2; **b,** 2.24×10^3; **c,** 1.14×10^3; **d,** 3.24×10^2; **e,** 1.70×10^2 N m^{-1}.
ii: a, 24.9; **b,** 13.9; **c,** 9.30; **d,** 3.33; **e,** 1.28 kJ mol^{-1}.
6. n_1/n_0: **a,** 1.9×10^{-9}; **c,** 5.5×10^{-4}; **e,** 0.356.
7. $\Delta E = 2.36$ kJ mol^{-1}; $n_1/n_0 = 0.386$.
8. i, 742 N m^{-1}; **ii,** 7.82 kJ mol^{-1}.
9. i, 2.39×10^{-2} J mol^{-1}; **ii,** $n_1/n_0 = 0.999991$.

Chapter 2

1. −5053.0 kJ mol^{-1}.
2. 20.89 kJ.
3. −200.8 kJ mol^{-1}.
4. i: (**1**), −4237.2; (**2**), −4227.1; (**3**), −4111.7; (**4**), −4200.0 kJ mol^{-1}.
ii: (**1**), 91.2; (**2**), 82.1; (**3**), 247.6; (**4**), 339.1 kJ mol^{-1}.
iii: a, 9.1; **b,** −91.5 kJ mol^{-1}.
5. i, 68.78; **ii,** −432; **iii,** −311.4; **iv,** −597.2; **v,** −37.08 kJ mol^{-1}.
6. i, −546.61; **ii,** −9.71; **iii,** 27.5; **iv,** −83.41; **v,** −180.5 kJ mol^{-1}.
7. i, −57.9; **ii,** −6.6; **iii,** 120.9; **iv,** −383.1; **v,** 51.2; **vi,** 19.7 kJ mol^{-1}.

Chapter 3

1. i, 21.82; **ii,** 14.075; **iii,** −3.5; **iv,** −1.8; **v,** 173.0; **vi,** 333.4; **vii,** 362.7; **viii,** −98.1 J K^{-1} mol^{-1}.
2. i: a, 58.8; **b,** −149.81; **c,** −80.78; **d,** −44.36; **e,** −5.66; **f,** −98.94 J K^{-1} mol^{-1}.
ii: a, 209; **b,** −889.6; **c,** −50.79; **d,** −228.6; **e,** 16; **f,** −16.44 kJ mol^{-1}.
3. i, −52; **ii,** −0.9; **iii,** −39.5; **iv,** −497.8 kJ mol^{-1}.
4. i, 21.82; **v,** 146.4; **vi,** 280.2; **vii,** 282.9; **viii,** −71.5 J K^{-1} mol^{-1}.
5. i, 20.5; **ii,** 129; **iii,** −173; **iv,** 297 J K^{-1} mol^{-1}.
6. a: i, 5.9; **ii,** −11.5 kJ mol^{-1}.
b: i, −2.0; **ii,** −19.4 kJ mol^{-1}.

7. **i,** −9.43; **ii,** 3.87; **iii,** −12.00; **iv,** 3.05; **v,** −33.0; **vi,** −17.1 kJ mol^{-1}.
8. **i,** −9.6; **ii,** 0.2; **iii,** 2.70; **iv,** −22.6; **v,** 2.8; **vi,** 15.8 kJ mol^{-1}.

Chapter 4

1. 4.48.
2. 2.21.
3. **i,** 0.313; **ii,** 9.23 atm.
4. **i,** −4.9 kJ mol^{-1}; **ii,** 0.733; **iii,** 423.5 °C; **iv,** 414.4 °C.
5. −52.0 kJ mol^{-1}.
6. **i,** 383.9 K, $K_P^\circ = 0.112$, $K_c^\ominus = 3.56 \times 10^{-3}$;
ii, 485.6 K, $K_P^\circ = 9.29$, $K_c^\ominus = 0.233$;
iii, $\Delta_r H^\circ = 67.4$ kJ mol^{-1}, $\Delta_r S^\circ = 157$ J K^{-1} mol^{-1}.
7. **i,** $\Delta_r U^\circ = -131.0$ kJ mol^{-1}; **ii,** $\Delta_r H^\circ = -135.7$ kJ mol^{-1}.

Chapter 5

1.

	ΔG°/kJ mol^{-1}	K_P°	$\Delta G^\ominus$/kJ mol^{-1}	$K_c^\ominus$
a	0	1	170	1.6×10^{-30}
b	−78	4.6×10^{13}	22	1.6×10^{-4}
c	−21	4.8×10^{3}	51	1×10^{-9}
d	73	1.6×10^{-13}	−13	200
e	−94	2.9×10^{16}	25	5×10^{-5}

2. **i,** 5.06; **ii,** 3.45.
3. **i,** 94.6%; **ii,** 18.3%; **iii,** 24.0%.
4. **a,** pH = 3.85, $\alpha = 0.7\%$
b, pH = 3.24, $\alpha = 2.9\%$
c, pH = 2.74, $\alpha = 9.0\%$.
5. **i,** $\alpha = 75\%$; **ii,** pH = 4.28.
6. **a,** pH = 10.9, $\alpha = 7.9\%$
b, pH = 9.5, $\alpha = 0.3\%$
c, pH = 8.3, $\alpha = 0.02\%$.
7. **i,** 3.40; **ii,** 5.94; **iii,** 7.9.
8. **a,** $pK_{AH} = 2.89$, $\alpha = 14.8\%$
b, $pK_{AH} = 3.84$, $\alpha = 5.3\%$
c, $pK_{AH} = 4.83$, $\alpha = 1.7\%$.
9. **i,** 81% neutral, 19% cationic, ~0% anionic;
ii, 83% neutral, 17% anionic, ~0% cationic.
10. **a,** $pK_{BH} = 9.50$, $\alpha = 7.6\%$
b, $pK_{BH} = 8.09$, $\alpha = 1.6\%$
c, $pK_{BH} = 4.86$, $\alpha = 0.04\%$.
11. $\rho = 0.86$; $pK_{AH} = 10.72$.

Chapter 6

1. $k = 2.47 \times 10^{-3}\,s^{-1}$, $\Delta G^{\ominus\ddagger}_{298\,K} = 87.9\,kJ\,mol^{-1}$.
2. $k = 5.42 \times 10^{-4}\,s^{-1}$, $\Delta G^{\ominus\ddagger}_{703\,K} = 221\,kJ\,mol^{-1}$.
3. $k = 1.25 \times 10^{-4}\,dm^3\,mol^{-1}\,s^{-1}$.
4. $k_P = 1.01 \times 10^{-6}\,Torr^{-1}\,min^{-1}$; $k_c = 4.02 \times 10^{-4}\,dm^3\,mol^{-1}\,s^{-1}$; $\Delta G^{\ominus\ddagger}_{385\,K} = 120\,kJ\,mol^{-1}$.
5. $k = 1.02 \times 10^{-3}\,dm^3\,mol^{-1}\,s^{-1}$.
6. $E_a = 102\,kJ\,mol^{-1}$; $A = 2.1 \times 10^{16}\,s^{-1}$; $\Delta H^{\ominus\ddagger} = 99\,kJ\,mol^{-1}$; $\Delta S^{\ominus\ddagger} = 60\,J\,K^{-1}\,mol$.
7. $E_a = 8.63\,kJ\,mol^{-1}$; $A = 2.19 \times 10^{10}\,dm^3\,mol^{-1}\,s^{-1}$; $\Delta H^{\ominus\ddagger} = 6.48\,kJ\,mol^{-1}$; $\Delta S^{\ominus\ddagger} = -54\,J\,K^{-1}\,mol^{-1}$.
8. $\Delta H^{\ominus\ddagger} = 104\,kJ\,mol^{-1}$; $\Delta S^{\ominus\ddagger} = -113\,J\,K^{-1}\,mol^{-1}$.
9. $E_a = 130\,kJ\,mol^{-1}$; $A = 5.62 \times 10^{9}\,s^{-1}$; $\Delta H^{\ominus\ddagger} = 125\,kJ\,mol^{-1}$; $\Delta S^{\ominus\ddagger} = -72\,J\,K^{-1}\,mol^{-1}$.
10. $\Delta H^{\ominus\ddagger} = 72\,kJ\,mol^{-1}$; $\Delta S^{\ominus\ddagger} = -85\,J\,K^{-1}\,mol^{-1}$.

Chapter 7

1. $k_1 = 3.28 \times 10^{-5}\,s^{-1}$; $k_{-1} = 1.87 \times 10^{-5}\,s^{-1}$.
2. $k_1 = 5.2 \times 10^{-6}\,s^{-1}$; $k_{-1} = 4.1 \times 10^{-6}\,s^{-1}$.
4. $K = 3.57 \times 10^{3}\,dm^3\,mol^{-1}$; $\Delta_r H^\circ = -136\,kJ\,mol^{-1}$; $\Delta_r S^\circ = -194\,J\,K^{-1}\,mol^{-1}$.
5. $k_X = 3.53 \times 10^{-4}$, $k_Y = 1.08 \times 10^{-4}$, $k_Z = 1.08 \times 10^{-4}\,dm^3\,mol^{-1}\,s^{-1}$.
6. $\Delta H^{\ominus\ddagger}$(enant.) $= 81\,kJ\,mol^{-1}$.
7. **i,** $E_a(k_1) = 233\,kJ\,mol^{-1}$, $A(k_1) = 2.7 \times 10^{13}\,s^{-1}$; **ii,** $E_a(k_2) = 239\,kJ\,mol^{-1}$, $A(k_2) = 1.4 \times 10^{13}\,s^{-1}$.
8. $k_{(rac)} = 2.38 \times 10^{-5}\,s^{-1}$; $k_{(enant.)} = 1.19 \times 10^{-5}\,s^{-1}$; $\Delta G^{\ominus\ddagger}_{369\,K}$(enant.) $= 126\,kJ\,mol^{-1}$.
10. **i,** $k_{obs} = k_2$; **ii,** $k_{obs} = k_2 K[NH_2OH]$.

Chapter 8

1. 2.06 and $1.9\,dm^3\,mol^{-1}\,s^{-1}$.
2. $0.30\,dm^3\,mol^{-1}\,s^{-1}$.
4. $k_{obs} = 1.1 \times 10^{-8}\,s^{-1}$, pH = 5.54, $k_{min} = 7.8 \times 10^{-10}\,s^{-1}$.
5. $k^{H}_{cat} = 2.0 \times 10^{-3}$ and $k^{D}_{cat} = 1.8 \times 10^{-3}\,dm^3\,mol^{-1}\,s^{-1}$.
8. $k_0 = 8.5 \times 10^{-4}\,s^{-1}$; $k_H = 2.0 \times 10^{-4}\,dm^3\,mol^{-1}\,s^{-1}$; $k_{OH} = 49\,dm^3\,mol^{-1}\,s^{-1}$.
10. $k_H = 2.3 \times 10^{3}\,dm^3\,mol^{-1}\,s^{-1}$; $k_{OH} = 3.3\,dm^3\,mol^{-1}\,s^{-1}$.

Chapter 9

1. ^{2}HF: 2801.3 and $2926.2\,cm^{-1}$;
^{2}HCl: 2040.6 and $2114.0\,cm^{-1}$;
^{2}HBr: 1809.7 and $1873.3\,cm^{-1}$.
2. 2023.3, 2344, 2641.2, 1911, and $1443.8\,cm^{-1}$.

3. **i,** 13.2; **ii,** 10.6.
4. $\bar{\nu}_{XD}$ (calc.) = 2524 and 2342 cm^{-1}.

Chapter 10

3. $\alpha = 0.64$; $k_{cat} = 1.56\ dm^3\ mol^{-1}\ s^{-1}$.
4. 0.62.
5. 0.90.
6. $\rho = -1.6$; $k_H(4\text{-}CF_3) = 9.5 \times 10^{-3}\ dm^3\ mol^{-1}\ s^{-1}$.
7. 0.58.
8. −0.35.
9. −3.34.
10. $\rho = -3.7$, $r^- = 0.7$.
11. −3.22.

Index

('T' indicates a table or list of values)